Recent Sedimentary Carbonates

In Two Parts by

J. D. Milliman · G. Müller · U. Förstner

Part 1

J. D. Milliman

Marine Carbonates

With 94 Figures and 39 Plates

Springer-Verlag Berlin · Heidelberg · New York 1974

John D. Milliman
Woods Hole Oceanographic Institution
Woods Hole, MA 02543/USA

For the cover plates V (a) and VII (a) D of this volume have been used

ISBN 3-540-06116-9 Springer-Verlag Berlin · Heidelberg · New York
ISBN 0-387-06116-9 Springer-Verlag New York · Heidelberg · Berlin

 Library of Congress Catalog Card Number 73-82359. Printed in Germany. Typesetting, printing and bookbinding: Universitätsdruckerei H. Stürtz AG, Würzburg.

To my father, E. B. Milliman

and my teacher, J. E. Hoffmeister

Foreword

Few fields of research in the earth sciences have produced as much data and literature as the study of carbonate sediments and rocks. The past 25 years in particular, have seen a significant increase in studies concerning modern marine and freshwater carbonates. With the present worldwide interest in oceanographic research, marine carbonates have received the bulk of the attention, particularly with respect to shallow-water sediments. However, in terms of the variety of environments, compositions and modes of formation, non-marine carbonates probably encompass a wider spectrum than do marine types.

Our purpose is to present a two-volume treatise on carbonate sediments and rocks, both marine and non-marine. We have confined ourselves to the discussion of modern (Holocene) environments, sediments and components, assuming that the compilation of these data will not only be relevant to those working with modern carbonates but will also serve as a necessary reference source for those interested in ancient analogs. The first volume, by MILLIMAN, deals almost exclusively with marine environments, while the second volume, by MÜLLER and FÖRSTNER, will concentrate on the non-marine carbonates.

JOHN D. MILLIMAN
GERMAN MÜLLER
ULRICH FÖRSTNER

Preface

This book is an attempt to synthesize our present-day knowledge about calcium carbonate in the marine environment, its composition, sedimentation and diagenesis. Because of the rapid increase in marine carbonate research in the past 25 years, the need for such a book is obvious. The classic texts on carbonate rocks (SCHMIDT, PIA and CAYEUX) are not only out of print, but with respect to marine carbonate sedimentation, are largely out of date. Discussion of modern carbonate sedimentation has been treated in numerous review articles and in several books, but for the most part these discussions have been restricted to tropical reefs and banks, often limited to the Caribbean area (for example, see BATHURST, 1971). Similarly, recent advances in carbonate petrography have not been summarized. Books by MAJEWSKI (1969) and HOROWITZ and POTTER (1971) tend to emphasize ancient rather than modern skeletal components.

Perhaps the most obvious need in such a synthesis is in documenting the mineralogy and chemistry of carbonate components and marine limestones. A. P. VINOGRADOV's massive compilation of the compositions of various marine organisms was based mainly on pre-World War II data, and does not take into account the rapid advances in conceptual models and instrumentation that have occurred during the past 25 years. New data on the mineralogy, minor and trace elements and stable isotopes have been published at an ever-increasing rate.

With these needs in mind, I have tried to summarize the available information concerning marine carbonates. The view is from the eyes of a marine geologist; I have not treated ancient limestones nor freshwater carbonates and subaerial diagenesis. Discussions of these various subjects, however, are available in other recent books (BATHURST, 1971; FÜCHTBAUER and MÜLLER, 1970; G. MÜLLER, in preparation).

The book is divided into four parts, each dealing with a specific aspect of marine carbonates. Part one discusses calcium carbonate and its stability within the marine environment (Chapter 1) and also treats the various methods used in carbonate study (Chapter 2). The second part concerns the ecology, calcification, petrography and composition of various skeletal and non-skeletal carbonate components (Chapters 3–5), while the third part deals with the distribution of these components in shallow seas (Chapter 6), shelf waters (Chapter 7), and the deep sea (Chapter 8). Part four traces the diagenetic alteration of carbonates (within the marine environment), through degradation (Chapter 9), cementation (Chapter 10), and dolomitization (Chapter 11).

A special note should be made concerning the composition tables given in Chapters 3 and 4. Complete listings of all analytical data are not given, simply because the amounts of published data are too extensive and unwieldy. Instead I have presented average compositions of representative genera. The range of values used in compiling the averages are given in parentheses below the average values. If the composition of a genus varies greatly between different published analyses (probably the result of an analytical error or different environmental or physiological factors), averages of the various analyses have been listed separately. In some skeletal groups (notably coralline algae and serpulids) sparse by published data have been supplemented by unpublished analyses performed by the writer. Otherwise all the data listed in the various tables have been taken exclusively from the literature; the interested reader is referred to those references listed in each table caption.

This book could not have been written without the help, encouragement and critical reading of many individuals. Much of the book was written during the tenure of a six-month Alexander von Humboldt-Stiftung Stipendium at the Laboratorium für Sedimentforschung, Universität Heidelberg; I thank Professor GERMAN MÜLLER for supplying me with the space and facilicies to do this research. H. G. MULTER, JENS MÜLLER, C. H. MOORE, JR., C. D. GEBELEIN, W. R. COBB, and S. HONJO supplied valuable illustrations and photographs. K. O. EMERY, M. R. CARRIKER, T. J. M. SCHOPF, JENS MÜLLER, J. C. HATHAWAY, W. H. ADEY and D. P. ABBOTT, S. HONJO, F. T. SAYLES, J. L. BISCHOFF, J. I. TRACEY, H. S. LADD and P. R. SUPKO all read and commented on parts of the manuscript, but, of course, final responsibility for any errors in the text remains mine.

DONALD SOUZA did all the drafting, and FRANK MEDEIROS much of the photographic work. Typing was done mostly by ANNE RILEY and ANITA TUCHOLKE, with help from BETTY HEMENWAY and ANNE COLLINS.

Finally, I would like to express my gratitude to ANN, CHRISTOPHER and HEATHER, who gave their constant support and understanding throughout the three years of research and writing.

Woods Hole, Massachusetts
October, 1973

JOHN D. MILLIMAN

Contents

Part I. Introduction

Chapter 1. Carbonates and the Ocean

Carbonate Mineralogy

Carbonate minerals include those minerals composed of $CO_3^=$ and one or more cations. The various cations that are usually incorporated in carbonates are listed in Table 1. Most carbonate minerals are either rhombohedral or orthorhombic in

Table 1. Atomic weights and ionic radii of common elements contained within carbonates. Data from MASON (1962)

Element	(Common ionic state)	Atomic weight	Ionic radius (Å)
B		10.82	0.23
Fl	(−1)	19.00	1.36
Na	(+1)	22.99	0.95
Mg	(+2)	24.32	0.66
S	(+6)	32.07	0.30
Cl	(−1)	35.46	1.81
K	(+1)	39.10	1.33
Ca	(+2)	40.08	0.99
Mn	(+2)	54.94	0.80
Fe	(+2)	55.85	0.74
Ni	(+2)	58.71	0.69
Co	(+2)	58.94	0.72
Cu	(+2)	63.54	0.72
Zn	(+2)	65.38	0.74
Sr	(+2)	87.63	1.12
Ba	(+2)	137.36	1.34
Pb	(+2)	207.21	1.20

crystal habit. Rhombohedral carbonate minerals have six-fold coordination (2 cations for every oxygen), while orthorhombic minerals have nine-fold co-ordination (3 cations for every oxygen). Smaller cations, such as Mg, Fe, Mn, Zn and Cu, are energetically favored in six-fold coordination, while larger cations, such as Sr, Pb and Ba, are favored in the larger orthorhombic structure. Calcium occupies a unique position in that its ionic radius (0.99 Å) is intermediate between the small and large cations, and as such can form either rhombohedral (calcite) or orthorhombic (aragonite) carbonates. Aragonite is the high pressure polymorph of calcium carbonate (JAMIESON, 1953; MACDONALD, 1956; CRAW-

FORD and FYFE, 1964) (Fig. 1), and at present earth-surface conditions is metastable. Calcium also can form a highly unstable spherulitic hexagonal polymorph, vaterite, sometimes referred to as μ-$CaCO_3$ (JOHNSTON and others, 1916); this polymorph rarely occurs in nature. Other common calcite-type rhombohedral carbonates include magnesite ($MgCO_3$), rhodochrosite ($MnCO_3$) and siderite ($FeCO_3$). Aragonite-type orthorhombic carbonates include strontianite ($SrCO_3$), witherite ($BaCO_3$) and cerrusite ($PbCO_3$).

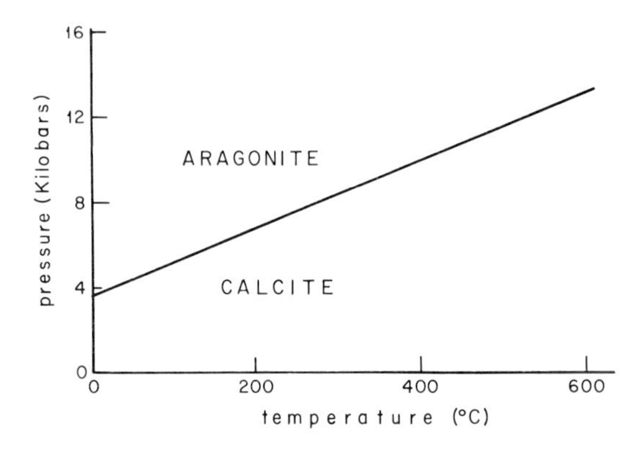

Fig. 1. Equilibrium curve for $CaCO_3$. Aragonite is the high-pressure polymorph. After JAMIESON (1953) and CLARK (1957)

Carbonate minerals are seldom pure; rather they tend to contain other cations within their lattices. Aragonite-type minerals show preferential substitution with larger cations, while calcite-type minerals prefer smaller cations. For example, calcite rarely contains more than 2000 to 3000 ppm Sr, but aragonite can contain up to 3.3% (HUTTON, 1936). The amount of substitution of large cations into aragonite structures, however, seldom exceeds a few percent at earth-surface conditions. In contrast, complete Mg-Fe and Fe-Mn transitions can occur in the calcite group (DEER and others, 1962; PALACHE and others, 1963; GOLDSMITH, 1959). The crystal structure of calcite itself, however, is such that complete solid-solution substitution by other cations is not common at normal earth-surface temperatures and pressures (GRAF, 1960).

Generally, the concentrations of Fe and Mn in calcite are less than 5 mole % (DEER and others, 1962; PALACHE and others, 1963). Although thermodynamic calculations suggest that similarly small amounts of Mg should substitute with calcium at earth-surface conditions (GOLDSMITH, 1959), calcites with more than 20 mole % $MgCO_3$ are common in modern oceanic sediments (see below). The great availability of Mg in oceanic waters accounts for this difference. Those calcites containing more than 1% Mg (4 mole % $MgCO_3$) are termed high-magnesium (or magnesian) calcites (CHAVE, 1952).

Dolomite is a rhombohedral carbonate which contains alternating layers of calcium and magnesium ($(Ca, Mg)CO_3$). Both iron and manganese can substitute for the magnesium to form ankerite ($Ca(Mg, Fe, Mn)(CO_3)_2$). Because magnesium ions have smaller ionic radii than calcium ions, the lattice spacings of pure dolomite

Table 2. Petrographic properties of calcite, aragonite and dolomite. Data from WAHLSTROM, 1955; DEER and others, 1962; and PALACHE and others, 1963

	Calcite $CaCO_3$	Aragonite $CaCO_3$	Dolomite $(CaMg)(CO_3)$
crystal system	rhombohedral uniaxial negative	orthorhombic biaxial negative	rhombohedral uniaxial negative
other cations	Mg, Fe, Mn, Zn, Cu	Sr, Ba, Pb, K	Fe, Mn, Zn, Cu
cleavage	$(10\bar{1}1)$		$(10\bar{1}1)$
dominant twinning	glide $(01\bar{1}2)$	(110) (010)	glide $(02\bar{2}1)$
color	colorless	colorless	colorless
specific density	2.72	2.94	2.86
n_0 n_e	1.658 1.486	n_x 1.530 n_y 1.681 n_z 1.685	1.679 1.502
birefringence	0.172	0.155	0.177
hardness	3	3.5–4	3.5–4

are considerably smaller than those of calcite (for example, 2.84 Å versus 3.04 Å in the 211 plane).

At present earth-surface temperatures calcite (including magnesian calcite), aragonite and dolomite are the common carbonate minerals. Many of their petrographic properties are listed in Table 2.

The Oceans as a Chemical System

Sea water contains nine major elements: sodium, magnesium, calcium, potassium, strontium, chlorine, sulfur (predominantly as sulfate), bromine and carbon (primarily as bicarbonate and carbonate). These nine elements contribute more than 99.9% of the total dissolved salts in the ocean (Table 3). The constancy of the ratios of the major elements throughout the oceans is well-known (DITTMAR, 1884), but variations in these ratios occur in nearshore areas (by introduction of river effluents) (Table 4) as well as in areas of high biologic activity and chemical precipitation. The major cations are present in ionic form, while a significant proportion of the anions (sulfate, bicarbonate and carbonate) are bonded to cations GARRELS and THOMPSON, 1962).

Terrestrial erosion contributes most of the major cations found in ocean water and also transfers CO_2 from the atmosphere to the oceans as HCO_3^-. The weathering cycle can be expressed in the following terms:

$$\underset{\text{(K-feldspar)}}{4\,KAlSi_3O_8} + 22\,H_2O + 4\,CO_2 = 4\,K^+ + \underset{\text{(Kaolinite)}}{Al_4Si_4O_{10}(OH)_8} + 8\,H_4SiO_4 + 4\,HCO_3^-$$

$$CaCO_3 + H^+ + HCO_3^- = Ca^{+2} + 2\,HCO_3^-$$

Table 3. Major cations and anions, together with some selected minor cations in sea water. Data from SVERDRUP and others (1942), CULKIN (1965) and GOLDBERG (1965)

	mg/kg	Dissolved solids %	Milli-moles/l	Atoms element / Atoms calcium
Cations				
Na^+	10760	30.6	468	45.3
Mg^{+2}	1294	3.7	53.2	5.15
Ca^{+2}	413	1.2	10.3	1
K^+	387	1.1	9.95	0.96
Sr^{+2}	8			
Ba^{+2}	0.015			
Fe^{+2}	0.01			
Mn^{+2}	0.002			
Cu^{+2}	0.003			
Zn^{+2}	0.01			
P	0.01			
Anions				
Cl^-	19353	55.2	545	
$SO_4^=$	2712	7.7	28.2	
HCO_3^-	142	0.4	2.3	
Br^-	67	0.2	0.84	
$B(H_2BO_3^-)$	26	0.07	0.43	
	35162	99.9*	1118.3	

(KRAUSKOPF, 1967; SIEVER, 1968). Most workers have considered the oceans to be in a steady state equilibrium, where the dissolved solids being introduced are equal to the solids being precipitated in the sea. By knowing the concentrations of dissolved species presently in sea water and river water and by estimating the mean annual discharge of river waters into the ocean, the estimated mean residence

Table 4. Composition of average river water. After LIVINGSTONE (1963)

	mg/kg	Dissolved solids %	Milli-moles/l
Na^+	6.3	5.3	0.274
Mg^{+2}	4.1	3.4	0.169
Ca^{+2}	15.0	12.5	0.374
K^+	2.3	1.9	0.059
Sr^{+2}	—	—	—
SiO_2	13.1	10.9	0.218
Cl^-	7.8	6.5	0.220
$SO_4^=$	11.2	9.4	0.117
HCO_3^-	58.4	48.7	0.954
Br^-	—	—	—
$H_2BO_3^-$	—	—	—
Totals	118.2	98.6	2.385

Table 5. Residence times of major dissolved solids in the ocean. Data from LIVINGSTONE (1963) and MACKENZIE and GARRELS (1966a)

	Amount in oceans ($\times 10^{18}$ kg)	Amount delivered by rivers annually ($\times 10^{10}$ kg)	Residence time (assuming entire influx from rivers) (years)
Cl^-	26.1	25.4	1.0×10^8
Na^+	14.4	20.7	0.7×10^8
$SO_4^=$	3.7	36.7	1.0×10^7
Mg^{+2}	1.9	13.3	1.0×10^7
Ca^{+2}	0.6	48.8	1.2×10^6
K^+	0.5	7.4	6.8×10^6
HCO_3^-	0.19	190.2	1.0×10^5
SiO_2	0.008	42.6	1.9×10^4

time of the dissolved solids can be calculated (Table 5). Such calculations tend to be in error since they fail to consider other sources for the dissolved ions, such as volcanic eminations, air-sea interactions and diagenesis of suspended river sediment. Still the general trend can be seen: silica and bicarbonate have very short residence times (10^4–10^5 years) and chlorine and sodium have extremely long residence times (10^7–10^8 years). The implications of the short residence times of silicon and carbonate will be discussed later in this chapter and elsewhere in the book.

The CO_2 System in the Oceans

The carbon dioxide system in the oceans has received considerable attention during the past 40 years. Much of our basic understanding of this complicated system came in the 1930's through studies by BUCH, WATTENBERG, HARVEY, GRIPENBERG, REVELLE and their co-workers. More recent studies by HOOD, PARK, PYTKOWICZ, BROECKER and co-workers on the CO_2 system, and by SILLÉN, GARRELS and MACKENZIE on the silicate system (and its relation to pH and the CO_2 system) have greatly added to our knowledge. Excellent reviews can be found in REVELLE and FAIRBRIDGE (1957), HOOD and others (1959), HARVEY (1960), SKIRROW (1965), MACINTYRE (1965), LI (1967), SPENCER (1965) and SIEVER (1968).

Carbon dioxide is the fourth most abundant gas in the atmosphere, amounting to some 3×10^{-4} atmospheres. If the CO_2 concentration in seawater were present in the same 1:2400 ratio with nitrogen as it is in the atmosphere, it would be only a minor component of sea water. The dissolved CO_2 in sea water, in fact, is about in equilibrium with the atmosphere (KANWISHER, 1960), but the total CO_2 is much greater (Table 6). To a large extent this increased ΣCO_2 concentration is related to the dissociation of CO_2 in sea water:

$$CO_{2\,(g)} \rightleftharpoons CO_{2\,(l)} + H_2O$$

$$\upharpoonleft\downharpoonright$$

$$H_2CO_3 \rightleftharpoons H^+ + HCO_3^- \rightleftharpoons H^+ + CO_3^= .$$

As a result of these reactions, only a small percentage of the total CO_2 in sea water is present as aqueous CO_2.

In quiet water the diffusion of gaseous CO_2 into the ocean is relatively slow; similarly, the hydration of CO_2 is very slow (KERN, 1960). Uptake of CO_2 increases with increasing water turbulence. For example, WEYL (1958) found that it requires 210 min for an unstirred sample of water to reach equilibrium with the atmosphere, but only 0.8 minutes for a stirred sample. It is because of high turbulence (together with cold temperatures) that a major influx of CO_2 into the oceans occurs in the stormy polar seas (BROECKER and others, 1960). The residence time of atmospheric CO_2 is about 7 years; once in the oceans, the mean residence time is between 600 and 1000 years (CRAIG, 1957b; BROECKER and others, 1960).

Table 6. Saturation avl values of atmospheric gases in sea water (chlorinity $= 20^0/_{00}$), where $P_{gas} = 1$ atm. Note the high solubility of CO_2 relative to O_2 and N_2. After SVERDRUP and others (1942)

	O_2 (ml/l)	N_2 (ml/l)	CO_2 (ml/l)
0 °C	49.24	23.00	1715
12 °C	36.75	17.80	1118
24 °C	29.38	14.63	782

The dissociation of H_2CO_3 and HCO_3^- can be expressed by the following equations:

$$K_1 = \frac{a_{H^+} \times a_{HCO_3^-}}{a_{H_2CO_3}}, \quad (1)$$

$$K_2 = \frac{a_{H^+} \times a_{CO_3^=}}{a_{HCO_3^-}} \quad (2)$$

where a = the activity coefficients of the various ions. For practical purposes, the equations can be modified with respect to sea water into the following equations:

$$K_1' = \frac{a_{H^+} \times C_{HCO_3^-}}{a_{H_2CO_3}}, \quad (3)$$

$$K_2' = \frac{a_{H^+} \times C_{CO_3^=}}{C_{HCO_3^-}} \quad (4)$$

where C = concentrations of the various ions and aH^+ is assumed to be the measured pH. Formulas 3 and 4 employ apparent dissociation constants (K') rather than true thermodynamic constants. As a result, K' values change with temperature and pressure (as do true thermodynamic constants) as well as with salinity (Fig. 2).

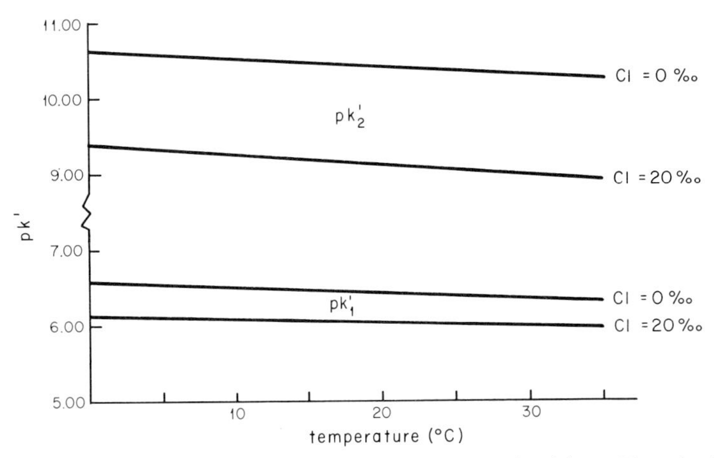

Fig. 2. Variation of K_1' and K_2' with temperature and chlorinity; $pK' = -\log K'$. Data from LYMAN (1956)

pH

The term pH is an expression of the negative logarithm of hydrogen ion activity:

$$pH = -\log a_{H^+}.$$

At neutrality and a temperature of 25 °C, pH = 7. Using Eq. (3) and (4), pH can be related to the carbon dioxide system:

$$pH = pK_1' + \log \frac{C_{HCO_3^-}}{a_{H_2CO_3}}, \quad (5)$$

$$pH = pK_2' + \log \frac{C_{CO_3^=}}{C_{HCO_3^-}}. \quad (6)$$

Thus the pH of sea water is inter-related to the activities and concentrations of H_2CO_3, HCO_3^- and $CO_3^=$. The dissociation of these CO_2 species and the particulary rapid dissociation of H_2CO_3 cause sea water to be a buffered solution. At pH's lower than 7.5, the main CO_2 species are H_2CO_3 and HCO_3^- (Fig. 3) and Eq. (5) approximates the system. For pH's higher than 7.5, the dominant species are HCO_3^- and $CO_3^=$, and Eq. (6) is applicable (PARK, in HOOD and others, 1959).

In recent years, several workers have concluded that the basic buffering capacity within the oceans is related more to the silica system than to the CO_2 system (SILLÉN, 1961, 1967; GARRELS, 1965; MACKENZIE and GARRELLS, 1966a, 1966b). The reaction can be expressed as:

$$\text{X-ray amorphous silicates} + HCO_3^- + SiO_2 = \text{cation} - \text{Al silicates} + CO_2 + H_2O$$

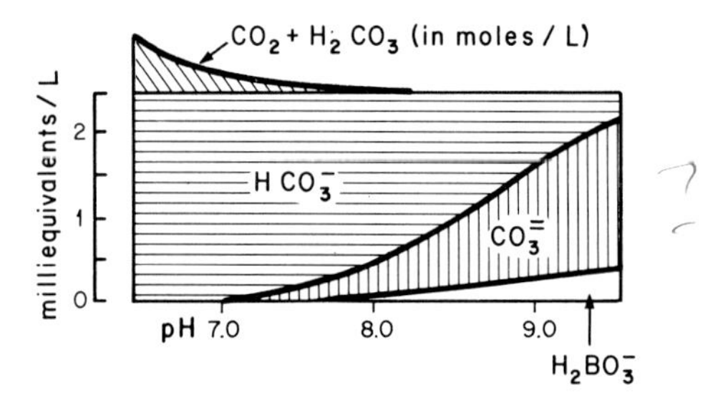

Fig. 3. Variation of CO_2, H_2CO_3, HCO_3^-, $CO_3^=$, and $H_2BO_3^-$ in sea water with changing pH. After HARVEY (1960)

This "reverse weathering" prevents alkali metals from accumulating, regulates the large HCO_3^- and SiO_2 contributions from the rivers (Tables 4, 5), and perhaps also regulates the CO_2 balance within the atmosphere (MACKENZIE and GARRELS, 1966a, 1966b). This process, however, would invoke large-scale diagenetic alterations of marine clays, an occurrence which has not been observed in modern oceanic sediments (BISCAY, 1965; GRIFFIN and others, 1968). Such alterations may occur in restricted sedimentary basins (SIEVER, 1968), but whether these alterations are sufficient to explain the predicted system is not known. Perhaps the silica system is completely or partly responsible for long-term equilibrium within ocean waters, but the CO_2 system must be considered a prime factor in short-term pH variations (PYTKOWICZ, 1965).

Alkalinity

Although the term alkalinity has been defined in numerous ways (including buffer capacity, excess base and titration alkalinity), in its simplest meaning alkalinity can be thought of as the dominance of base effectiveness over acid effectiveness; that is, the bases HCO_3^-, CO_3 and $HBO_3^=$ are more effective in sea-water than the potentially strong acids, such as $Na(H_2O)_x^+$, $Ca(H_2O)_x^{+2}$, and $Mg(H_2O)_x^{+2}$ (RAKESTRAW, 1949). Chemically, the alkalinity of sea water can be defined as:

$$\text{Alk. (eq/l)} = C_{HCO_3^-} + C_{CO_3^=} + C_{H_2BO_3^-} + (C_{OH^-} - C_{H^+}). \tag{7}$$

Within the pH range of 5.5 to 8.5, the term $(C_{OH^-} - C_{H^+})$ is negligible, so that the equation can be simplified:

$$\text{Alk. (eq/l)} = C_{HCO_3^-} + 2\,C_{CO_3^=} + C_{H_2BO_3^-}. \tag{8}$$

The term carbonate alkalinity (CA) is defined as

$$\text{CA} = C_{HCO_3^-} + 2\,C_{CO_3^=} = \text{Alk} - C_{H_2BO_3^-}.$$

As pH falls, boric acid content also decreases and carbonate alkalinity approaches alkalinity. BUCH (1933) and LYMAN (1956) have provided tables by which carbonate alkalinity can be determined. Thus by measuring the alkalinity, the carbonate alkalinity can be calculated and the sum total of $C_{HCO_3^-}$ and $C_{CO_3^=}$ determined.

Alkalinity can be expressed analytically as the amount of acid required to neutralize sea water (pH = 7.0); this term commonly is called titration alkalinity. In normal sea water, the titration alkalinity varies from 0.18 to 0.25×10^{-3} equivalents/liter, depending upon the absolute chlorinity of the water. The term specific alkalinity (SA) refers to alkalinity relative to a constant chlorinity:

$$SA = \frac{\text{titration alkalinity} \times 10^3}{Cl\ ^0/_{00}}. \tag{9}$$

Specific alkalinity varies with both latitude and water depth, but the range of values is generally between 0.120 and 0.130. Lowest alkalinities are usually found in tropical surface waters, where carbonate precipitation has depleted the $CO_3^=$. In deep waters, where dissolution occurs, alkalinity values are high.

Variations in the CO_2 in Nature

The CO_2 system in the ocean is dependent upon five major parameters: salinity, temperature, pressure, biologic activity and carbonate precipitation (or dissolution). Increased temperature will increase reaction rates of the CO_2 species, but also will decrease the solubility of gaseous CO_2 in sea water. Thus, in an open system (where CO_2 can enter or escape) pH will increase with increasing temperature; in a closed system (such as the deep sea), in which CO_2 cannot readily escape, pH will fall with increasing temperature. The decrease in pH is in response to increases in K_1' and K_2' which cause further dissociations and the release of hydrogen.

The solubilities of CO_2 and H_2CO_3 are inversely related to salinity, while the solubilities of HCO_3^- and $CO_3^=$ are directly related. Pressure increases the solubility of CO_2 and the dissociation constants increase at the following rates:

$$\Delta pK_1' = -0.48 \times 10^{-4}\,\Delta z, \tag{10}$$

$$\Delta pK_2' = -0.18 \times 10^{-4}\,\Delta z \tag{11}$$

where z = depth in meters (SVERDRUP and others, 1942).

Biologic activity (photosynthesis and respiration) has perhaps the greatest effect upon the CO_2 system in the oceans:

$$H_2O + CO_2 \underset{\text{respiration}}{\overset{\text{photosynthesis}}{\rightleftharpoons}} \underset{\text{(organic matter)}}{HCOH} + O_2.$$

During photosynthesis, the partial pressure of CO_2 decreases and pH increases. During respiration, CO_2 increases and pH decreases. Such variations occur in both diurnal and yearly cycles within shallow waters. In shallow waters pH values generally range from 7.5 to 8.5, although in some shallow tidal pools ranges may be as great as 6 to 11 (REVELLE and EMERY, 1957). In the deeper aphotic zones of the ocean, photosynthesis is negligible. Respiration of zooplankton and the oxidation of settling biogenic debris result in the consumption of oxygen and the production of CO_2. Thus it is not surprising that minimum pH values within the water column often occur at depths coincident with oxygen minimum layers, generally 300 to 600 meters.

The CO_2 balance also varies with the precipitation of calcium carbonate:

$$Ca^{+2} + HCO_3^- \rightleftharpoons CaCO_3 + H^+ .$$

When carbonate is precipitated, the hydrogen ion concentration increases (pH decreases); with dissolution, pH increases.

Calcium Carbonate in Sea Water

The thermodynamic solubility product of calcium carbonate (K_{sp}) is the ionic activity product of calcium and carbonate in any solution which is in equilibrium with calcium carbonate:

$$K_{sp} = a_{Ca^{+2}} \times a_{CO_3^=} .$$

In an aqueous solution at $T = 20$ °C, and $P = 1$ atm., K_{sp} of calcite is equal to about 4 to 5×10^{-7} moles^2/liter (see LI, 1967, for a discussion). K_{sp} is a function of temperature and pressure (not salinity). However for this constant to be applied in sea water, one must know the activities of calcium and carbonate in sea water, which are related to temperature, pressure *and* salinity.

For greater ease in computing the solubility of calcium carbonate in sea water, the apparent solubility product (K'_{sp}) is commonly used. The $CO_3^=$ concentration can be determined by knowing the alkalinity and pH. K'_{sp} is a function of temperature, pressure and salinity, and calculations of the K'_{sp}'s of calcite and aragonite in sea water have been made over a fairly wide range of environments (WATTENBERG, 1936; SMITH, 1941; MACINTYRE, 1965). The ranges of calculated values are summarized in Figs. 4 and 5.

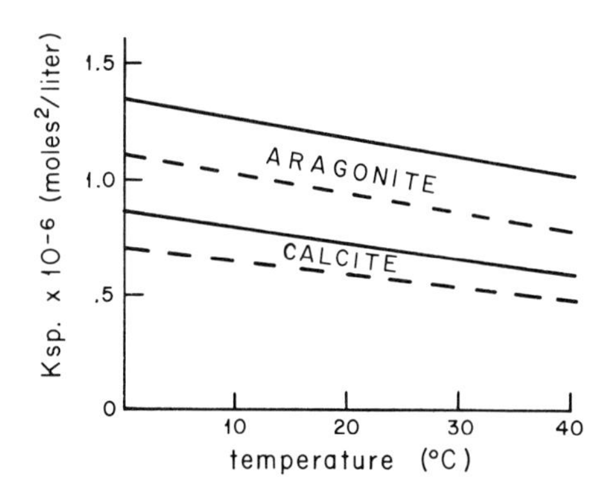

Fig. 4. Variation of K'_{sp} of aragonite and calcite with temperature at Cl = 19‰. The dashed lines are calculated K'_{sp} values based on LYMAN's (1956) K_2; the solid lines are calculated on the basis of Buch's (1938) K_2 value. Most calculated and observed values fall within these fields. After LI (1967)

Calculated K'_{sp} products are several hundred times smaller than the ionic products of calcium and carbonate in sea water, which would imply that the oceans are greatly over-supersaturated with calcium carbonate and that precipitation occurs throughout. We know that this is not the case, and the discrepency results from the fact that activity products of calcium and carbonate are much smaller in sea water than one would intuitively imagine. Decreased activity coefficients (because the ionic strength of sea water is than 1) together with ionic complexing (especially with $SO_4^=$ and Mg^{+2}) cause this decrease in the activity products (GARRELS and THOMPSON, 1962). Still, the activity products are sufficiently high to suggest that surface waters in warm climates are supersaturated with respect to both calcite and aragonite. However, because the solubility of $CaCO_3$ increases greatly with increasing water depth (pressure) and decreasing temperature and pH, most ocean water is undersaturated with respect to one or both polymorphs. Calcium carbonate dissolution is discussed at greater length in Chapters 8 and 9.

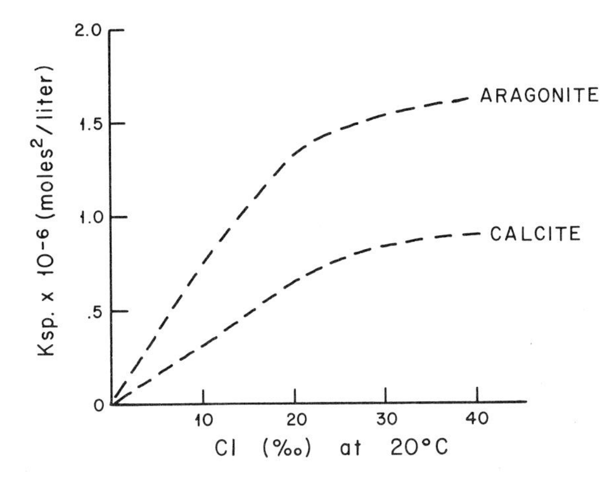

Fig. 5. Variations in the K_{sp} of aragonite and calcite with varying chlorinities, at 20 °C. After SMITH (1941) and WATTENBERG (1936)

Calcite, being the stable polymorph of calcium carbonate, should be the ultimate phase to survive. Observations, however, show that aragonite is the most common phase to be precipitated inorganically from sea water (for example, see GEE and others, 1932; REVELLE and FAIRBRIDGE, 1957). Many cations present in sea water can inhibit the precipitation of calcite. Among these (and with decreasing inhibition effects) are Cu, Zn, Ni, Mn and Mg (KITANO and others, 1969). Although magnesium has the least effect, it is far more abundant in sea water than the other elements and therefore plays the major role in inhibiting calcite nucleation (MONAGHAN and LYTLE, 1956; ZELLER and WRAY, 1956; LIPPMANN, 1960; CLOUD, 1962a; KITANO, 1962; KITANO and HOOD, 1962; USDOWSKI, 1963; SIMKISS, 1964; BISCHOFF, 1968a, c). LIPPMANN (1960) has speculated that this inhibition is caused by the hydration of magnesium ions; as calcite grows, incorporated magnesium, tends to remain attached to the hydroxyl ions and thus interfere with crystal growth. Crystal growth of aragonite, on the other hand, is not inhibited by magnesium.

Some cations, notably Sr, Ca and Pb, favor aragonite precipitation (ZELLER and WRAY, 1956; BISCHOFF, 1968b), but whether this is the result of calcite inhibition is not known. Elevated pH and temperature also favor aragonite precipitation (MURRAY and IRVINE, 1891; ZELLER and WRAY, 1956; WRAY and DANIELDS, 1957; GOTO, 1961; KINSMAN and HOLLAND, 1969) (Table 7). During experiments in which calcium carbonate was inorganically precipitated from sea water, KINSMAN and HOLLAND found that at temperatures between 15 and 17 °C 50 to 100% of the precipitate was aragonite; at 30 °C the entire precipitate was aragonite.

Table 7. Factors favoring aragonite and calcite precipitation

Factors favoring aragonite	Factors favoring calcite
1. Mg in solution	1. absence of Mg
2. high temperature	2. low temperature
3. high pH	3. low pH
4. sodium succinate, chondroitin sulfate	4. $SO_4^=$
5. Sr, Ba, Pb	5. $NaCO_3$; $(NH_4)_2CO_3$
	6. presence of many organic compounds, such as sodium citrate and sodium malate

Until recently, most workers agreed that the inhibiting effects of Mg would preclude the inorganic precipitation of calcite in sea water (for example, see CLOUD, 1962a; DE GROOT, 1965). However, experimental data show that magnesian calcites containing 10 to 15 mole % $MgCO_3$ have similar solubilies to aragonite (Fig. 6). The preponderance of magnesian calcite lutites and cements

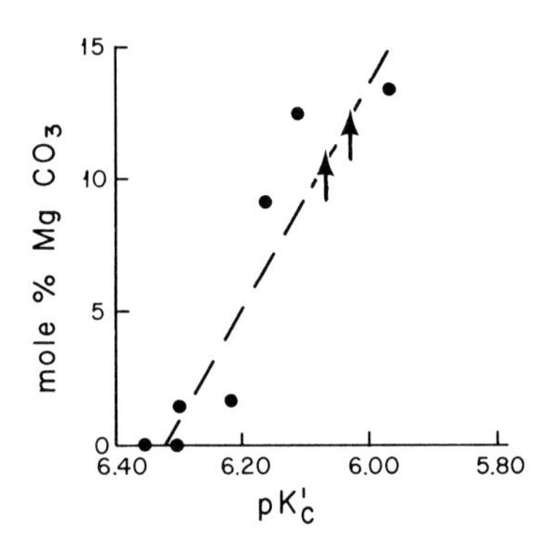

Fig. 6. Apparent solubilities of calcites with varying $MgCO_3$ content, in seawater at a temperature of 30 °C. Arrows indicate aragonite solubilities relative to magnesium calcite. $pK'_c = -\log_{10} [(Ca^{+2})(CO_3^=)]$. After WEYL (1967)

within this Mg-composition range suggests that such carbonates may be inorganically precipitated. In fact, they may be as abundant or even more abundant than inorganically precipitated aragonite (see Chapters 6, 8 and 10).

Low Mg concentrations within the precipitating liquid, low pH's and low temperatures are conducive for calcite precipitation (KITANO and HOODS, 1962; BROOKS and others, 1950; KINSMAN and HOLLAND, 1969). BROOKS and others precipitated a probable magnesian calcite (6 to 10 mole % $MgCO_3$) at 10 °C, and

KINSMAN and HOLLAND (1969) found magnesian calcite (8 to 10 mole % $MgCO_3$) is precipitated from sea water at temperatures lower than 16 °C. Other workers have precipitated magnesian calcite by decreasing the rate of precipitation by adding organic compounds (such as pyruvate, citrate and malate) which form more soluble calcium ion complexes (KITANO and KANAMORI, 1966; KITANO and others, 1969; TARUTANI and others, 1969) or by adding inorganic coupounds, such as Na_2CO_3 or $(NH_4)_2CO_3$ (LUCAS, 1948; BARON and PESNEAU, 1956; TOWE and MALONE, 1970). BISCHOFF (1968b) suggested that $SO_4^=$ inhibits aragonite precipitation and therefore may favor magnesian calcite precipitation. Creation of supersaturation under extreme physical and chemical conditions can also cause magnesian calcite precipitation (GLOVER and SIPPEL, 1967; CONRAD, 1968).

Chapter 2. Methods

A complete description of a carbonate sediment (or rock) should include measures of its texture, optical petrography, mineralogy and chemistry[1]. Each of these four parameters is discussed in this chapter. Further discussion can be found in standard petrography books (such as KRUMBEIN and PETTIJOHN, 1938; MÜLLER, 1967), as well as in papers quoted throughout this chapter.

Texture

The composition of terrigenous grains in large part is determined by the grain size of the sediment, which in turn is dependent upon the agents of transportation and deposition. In contrast, carbonate grains generally are products of the environment in which they are deposited. Being predominantly biogenic, the type of carbonate grain or the way in which it is formed can greatly influence the textural parameters of a sediment. For instance, mud-ingesting animals can produce a sufficient quantity of fecal pellets to transform a carbonate mud into a muddy sand. Similarly, a *Halimeda* meadow in a quiet lagoon may produce a sediment much coarser than would be deposited by terrigenous sedimentation. Thus, whereas size frequency data can give one a basic understanding of the environment of terrigenous deposition, similar parameters must be used with extreme care in describing carbonate sediment.

Many investigators have found that a complete size analysis of a carbonate sediment may require more time and effort than the results warrant. A simple tabulation of the various grain size catagories (gravel, greater than 2000 microns; sand, 62 to 2000 microns; silt and clay, less than 62 microns) may supply a sufficient amount of textural information about a carbonate sediment. These data can be acquired during the separation of the sediment into the various size fractions used in subsequent petrographic and geochemical analyses (Fig. 7). Some workers (GINSBURG, 1956; PURDY, 1963; MILLIMAN, 1967) have combined very fine sand (62–125 microns) and silt and clay into a single category, on the assumption that particles finer than 125 microns are too small for standard petrographic identification, and therefore should be grouped together.

1 Although primary sedimentary structures may have important environmental significance in the study of carbonate rocks (PETTIJOHN and POTTER, 1964), relatively few modern marine carbonates have been investigated with regard to their sedimentary structures. One notable exception is the report on the megastructures on Great Bahama Bank (BALL, 1967). Probably the lack of studies on smaller sedimentary structures is due to the difficulty in obtaining the required three-dimensional prospective.

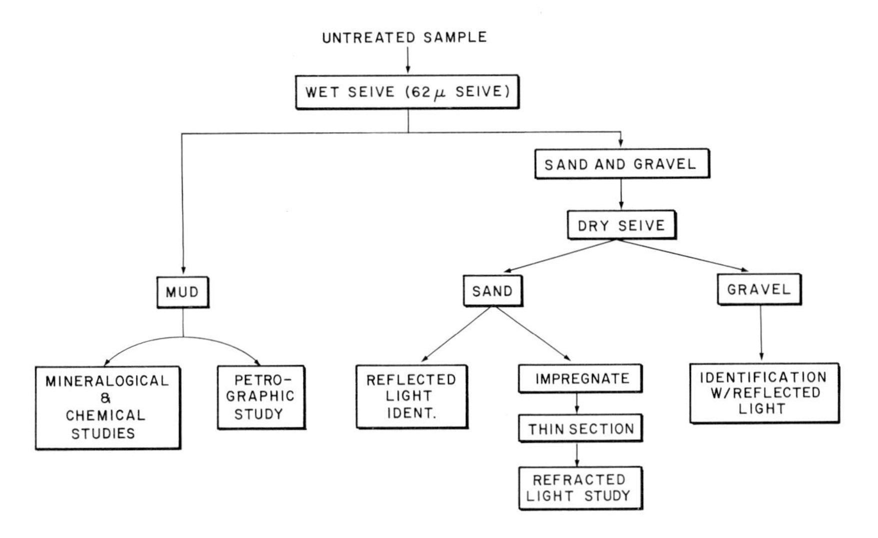

Fig. 7. Flow chart for textural and petrographic analyses

This is not to say that more detailed textural analyses cannot be used effectively in carbonate investigations. For example, sorting and skewness coefficients permit a rapid differentiation between beach and reef sands. The study of beach sands at Alacran Reef by FOLK (1967) is one example of how careful use of textural statistics can be utilized for the fuller understanding of carbonate sedimentation. Similarly, complete silt and clay analyses by standard pipette methods (see MÜLLER, 1967) may be extremely valuable in studying fine grained carbonate sedimentation.

Petrography

The components of an unaltered carbonate sediment dictate its mineralogy and chemistry as well as the texture. Thus few carbonate studies are complete without some microscopic investigation.

Carbonate sediments can be studied with reflected or transmitted light, or by electron microscopy. Identification keys for both reflected and transmitted light are presented in Appendices A and B, and the general properties are discussed in Chapter 3 and 4. The type of analysis depends upon the size and degree of alteration of the particles involved:

1. Many carbonate particles, especially small, reworked or encrusted grains, can be identified only by characteristics defined under transmitted light. Point-counting assures a reasonable estimate of total composition. If a grain count is required, one should include grain voids. If a weight percent is needed, grain voids should not be counted. In the former method, planktonic foraminifera within a given sample may constitute 70% of the sand-size grains, whereas using the latter method, they may only represent 25% of the sediment weight. Most carbonate

sediments contain five to eight major components, and 300 point counts generally is sufficient to assure reasonable accuracy (GINSBURG, 1956; PURDY, 1963).

2. The investigation of carbonate particles under reflected light is suited to those sediments containing unaltered grains or to those sediments requiring species identification or the differentiation of relict from modern components (for example, see SARNTHEIM, 1971). A comparison between loose grain counts and point counts can be tedious and mis-leading. One can divide the loose grains into various size intervals by sieving, and back-calculate the mean composition of the sediment; but one is still left with the problem of converting grain counts into weight percentages.

The study of gravel-size components is restricted by the sheer size of the individual grains. Only a few such grains can fit onto the standard petrographic slide. Unless one makes over-size slides, he must perform loose grain counts in order to identify enough grains to insure some degree of statistical accuracy. As with loose grain sand studies, the gravel fraction can be divided into various size grades; each fraction is then identified and the average composition computed by knowing the composition and weight percentage of each size interval.

Table 8. Approximate lower size limits (in microns) at which various skeletal components can be recognized in both reflected and refracted light. Adapted from LEIGHTON and PENDEXTER (1962)

	Reflected	Refracted
Mollusk	250 μ	125 μ
Coral	250	250
Foraminifera	62	62
Bryozoans	250	125–250
Barnacles	500	125–250
Echinoids	125	62
Halimeda	125	62–125
Coralline algae	500	125
spicules	<62	<62

3. Most carbonate particles can be identified down to sizes of about 125 microns (Table 8). The ease of identification of finer particles depends upon the type of grain. Spicules and micro-organisms are fairly easy to identify. Most fine comminuted particles, however, are difficult if not impossible to study; thus the distribution of these finer components within most sediments is poorly documented. MATTHEWS (1966) and MOLNIA and PILKEY (1972) have used immersion oils in the identification of smaller components. But even this method has serious limitations, and mineralogical and chemical studies have proved to be extremely helpful (MATTHEWS, 1966). Recent advances in electron microscopy may provide the ultimate answer in identifying of these finer carbonate particles. Using a scanning electron microscope, STIEGLITZ (1972) was able to identify most grains (in Bimini Lagoon sediment samples) coarser than 16 microns and many grains finer than 16 microns. The interested reader is referred to STIEGLITZ' Table 2 and the many illustrations presented in this valuable paper.

Sediment Impregnation

Loose sediments requiring study under transmitted light must be impregnated before thin sectioning. Sands are placed in containers (ice-cube trays are cheap and allow for easy removal of the hardened sample cubes), the impregnating medium is added, and the containers are placed in an evacuating jar. After 5 to 10 minutes under vacuum, the jar is brought back to atmospheric pressure, forcing the resin into the evacuated pore spaces. The less viscous the impregnating medium, the easier it can flow into the voids. This procedure is repeated two or three times, until further evacuation produces relatively few air bubbles escaping from the sample. Depending upon the amount of catalyst added to the impregnating medium, the cubes will take two to twenty-four hours to harden. These cubes then can be sliced and thin-sectioned exactly the way one could treat a rock sample.

Normal atmospheric pressure may not be sufficient to force the impregnating medium into the pore spaces of finer sediments. In this case a pressure bomb may be required. GINSBURG and others (1966) have described the "Shell Method" of impregnating cores; this technique also can be applied in the impregnation of finer sediments.

Peels

The thin section study of carbonates involves two problems which can hinder complete understanding of the sediment. First, the actual time and expense involved in thin-section production can be great. Second, the thickness of a thin-section (5 to 10 microns) may be considerably thicker than many of the smaller crystals present within the grains, resulting in the crystals appearing as an amorphous or cryptocrystalline mass.

The advent of cellulose and acetate peels (BUEHLER, 1948) offered a solution to both problems. Not only is the peel method fast and efficient, but peel impressions represent relatively thin topographic surfaces, thus allowing the detection of fine-grain petrographic characteristics. Mineral stains (see p. 20) also can be used on peels (KATZ and FRIEDMAN, 1965; DAVIES and TILL, 1968).

The peel method involves making a negative impression of an etched smooth slab with a surface-conforming material. Early techniques involved applying a liquid over the slab; upon hardening it was peeled from the surface (BUEHLER, 1948; BISSELL, 1957). Subsequent studies found that acetate sheets are neater and faster (15 minutes versus 12 to 24 hours) than the liquid method (LAND, 1962; MCCRONE, 1963).

The following technique was given by MCCRONE (1963) for the making of an acetate peel:

1. Grind the rock section to a very smooth surface. The finishing abrasive should be about no. 1000.

2. Expose the polished surface to a mild acid for a short period of time in order to achieve a slight etching. Wash and dry.

3. Cover slab with acetone and apply the acetate film over the slab, starting

from the center and moving out to the sides (this technique will eliminate most air bubbles).

4. Dry for about 15 minutes and then gently peel off and mount between glass slides.

HONJO (1963) states that ethyl cellulose, about 100 microns thick, is an excellent material for the peel. To make this material, add 10 g of ethyl cellulose powder to 100 ml of trichloroethylene and stir thoroughly. Pour onto a glass plate whose borders have been edged with tape, and let dry for 5 minutes. The dry sheet then can be removed and used.

Mineralogy

Knowing the mineralogy of a carbonate sediment can give added insight into both its petrographic and chemical composition and the environmental conditions under which it was deposited. Two major types of mineralogical determinations are used: organic and inorganic stains and X-ray diffraction. Both techniques have their use in modern carbonate studies, and each will be discussed below.

Staining

Knowing the mineralogy of individual carbonate grains (or even carbonate fillings) can be valuable, especially in documenting diagenesis. Microprobe analysis combines electron microscopy and quantitative accuracy, but until recently the time and expense of this method precluded its use by most workers. A more economical and practical method of determining mineralogy involves treating the sample with organic and inorganic stains. A discussion of the various staining techniques used in differentiating calcite, aragonite and dolomite was given by FRIEDMAN (1959). WARNE (1962) expanded the staining techniques to include all major carbonates plus gypsum and anhydrite.

Calcium carbonate can be differentiated from dolomite by two common methods. When boiled in a solution of Titian Yellow (30% NaOH), dolomite turns orange-red, while calcite and aragonite remain colorless. In the other method, Alizarin red-S in a slightly acid solution will turn calcite red, while leaving dolomite colorless; this difference in color is related to the rapidity with which pure $CaCO_3$ will react with dilute acids (DICKSON, 1966).

Aragonite and calcite also can be differentiated from one another by staining. After boiling the sample in a 0.1 N solution of $Co(NO_3)_2$ (Meigen's solution) for 20 minutes, aragonite will turn purple and calcite remain colorless (Friedman, 1959); however, magnesian calcite also may turn purple. In another method, aragonite turns black when exposed to Feigel's solution; calcite remains colorless. If the slide is then treated with Alizarin red-S (in a 30% NaOH solution) magnesian calcite will turn purple, while calcite remains colorless (FRIEDMAN, 1959) (Fig. 8). Many workers, however, have found that the differentiation between calcite and magnesian calcite is poor. Perhaps a more reliable technique involves the use of Titian Yellow (also called Clayton Yellow), which is specific for magnesium; when this stain is added to EDTA, magnesian calcite turns red and calcite remains colorless (WINLAND, 1971).

Fig. 8 shows the sequence of calcium carbonate staining as suggested by FRIEDMAN (1959). The formula for Alizarin red-S is: 0.1 g Alizarin red-S in 100 cc of 0.2% cold HCl. Feigel's solution is made by adding solid Ag_2SO_4 to 11.8 g of $MnSO_4 \cdot 7\,H_2O$ in 100 cc of distilled H_2O and boil. Cool and add two drops of dilute NaOH. Filter and store solution in dark bottle.

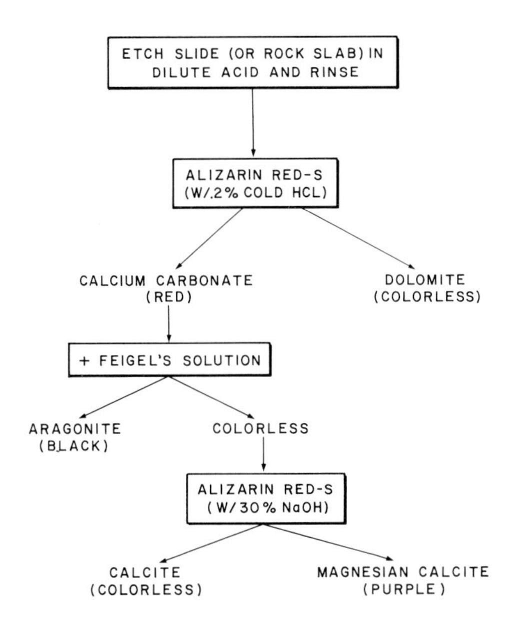

Fig. 8. Flow sheet for carbonate differentiation using the Alizarin red-S — Feigel's solution technique. After FRIEDMAN (1959)

The Clayton Yellow stain for magnesian calcite is made by adding 0.5 g of Titian Yellow, 0.8 g of NaOH and 2 g of EDTA to 500 ml of H_2O. The carbonate (loose grain or thin section) is etched with dilute acetic acid for 30 seconds and the excess acid blotted away. The slide (or grains) is then immersed in the staining solution for 30 minutes, excess blotted away and the slide is then dried by rapid warming. A few drops of mineral oil helps intensify the stain for petrographic study. WINLAND (1971) has noted that the stain tends to fade to reddish brown with time; the original stain can be regained by re-exposure to the staining solution.

X-ray Diffraction

X-ray diffraction techniques serve two important purposes in carbonate mineralogy studies. First, one can determine quantitatively the mineral composition of a carbonate component or sediment. Second, one can calculate the Mg (and thus the Ca) concentrations in calcites and dolomites. The principles of X-ray diffraction are well documented (for example, see CULLITY, 1956 or KLUG and

Table 9. Dominant X-ray diffraction peaks for aragonite, calcite and dolomite, in order of decreasing intensities. 2 Θ angles are for Cu-Kα radiation. Data from Joint Committee on Powder Diffraction Standards, 1970

dÅ	Degrees 2 Θ	I/I_1	hkl
	Aragonite		
3.396	26.24	100	111
1.977	45.90	65	221
3.273	27.25	52	021
2.700	33.18	46	012
2.372	37.93	38	112
2.481	36.21	33	200
1.882	48.36	32	041
2.341	38.45	31	130
1.877	48.50	25	202
1.742	52.53	25	113
	Calcite		
3.035	29.49	100	104
2.285	39.43	18	113
2.095	43.18	18	202
1.913	47.53	17	108
1.875	48.55	17	116
2.495	36.00	14	110
3.860	23.04	12	102
	Dolomite		
2.886	30.98	100	104
2.192	41.18	30	113
1.781	51.30	30	009
1.786	51.15	30	116
1.804	50.60	20	018
2.015	44.99	15	202
2.670	33.56	10	006
2.405	37.39	10	110

ALEXANDER, 1954) and the interested reader is referred to these texts. The lattice spacings and the 2θ angles (using $K\alpha$ radiation and Cu target) for aragonite, calcite and dolomite are listed in Table 9.

Quantitative Mineralogy

Many workers have used X-ray diffraction data to determine the amount of calcite and aragonite within a sediment. The simplest method of quantification involves the measuring of peak heights (LOWENSTAM, 1954a; TUREKIAN and ARMSTRONG, 1960) (Fig. 9). Any one type of calcite (for example, oyster calcite) tends to display a relatively constant peak height. But different calcites (for example, coralline algae versus oyster calcite) can display markedly different peak heights (Fig. 10). Therefore quantifying the mineralogy of a mixed carbonate by using peak height analysis can result in considerable error.

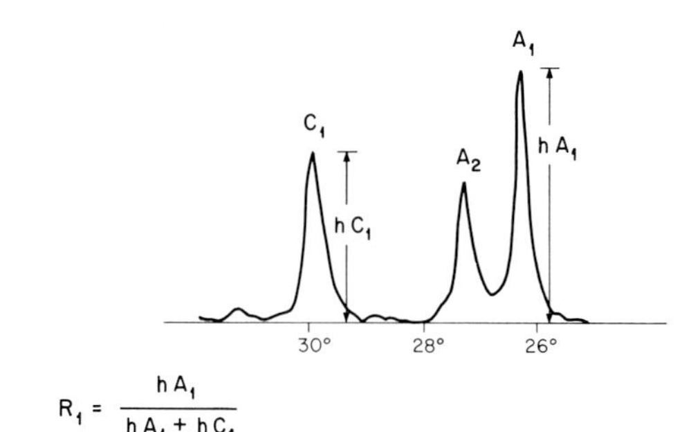

A $$R_1 = \frac{h A_1}{h A_1 + h C_1}$$

B $$R_2 = \frac{a A_1}{a A_1 + a C_1} = \frac{(hA \cdot wA)\,1/2}{1/2\,(hA \cdot wA) + 1/2\,(hC \cdot wC)} = \frac{hA\,wA}{hA\,wA + hC\,wC}$$

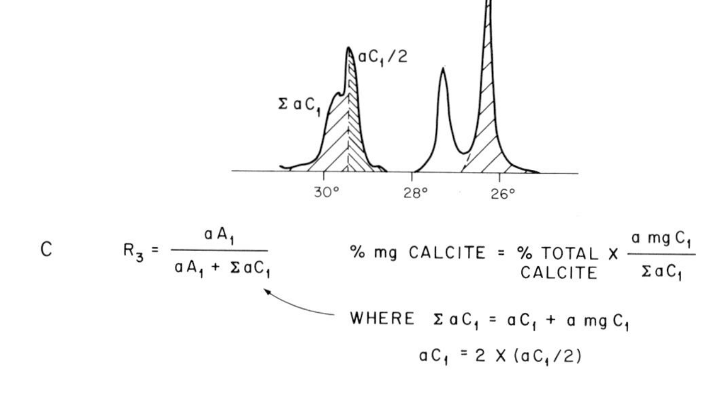

C $$R_3 = \frac{a A_1}{a A_1 + \Sigma a C_1}$$

$$\%\ \text{mg CALCITE} = \%\ \text{TOTAL CALCITE} \times \frac{a\ \text{mg}\ C_1}{\Sigma a C_1}$$

WHERE $\Sigma a C_1 = a C_1 + a\ \text{mg}\ C_1$

$a C_1 = 2 \times (a C_1/2)$

Fig. 9 A–C. Calculation of carbonate mineralogy by X-ray diffraction. In peak height analysis (A), the simple ratio $h_A/(h_A + h_C)$ is calculated. A geometric calculation can be used (B) with simple calcite-aragonite mixtures, in which both peaks are assumed to be triangles, and the area of each is calculated. The area is the product of the height and the width; in this illustration, the width was taken 1 cm above the background level. In practice, any width base can be used, as long as its location relative to the peak height remains constant. Where the calcite curve is composed of two or more types of calcite, a more complex analysis is required (C). The intensity of the free half of major peak (the right side of the low Mg calcite curve) is calculated and multiplied by two. This intensity is then compared to the total intensity in order to differentiate between it and the other calcites present. The integrated peak intensity is calculated by either planimeter analysis or by weighing

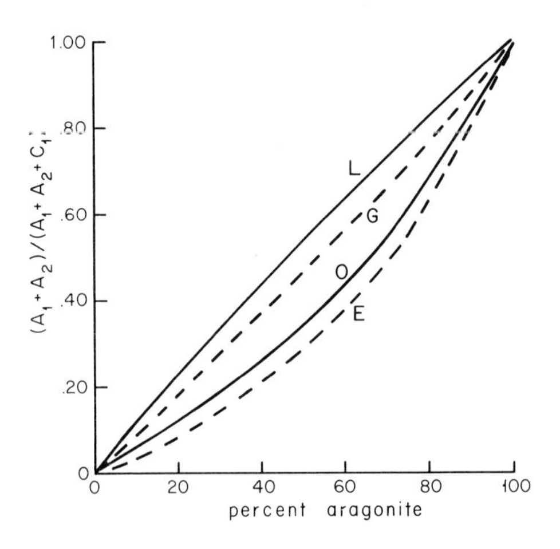

Fig. 10. Variation in $(hA_1+hA_2)/(hA_1+hA_2+hC_1)$ curves with various calcite standards. L = *Lithophyllum;* G = *Goniolithon;* O = *Ostrea;* E = Echinoid test. In all cases the aragonite was from the same mollusk shell

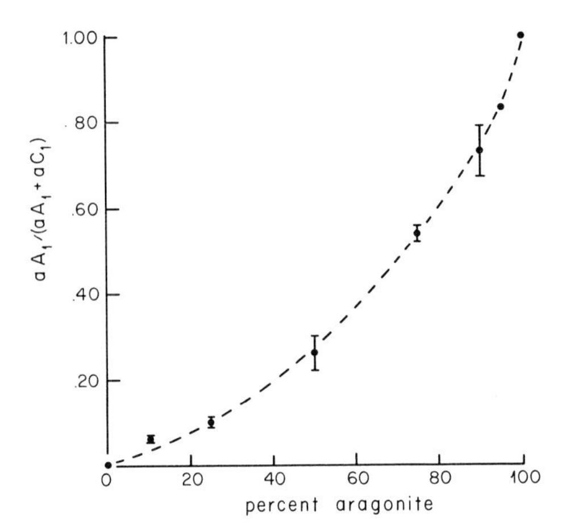

Fig. 11. Standard curve for aragonite determination, using peak area analysis. Each data point represents 8 to 10 analyses of various calcite and aragonite standards. In most instances the standard deviation (represented by vertical bars) is considerably less than 5% (absolute)

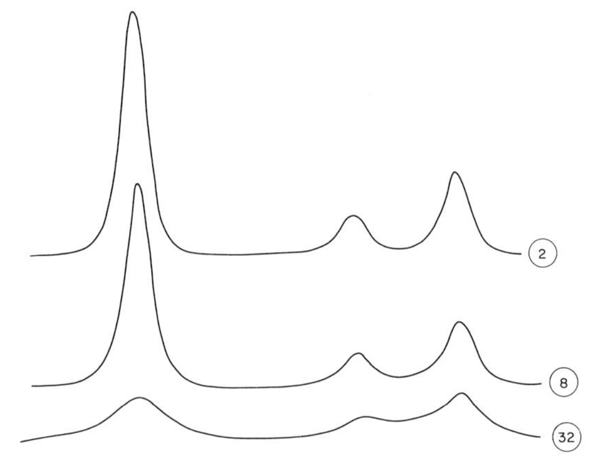

Fig. 12. X-ray diffractograms (A_1, A_2, C_1 peaks, respectively), showing the effect of extended grinding on the peak intensity of a 50-50 mixture of mollusk aragonite and oyster calcite. Grinding times (circled numbers) are in cumulative minutes; in each instance in scales are the same

A more accurate method of quantifying diffraction data is to measure the total peak intensity. Intensity is directly related to peak area, which can be calculated by planimeter (NEUMANN, 1965), by cutting out and weighing peak areas (PILKEY, 1964), or in simple mixtures, by geometric analysis (Fig. 9). The accuracy of peak-intensity analysis appears to be about 5% (CHAVE, 1962; Fig. 11). DAVIES and HOOPER (1963) claim that by careful integration of intensities, mollusk shell mineralogies can be determined to within an accuracy of 1%.

Several errors in the application of X-ray diffraction techniques, however, can seriously affect the quality of results; among these are particle size, the effects of grinding, $MgCO_3$ content within the calcite, and the crystallinity of the carbonate components:

1. The probability of the component crystallites within a sample displaying random orientation increases as particle size decreases. Below a certain size, however, the particles tend to become increasingly amorphous to X-rays (Fig. 12).

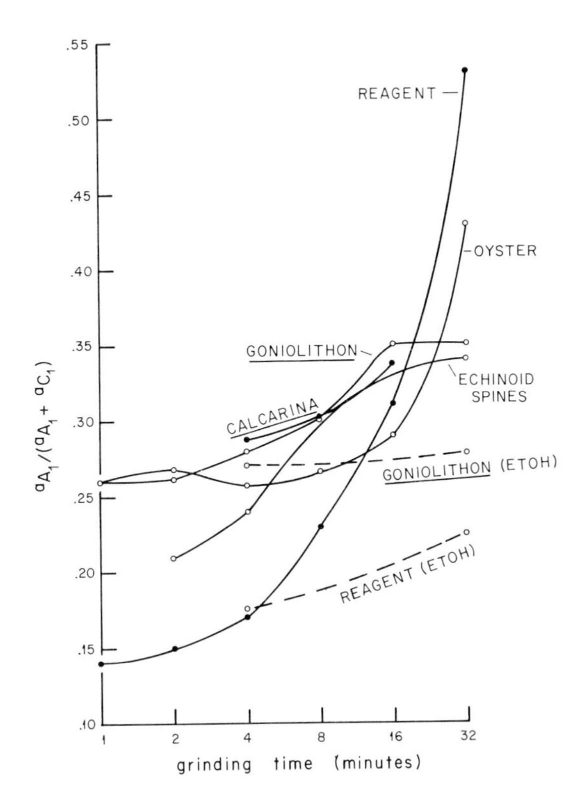

Fig. 13. Variation with grinding time of the $aA_1/(aA_1 + aC_1)$ ratio for mixtures of various calcites with a molluscan aragonite. Each mixture contained 50% aragonite and 50% calcite. Some of the samples were cooled during grinding with ethanol (ETOH), which appreciably decreased the degradation of the calcite

Calcite peak intensity decreases more rapidly with particle degradation than does aragonite intensity, although the rate of decrease varies with the type of calcite (Fig. 13). This variation in calcite intensity reduction in part may depend upon the variable utilization of organic matter within different skeletons as a lubricant during grinding.

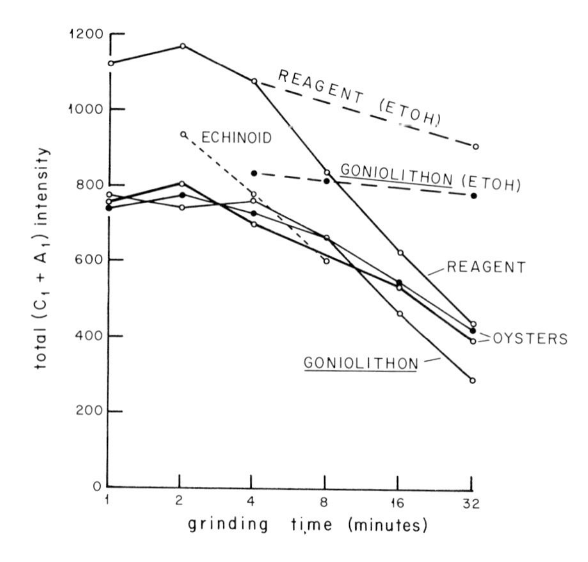

Fig. 14. Variations in total intensity of the aA_1 and aC_1 of several mixtures of 50% calcite and 50% aragonite. The actual patterns of the oyster calcite and aragonite mixture are shown in Fig. 12. The use of ethanol (ETOH) as a coolant will retard peak destruction

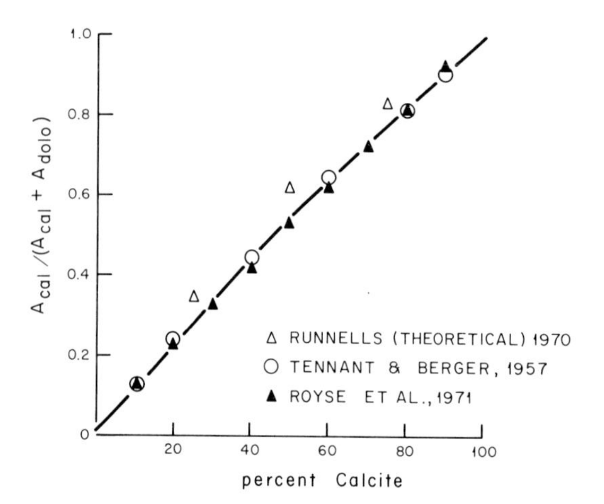

Fig. 15. Comparison of the integrated area of calcite relative to dolomite $(aC_1)/(aC_1+aD_1)$ with the percent calcite in the mixture. Observed values are somewhat lower than Runnell's predicted values

With increased grinding time, the total calcite + aragonite peak intensity decreases (Fig. 14) and the aragonite/calcite intensity ratio increases (Figs. 12 and 13). Using a pulverized fine sand sample, 250 mg in weight, the writer has found that 4 minutes of dry grinding provides an optimal peak intensity for X-ray analysis (Figs. 13 and 14). If one uses ethanol (ETOH) or other coolants while grinding, a longer grinding time is required.

In addition to these errors, over grinding can produce sufficient heating to alter aragonite into calcite (Goodell and Kunzler, 1965). However the high pressures caused by mechanical grinding also can alter calcite into aragonite (Burns and Bredig, 1956; Dachille and Roy, 1960; Jamieson and Goldsmith, 1960). Both these reactions, however, become quantitatively important only after extensive and vigorous grinding.

Table 10. Theoretical and observed peak intensities for various calcites and dolomites. Theoretical intensities are based on the solid-solution variation factor described by RUNNELLS (1970). Observed intensities are computed relative to oyster calcite, which is assumed to be 100. Standard deviations are in parentheses (unpublished data from the writer and B. D. BORNHOLD)

Calcite (dolomite)	Mole percent $MgCO_3$	Theoretical peak intensity	Observed peak intensity	Number of analyses
Reagent grade	1	100	221 (±56)	10
Oyster (*Crassostrea virginica*)	1	100	100 (±12)	13
Coccoliths	1	100	125 (±12)	9
Planktonic foraminifera	1	100	92 (±7)	9
Serpulid (*Serpula vermicularis*)	6	95	92 (±8)	12
Echinoid spines	8.5	93	107 (±7)	9
Alcyonarian spicules	9.5	92	91 (±6)	9
Coralline algae (*Lithothamnium*)	9.5	92	103 (±8)	9
Echinoid plates	11.5	90.5	142 (±14)	11
Coralline algae (*Lithophyllum*)	13.0	89	111 (±12)	6
Foraminifera (*Calcarina*)	16.5	86	111 (±3)	9
Coralline algae (*Jania*)	16.5	86	89 (±6)	8
Coralline algae (*Goniolithon*)	19.0	84	101 (±12)	12
Coralline algae (*Porolithon*)	20.5	83	100 (±13)	9
Coralline algae (*Lithophyllum*)	20.0	83	101 (±9)	6
Marine dolomite	50	62	99 (±13)	12
Dolomite crystal	50	62	270 (±48)	8

2. Calculations of solid-solution variations suggest that the replacement of calcium ions by magnesium, iron or manganese cations can greatly reduce the peak intensity of various carbonates. RUNNELLS (1970) calculated that dolomite should have only 61.5% the peak intensity of pure calcite. RUNNELS concluded that this factor may have serious implications in the quantification of X-ray data, but empirical evidence shows that the integrated intensity of dolomite is only slightly less than that of calcite (Fig. 15). Magnesian calcites have intensities that approximate (with the limits of error) those of pure calcites (Table 10). In fact the intensities of calcites with similar compositions (for example oyster calcite versus reagent grade calcite) vary more than do the intensities of calcities with different compositions. This observation is important for several reasons: First it indicates that most marine calcites (with the exception of echinoid plates) have generally similar peak intensities, thereby facilitating quantitative X-ray diffraction analysis (Table 10). Second, it suggests that reagent grade calcite should not be used as a calcite standard in X-ray diffraction analysis, since its peak intensity does not correspond with marine calcites and dolomites.

3. Probably the most important parameter which determines carbonate peak intensity is the crystallinity of the particular material. Those calcites with large crystal size, such as reagent grade calcite, some dolomite crystals and echinoid plates (TOWE, 1967), show the greatest peak intensities. Most marine calcites, on the other hand, display lower peak intensities. Generally all aragonites have similar intensities; the intensity of the 3.48 Å peak is approximately 0.35 that of the (211) calcite (oyster) peak.

$MgCO_3$ Content in Carbonate

As mentioned in Chapter 1, magnesium substitution decreases the lattice spacing in the calcite crystal. Calcite has a lattice spacing of 3.04 Å, while dolomite has a (211) spacing of 2.84 Å. Assuming a reasonably uniform distribution of magnesium throughout the crystal, one can determine the amount of magnesium within a calcite by measuring the shift of the calcite peak position towards that of dolomite (CHAVE, 1952). Several problems, however, are involved in relating the shift in the calcite peak to an absolute magnesium composition:

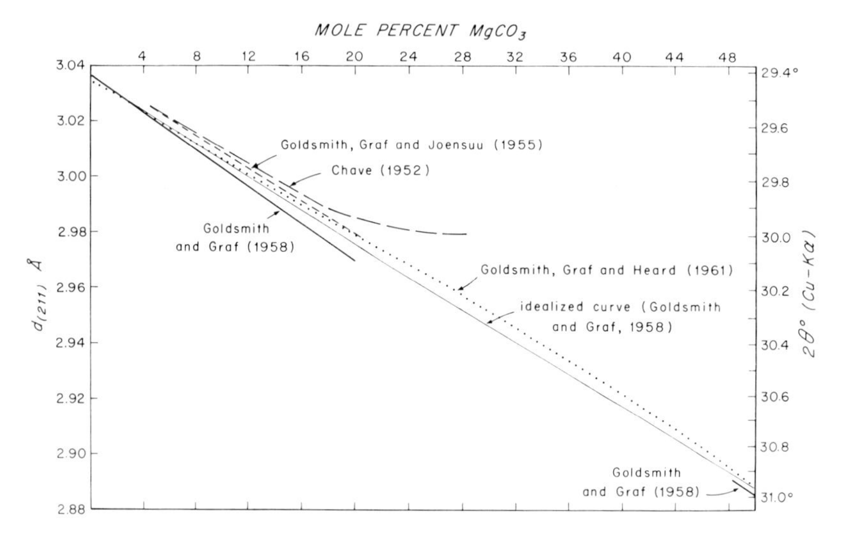

Fig. 16. Various curves relating the mole percent $MgCO_3$ in a calcite (or dolomite) with the position of the d(211) peak. The GOLDSMITH, GRAF and HEARD (1961) curve comes the closest to the idealized curve

1. Choice of an internal standard. The exact location of the diffraction peak on the recorder tracing depends upon the packing of the powder in the sample holder and the alignment of the X-ray machine. Recorded peaks often can vary by more than 0.10° 2θ from their actual values. Thus an exact measurement of the calcite peak requires the use of an internal standard, whose absolute lattice spacing is known. Fluorite (3.153 Å, 28.30° 2θ using Cu—$K\alpha$ radiation), quartz (3.343 Å, 26.66° 2θ), and halite (2.821 Å, 31.71° 2θ) often are used as internal standards since their peaks fall reasonably close to the main calcite peak. Aragonite (3.40 Å, 26.22° 2θ), if present, can be used. If an internal standard must be added, halite is a good choice, since it dissolves in water, leaving the sample relatively pure and capable of further chemical analysis.

2. Choice of lattice constant—composition curve. Five curves relating the shift of the lattice spacing to composition have been proposed (Fig. 16) and the spread between these various curves is surprisingly great. For example, a peak at 29.90° 2θ could represent an $MgCO_3$ concentration between 15 and

21 mole %, depending upon which lattice constant-composition curve is used. The curves proposed by CHAVE (1952) and GOLDSMITH and others (1955) were derived from values obtained from calcitic organisms, and were based on the premise that the distribution of $MgCO_3$ is unimodal and symmetrical. As will be seen in the following chapters, for at least some organisms this assumption is incorrect. The GOLDSMITH and GRAF (1958) curve has been used by many geologists, but a later curve (GOLDSMITH and others, 1961) appears to be more accurate (J.R. GOLDSMITH, 1970, written communication).

3. Distribution of $MgCO_3$ within the calcite. If the calcite curve is symmetrical or if Mg variations among the crystallites are small, the modal position of the X-ray peak will be coincident with the true $MgCO_3$ content. Since most calcites have X-ray curves which are more or less symmetrical, the peak shift method is generally valid. In some carbonates, however, internal distribution of $MgCO_3$ content can show considerable variation. For example, X-ray diffraction curves in many genera of coralline algae are markedly asymmetrical, so that the modal estimate generally underestimates the actual $MgCO_3$.

In most instances, if one knows the total Mg content or the mole % $MgCO_3$, the other value can be computed, since the Mg within a carbonate grain is usually confined to the $MgCO_3$ phase. However in coralline algae, Mg also can be present as brucite ($Mg(OH)_2$); thus the total Mg concentration within coralline algae generally is much greater than that present as $MgCO_3$ (MILLIMAN and others, 1971).

Chemical Composition

Calcium Carbonate Determination

A number of ways of determining the quantity of calcium carbonate in a sample have been developed. Among the more widely used methods are gasometry (HULSEMANN, 1966), acid-leaching (TWENHOFEL and TYLER, 1941), EDTA (BISQUE, 1961), combustion at 1000° (SCHOPF and MANHEIM, 1967) and atomic absorption (SIESSER and ROGERS, 1971). The type of analysis to be used depends upon the accuracy required, the nature of the sample and the analytical speed desired. Acid leaching is fast (100 samples per day) and it is reasonably accurate in sand and gravel-size sediments. On the other hand, acid can leach clay minerals and other clay particles probably escape during decanting. As a result, carbonate values of fine-grained sediments obtained by acid leaching tend to be higher than values obtained by standard gasometric analysis (SIESSER and ROGERS, 1971). EDTA analysis is time-consuming, as only about 10 samples can be analyzed per day.

Combustion techniques are extremely advantageous in the analysis of whole carbonate organisms. Heat treatments at 100 and 500 °C will give measures of the water and organic content of the material; a temperature of 1000 °C will drive off the CO_2 incorporated in the $CO_3^=$ (SCHOPF and MANHEIM, 1967). This method is also applicable in carbonate sediments containing minor amounts of clay or hydrated sediments. Gasometry is relatively accurate, results generally being reproducable within 0.1 to 0.2%. On the other hand, the analysis is slow

(10 to 15 samples per day) and accuracy is dependent upon atmospheric pressure, temperature and the skill of the analyst.

Elemental Analysis

Most marine carbonates contain one major cation (Ca^{++}), two minor cations (Mg^{++}, Sr^{++}) and a number of prominent trace cations (Fe^{++}, Mn^{++}, Ni^{++}, Na^{+}, K^{+} and others). The major anion is $CO_3^{=}$, but significant amounts of $SO_4^{=}$, P_2O_5, OH^{-} and Fl^{-} also may be present. The distribution of minor and trace elements within carbonates can be of both biologic and diagenetic interest. Many applications are discussed in subsequent chapters of this book.

The ease and precision of elemental analysis has increased greatly during the past 30 years. Before World War II most analyses were performed by wet chemical methods; now X-ray fluorescence (BIRKS, 1959) emission spectrography (NACHTRIEB, 1950; BURAKOV and IANKOVSKII, 1964; American Society for Testing Materials, 1971), flame photometry (DEAN, 1960) and recently, atomic absorption spectrophotometry (SLAVIN, 1968; PERKIN-ELMER, 1966) are used by most analysts. Atomic absorption is especially easy for non-chemists, but recent data suggest that there may be subtle differences between the results of atomic absorption analyses and those obtained by other methods (L.S. LAND, 1971, oral communication).

Stable Isotopes

The isotopic composition of a solid precipitated from solution is dependent upon the isotopic composition of the precipitating medium (UREY, 1947). In the $CaCO_3$—CO_2—H_2O system, the stable isotopes of oxygen and carbon, O^{18} and C^{13}, are incorporated into the carbonates by the following reactions:

$$H_2O^{18}(l) + 1/3\, CaCO_3^{16}\,(s) \rightleftharpoons H_2O^{16}(l) + 1/3\, CaCO_3^{18}\,(s)$$

$$C^{13}O_3^{=}(aq) + C^{12}O_3^{=}\,(s) \rightleftharpoons C^{12}O_3^{=}(aq) + C^{13}O_3^{=}\,(s)$$

(GRAF, 1960). Thus, knowing the distribution of stable isotopes within a carbonate can provide a valuable insight into the nature of the medium from which the carbonate was precipitated (MCCREA, 1950; UREY and others, 1951). Excellent reviews on these two stable isotopes have been given by GRAF (1960) and DEGENS (1967a).

Both O^{18} and C^{13} are present on the earth's crust in relatively small amounts. The ratio of O^{16} to O^{18} in air is about 480/1 (EPSTEIN, 1959), but in other systems this ratio can vary by as much as 10% (GRAF, 1960). The ratio of C^{12} to C^{13} is about 89/1, but the ratio varies greatly within different environments.

Isotope contents are determined with a high-precision mass spectrometer. Generally the content of both stable isotopes is reported as a deviation from a standard sample:

$$\delta O^{18} = 1000\left[\left(\frac{O^{18}/O^{16}\text{ sample}}{O^{18}/O^{16}\text{ standard}}\right) - 1\right]$$

$$\delta C^{13} = 1000\left[\left(\frac{C^{13}/C^{12}\text{ sample}}{C^{13}/C^{12}\text{ standard}}\right) - 1\right].$$

Oxygen isotope data are reported relative to either the Chicago Belemnite Standard (PDB) (UREY and others, 1951) or Standard Mean Ocean Water (SMOW) (CRAIG, 1961). These two standards are related as follows:

$$\delta O^{18}_{SMOW} = (\delta O^{18}_{PDB} \times 1.03) + 29.5$$

(DEGENS, 1965). Both oxygen and carbon isotope values are usually reported relative to the PDB standard (CRAIG, 1953, 1957a).

Assuming that a carbonate is precipitated in equilibrium with the ambient environment, the stable isotope composition of the carbonate will be dependent upon the mineral phase of the carbonate and upon the salinity and the temperature of the water. Non-equilibrium deposition can occur in those organisms which incorporate metabolic oxygen or carbon, or which can preferentially fractionate the isotopes (CRAIG, 1953; EPSTEIN and others, 1951, 1953; KEITH and WEBER, 1965). This effect of metabolic fractionation is especially critical in the deposition of C^{13}. Photosynthesizing plants preferentially incorporate C^{12} into their plant tissue (CRAIG, 1953; PARK and EPSTEIN, 1960); organic matter commonly displays δC^{13} values of -25 to $-30^0/_{00}$, and methane can have ratios as low as $-80^0/_{00}$ (HATHAWAY and DEGENS, 1969). This fractionation can have considerable effect upon the δC^{13} content of the surrounding medium and any carbonates precipitated within such an environment.

Mineralogy can play an important role in the fractionation of both oxygen and carbon isotopes (SHARMA and CLAYTON, 1965). The O^{18} content of aragonite at 25 °C is $0.6^0/_{00}$ higher than in coprecipitated calcite; in magnesian calcite, O^{18} values increase at the rate of $0.06^0/_{00}$ for each mole % of $MgCO_3$ (TARUTANI and others, 1969). Calcite is enriched in C^{13} relative to the HCO_3^- in the precipitating medium by $0.9^0/_{00}$; aragonite is enriched by $1.8^0/_{00}$ relative to calcite (ROBINSON and CLAYTON, 1969). The pH at which the carbonate is precipitated also can influence both the C^{13} and O^{18} values (EPSTEIN, 1959; DEUSER and DEGENS, 1967).

Oxygen and carbon isotope content also is dependent upon salinity. In open ocean waters the salinity remains more or less constant and thus variations in O^{18} and C^{13} due to salinity changes are minimal (EMILIANI, 1966). But in areas influenced by melt waters, river discharge or high rates of evaporation, the δO^{18} content can be affected strongly. Fresh waters exhibit considerably lower O^{18} and C^{13} ratios than do marine waters. Land runoff derives most of its CO_2 from plant respiration and humus decomposition, both of which are low in O^{18} and C^{13} (VOGEL, 1959). On the other hand, evaporation preferentially removes O^{16}, thus enriching surface waters with O^{18} (EPSTEIN and MAYEDA, 1953). Beyond salinities of about $50^0/_{00}$, however, δO^{18} values generally remain between $+3$ and $+5^0/_{00}$ (Fig. 17).

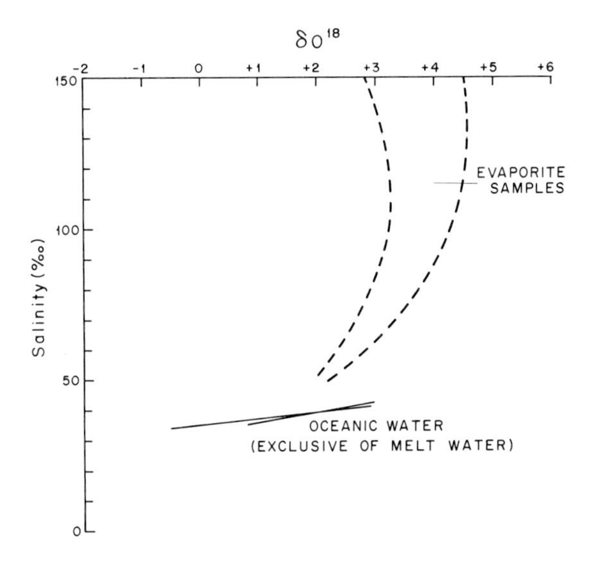

Fig. 17. Variation of O^{18} content in oceanic water with salinity. After EPSTEIN and MAYEDA (1953), LOWENSTAM and EPSTEIN (1957) and LLOYD (1966)

The preferential evaporation of O^{16} and C^{13} from surface waters decreases with increasing temperature; thus the δO^{18} and δC^{13} ratios increase. If salinity is held constant, the δO^{18} content can be used as a reliable indicator of the temperature at which the carbonate was precipitated (MCCREA, 1950; UREY and others, 1951; EPSTEIN and others, 1951, 1953). The variation of δO^{18} with temperature is shown in Fig. 18. The rate of change in δC^{13} is much smaller (EMRICH

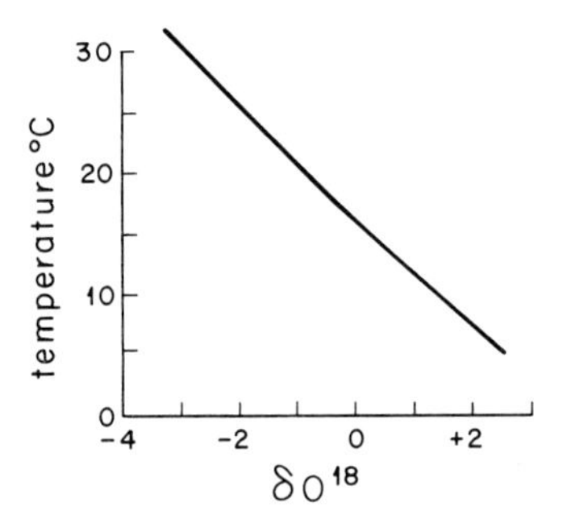

Fig. 18. Variation of O^{18} with temperature; $S = 35^0/_{00}$. After EPSTEIN and MAYEDA (1953)

and others, 1970) and thus cannot be used as a reliable indicator of temperature. The variation of δO^{18} and δC^{13} values in various marine carbonates is shown in Fig. 19 and is discussed further in the following chapters.

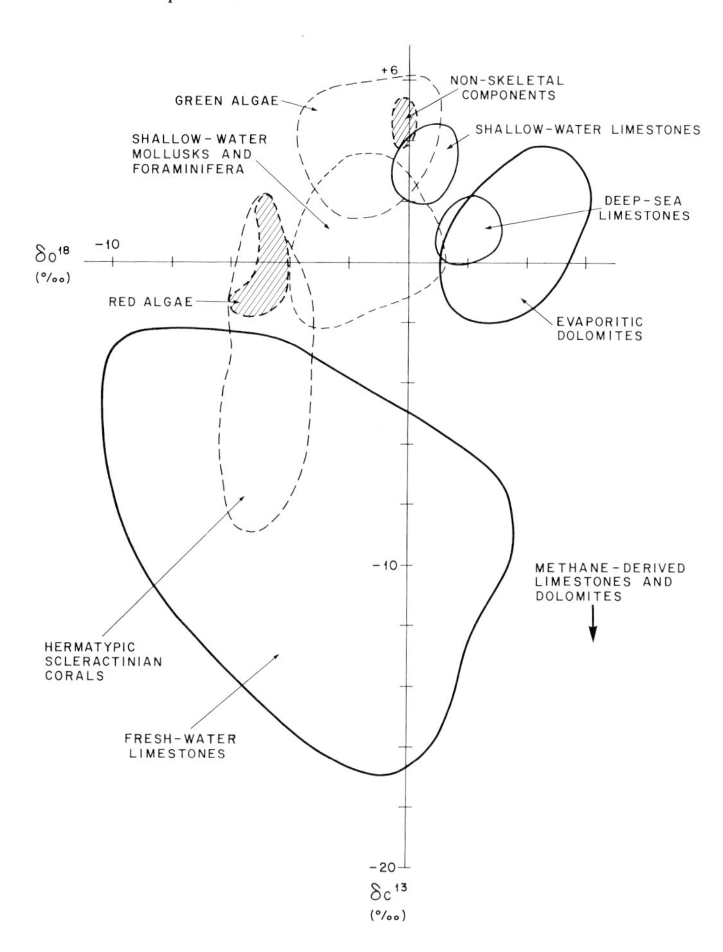

Fig. 19. Distribution of δO^{18} and δC^{13} values in various types of marine carbonates. Many of the data are derived from values reported elsewhere in this book

Part II. Carbonate Components

The carbonate particles within a sediment can be differentiated into organic and inorganic components. This classification, however, presents difficulties because of the uncertainty concerning the origin of many of the "inorganic" components, such as oolite and grapestone. To rectify this problem, ILLING (1954) classified grains as skeletal and non-skeletal. Non-skeletal components are defined as grains which do not appear to have been precipitated as skeletal tests. This does not necessarily mean, however, that they could not have once been skeletal nor that they have an inorganic origin; it only signifies that in their present state no skeletal origin can be ascertained.

The general ecologic, petrographic and compositional characteristics of various non-skeletal and skeletal carbonate components are discussed in Chapters 3 and 4. Mineralogical and chemical trends between various groups are summarized in Chapter 5.

Chapter 3. Non-Skeletal Components

Lithoclasts

Carbonate lithoclasts are defined as fragments of limestone (FOLK, 1959). ILLING (1954) used the term "derived grains" to describe such fragments. Generally the presence of lithoclasts within a sediment suggests the proximity of an outcrop, either subaerial or subaqueous. Most lithoclasts are recognized by their generally well-developed matrix and by grain boundaries which may cut constituent particles. Lithoclasts from older outcrops may contain particles which are compositionally different from those occurring in ambient sediments. In contrast, younger lithoclasts can contain components similar to those in neighboring sediments. Mineralogy commonly depends upon the age and source of the rock. Late Quaternary lithoclasts in continental shelf sediments off the eastern United States are composed mostly of aragonite and magnesian calcite, but most older lithoclasts, such as those derived from Miocene outcrops, are calcitic (MILLIMAN, 1972). In some sediments lithoclasts may represent the major source of calcitic grains; for example, the carbonate-rich sediments leeward of Andros gain most of their calcite from land-derived lithoclasts.

Pelletoids

Pelletoids are round, oblong or cylindrical (rod-like) carbonate grains that have a disoriented or cryptocrystalline granular texture (Plate I). Pelletoids range in length from 0.2 to 2 mm, but are usually between 0.2 and 0.6 mm in length with diameters between 0.1 and 0.4 mm. Colors are generally white, cream or tan, occasionally black or gray. The surface may be dull and pitted (especially in lower-energy environments) or smooth and polished (in high-energy environments).

Most pelletoids are fecal in origin, but others may be altered or recrystallized ooids (Plate II f). Some even may be severely altered coralline algal fragments (WOLF, 1962, 1965). Because of the diverse origin of these particles, the writer prefers the term "pelletoid" since it is descriptive and does not connote origin. Fecal pellets can be recognized by an internal texture of unoriented silt and clay-size grains, bound together by an organic and carbonate matrix. Fragments of larger grains also may be incorporated into the pellet. Microscopic algae and organic blobs can consitute and important part of the matrix (KORNICKER and PURDY, 1957; PURDY, 1963). FOLK and ROBLES (1964) estimated that 50 to 90% of the solid material in fecal pellets of the gastropod *Batillaria minima* at Alacran

Reef is organic material. This estimate is considerably higher than for most other fecal pellets (FRANKENBERG and others, 1967), but indicates the high organic content that can be present. Other dark "blebs" which occur in gray pellets may represent chitinous fragments (ILLING, 1954) or altered organic matter (GINSBURG, 1957; PURDY, 1963).

The shape and composition of fecal pellets can have important phylogenetic and taxonomic significance (KORNICKER, 1962). For exampe, ABBOTT (1958) used the size and shape of fecal pellets to aid in the identification of various gastropod species. However, because of the large number of animals that produce fecal pellets (MOORE and KRUSE, 1956, and MANNING and KUMPF, 1959, have listed more than 100 species in south Florida alone) and because most pellets are unsculptured ovoids (MOORE, 1939), identification of the pellet-producing organism is often difficult or impossible.

The best pellet producers are fine-grained deposit feeders (MOORE, 1939). Some organisms, such as holothurians and numerous crustaceans, produce loosely cohesive strings and rods that disintegrate soon after excretion. Recognition of these in bottom sediments often is difficult. GINSBURG (1957) concluded that nereid worms and crustaceans (especially *Callianasa*, SHINN, 1968c) are the most important pellet producers in south Florida. THORP (1936a), ILLING (1954) and NEWELL and RIGBY (1957) speculated that the fecal pellets in the Bahamas are mainly molluscan and crustacean, but CLOUD (1962a) found that the burrowing and mud-ingesting polychaete *Armandia maculata* is the most prolific pellet producer on the inner banks. The gastropod *Batillaria minima* is the dominant intertidal pellet producer (KORNICKER and PURDY, 1957; SHINN and others, 1969). Similar Cerithid gastropods are major pellet producers in the Persian Gulf (ILLING and others, 1965). Some fecal pellets are shown in Plate I, but the interested reader should refer to the numerous papers of MOORE (1939 and references therein) and to MANNING and KUMPF (1959) for further descriptions, as well as the annotated bibliography by HÄNTZSCHEL and others (1968) for further references.

Fecal pellets are produced in all environments, often in large quantities. For instance, FRANKENBERG and others (1967) estimated that the shrimp *Callianasa major* produces about 450 pellets per day per burrow. However pellets are preserved only in special environments. Carbonate sands and muds from reefs and lagoons can be aggregated into pellets, but these pellets generally are loosely bonded and easily decomposed (STARK and DAPPLES, 1941; GINSBURG, 1957). Fecal pellets are a dominant sedimentary component in subtidal and lower intertidal zones off low energy coast lines, but because of their friable nature, they are easily disaggreated (SHINN and others, 1969; ILLING and others, 1965; VON DER BORCH, 1965). In the deep sea fecal pellets are common, but they seldom are preserved.

Preservation of fecal pellets within a sediment generally requires intragranular lithification (see Chapter 10) of the individual pellets. Apparently the first step in this process occurs during dehydration within the gut of the excreting organism (MOORE and KRUSE, 1956). Soon after deposition, however, the pellets must harden or be subjected to rapid decomposition. PURDY has suggested that intragranular lithification is related to the decomposition of the organic mucus within

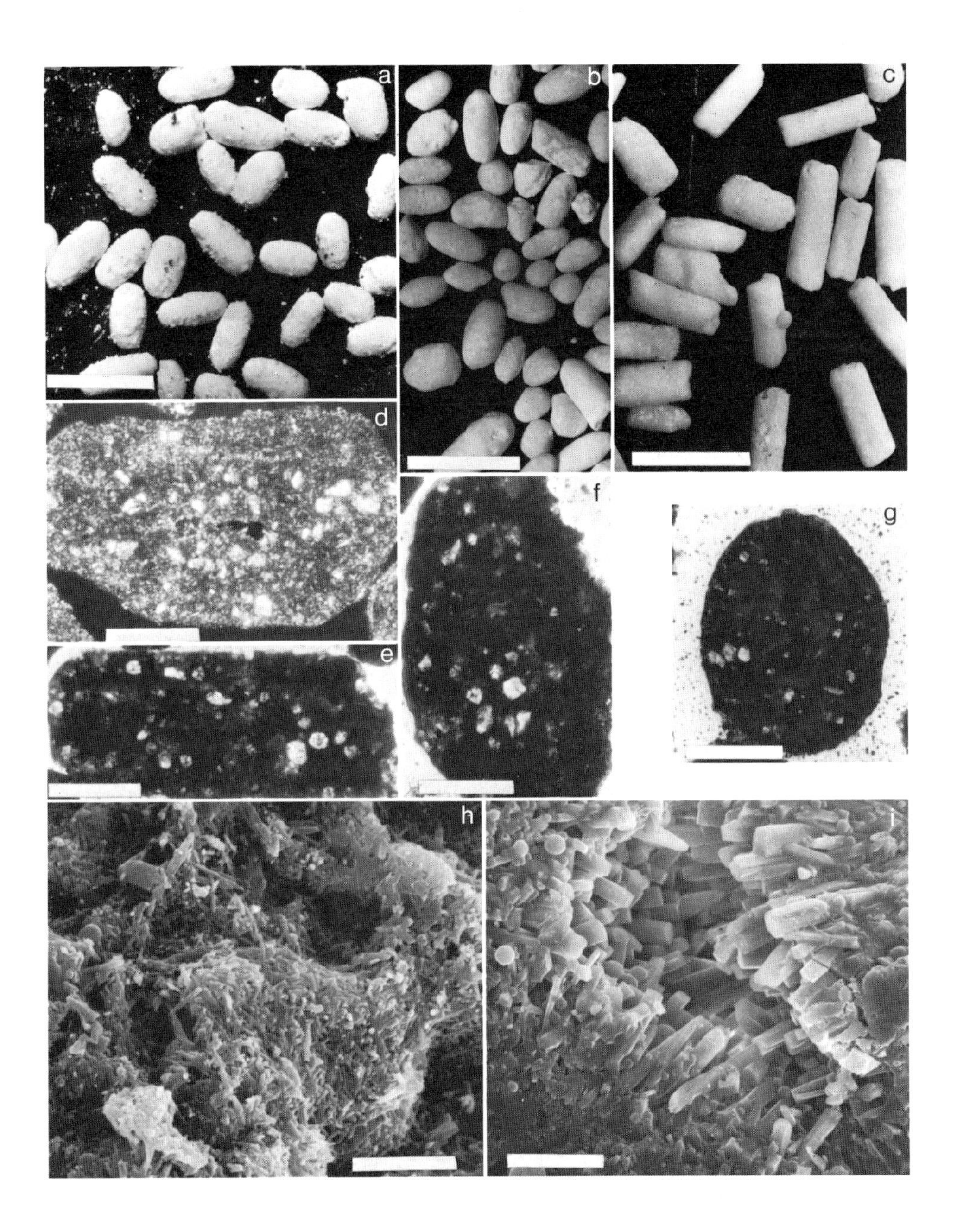

Plate I. Pelletoids. (a) Soft, unlithified pelletoids. Reflected light; scale is 1 mm. (b) Hard pelletoids. Reflected light; scale is 1 mm. (c) Hard cylindrical pelletoids. Reflected light; scale is 1 mm. (d)–(g). Refracted-light photomicrographs of pelletoids, showing the opaque nature of these grains. (d) is under polarized light. Scales are all 250 microns. (h) Micro-fabric of an unlithified pelletoid. Note the very small aragonite needles comprising the matrix, and the relatively large amount of pore space. SEM; scale is 8 microns. (i) Pore space within a lithified pelletoid. The intragranular cement is composed of large aragonite crystals. SEM; scale is 25 microns

Table 11. Location and type of non-skeletal carbonates found in modern carbonate sediments

Location	Latitude	Climate	Salinity	Non skeletal grains reported				Ref.
				Oolite	Pelletoids	Aggregates	Cryptoxlline grains	
Golfe de Bou Grara, Tunesia	33° N	temperate	supersaline	?	?	×		1
Lagoons on Texas Gulf Coast	27° N	temperate	supersaline	×		×		2
Great Bahama Bank	25° N	tropical	normal-supersaline	×	×	×	×	3
Persian Gulf	24° N	subtropical	supersaline	×	×	×	×	4
Batabano Bay, Cuba	22° N	tropical	supersaline	×	×	×	×	5
Gulf of Suez, Red Sea	29° N	tropical	supersaline	×				6
Southeastern Bahamas	22° N	tropical	normal	×	×	×	×	7
Pedro Bank	17° N	tropical	normal	×				8
Serrana Bank	14° N	tropical	normal	×	×	×	×	9
Malden Atoll, Pacific	4° S	tropical	supersaline	×				10
Shark Bay, Australia	26° S	subtropical	supersaline	×	×	×	×	11

1. Unpublished data.
2. RUSNAK, 1960; DALRYMPLE, 1964, 1965.
3. ILLING, 1954; NEWELL and RIGBY, 1957; PURDY, 1963.
4. HOUBOLDT, 1957; KINSMAN, 1964a; EVANS, 1966; KENDALL and SKIPWITH, 1969.
5. DAETWYLER and KIDWELL, 1959.
6. WALTHER, 1888; ROTHPLETZ, 1892; SASS et al., 1971.
7. DORAN, 1955; MILLIMAN, 1967.
8. ZANS, 1958.
9. MILLIMAN, 1969b.
10. J. I. TRACEY, JR., personal communication.
11. DAVIES, 1970a.

the pellets by bacteria. He also suggested that the specific composition of the mucus may be critical in the lithification. This would explain why only those pellets of certain animals are preserved.

Such lithification occurs in shallow waters which are greatly supersaturated with respect to calcium carbonate (Table 11). Equally important is the presence of a low-energy environment; for example, pellets form a prominent sedimentary component in the interior portions of the Bahama Banks (THORP, 1936a; ILLING, 1954; NEWELL and RIGBY, 1957; CLOUD, 1962a; PURDY, 1963). Both CLOUD and PURDY have estimated that pellets consitute about 30% of the total sediment leeward of Andros Island and over 75% of the sand fraction. In contrast, pellets are seldom abundant in higher energy sedimentary environments (Table 49).

Ooids

An ooid is defined as a "... grain which displays one or more regular lamellae formed as successive coatings around a nucleus ...", and in which "... the constituent crystals of the lamellae must show a systematic crystallographic orientation with respect to the grain surface" (NEWELL and others, 1960). An oolite refers to a sediment composed of ooids. Pisolites have the same characteristics as ooids but consist of grains greater than 2 mm in diameter. A recent discussion of term usage has been given by TEICHERT (1970).

Individual laminae generally range from 3 to 15 microns in thickness (Plate II); some ooids contain as many as 175 to 200 lamellae (PURDY, 1963). Component crystals average from 1 to 4 microns in length (SHEARMAN and others, 1970; LOREAU, 1970). The c-axis of the crystals can be oriented tangential to the grain surface (true of most aragonitic ooids) or radially (most calcitic ooids). RUSNAK (1960) found that individual laminae of the Laguna Madre ooids can have either orientation. In addition, ooids can have layers or inclusions of unoriented crystals. However those grains which display completely unoriented crystals, by definition, can not be considered ooids (contrary to RUSNAK, 1960), but rather as pelletoids (or pseudo-oolite, CAROZZI, 1960).

The diameter of most individual ooids generally ranges from 0.2 to 0.6 mm (Plate II) and are characteristically round, but their shape ultimately depends upon the shape of the grain around which the ooid has nucleated. Ooids can nucleate around nearly any available grain (usually sand size) regardless of its composition. In terrigenous sediments the nuclei tend to be non-carbonate, in carbonate areas they are mostly carbonate.

The tan to cream color of most ooids together with the pearly luster (the result of mechanical abrasion; NEWELL and others, 1960) and the ovoid to spherical shape are diagnostic properties under reflected light. Ooids are easily recognized in thin section by their lamellae and the pseudo-uniaxial interference figure under crossed nicols. ILLING (1954) differentiated true (or mature) ooids from superficial ones on the basis of the number of layers around the nucleus. CAROZZI (1957) suggested that superficial ooids be restricted to those ooids containing only one lamina. However the justification in restricting the definition of this term seems artificial since the number of laminae does not necessarily reflect the ability of an ooid to precipitate (PURDY, 1963).

Ooids contain perforating blue-green and green algae, together with lesser amounts of fungi and bacteria. The absolute amount of organic matter in ooids is not great; NEWELL and others (1960) report about 0.1% in the Bahama ooids. But the organic matter has a relatively low density (compared to aragonite) and thus occupies a far greater volume than chemical data would suggest. Furthermore, organic constituents concentrate along laminae or in unoriented aragonitic layers (BATHURST, 1967a; SHEARMAN and others, 1970), thus being a visually prominent part of the ooid. The characteristic tan to brown color of modern ooids is a direct result of the organic matter distributed throughout the ooid (SHEARMAN and others, 1970). NEWELL and others (1960) reported that the dominant algae within the Bahaman ooids are the blue-green *Entophysalia deusta* and the green *Gomontia polyrhiza*. Both species are known to bore carbonate and in some instances perforate the ooid lamellae. More generally, however, the algae tend to coat and congregate around the lamellae rather than perforating them.

Modern ooids occur only in rather restricted subtropical and tropical environments. In subtropical areas the waters are supersaline, while in subtropical areas the waters can be either normal or supersaline (Table 11)[1]. Within these environments oolitic-rich sediments are restricted to shallow depths, generally less than 2 meters, although ooids can form at considerably greater depths (up to 10 to 15 meters; NEWELL and others, 1960, MILLIMAN, 1967, 1969b). Shallow-water ooids form mainly in agitated environments, often in areas influenced by tidal currents or wave action. Ooids forming in non-agitated environments tend to be superficial or have incomplete discontinuous layers around the nucleus, as well as having a dull, chalky appearance (FREEMAN, 1962; BATHURST, 1967a).

Aggregates and Cryptocrystalline Lumps

The term "aggregate" describes two or more fragments joined together by a cryptocrystalline matrix which constitutes considerably less than half the grain (Plate III). The matrix consists of disoriented aragonite needles, usually less than 10 microns long, or aragonitic discs, about 5 to 10 microns in diameter (Plate IIIe, f). Under refracted light this matrix, which contains a large amount of organic matter (PURDY, 1963), exhibits a tan to brownish color.

Although previous workers (SORBY, 1879; VAUGHAN, 1919) recognized the presence of such grains, it was ILLING (1954) who first realized that these grains were of modern origin. ILLING termed these grains "lumps" and defined several types: 1. The term "grapestone" was reserved for those lumps from which round grains protrude, thus resembling a bunch of grapes. 2. "Botryoidal lumps" were grapestones with superficial oolitic coatings. PURDY (1963) has argued that

1 Modern ooids have been reported in the sediments of the Florida Keys (GINSBURG and others, 1963), but little else is known about these ooids. The ooids off north Africa (LUCAS, 1955; HILMY, 1951) appear to be relict grains that were precipitated during the Holocene transgression (FABRICIUS and others, 1970). It is possible, however, that ooids are presently forming in some restricted lagoons along the north African coast, such as Golfe de Bou Grara, in southern Tunisia.

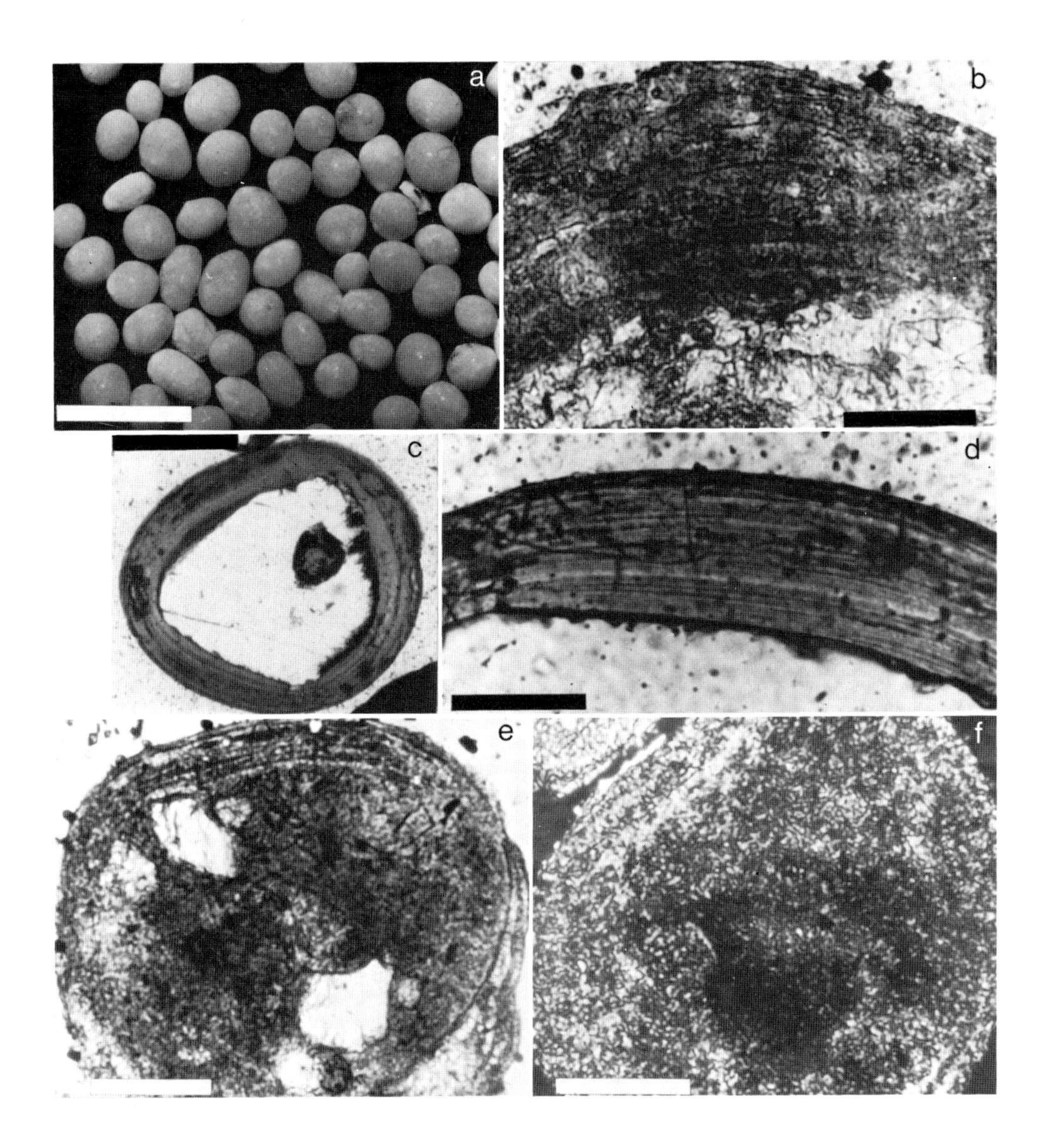

Plate II. Ooids. (a) Ooids from Cat Cay, Great Bahama Bank. Reflected light; scale is 1 mm. (b) and (d) Bands of laminae surrounding a nucleus. Refracted light; scale is 250 microns. (c) Superficial ooid coating a quartz grain. Refracted light; scale is 250 microns. (e) Superficial ooid surrounding a pelletoid. The laminae are poorly defined on the lower left. Refracted light; scale is 100 microns. (f) Cryptocrystalline ooid, closely resembling a pelletoid. It is because of such occurrences that the term "pelletoid" is preferred to "fecal pellet". Polarized refracted light; scale is 100 microns

botryoidal lumps should be treated as ooids with grapestone nuclei. 3. "Encrusted lumps" had smoother outer surfaces than grapestones, and often hollow interiors. Many workers have found it difficult to distinquish between ILLING'S grapestone and encrusted lumps. DAETWYLER and KIDWELL (1959) grouped these two components into "constituent grains". PURDY (1963) defined grapestones as having hollow interiors, thus overlapping with ILLING'S encrusted lumps. Because grapestone and encrusted lumps represent a continuum rather than discrete grain types, the writer prefers the inclusive term "aggregates".

Several other types of aggregates occur in modern carbonate sediments. ILLING (1954) described "friable aggregates" in Bahamian sediments as being "... irregularly shaped, loosely bound aggregates of silt particles ... (with) a chalky white texture that are easily disrupted with a pin point". PURDY identified these grains as "mud aggregates". ILLING and PURDY both found these grains in all stages of cementation, with the final stage being either a grapestone or cryptocrystalline lump. PURDY (1963) defined "organic aggregates" as grains cemented by organic slime or by a binding skeletal matrix. Where the encrusting organism is recognizable, such as coralline algae, serpulids or encrusting foraminifera, the grain can be classified as skeletal. Only small portions of mud aggregates and organic aggregates are found generally in non-skeletal sediments, but locally they can be abundant.

Cryptocrystalline grains (lumps) have been defined by PURDY (1963) as grains that have been altered, filled and encrusted (inorganically) to the point where their original source cannot be recognized. The term "lump" (MILLIMAN, 1967) is essentially the same, and refers to grains in which the matrix constitutes more than 50% of the grain (as opposed to aggregates). In non-skeletal sediments, cryptocrystalline lumps often are composed of altered aggregates, while in skeletal-sediments, *Halimeda* and foraminifera (especially peneroplids and miliolids) are major sources of cryptocrystalline lumps (PURDY, 1963). In many instances the only petrographic difference between aggregates and cryptocrystalline lumps is the greater amount of matrix found in the latter. Aggregates and cryptocrystalline lumps generally form in sandy sediments, although they are exposed to lower current and wave energies than sediments found in oolitic areas.

→

Plate III. Aggregates and Lumps. (a) Aggregates. Reflected light; scale is 1.5 mm. (b) Aggregate from Hogsty Reef lagoon. Individual particles (many of which are pelletoids) are cemented together, but in this particular aggregate, matrix is lacking. Polarized refracted light; scale is 250 microns. (c) Close-up of the cemented contact between two constituent grains within an aggregate. Refracted light; scale is 50 microns. (d) Broken edge of aggregate. Of particular interest is the presence of disk-like carbonate "crystals". These are seen in other shallow-water non-skeletal carbonates and cements (as seen in many of the following plates). SEM; scale is 25 microns. (e) Matrix of aggregate. Main components are the disk-like carbonate grains; preliminary micro-probe analyses suggest that these are high-strontium aragonite. Also present are some diatoms. SEM; scale is 8 microns. (f) Refracted-light photomicrograph, showing the matrix within an aggregate. Scale is 50 microns. (g) A lump is defined as a particle in which more than half the grain is matrix. Such a particle is pictured here. Refracted light; scale is 250 microns

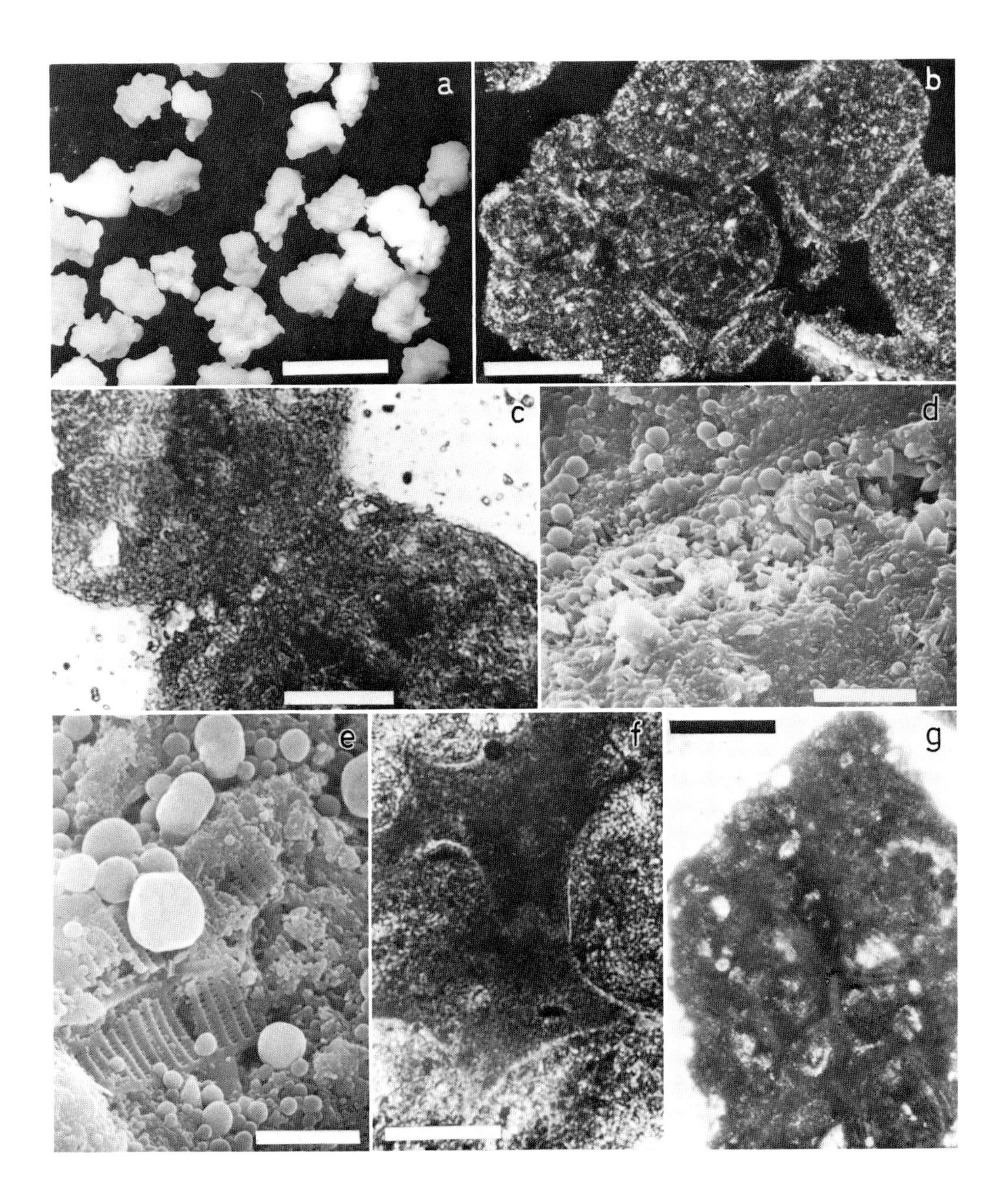

Plate III.

Table 12. Composition of modern marine non-skeletal carbonate grains. Most data are based on unpublished analyses by the writer. Other data are from WALTHER (1888), ZELLER and WRAY (1958), NEWELL and others (1960), KAHLE (1965), MILLIMAN (1967), KINSMAN (1969b), MÜLLER (1970), TILL (1970), and F.T. MANHEIM (unpublished data)

Carbonate component	Org. matter %	Mineral.	Percent				Parts per million									
			Ca	Mg	Sr	Na	Fe	Mn	K	Cu	Ba	Si	P	S	Al	Cr
Pelletoids																
Great Bahama Bank [a]	4.7	87% Arag.		0.41	0.80	0.23	1290	80	<20	<50						
Great Bahama Bank (rexl)	2.3	96% Arag.		0.19	1.00	0.23	274	10	7	10						
Oolite																
Great Bahama Bank	1.9	99% Arag.	38.2	0.08 (0.05–0.19)	0.99 (0.93–1.06)	0.24 (0.21–0.28)	300 (95–500)	3	11	<20	<65	376	70	1000	<300	
Great Inagua Island, Bahamas	2.5	88% Arag.		0.32	0.94	0.44	350	7	<30	<20						
Bimini Lagoon, Bahamas[a]		97% Arag.	37.8	0.69	0.96		14	5			9	730			<100	<3
Persian Gulf		99% Arag.		0.12 (0.06–0.24)	0.97 (0.90–1.06)											
Suez, Red Sea	0.34	Arag.	37.8	0.05		0.34										
Aggregates																
Great Bahama Bank		95–96% Arag.		0.40 (0.34–0.44)	0.92 (0.92–0.93)	0.32										
Serrana Bank	2.6	88% Arag.		0.63	0.97	0.39	285	47	<10	<30						

[a] 87% oolite.

Composition of Non-Skeletal Carbonates

The composition of unrecrystallized pelletoids is directly related to the composition of their constituent fine-grained sediments, generally aragonite and magnesian calcite. Organic matter tends to be higher in unrecrystallized pellets than in other non-skeletal grains, about 5% (Table 12). Recrystallized pellets contain less organic matter and are composed almost entirely of aragonite. The chemical composition of pelletoids is similar to that of the other non-skeletal fragments (Table 12). Modern ooids are dominantly aragonite (a point first recognized by WALTHER, 1888). A nucleus with a different composition may affect the ooid's total mineralogy, depending upon the size of the nucleus. Organic carbon in ooids is less abundant than in other non-skeletal fragments, but tends to be higher than the 0.13% reported by NEWELL and others (1960) (Table 12). Strontium values are higher than in any other marine carbonate, averaging close to 1%. Magnesium is very low, often less than 0.1%. Trace elements also tend to be low.

Ooids in Laguna Madre (Texas) contain both magnesian calcite and aragonite. The magnesian calcite is restricted to the radially oriented crystal layers and the aragonite to the tangentially oriented and micritic crystals (FRISHMAN and BEHRENS, 1969). The magnesian calcite contains 13 to 15 mole % $MgCO_3$, while the aragonite contains about 1.1% Sr (L.S. LAND, 1970, written communication). The reason for the variable composition and crystal structure in the Laguna Madre ooids is not known.

Fluorine content in modern ooids averages 1600 ppm, a concentration that is almost twice as high as any other marine carbonate yet analyzed (CARPENTER, 1969). Available isotope data suggest that the O^{18}/O^{16} ratio in non-skeletal fragments is in equilibrium with the ambient environment. The δC^{13} ratio usually ranges between +4 and +5‰ (DEUSER and DEGENS, 1969; LLOYD, 1971).

Aggregates have similar compositions to other non skeletal fragments. Magnesian calcite (usually less than 10% of the total carbonate) may be unrecrystallized deteritus or may represent an prior stage of cementation (WINLAND and MATTHEWS, 1969; see 50). When calculated on a 100% aragonite basis, the concentrations of Sr, Mg and various trace elements are similar to those found in ooids and pelletoids. Because of the difficulty in separating them from other grains, no analyses have been made on cryptocrystalline lumps, but sediments abundant in these lumps have compositions similar to other non-skeletal grains (MILLIMAN, 1967).

Origin of Non-Skeletal Carbonates

Non-skeletal carbonate grains (ooids, pelletoids, aggregates and cryptocrystalline lumps) have many similar compositional characteristics (Table 12). Most grains contain cryptocrystalline aragonite cement. Organic matter generally is concentrated within the cement or along individual laminae. The main difference between the various grains are shape, the arrangement of constituent aragonite needles and the relative importance of the matrix.

Ooid precipitation is most rapid in shallow aggitated environments. The high degree of aggitation undoubtedly explains the evenly coated laminae found in ooids; BATHURST (1968) suggests that aggitation also may prevent ooids from cementing together. Ooids which have not been exposed to such high energy conditions tend to display uneven or thin laminations, random crystal orientation and rough surface textures (FREEMAN, 1962; BATHURST, 1967a; DONAHUE, 1969). Aggregates and cryptocrystalline lumps occur in deeper waters or in lower energy environments bankward of ooid deposits. Such sediments often are covered mucilagenous algal mats. Aggregate grains do not contain discrete laminae, but rather a disoriented cement matrix, perhaps the result of being precipitated in a non-aggitated environment. Because of their initially friable nature, pelletoids are usually found in even lower energy environments, often in muddy sediments bankward of the aggregate facies (Fig. 59).

Thus the petrographic and morphologic differences between the various grain types may reflect the various environmental conditions of precipitation (PURDY, 1963; KENDALL and SKIPWITH, 1969). RUSNAK (1960) also has suggested that petrographic differences may reflect differences in the rate of precipitation. The *mechanism* of non-skeletal carbonate precipitation, however, is still debated.

Theories concerning oolite formation are numerous and diverse, due in part to the highly variable definitions of "ooid" and the wide range of oolitic sediments and rocks that have been investigated. Basically three types of theories have been proposed: algal, microbial and inorganic precipitation. Other theories, such as ooids being fossil insect eggs (VIRLET-D'AOUST, 1857) or aeolian aggregates (MATTHEWS, 1930) have been discarded. Excellent reviews on ooid formation have been given by LINCK (1903), BROWN (1914), EARDLEY (1938), ILLING (1954), NEWELL and others (1960) and BATHURST (1967a, 1971).

ROTHPLETZ (1892) and WETHERED (1895) were impressed with the amounts of algal tubules found within ooids, and proposed that ooids were algally precipitated. This theory was favored by many workers at the turn of the last century. In subsequent years, however, the algal theory fell into general disrepute and had only a few scattered supporters (such as DANGEARD, 1936, and NESTEROFF, 1956). It has been known for a long time (see HARRIS and JUKES in WETHERED, 1895) that many of the algae within ooids are boring varieties and it was argued that these forms probably represent post-depositional alterations rather than syngenetic additions. Ooids (and pisolites) form in caves and steam boilers and have been precipitated in the laboratory (MONAGHAN and LYTLE, 1956; DONAHUE, 1965), thus demonstrating that algae are not critical for precipition.

The most popular theory in recent years has concerned inorganic precipitation. LINCK (1903), as a result of extensive laboratory experiments, suggested that ooids can precipitate inorganically. Many subsequent workers have supported this theory (EARDLEY, 1938; ILLING, 1954; NEWELL and others, 1960; RUSNAK, 1960; MILLIMAN, 1967), but the exact mechanism of inorganic precipitation is not clear. ILLING (1954) suggested that upwelling and subsequent warming of deep ocean waters on shallow banks can produce supersaturations sufficiently high to initiate carbonate precipitation. If the warming of upwelled water were critical, however, one should detect significant increases in water temperatures across shallow banks, as well as high concentrations of ooids on the windward sides of such banks.

In at least some areas, neither assumption holds. Water temperatures at several small Caribbean atolls remain constant across both the reef flats and lagoons, and non-skeletal sediments are most prominent in the leeward parts of the atolls, not the windward sections (MILLIMAN, 1967, 1969 b).

Other workers have suggested that supersaline conditions can result in sufficiently high supersaturations with respect to calcium carbonate to favor precipitation (SMITH, 1940; CLOUD, 1962a; BROECKER and TAKAHASHI, 1965). While non-skeletal precipitation in sub-tropical areas is limited to supersaline environments, in the tropics similar sediments can form in waters with normal salinities (Table 11). This suggests that normal tropical waters are sufficiently supersaturated with respect to calcium carbonate to facilitate non-skeletal precipitation. It does not, however, explain why such precipitation occurs only in certain areas.

VAUGHAN (1914) suggested that ooid precipitation involved a biochemical decomposition of organic material by denitrifying bacteria, resulting in a gelatinous carbonate precipitate, which later could be molded into oolitic spherules by water aggitation. Microbial processes have also been suggested by BLACK (1933a), MONAGHAN and LYTLE (1956), OPPENHEIMER (1960) and PURDY (1963). DALRYMPLE (1965) concluded that the aragonitic cryptocrystalline lumps found in Baffin Bay (Texas) sediments were "algal micrites" formed in blue-green algae by bacterial activities. One main stumbling block in the bacterial theories, however, is understanding how bacterial activity could provide the necessary micro-environments conducive for carbonate precipitation in the high energy sandy environments in which ooids occus. Earlier workers had suggested that ooids nucleated in muds and later were exposed to high energies, but carbon-14 dates show that the outer layers of modern ooids are contemporaneous in age, indicating that they are forming in the present sandy environments (NEWELL and others, 1960; MARTIN and GINSBURG, 1966). BATHURST (1967b) has suggested that Bahamian ooids mostly remain buried within migrating oolitic dunes and are not exposed to the ambient sea water for more than 5% of their "life". Not only would this situation offer the isolation required for bacterial activity, but it may also explain how ooid laminations form. Each lamination may represent one exposure to the precipitating medium, followed by subsequent burial, during which algae can aggregate around the ooid's surface.

Recent studies of the amino acids (MITTERER, 1968) and δC^{13} content (DEUSER and DEGENS, 1969) in ooids suggest an even more direct biochemical process. DEGENS (in DEUSER and DEGENS, 1969) has suggested a template theory, by which calcium and carbonate ions attach themselves to available organic templates, resulting in slow ooid accretion. This theory is supported by the fact that laminae often are coated with mucilagenous sheaths (SHEARMAN and others, 1970). An organic control of non-skeletal precipitation also has been suggested by KINSMAN (1969 b), who found Sr values in ooids to be some 18% higher than values predicted by physiochemical arguments. KINSMAN suggested that the strontium concentration may be enriched during migration of sea water through the mucilagenous sheaths.

Bacterial theories are even more plausible in explaining the origin of aggregates. TAFT and HARBAUGH (1964) and BATHURST (1967a) report brown and green algal slimes and mats which cover Bahamian grapestone sediments. These mats immobilize the sediment and offer the isolation necessary for microbial activities. The concentration of organic material within grapestone cement (PURDY, 1963) also may indicate a microbial process. WINLAND and MATTHEWS (1969) found that initial grapestone precipitation occurs around muciliagenous algae as magnesian calcite; subsequent precipitates are aragonite and the original magnesian calcite also alters to aragonite. MONTY (1967) has described the aggregation of particles by blue-green algae. The algae can precipitate microcrystalline carbonate which binds the aggregates into a solid mass. Similar processes may occur during the formation of other non-skeletal fragments.

Any theory of non-skeletal precipitation must also explain why such grains are found only within certain environments. Why are non-skeletal grains common in the Bahamas and yet totally absent in Florida and most other shallow-water tropical areas? The answer to this question may involve the availability (or the unavailability) of skeletal sediments. The non-skeletal sediments in some Caribbean atolls are restricted to relatively shallow (5 to 15 m) lagoons which are either adjacent to non-productive reefs (such as Hogsty Reef) or distant from productive reefs (such as at Serrana Bank) (MILLIMAN, 1967, 1969 b). Because the productivity of lagoonal benthos in these areas is low, the rate of skeletal sedimentation is minimal. The lack of skeletal production may result in ambient waters remaining supersaturated with respect to calcium carbonate (since the organisms are not removing it from solution) thus increasing the possibilities of inorganic precipitation. Perhaps the lack of sediment accumulation also allows potential nuclei a longer exposure time to these supersaturated waters. The generally "old" ages of modern non-skeletal sediments (often more than 500 years old) and the very slow rate of ooid precipitation infer that slow sedimentation may be a critical factor in non-skeletal precipitation.

In conclusion, one is struck with the general compositional similarities of the various non-skeletal carbonate grains. These similarities suggest a common origin, while petrographic differences infer environmental control on the exact type of cemented grain produced. Although the formation of non-skeletal grains is still not fully understood, biochemical processes seem to be important. The fact that ooids, pelletoids and aggregates are restricted to certain environments, however, suggests an inorganic control on the precipitation.

Chapter 4. Skeletal Components

Skeletons serve two primary biologic functions: they provide structural support (and muscle attachment) and protection for the organism. Skeletal components can also serve other purposes, such as in attachment of the organism to the substrate (for example the basal plate of cirripeds) or in feeding (the radulae of gastropods). Skeletons can be comprised of many minerals (including magnetite, goethite and fluorite) but most skeletons are composed of calcium carbonate, opaline-silica or calcium phosphate. Opaline-silica skeletons generally are restricted to the more primitive plants (such as diatoms) and animals (sarcodinans and poriferans), while calcium phosphate skeletons are best represented in the phyla Arthropoda and Vertebrata.

The most commonly occurring skeleton composition and by far the most important in terms of sediment deposition is calcium carbonate. Many different plant and animal phyla precipitate calcium carbonate either as hard tests or as disjunct plates. Many of these phyla are dicussed in the following sections.

Plants

Cyanophyta (Blue-green Algae)

Blue-green algae are not considered to precipitate calcium carbonate in the marine environment, although, as will be discussed below, there is some evidence to the contrary. However, even if blue-green algae were not direct skeletal contributors, their role in both modern and ancient carbonate sedimentation merits their discussion in this chapter. Blue-green algae have numerous important functions within the carbonate system: 1. They are probably the most important agent in the boring and biologic destruction of shallow-water carbonate substrata (see Chapter 9 for a discussion); 2. Blue-green algal mats provide food for intertidal and subtidal organisms (BATHURST, 1967c); 3. Algal mats can stabilize the bottom, thereby influencing the rate of sediment accumulation and the size of the ambient particles (GEBELEIN, 1969; NEUMANN and others, 1970). In addition, the mats can create macro- and micro-environments which can have profound effects on carbonate precipitation and diagenesis; 4. Finally, blue-green algae are critical in the formation of stromatolitic features.

The term stromatolite was first used by KALKOWSKI (1908) to describe laminated structures of probable blue-green algal origin. A more complete definition has been offered by LOGAN and others (1964): "Algal stromatolites are laminated structures composed of particulate sand, silt and clay-sized sediment, which have formed by the trapping and binding of detrital sediment particles by an algal film." Unlayered or poorly laminated algal mats may be considered marginal stromatolites. Oncolites are defined as unattached stromatolites (PIA, 1927). These definitions are important to keep in mind, for numerous workers have used the terms stromatolite and oncolite erroneously in referring to structures that have been formed by other organisms, such as coralline algae. A complete review of terminology as well as a general discussion of stromatolites has been given by HOFMANN (1969).

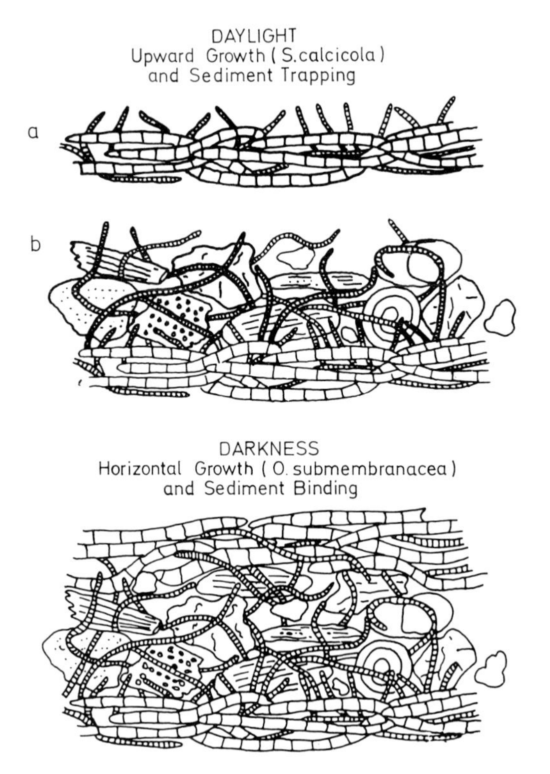

Fig. 20. Schematic representation of the day-night accretion of stromatolites. From GEBELEIN (1969)

Layers are defined by a couplet of laminae consisting of a dark, organic-rich lamina and a light, carbonate-rich lamina. BLACK (1933b) offered three possible explanations for these laminations: 1. rhythmic variations in filament growth with relation to sedimentation, 2. alteration of algal species, 3. sedimentary lamination of mineral particles between filaments. GINSBURG (1960) concluded that the first alternative is the most likely, and subsequent studies have substantiated this conclusion. Apparently laminar accretion occurs in two discrete steps. First vertically growing algal filaments trap and bind the sediment that accumulates on the mat. Then a horizontally growing layer (often a different algal species) is added (HOMMERIL and RIOULT, 1965; GEBELEIN, 1969) (Fig. 20). The time interval between the accretion of these two laminae depends greatly upon the surrounding environment and the algal species involved. In intertidal zones exposed to diurnal floodings, two layers can form in one day, the sediment part of each couplet being deposited during high tide (GEBELEIN and HOFFMAN, 1968). MONTY (1965) and GEBELEIN (1969) found that layers in subtidal environments can form daily; clearly the controlling factor is the availability of sunlight. Most of the vertical accretion occurs during the day, when *Schizothrix calcicola* grows as much as 1 mm. At night the slower growing *Oscillatoria submembranacea* accretes a thin horizontal layer that seals the couplet (GEBELEIN, 1969) (Fig. 20). GEBELEIN has calculated that if such daily accretion rates were continuous, 2 to 3 cm-thick stromatolites could form within a month. However, MONTY (1967) has shown that such structures do not necessarily accrete continuously, so that the number of layers within a *Schizothrix* mat gives only a minimum age. In addition, formation of new layers (and therefore vertical accretion) can be dependent upon seasonal growth patterns. For instance MONTY (1965) found that subtidal mats in the Bahamas may grow only in late summer.

Mats containing other blue-green algal species and mats growing in other environments do not necessarily have such rapid rates of accretion. The supratidal *Scytonema* accretes every week (*S. crustaceus*) or every few week (*S. myochrous*), and *Rivularia biasolettiana* apparently forms a new layer every year (MONTY, 1967). Something other than diurnal or tidal cycles must control these laminar accretions. Perhaps it is related to periodic wettings (such as during high tides or storms) or to seasonal algal growth.

Early workers tended to assign "form genus" to the various stromatolitic structures found in the geologic record. Thus, for example, stromatolites with hemispherical, cabbage-like features were assigned to the "genus" *Cryptozoon*. JOHNSON (1951) has presented a synopsis of this type of classification.

When modern analogues of ancient stromatolites were studied (BLACK, 1933b; GINSBURG and others, 1954; LOGAN, 1961; MONTY, 1967; GEBELEIN, 1969) it was found that stromatolites and algal mats often are composed of numerous binding species, in some instances more than 20. While the stromatolitic microstructure may depend upon the blue-green algae present, the gross morphologic form of the stromatolite depends strongly upon environmental factors, such as sediment supply, water depth and current speed (C. D. GEBELEIN, 1972, written communication). In place of the bionomial classification. LOGAN and others (1964) proposed a classification based strictly upon morphologic form. Three main grouth types were recognized. 1. laterally linked hemispheroids (LLH) which contain horizontally continuous layers; 2. vertically stacked hermispheroids (SH) in which the laminae between domes are not connected; and 3. spheroidal structures (SS) which are equivalent to oncolites. Some of the various types are shown in Plate IV.

Many of the differences between these various morphologic forms can be explained by the environment of accretion (HOFFMAN and others, 1968). LOGAN and others (1964) and DAVIES (1970b) found that LLH forms are most common in supratidal and intertidal environments. SH structures are restricted mostly to protected arid areas with considerable tidal range in which sediment preferentially accumulates on the domes. Oncolites, being unattached to a firm substrate, infer a turbulent environment; the degree of turbulence determines the ultimate morphology of the oncolite. In subtidal areas these generalizations may not hold. For instance, GEBELEIN (1969) has shown that in Bermuda subtidal algal mats form in areas where currents range from 15 to 20 cm/sec and algal bisquits (low-energy oncolites) in areas with currents between 1 and 11 cm/sec.

Although *Schizothrix calcicola* is the most prominent species in most modern algal stromatolites (SHARP, 1969) morphologic form can depend upon other species of blue-green algae. For instance, on Andros Island, *Scytonema* is the dominant species in supratidal and upper intertidal mats, while *Rivularia* dominates the intertidal and *Schizothrix* in the subtidal (MONTY, 1967).

Supratidal stromatolites are normally compact and generally hardened ("lithified"), while subtidal stromatolites often are non-compact and can be removed by artificially-induced agitation. The amount of agitation required to remove some subtidal algal mats, however, may be considerable. NEUMANN and others (1970) found that some mat surfaces can withstand velocities up to 100 cm/sec, or five times the current velocity required to erode the loose sediment within the mat. The main cause of compaction of intertidal and supratidal mats has been assumed to be alternating desication and wetting by periodic floodings and ebbings of sea level. LOGAN (1961) reported aragonite needles growing in the pore spaces of some stromatolites, but concluded that the needles were the result of diagenetic alteration rather than direct biogenic precipitation. Although the binding of bottom sediments by blue-green algae may be critical in the lithification of those sediments (DALRYMPLE, 1965; GEBELEIN and HOFFMAN, 1971), direct evidence of the precipitation of calcium carbonate by marine blue-green algae is lacking. MONTY (1965, 1967) reported small (mostly finer than 4 microns) crystals of magnesium calcite (5 to 10 mole % $MgCO_3$) forming sheaths around algal filaments; he concluded that these were biologically precipitated. Such occurrences, however, may be limited to fresh and brackish water conditions; GEBELEIN (1969) found no evidence of carbonate precipitation in the subtidal stromatolites at Bermuda nor in other localities (C. D. GEBELEIN, 1972, written communication).

Chrysophyta (Coccolithophorids)

Coccolithophorids are flagellated yellow-green algae belonging to the phyllum Chrysophyta. Individual plants consist of round coccospheres, on which are numerous calcareous plates, called coccoliths. The coccosphere averages about 10 to 100 microns in diameter, and the coccoliths 2 to 20 microns (Plate V). Each coccolith is composed of very small calcite crystals, 0.25 to 1 micron in diameter (BATHURST, 1971). The composite c-axis of the coccolith is oriented parallel to the direction of elongation (WATABE, 1967). Coccolith formation takes place within the cell wall of the coccosphere, apparently in less than one hour. Calcification is dependent upon photosynthesis (PAASCHE, 1962; WILBUR and WATABE, 1967), although slow calcification can occur in the absence of light (PAASCHE, 1966; ISENBERG and others, 1967). The biologic response to light and the presence of chlorophyl within their cells suggest that coccolithophorids are autotrophs. However, deeper-water forms possess flagella, indicating that these organisms

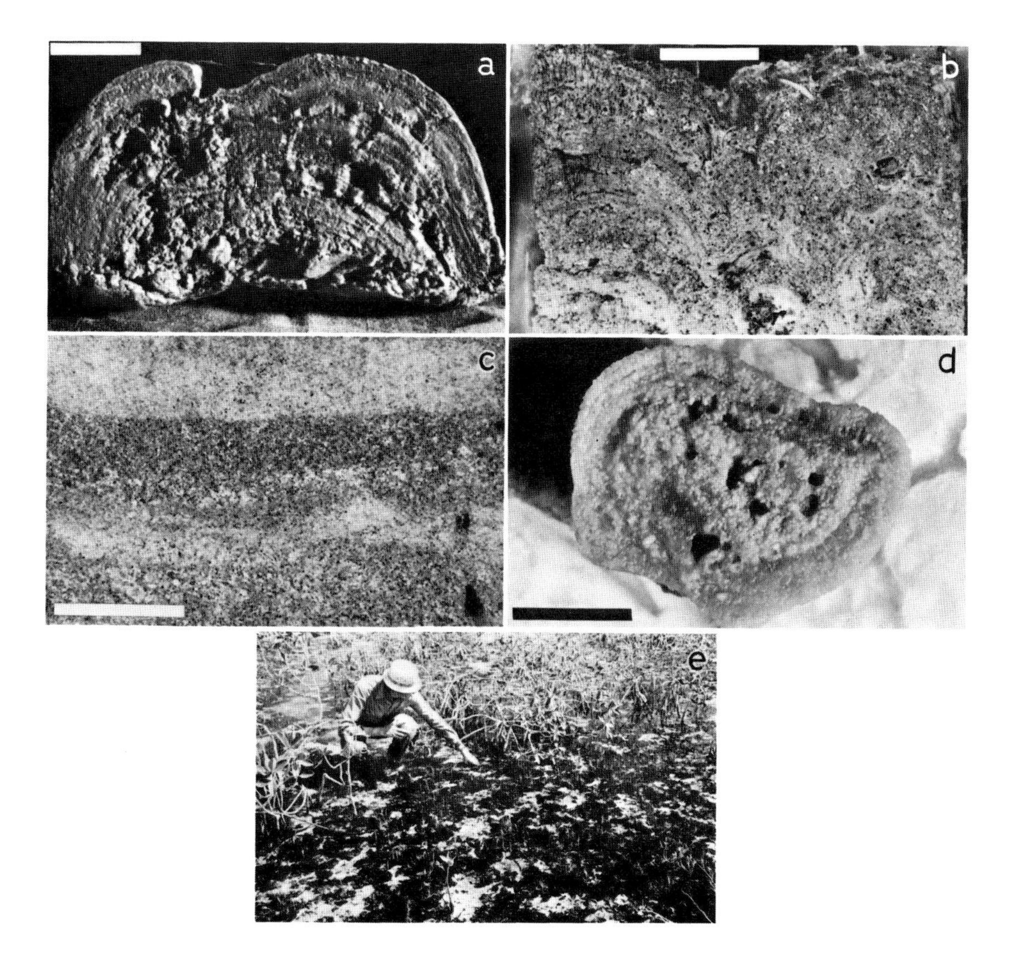

Plate IV. (a) SH stromatolite, lower intertidal zone, Gladstone Embayment, Shark Bay, Western Australia. This low domal form is elongate in the plane parallel to current direction. Scale is 5 cm. (b) SH-LLH stromatolitic domes, Hamelin Pool, Shark Bay. Lateral linkage occurs at the surface and at one level in about the middle of the structures. Scale is 2.5 cm. (c) Flat laminated algal sediments, high intertidal zone, Shark Bay. Laminae are thin, often discontinuous. Individual laminations are marked by changes in grain size or color (usually changes in organic content). Scale is 2 cm. (d) SS oncolite, thin grass bed with algal mat on surface, 1 km east of Joulters Cay, Bahamas. Water depth is 1 m. Core of oncolite consists of SH stromatolite. This form became detached and concentric laminae formed over the core. Scale is 1 cm. (e) Thick (5 cm) *Scytonema* algal mats and red mangroves growing on a tidal flat surface at SW Andros Island, Bahamas. These areas, which are covered by at least a thin layer of water most of the time, border both the ponds and algal flats of Andros tidal flats. Photographs courtesy of CONRAD GEBELEIN

may also be heterotrophic (BERNARD, 1964). A comprehensive review of coccolithophorid biology and calcification has been given by PAASCHE (1968).

Although coccoliths were recognized in sediments more than 100 years ago, it was not until the development of the electron microscope that these nannofossils were widely studied. Because of the relatively short geologic duration of individual species, coccoliths (as well as extinct forms such as rhabdoliths, pentaliths and discoasters) have proved to be excellent stratigraphic indicators. BLACK (1971) has presented a useful series of photomicrographs of the coccoliths from various coccolithophorid families.

Coccoliths composed of simple calcite crystals are called holococcoliths. Heterococcoliths are more complex and diverse, with four prominent forms: caneoliths, scapholiths, cyrtoliths and placoliths (HALLDAL and MARKALI, 1955; BLACK, 1963). Cyrtoliths and placoliths, being more complex and massive, are more resistant to solution and therefore are more prominent in deep-sea sediments (MCINTYRE and MCINTYRE, 1971). The soluble holococcoliths rarely are preserved in the deep sea (GAARDER, 1971). SCHNEIDERMANN (1971) reports that both caneoliths and cyrtoliths decrease in Atlantic Ocean sediments at depths greater than 3000 m, and disappear below 5000 m. Placoliths of the species *Cyclococcolithina leptopora* are the most resistant.

Coccolithophorid concentrations in surface waters commonly range from 50000 to 500000 cells per liter (BERNARD and LECAL, 1953) but concentrations as high as 13800000 per liter have been reported from Norwegian fjords (BIRKENES and BRAARUD, 1952). Although surface populations of coccolithophorids are greater in higher latitudes, coccolith-rich sediments generally are restricted to lower latitudes (BLACK, 1965).

In the Atlantic Ocean most of the fine-grained carbonate is probably composed of coccolith debris, amounting to between 5 and 20% of the total sediment and sometimes more (MCINTYRE and BÉ, 1967). BRAMLETTE (1958) estimated lower concentrations in the Pacific, but pointed out that during the Tertiary and pre-Tertiary, when ocean temperatures were warmer, coccoliths were much more important. In modern enclosed basins, such as the Black Sea and the Mediterranean Sea, coccoliths are still the dominant carbonate component (BERNARD and LECAL, 1953; BUCKRY and others, 1970; GAARDER, 1971). Coccoliths also can be abundant in shallow-water deposits (see SCHOLLE and KLING, 1972), although they generally are far less common than in deeper water sediments.

Modern coccoliths are thought to be calcitic, but aragonite and vaterite platelets have been precipitated experimentally (WILBUR and WATABE, 1963). The only aragonitic coccoliths found in nature occur in South African sedimentary strata (HART and others, 1966). Owing to the difficulties of separation, coccolith chemical analyses have been relatively few. The best published data are those of THOMPSON and BOWEN (1969) and THOMPSON (1972), who presented analyses of relatively pure coccolith cozes from various deep-sea sediments (Table 13). The calcite is very low in magnesium and strontium, but does contain relatively high concentrations of barium and manganese. CARPENTER (1969) reports fluorine concentrations of about 280 ppm. To what extent these various trace element concentrations might represent diagenetic replacements or clay mineral contamination is not known.

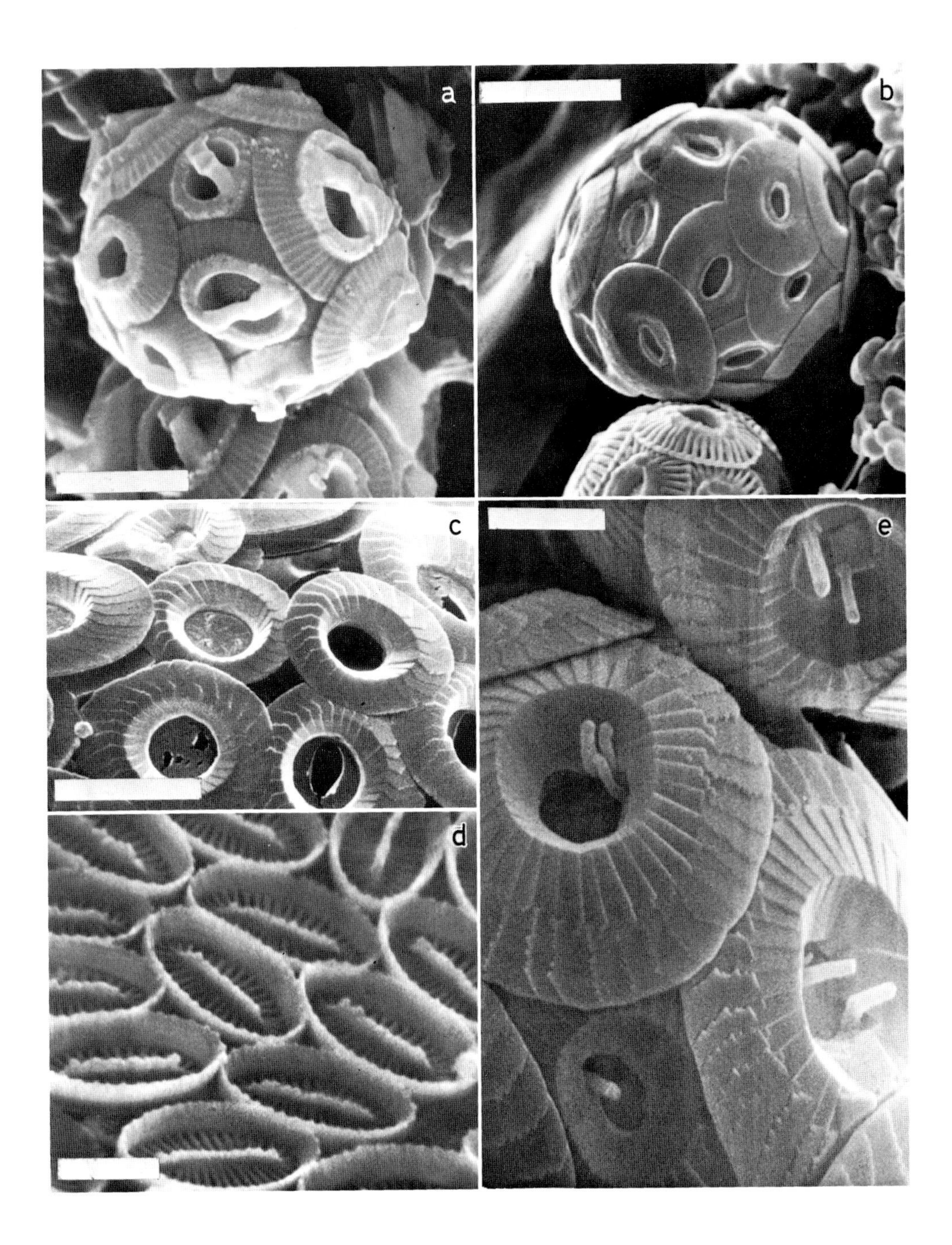

Plate V. Coccospheres and Coccoliths. (a) Coccosphere of the coccolithophorid *Geophyrocapea oceanica*. SEM; scale is 3 microns. (b) Coccosphere of the coccolithophorid *Syracosphaera hulberti*. SEM: scale is 3 microns. (c) Single coccoliths, *Cyclococcolithus mirabilis*. SEM; scale is 2 microns. (d) *Syracosphaera purchura*. SEM; scale is 1 micron. (e) *Cyclococcolithus* sp. SEM; scale is 1 micron. Photomicrographs courtesy of SUSUMU HONJO

Table 13. Average chemical composition of coccolith oozes from the Atlantic, Pacific and Indian Oceans. Mg and Sr contents determined on dissolved carbonate portion. Regional variation in various trace elements probably reflects the presence of insoluble clays rather than the coccoliths themselves. Indian Ocean coccolith oozes are mid-Tertiary, while those from the Pacific and Atlantic are primarily late Tertiary and Quaternary. Data from THOMPSON and BOWEN (1969) and G. THOMPSON (unpublished data). Range of values in parentheses

Mineralogy	Atlantic calcite	Pacific calcite	Indian calcite
Percent	2.2 (2.1–2.4)	3.6 (2.9–4.3)	8.8 (8.8–13.0)
Insol.			
Mg	0.11 (0.10–0.12)	0.17 (0.14–0.22)	0.13 (0.10–0.16)
Sr	0.15 (0.14–0.16)	0.16 (0.15–0.17)	0.21 (0.18–0.22)
Parts per million			
Fe	> 1000	1200 (500–1700)	> 2000
Mn	265 (245–280)	500 (290–930)	1500 (950–2000)
Ba	175 (135–195)	510 (485–530)	160 (110–240)
Cu	13 (10–15)	45 (22–80)	50 (16–105)
Zn	5 (4–6)	10 (5–17)	28 (18–35)
Pb	< 2	< 2	4 (tr-9)
B	35 (30–40)	13 (10–16)	28 (24–32)
Ni	4	12 (3–25)	45 (32–70)
Cr	5 (5–6)	3 (1–4)	4 (tr-10)
Co	4 (3–5)	6 (4–9)	20 (15–28)
Li	6 (4–7)	4 (2–6)	7 (6–7)
V	< 5	10 (8–12)	4 (tr-10)
Y	14 (11–15)	26 (19–30)	33 (18–40)

Rhodophyta (Red Algae)

Red Algae, so termed because of their reddish to purple color, are the most cosmopolitan carbonate-producing benthic algae. Coralline algae (order Cryptonemiales, Family Corallinaceae) precipitate carbonate both within and between the cell walls, thus resulting in a rigid skeletal structure (JOHNSON, 1961). Some members of the Order Nemalionales (Family Chaetangiaceae) also precipitate calcium carbonate, but calcification is not as complete as in the corallines; for all practical purposes, these algae can be ignored as significant carbonate contributors. LEMOINE (1911), FOSLIE and PRINZ (1929), PIA (1926), FRITSCH (1945) and recently ADEY and MACINTYRE (1973) have presented comprehensive discussions on calcareous red algae. Much of the following discussion is based upon these papers, together with the excellent but somewhat dated review by J. H. JOHNSON (1961).

Coralline algae consist of vegetative tissue (thalli) and reproductive structures called conceptacles. The thallus is separated into three tissues: the epithallus, above the lateral meristem, the perithallus, which constitutes the upper thallus below the meristem, and the hypothallus, which constitutes the lower hypothallus. Hypothallial cells are generally larger than perithallial cells. The spore-bearing cells (sporangia) commonly are collected in large cavities, called conceptacles.

The character of the cells, the relative development of the various tissues and the type of sporangia and conceptacles are used in determining subfamilies and genera. Most genera contain all three major types of tissue, although some genera may lack perithallium or sometimes epithallium (W.H. ADEY, 1973, oral communication). An extremely helpful key to the identification of corallines has been presented by ADEY and MACINTYRE (1973).

Coralline algae are separated into two distinct subfamilies, Melobesoideae and Corallineae[1]. The former are commonly crustose and massive, and by far the more prominent carbonate contributors. JOHNSON (1961) recognized three types of crustose growth forms: 1. rounded heads of tightly packed branches, restricted to high energy environments, 2. long slender branches that grow in environments with less severe currents and waves, and 3. encrusting species which generally are limited to quiet waters. This catagorization is somewhat simplistic, since, for example, many of the most massive and encrusting species (such as *Porolithon onkodes*) are major contributors to high energy areas (such as algal ridges).

Some branching erect corallines can thrive in sand or mud as well as on hard bottoms, while encrusting species generally require hard substrata upon which to grow. Massive encrustations on shells or other nuclei can result in algal nodules (termed rhodolites by BOSELLINI and GINSBURG, 1971). Generally algal nodules are restricted to unstable bottoms, where wave action or strong currents move them occasionally (ADEY and MACINTYRE, 1973). Such nodules display one or more distinct growth forms, such as branching, columnar, globular or laminar. In shallow-water tropical environments any one growth form often represents a mixture of two or more coralline algal species and sometimes genera (BOSELLINI and GINSBURG, 1971).

Corallineae, represented by three common genera, *Amphiroa*, *Corallina* and *Jania*, consist of articulated branches with calcified portions separated from one another by uncalcified or poorly calcified internodes (genicula). These species grow in "small branching tufts, usually less than 10 cm high, formed of fine stems which are composed of cylindrical articles or flattened leaf-like segments" (JOHNSON, 1961). Unlike the erect and crustose corallines, Corallineae are usually limited to sheltered environments.

Ecology, Growth Rates and Sedimentary Distribution

Contrary to the belief of many geologists, coralline algae are not limited to shallow tropical waters; they also can grow in polar seas as well as at relatively great depths. Among the common crustose genera, *Porolithon*, *Neogoniolithon* and *Lithoporella* are restricted to the tropics (the former two genera, together with *Lithophyllum*, are the dominant tropical genera; ADEY and MACINTYRE, 1973). *Lithophyllum* also occurs in temperate waters, while *Lithothamnium* is found in all climates. JOHNSON (1961) pointed out that the diversification of *Lithothamnium* increases in higher latitudes relative to the number of other species. Among the

1 ADEY and JOHANSEN (1972) and ADEY and MACINTYRE (1973) have suggested another classification with three crustose subfamilies: Lithophylloideae, Mastophorideae and Melobesoideae.

Table 14. Deposition rates of calcium in various calcareous red algae. Generic averages are derived from data presented in GOREAU (1963); ranges of values are in parentheses

	Calcium deposition in light (μg Ca/mgN/hr)	Deposition in light / Deposition in dark
Galaxaura	1198 (46–4318)	1.68 (0.40–2.34)
Liagora	3932 (3551–4314)	1.86 (1.83–1.89)
Amphiroa	1335 (147–2465)	1.25 (0.93–1.63)
Jania	390 (59–722)	2.18
Goniolithon	1435	1.55
Lithothamnium	186 (31–341)	1.30 (1.09–1.51)

articulates *Amphiroa* is limited to the tropics and *Jania* and *Corallina* are generally globally dispersed.

In the tropics most branching species are restricted to shallow depths. Encrusting species dominate in deeper environments, although the crusts become thinner with increased water depth. Generally in the tropics at depths greater than 100 meters (or in cryptic environments) the dominant species belong to the genus *Lithothamnium* (JOHNSON, 1961; W.H. ADEY, 1973, oral communication). At higher latitudes this genus is also important at much shallower depths (ADEY, 1966) and *Leptophytum* is characteristic of deeper waters (W.H. ADEY, 1973, oral communication). Living corallines have been dredged from depths as great as 300 meters in the tropics (JOHNSON, 1961), but the maximum depth of growth is significantly less in higher latitudes, generally 50 to 100 meters (LEMOINE, 1940). This decrease in the maximum depth with increasing latitude may be in response to the lower average angle of incidence of the sun and the generally reduced clarity of the water. Both of these factors would tend to decrease the compensation depth of photosynthesis.

Relatively little work has been done on the growth rate of calcareous red algae. Data presented by GOREAU (1963) show that species from the order Nemalionales (notably the genera *Galaxaura* and *Liagora*) calcify more quickly than do most coralline algae (Table 14). Of the corallines, *Amphiroa* and *Neogonolithon* are more efficient producers of calcium carbonate than either *Jania* or *Lithothamnium*. All genera exhibit greater carbonate production in light than in dark, but the fact that calcification can occur in the dark (Table 14) suggests that red algae may utilize some other catalyst besides photosynthesis during calcification; perhaps the uptake of high energy organic phosphates may aid in calcification (PEARSE, 1972). The genus *Lithothamnium* shows the greatest relative precipitation of carbonate in the dark, which would explain its dominance in deeper waters[2].

Growth rates apparently increase with increasing water temperature (ADEY, 1970). In the Gulf of Maine, *Clathromorphum circumscriptum* grows at a rate of 3 mm per year (ADEY, 1965). Off western France, the crustose corallines *Lithophyllum incrustans* and *Lithothamnium lenormandi* grow 2 to 7 mm per year, and about $\frac{1}{2}$ to 1 mm per month during the summer (LEMOINE, 1940). HUVE (1954)

2 There is some question as to whether GOREAU correctly identified the *Lithothamnium* species involved in his studies. If the specimens were misidentified, then the results of his studied should be re-evaluated (W.H. ADEY, 1973, oral communication).

reported annual increase of 7 to 15 mm for *Lithophyllum tortuosum* in the western Mediterranean and SMITH (1970) found average growth rates of 150% per year for various corallines off southern California. BOSELLINI and GINSBURG (1971), on the other hand, have found annual growth rates for Bermudan algal nodules of only 0.4 mm, but these estimates are based on total algal encrustations on a ship wreck, and therefore only represent minimum estimates. Articulate corallines can have greater rates of growth. JOHNSON (1961) quotes rates of 3 to 4 cm per year in the shallow waters near France, and JOHNSON and AUSTIN (1970) and SMITH (1970) have observed somewhat similar rates.

Available data suggest that a number of species of deep-water coralline algae have slow growth rates (ADEY, 1970; 1973, oral communication). MCMASTER and CONOVER (1967) found the crust of a deep-water algal ball 4 to 5 cm in diameter to be 400±40 years old, and the core to be 1 500±100 years old, indicating an average annual rate of accretion of less than 0.1 mm. Similarly, VOGEL (1970) reports algal balls 15 cm in diameter that required 500 to 600 years to grow.

Coralline algae are prominent sediment contributors on most coral reefs, usually constituting more than 20% of the reef sediment and locally more than 50% (such as Johnson Island; EMERY, 1956). Because of the lack of hard substrate, corallines are less abundant in lagoonal environments, although they do contribute to patch reef sediments. Branching corallines, such as some species of *Neogoniolithon*, can form low-lying reefs in quiet lagoons, but these generally are only locally important (see p. 177). Corallines can also be important constituents in colder or deeper-water sediments. Many examples of polar communities have been cited by ADEY and MACINTYRE (1973). Usually such accumulations are limited to those areas with low rates of detrital sedimentation and with numerous hard substrates. Examples include the shallow current-swept shelves off Newfoundland, Norway, western France, the British Isles and the northeast Brazilian continental shelf.

The most critical role for coralline algae in most reef areas is not as sedimentary contributors nor as primary photosynthesizers (MARSH, 1970), but rather as encrusters and cementers. This role can be compared to the mortar used in cementing bricks (in this case coral and mollusk debris). In this manner, corallines can incorporate loose grains into hardened rock, and thereby construct stable bioherms. The most visually obvious algal encrustations occur on the windward edges of Indo-Pacific and some Caribbean (ADEY, studies in progress) coral reefs, where mounds, called algal ridges (also called *Lithothamnion* (sic) ridges even though the dominant genus is *Porolithon*) can reach thicknesses considerably greater than one meter. Other examples include the massive algal encrustations in the intertidal zones in the Mediterranean, the algal banks off Norway (FOSLIE, 1894) and the algal ridges along the seaward edges of many continental shelves (see below).

Petrography

In their unaltered state, coralline algae are one of the easiest carbonate components to recognize in thin section. The typical cellular structure, the brownish color under plain light, and the lack of any notable birefringence (due to the extremely small size of component crystals; tenths of microns according to BAAS-BECKING and GALLIHER, 1931), are all diagnostic properties (Plate VI). However, algal fragments can be rapidly "micritized", resulting in an apparently cell-less, dark brown opaque mass (WOLF, 1965). With practice these cryptocrystalline algal fragments can be recognized. Seldom should one confuse pelletoids with rounded cryptocrystalline coralline algae, although such confusion can occur in limestones (WOLF, 1962).

Articulated corallines can be distinquished under reflected light by their thin segmented branches. Encrusting algae often coat many grains in the gravel fraction of a sediment, thus hindering identification of grain. One or two washings with dilute HCL (the exact number of washings depends upon the particle size), however, usually will expose the nature of the grain's interior.

Composition

The Nemalionales (notably *Galaxaura* and *Liagora*) are aragonitic but contain some magnesian calcite (DAWSON, 1961). The few available data suggest that these algae contain relatively high Sr and low Mg concentrations (LOWENSTAM, 1963; Table 15). Coralline algae are thought to be totally magnesian calcite. ALEXANDERSSON (1969) reports an aragonitic coralline alga from the Mediterranean, but this may be a Nemalionales.

The magnesian calcitic corallines display a considerable range of magnesium concentrations (Table 15), which are partly dependent upon ambient water temperature (Fig. 21). Until recently is was thought that the magnesium contained within coralline algal skeletons was restricted to the magnesian calcite phase. SCHMALZ (1965) and WEBER and KAUFMAN (1965) found that *Goniolithon (Neogoniolithon)* is an exception to this rule, containing measurable amounts of brucite ($Mg(OH)_2$). However, more recent data show that most coralline algae contain an average of 1% (by weight) more magnesium than predicted from the shift in the X-ray diffraction peaks (MILLIMAN and others, 1971). This excess magnesium reaches as much as 3% in the tropical genera *Porolithon*, *Neogoniolithon* and *Amphiroa*. Much of the "excess" magnesium is present as distinct phases of higher magnesium calcite (concentrations of $MgCO_3$ greater than 25 mole %), but some may be present as brucite (Fig. 22). MOBERLY (1968 and references therein) found widely different concentrations of magnesium throughout coralline algal skeletons. He suggested that these variations in magnesium content

---

→

Plate VI. Coralline Algae. (a) Colony of branching coralline algae (*Lithothamnium*) from the Canary Islands. Scale is 1 cm; Photograph courtesy of J. MÜLLER. (b) Articulated coralline algae (*Jania*). Reflected light; scale is 500 microns. (c) SEM photomicrograph of the coralline algal cell structure. Scale is 50 microns. (d) Close-up SEM photo of the individual cells; note the small size of the magnesian calcite crystals. Scale is 5 microns. (e) and (f) Transverse cuts of branching coralline algae. Refracted light, plane and polarized; scales are 100 microns. (g) Longitudinal sections of coralline algae, showing the banding of cells. Refracted light; scale is 100 microns. (h) Longitudinal section of articulated coralline algae. Refracted light; scale is 100 microns

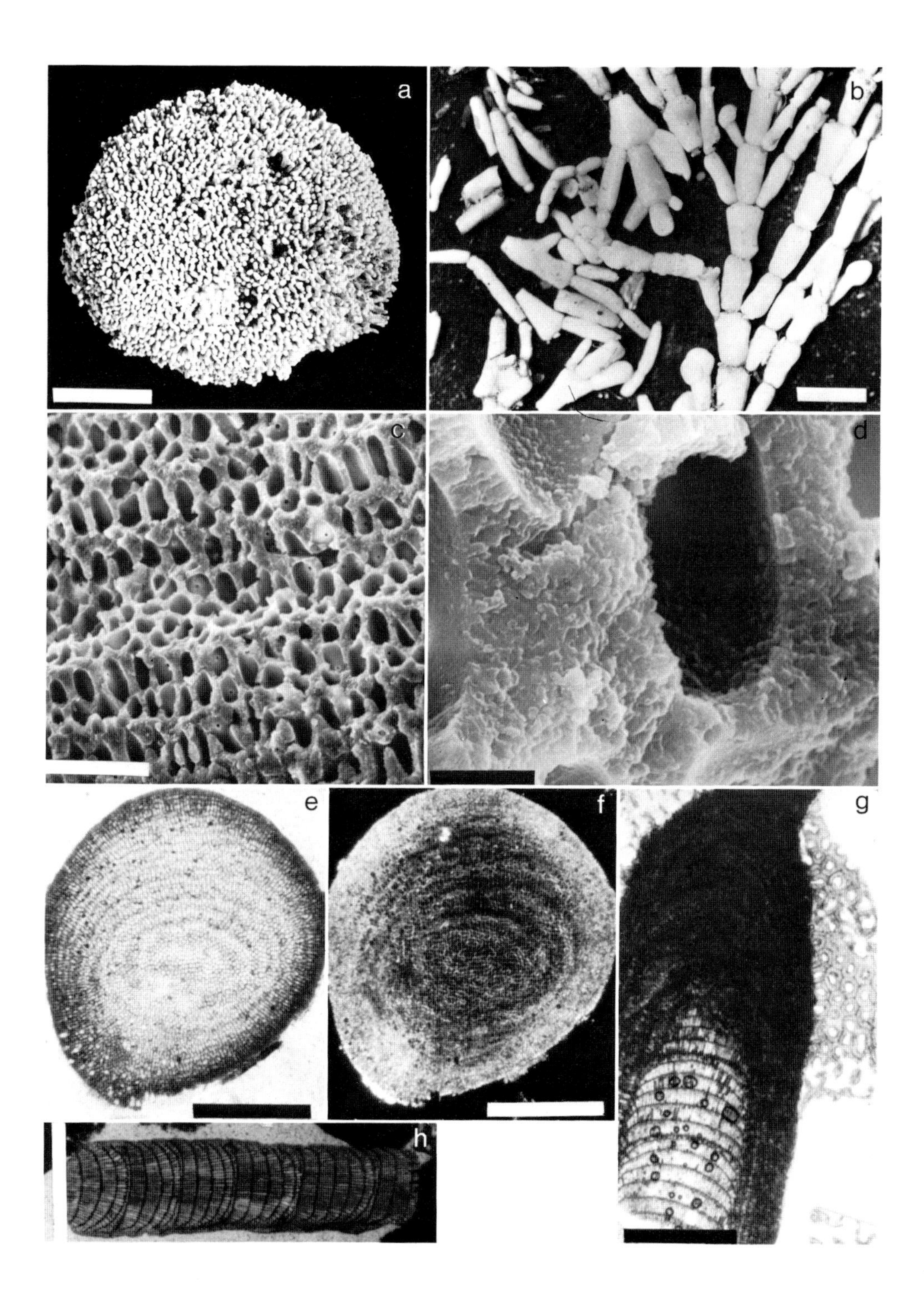

Plate VI.

Table 15. Chemical composition of articulate and crustose coralline algae. a) Are bulk analyses; b) are analyses with organics and soluble salts removed. Range of values are in parentheses. Most data are the results of unpublished analyses by the author. Other data are from VINOGRADOV (1953), GOLSMITH and others (1955), THOMPSON and CHOW (1956), SCHALZ (1965), KEITH and WEBER (1965), and MÜLLER and others (in press)

	Percent organic matter	Mineral.	Percent				ppm						
			Ca	Mg	Sr	Na	K	Fe	Mn	Cu	Zn	Ba	S
Nemalionales													
Galaxaura	12.7	Arag.	a)	0.78	0.95	3.5	4600	94	7	19	37		
Corallineae (articulates)													
Jania	23	MgC	a) 32.3	3.93	0.19	0.38	1270	834	76	83	36		
				(3.26–4.38)	(0.12–0.27)	(0.13–0.86)	(344–4410)	(63–1940)	(22–190)	(14–252)	(11–470)		
			b) 33.0	4.24	0.14	0.09	272	112	28	23	75		
			(32.3–34.4)	(3.14–4.30)	(0.12–0.16)	(0.08–0.10)	(192–426)	(77–148)	(20–42)	(12–42)	(26–117)		
Corallina		MgC	a)	4.28	0.21	0.57	1020	1132	113	43	63		
				(3.34–5.10)	(0.15–0.28)	(0.53–0.66)		(210–3800)	(17–392)	(29–64)	(48–90)		
Amphiroa		MgC	a) 27.0	5.40	0.20	0.53		350	37	35	38		3350
				(5.05–7.00)	(0.16–0.24)	(0.52–0.55)		(59–690)	(8–120)	(27–51)	(11–67)		
Cheilosporum		MgC	a) 32.8	4.20	0.18	0.21	340		16	14	30		
			(32.6–33.1)	(4.04–4.36)	(0.18–0.19)	(0.13–0.30)	(312–373)		(11–22)	(13–14)	(21–40)		
Melobesia (crustose)													
Lithothamnium		MgC	a) 32.1	3.48	0.24	0.65	2500	270	35	110	60		4000
				(1.74–5.00)	(0.11–0.32)	(0.22–0.78)	(230–4040)	(85–460)	(13–53)	(2–443)	(25–181)		
			b) 34.3	3.47	0.41	0.12							
				(2.39–4.11)	(0.30–0.56)	(0.08–0.16)							
Lithophyllum	2–5	MgC	a) 32.0	4.27	0.27	0.90	2980	248	46	58	86		2260
				(3.14–5.60)	(0.18–0.34)	(0.23–1.42)	(743–6900)	(43–847)	(11–94)	(7–120)	(26–282)		
			b) 32.6	4.03	0.31	16	200	60	15	15	38		
Goniolithon	4	MgC	a)	5.40	0.25	0.46	2700	35	35	25	45	8	2150
				(3.89–6.56)		(0.37–0.55)							
			b)	5.00	0.29	0.15	115	40	90		58		
				(4.31–5.38)	(0.15–0.33)	(0.08–0.20)							
Porolithon	2	MgC	a) 29.2	6.00	0.24	0.50	872	175	30	34	79		
			(29.0–29.4)	(4.25–7.20)	(0.15–0.35)	(0.36–0.88)	(514–1786)	(28–446)	(7–72)	(11–55)	(12–234)		
			b) 30.7	6.00	0.23	0.13	150	35	24	7	14		
				(5.90–6.10)	(0.23–0.24)		(123–172)	(33–37)	(3–64)	(1–12)	(9–19)		

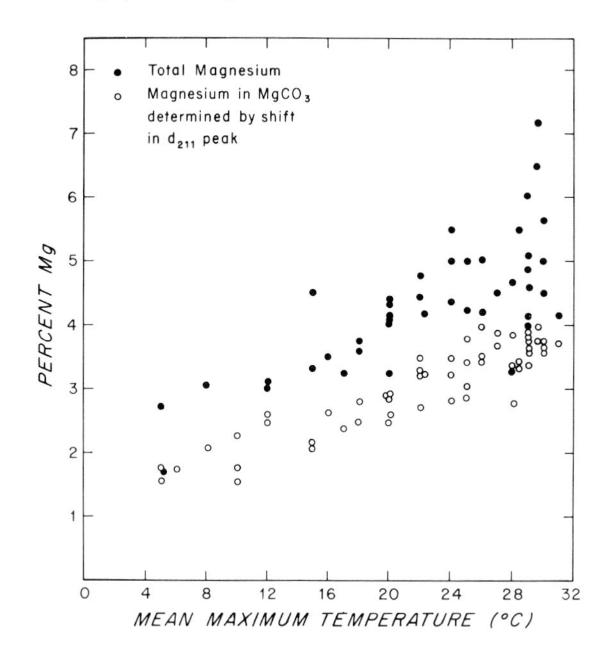

Fig. 21. Variation of total magnesium and the magnesium in $MgCO_3$ (as determined by the shift in the 211 peak) in various specimens of coralline algae. From MILLIMAN and others (1971)

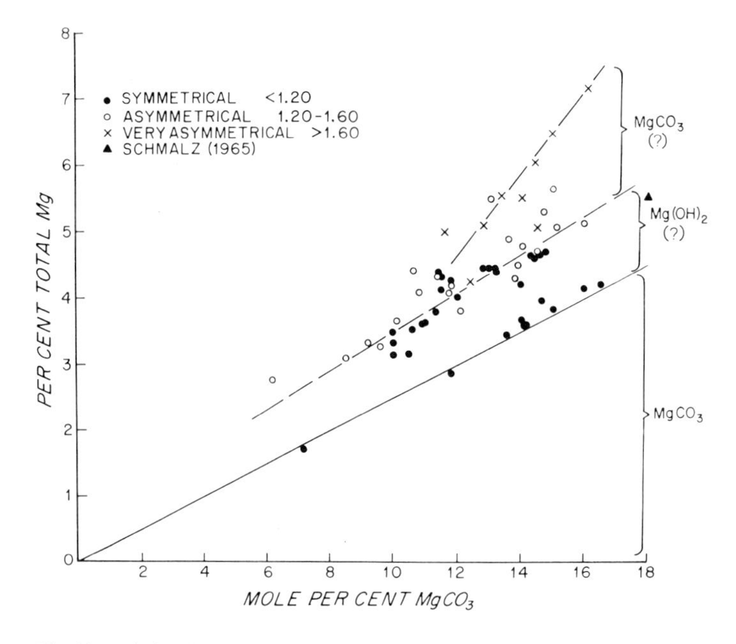

Fig. 22. Relation between peak symmetry and excess magnesium in coralline algae. The value to the far right is from *Goniolithon* sp., reported by SCHMALZ (1965). Mole % $MgCO_3$ was determined using the GOLDSMITH and GRAF (1958) lattice-constant versus composition curve. From MILLIMAN and others (1971)

are the result of periodic fluctuations in the ambient environment and corresponding changes in the organisms' metabolism. In temperate areas, seasonal variations in the temperature or growth rate may control magnesium incorporation (LEMOINE, 1911; HAAS and others, 1935; CHAVE and WHEELER, 1965), but such factors do not seem important in tropical areas, where internal variations in magnesium content are great but seasonal climatic variations are minimal. Perhaps the faster rate of calcification in the warmer areas results in less strict regulation of minor and trace element uptake.

Relatively large amounts of sulfur are found in coralline algae, implying that $CaSO_4$ may be a minor constituent of the carbonate skeleton. High sodium and potassium content reflect entrapped sea salts. Even the most severe boiling in water and hydrogen peroxide, however, does not totally remove these salts (Table 15), suggesting that salt entrapment must be on the near molecular level (see p. 130). Articulated coralline algae contain more organic matter, up to 20% versus less than 5%, than do most crustose species.

Table 16. Average stable isotope composition of some coralline algae. Data from CRAIG (1953), GROSS (1964), KEITH and WEBER (1965) and GROSS and TRACEY (1966)

	δO^{18}	δC^{13}
Jania	−2.1	
Corallina		−4.5
Amphiroa	−3.0 (−1.0 − −4.3)	−1.1 (−0.5 − −1.7)
Lithophyllum	−2.6 (−0.9 − −4.4)	+1.5 (+3.2 − −0.2)
Goniolithon	−3.6 (−0.9 − −5.4)	−0.8 (+2.5 − −3.6)
Porolithon	−5.3 (−4.6 − −6.0)	+0.3 (+2.2 − −1.5)

Red algae are considerably enriched in O^{16} and some are also enriched in C^{13}, indicating that metabolic CO_2 probably is utilized in the calcification process (CRAIG, 1953; KEITH and WEBER, 1965; GROSS, 1965). Some isotopic averages of various coralline algal genera are presented in Table 16.

Phaecophyta (Brown Algae)

One genus of brown algae is known to be calcareous. *Padina* a tropical plant, has a partly calcified body, composed of surficial bands of aragonite alternating with bands of organic matter (DAWSON, 1961). Because this plant is commonly encrusted with other organisms, it is very difficult to obtain accurate chemical analyses of the aragonite. The few available analyses indicate that *Padina padina* has between 0.8 and 0.9% Sr (about the same or slightly higher than green algae); Mg content, 0.4%, is higher than in other aragonitic skeletons (LOWENSTAM, 1963). VINOGRADOV (1953) reported even higher magnesium contents (1.8%), along with 3% Na, 1.4% K, and 0.8% Fe. *Padina* is common on tropical reef flats, but is probably only a minor carbonate contributor.

Chlorophyta (Green Algae)

Chlorophytes contain many of the same chlorophyl pigments as higher plants, resulting in a typical green color. Calcifying chlorophytes include two families: Codiaceae and Dasycladaceae. Members of the latter family are characterized by a central stem, usually cylindrical, from which branches develop, often in a radial pattern. Calcification entails the precipitation of a crust around the stem; inner portions may or may not be calcified. Most representatives of this family are extinct, the only prominent living members being *Acetabularia* and *Cymopolia*. The Codiaceans have a tubular thallus composed of intertwining and branching filaments (JOHNSON, 1961). Most of the common calcified green algae, including *Penicillus*, *Rhipocephalus*, *Halimeda* and *Udotea*, belong to this group.

Ecology, Calcification and Sedimentary Significance

Some codiaceans and dasycladaceans grow in temperate and subtropical climates, for example *Cymopolia* in the eastern Atlantic and *Halimeda tuna* in the Mediterranean. But most species are limited to tropical climates. In the western Atlantic, for instance, *Halimeda* is a prominent sediment contributor only south of Palm Beach, Florida (with the exception of Bermuda) (MILLIMAN, 1972). Dasycladaceans and most codiaceans grow in relatively quiet, shallow waters, usually in depths no greater than 15 to 20 meters. *Halimeda* is an exception in that it can thrive in a number of environments; it is prolific on shallow turbulent reefs as well as in deep lagoons and forereefs. Most erect forms can grow on loose sediments, while prostrate forms grow in large masses on hard substrate. GILMARTIN (1960), for example, found that most *Halimeda* at Eniwetok Atoll lagoon grows only on or near hard substrates such as coral patches. GOREAU and GRAHAM (1967) reported that *Halimeda copiosa* grows in depths exceeding 45 m off Jamaica, and as such, is a prominent sediment contributor along much of the reef front.

Most codiaceans, such as *Penicillus*, *Rhipocephalus* and *Udotea*, contain organic filaments sheathed with aragonite needles, averaging 5 microns in length and 0.3 microns in diameter (PIA, 1926; POBEGUIN, 1954; LOWENSTAM, 1955). Recently, PERKINS and others (1972) reported two other diagnostic crystal types within *Penicillus:* serrated crystals about 1 micron in length located in extracellular sheaths, and double-terminated crystals, 48 to 160 microns in length, which originate by extracellular calcification.

In contrast to the non-rigid codiaceans, *Halimeda* is composed of disjunct calcified segments of fused aragonite needles. *Halimeda* segments, which are joined to one another by narrow uncalcified nodal regions, consist of a central core of medullary filaments and surrounding cortex of lateral branches called utricles (HILLIS, 1959). The numerous aragonite needles within *Halimeda* segments usually are much smaller than those in other codiaceans, averaging 0.1 to 1 microns in length (BAARS, 1968).

Upon death, the organic nodal regions usually decompose and *Halimeda* plates separate from one another. However, the plates do not decompose further, but remain as distinct sedimentary fragments. *Halimeda* was first pictured as a prolific sediment contributor by JUDD (1904) and his coworkers at Funafuti Atoll. Subsequently CHAPMAN and MAWSON (1906) showed that *Halimeda* has been an important component in both recent and ancient carbonates throughout the tropical Pacific. It is one of the most common lagoonal constituents in Pacific lagoons, while in the Caribbean it is a prominent reef component (Plate VII).

CHAPMAN and MAWSON (1906) were also the first to suggest that other codiaceans could contribute to carbonate sediments; post-mortum decomposition of the organic filaments releases the aragonite needles to the sediment. Studies by LOWENSTAM and EPSTEIN (1957) and STOCKMAN and others (1967) indicate that these disjoint aragonite needles may contribute significant amounts of aragonite muds to shallow back-reef and bank sediments.

Chlorophyte calcification appears to be predominantly extracellular, beginning in the outer portions of the thallus and proceeding inwards (BAARS, 1968; PERKINS and others, 1972). Although the specific mechanism of calcification is not known (LEWIN, 1962), WILBUR and others (1969) found no organic matrix in *Halimeda*, suggesting possible physiochemical precipitation during photosynthesis. STARK and others (1969) have pictured a two-step calcification process: calcium ions are bound inorganically into the cell wall, followed by precipitation of the calcium carbonate. Polysaccharides may provide a critical link in the exchange of calcium with the precipitating skeleton (BÖHM, 1971). WILBUR and others (1969) reported that *Halimeda* adds daily layers to its growing end. Calcification of these layers, however does not begin until 36 to 48 hours later, and continues with increasing maturity of the segment.

Table 17. Deposition of calcium in various Codiacean algae. Generic averages are derived from data presented in GOREAU (1963). Ranges of values are in parentheses

	Calcium deposition in light (μg Ca/mg N/hr)	Deposition in light / Deposition in dark
Halimeda	334 (108–870)	1.14 (0.71–1.83)
Penicillus	472 (154–1377)	2.00 (1.43–2.53)
Udotea	176 (16–361)	1.88 (0.78–2.67)
Rhipocephalus	167 (113–221)	1.31 (0.96–1.67)

Although earlier studies by GOREAU (1961) had suggested that carbonate precipitation was directly related to photosynthesis, subsequent studies found that *Halimeda* can produce carbonate at almost the same rate in the dark as in the light (Table 17); this generally applies only for the more poorly calcified species (STARK and others, 1969). GOREAU speculated that calcification of *Halimeda* might be inhibited at higher light levels, thus explaining the prolific growth of *Halimeda* at relatively great depths (and low light levels).

All green algae are capable of reproducing at extremely rapid rates for short periods of time. Although calcification rates generally are slower than in coralline algae (compare Table 17 with Table 14), additions of one *Halimeda* segment per day are not unusual (COLINVAUX and others, 1965; H.G. MULTER, in preparation) and the genesis of complete *Halimeda* plants within 24 hours has been reported

Plate VII. Halimeda. (a) Some of the more prominent species of *Halimeda* found in the Caribbean: A–F, *H. simulans*, *H. scarba*, *H. opuntia*, *H. monile*, *H. incrassata*, *H. tuna*. Photograph courtesy of H. G. MULTER. (b) and (c) *Halimeda* grains under reflected light. Noteworthy characteristics are the spongy texture of the grains (upper right corner of b) and the surface expression of utricles (c). Scales are 1 mm and 750 microns, respectively. (d) The surface expression of the utricles is shown in greater detail in this photomicrograph. SEM; scale is 300 microns

Plate VIII. *Halimeda* grows rapidly over sporadic intervals, as is evidenced by this sequence of photographs of *H. incrassata* taken at 4 week intervals in an outdoor tank. (a) 2/6/67; (b) 3/4/67; (c) 4/2/67; (d) 5/17/67. Note the rapid growth between (b) and (c). Photograph and information courtesy of H. G. MULTER

Plate VIII. Continued

(Goreau, 1961). Finckh (1904) found that a single *Halimeda* plant in Funafuti lagoon produced more than 14 grams of carbonate in a 6-week period. Colinvaux and others (1965) also reported that *Penicillus* plants can reach nearly full growth after 1 to 2 weeks, and complete maturity within one month. However such rapid growth is sporadic, lasting 1 or perhaps 2 weeks, followed by long periods of non-growth (Plate VIII). Stockman and others (1967) found that the average life span of *Penicillus* plants in Florida Bay ranges from 1 to 2 months, suggesting that active growth and calcification occur in something less than 25% of the plant's life.

Petrography

Fresh *Halimeda* plates are easily recognized. The utricles, which are 20 to 60 microns in diameter, and the chalky white plates are diagnostic properties under reflected light. The small disoriented aragonite needles that constitute the *Halimeda* "skeleton" result in an isotropic image under transmitted light, becoming opaque in polarized light. The utricles give the plates a porous appearance that is easily identified (Plate IX a, b). Unfortunately *Halimeda* is one of the first carbonate components to alter diagenetically. The filling of utricles with crytocrystalline aragonite and the micritization of the plant walls result in a gradual transformation into a cryptocrystalline lump (Plates IX and XXXI).

Cymopolia is one of the most widely distributed calcareous green algae (Johnson, 1961), but does not appear to be a common sedimentary component. Outwardly segments are similar to *Halimeda*, but they are cylindrical and hollow instead of flat and filled (Plate IX).

Composition

Calcareous green algae are aragonitic. Strontium concentrations generally are higher than in other calcareous organisms, ranging from 0.77 to 0.97% in *Halimeda* plates (average of 0.88%). *Rhipocephalus*, *Penicillus* and *Cymopolia* have similar strontium contents (Table 18). Magnesium in *Halimeda* averages less than 0.1%; values are higher in *Penicillus* but this may be due to contamination from detrital magnesian calcite. However, electron photomicrographs show calcite (?) rhombs incorporated in with the aragonite needs of *Penicillus* plants (Plate IX f). The uniformity in size of these crystals and their positions relative to the aragonite needles suggest that they may be authigenic.

→

Plate IX. Green Algae. (a) and (b) *Halimeda* plate under plane and polarized refracted light. Note the swiss cheese-like character of the plate. Scale is 250 microns. (c) Close-up of *Halimeda* plate. Refracted polarized light; scale is 100 microns. (d) *Halimeda* plate in which the utricles are mostly filled with crytocrystalline aragonite. Refracted polarized light; scale is 100 microns. (e) *Cymopolia* plates grow around an organic stem. Reflected light; scale is 500 microns. (f) Aragonite crystals in the top of a *Penicillus* plant. The blocky crystals may be magnesian calcite; they do not appear to be the result of contamination, but may be authigenic. SEM; scale is 5 microns. (g) Aragonite needles in the stem of a *Halimeda* plant. SEM; scale is 3 microns

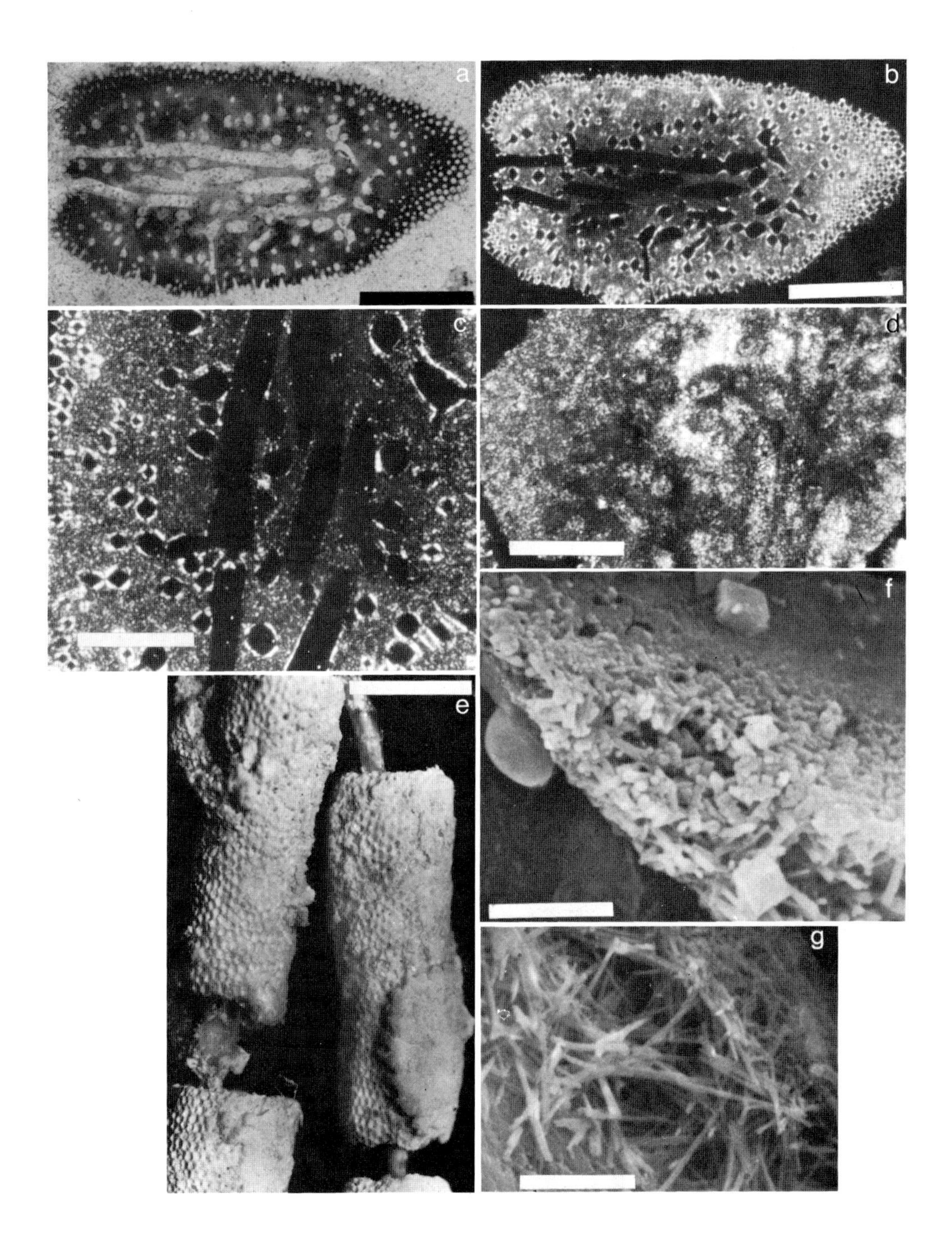

Plate IX.

Table 18. Chemical composition of various green algae (on an organic-free basis). Range of values in parentheses. Data are from VINOGRADOV (1953), THOMPSON and CHOW (1955), BOWEN (1956), ODUM (1957), MATTHEWS (1966), MILLIMAN (1967), MOBERLY (1968), KINSMAN (1969b), MÜLLER (1970), EDGINGTON and others (1970) and unpublished data of the author

	Percent organic matter	Mineral.	Percent				Parts per million						
			Ca	Mg	Sr	Na	Fe	Mn	P	Zn	Cu	K	U
Codiaceae													
Halimeda	8 (4.5–13.0)	Arag.	35.2 (34.7–35.6)	0.09 (0.04–0.27)	0.88 (0.77–0.97)	0.21 (0.18–0.26)	160[a] 3600[b] (1300–6800) 1900[c] (300–7700)	8	tr	30	30	210	1.7
Penicillus	24	Arag. tr. MgC (?)	32.9	0.37 (0.23–0.64)	0.87 (0.79–0.92)								2.2
Rhipocephalus		Arag.			0.80								
Dasycladaceae													
Cymopolia	17 (11–23)	Arag.	39.0	0.34 (0.32–0.37)	0.90 (0.89–0.91)	0.90 (0.89–0.92)	900 (675–1125)	17 (16–18)		16	11	100	1.8

[a] Author's analyses.
[b] Data from VINOGRADOV (1953).
[c] MOBERLY (1968).

Table 19. Isotopic compositions of various green algae. Data from CRAIG (1953), LOWENSTAM and EPSTEIN (1957), GROSS (1964), KEITH and WEBER (1965) and GROSS and TRACEY (1966)

	δO^{18}	δC^{13}
Acetabularia		+3.1
Halimeda	−2.2 (−0.6 − −4.0)	+2.9 (+0.8 − +5.5)
Udotea	−1.7 (−0.6 − −2.3)	+5.5 (+4.5 − +5.9)
Penicillus		
whole	−0.4	+2.9 (+2.2 − +3.7)
heads	+0.2 (−0.1 − +1.3)	+4.6 (+4.4 − +4.8)
stalks	−1.5 (−0.8 − −1.9)	+2.0 (+1.3 − +3.1)
Rhipocephalus		
whole	−1.1 (−0.9 − −1.4)	+4.4
heads	+0.6 (−0.1 − +1.5)	+5.3 (+4.3 − +5.9)
stalks	−1.1 (−0.5 − −1.9)	+1.3 (+0.1 − +2.0)

Halimeda and other green algae tend to be rich in C^{13} and poor in O^{18} (Table 19). The δC^{13} content within the organic carbon is strongly negative (−15 to 17‰) (CRAIG, 1953) suggesting that the δC^{13} enrichment within the carbonate is related to fractionation during photosynthesis. LOWENSTAM and EPSTEIN (1957) found that the heads of both *Penicillus* and *Rhipocephalus* tend to concentrate C^{13} and O^{18} relative to the stems (Table 19). Whether the heads and stems contain different trace element concentrations is not known.

Dinoflagellates

The existence of calcareous dinoflagellates has been known since the work of DEFLANDRE (1948). Recent studies have shown that these calcareous tests, which are 25 to 50 microns in diameter and possess spines 1 to 12 microns long, are not living plant cells, but rather resting cysts (WALL and DALE, 1968). WALL and DALE (1968) have observed that these cysts are widely distributed in the coarse silt fraction of both open ocean and shallow-marine sediments, but these microfossils probably account for only a small portion of the fine-size marine carbonates.

Dinoflagellate cysts are calcitic; the spines often are represented by single calcite crystals. Judging from the available data, the calcite is relatively pure, containing very little magnesium (WALL and others, 1970).

Invertebrates

Foraminifera

Foraminifera, belonging to the protozoan class Sarcodina, possess a secreted or agglutinated test. The shell not only acts as a chemical and physical buffer against the aqueous environment, but also serves as a "protective greenhouse" for symbiotic algae that live within the protoplasm (ANGELL, 1967; MARASZALEK and others, 1969). Foraminifera can be divided into arenaceous and calcareous forms. Those foraminifera with arenaceous (agglutinated) tests can be quantitatively important in nearshore and slope sediments, but generally their abundance is small compared to the calcareous forms. Calcareous foraminifera can be separated into two broad groups on the basis of wall structure (WOOD, 1949), porcellaneous and hyaline. Most porcellaneous foraminifera have imperforate tests, comprised of randomly oriented calcite crystals (although the surface layers may only be unoriented in two planes; TOWE and CIFELLI, 1967). This unoriented crystal arrangement, together with the possible incorporation of water in tiny voids (KENDALL and SKIPWITH, 1969), results in the porcellaneous foraminifera being optically opaque or having very low interference colors (such as first order brown and gray) under polarized light (Plate X c, d). Shell growth involves the addition of a new chamber onto the previous apertural face (REISS, 1958). Porcellaneous forms include many of the prominant Caribbean reef foraminifera, such as peneroplids and miliolids.

Hyaline tests are perforate and are composed of laminar sheets of calcite (in rare occurrences, aragonite) that cover both the outer and preceeding chambers. Two basic types of hyalines have been recognized under refracted light: the radial forms in which the c-axis of the calcite crystals are oriented normal to the wall, and granular types, which appear to have unoriented crystals. Under polarized light the radial types exhibit a diagnostic radial fibrous extinction (Plates X a, XI a). Electron microscope studies, however, show that the relationships between optical and actual crystal morphology in these two hyaline groups is not clear (TOWE and CIFELLI, 1967).

TOWE and CIFELLI (1967) have proposed that calcification in foraminifera occurs on organic templates that have active members (which form the templates for crystal nucleation) and passive members (which provide form and shape for crystals). The unoriented crystals in porcellaneous tests may be the result of a colloidal or structurally incoherent material in the calcifying matrix that induces spontaneous carbonate precipitation (TOWE and CIFELLI, 1967). The effect of symbiotic algae, which inhabit the protoplasm of many benthonic foraminifera (BOLTOVSKOY, 1963; HAYNES, 1965), is not known, but conceivably the algae aid in calcification in an analogous manner to that in corals (see below).

→

Plate X. Porcellaneous and Hyaline Foraminifera. (a) Porcellaneous benthonic foraminifera (peneroplids) with bright to dull white surface texture. Reflected light; scale is 2 mm. (b) The constituent magnesian calcite crystals within the test of porcellaneous foraminifera are exceedingly small. SEM; scale is 10 microns. (c) and (d) As a result of the small crystal size, porcellaneous foraminifera tend to be brown to opaque under refracted light. (c) is a transverse section and (d) a longitudinal section of a peneroplid. Scales in both photos are 250 microns. (e) Hyaline foraminifera (these are *Calcarina spengleri*) are particularly prominent sedimentary components on Indo-Pacific reef flats. Reflected light; scale is 1 mm. (f) The ultrastructure of the hyaline test consists of layers of magnesian calcite. SEM; scale is 10 microns. (g) As a result of the layered ultrastructure, hyaline foraminifera show a distinct radial fibrous structure under polarized refracted light. Scale is 250 microns

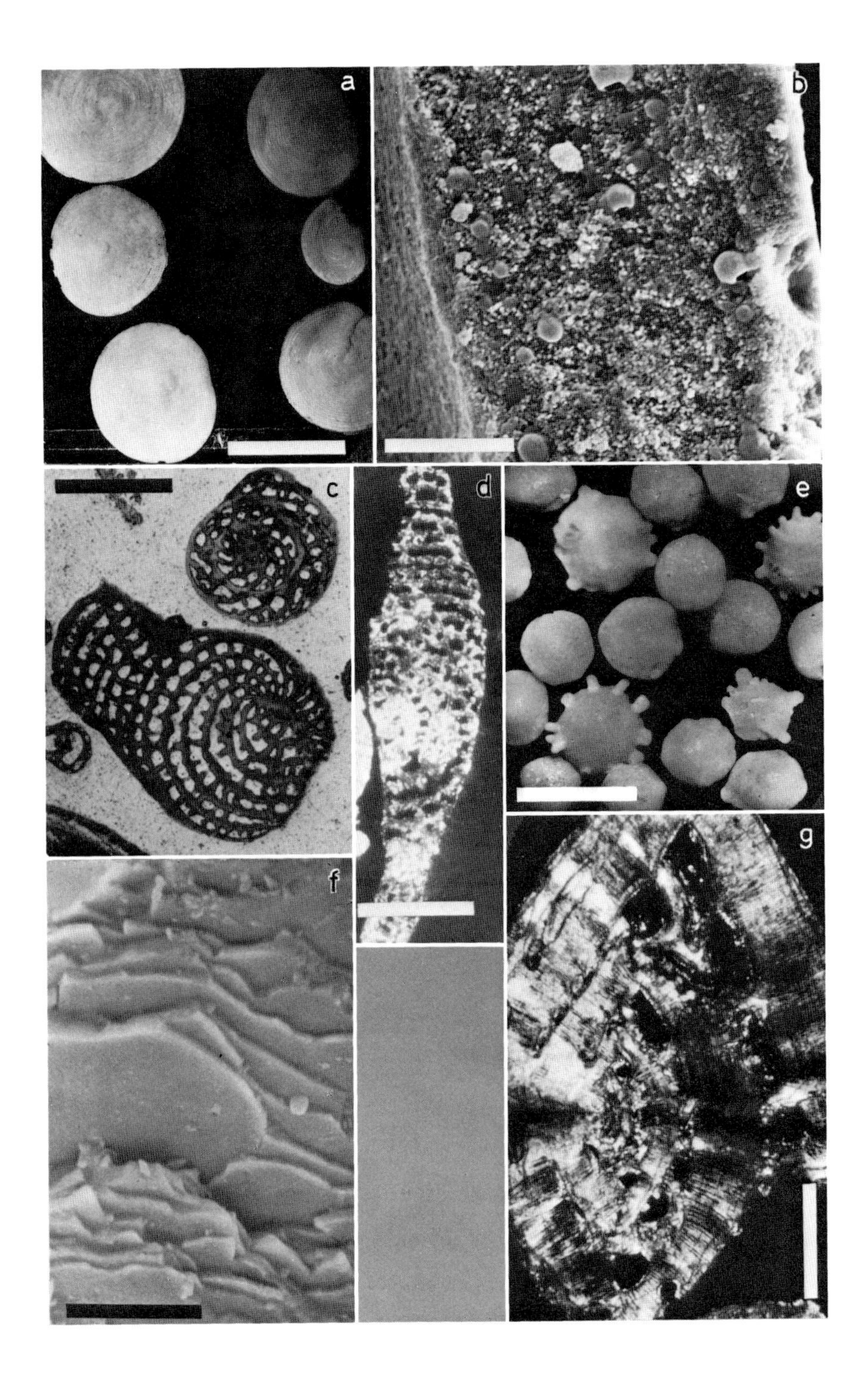

Plate X.

Ecology and Sedimentary Significance

Unlike most other carbonate components, foraminifera often remain preserved as entire individuals, and thereby can offer added insight into the age and depositional environment of a carbonate sediment. The following few paragraphs discuss some of the more important ecologic aspects of benthonic and planktonic foraminifera. More complete discussions have been given by PHLEGER (1960) and LOEBLICH and TAPPAN (1964).

Benthonic foraminifera can live on and in a wide variety of substrates; many shallow-water forms live on grass blades or on hard surfaces. The life span of an average individual appears to be about one year (WALTON, 1955). Available data suggest that the number of living individuals generally ranges between 1 and 100 individuals per m^2 (MURRAY, 1967), but even within relatively small areas the density concentrations of living individuals can change by orders of magnitude (LUTZE, 1968). Highest productivity occurs in nearshore areas adjacent to rivers and in mid shelf depths (WALTON, 1964). WALTON (1955) found that the highest living populations off Baja California occur in water depths between 50 and 90 meters; at depths between 500 and 600 meters, living populations are about 1/6 of those at 50 to 90 meters. Generally speaking the size of benthonic foraminifera increases and microarchitecture changes with both distance from shore and increasing water temperature (BANDY, 1960, 1964).

Benthonic foraminifera can be an important sedimentary component in shallow-water environments. The encrusting *Homotrema rubrum* (recognized by its red color and irregular cell structure, Plate XI) is characteristic of reef-flat sediments, especially those in the Caribbean and Bermuda (MACKENZIE and others, 1965). Peneroplids and miliolids predominate in Caribbean lagoonal deposits. An excellent review and taxonomic discussion of Caribbean reef foraminifera has been presented by WANTLAND (1967). Benthonic foraminifera are generally more abundant in Pacific reef sediments (see below). *Calcarina*, *Marginopora* and *Amphistegina* are common radial hyaline genera. CUSHMAN and others (1954) and COLLINS (1958) have presented descriptions of the reef foraminifera in the Marshall Islands and Great Barrier reef; these papers can serve as a general introduction to Indo-Pacific reef populations.

Living and dead benthonic foraminifera are not always represented by the same species within a sediment. Post-mortum transport (ILLING, 1950; MURRAY, 1966) and the mixture of recent and relict assemblages account for such discrepencies. An excellent example of the mixing of various assemblages occurs on the edges of many continental shelves, where shallow-water reef benthonics are mixed with deeper-water benthonics and planktonics. In outer shelf and upper slope sediments, benthonics gradually decrease and planktonics become predominant.

Most planktonic foraminifera precipitate their tests in the surface layers of the ocean, generally above the thermocline. Life spans probably do not exceed one year (BERGER and SOUTAR, 1967). However, studies by BÈ (BÈ and LOTT, 1964; BÈ, 1965; BÈ and HAMLEBEN, 1970) and ORR (1967) show that some planktonic genera (*Globigerinoides* and *Globorotalia*) can precipitate an outer "drusy" calcitic

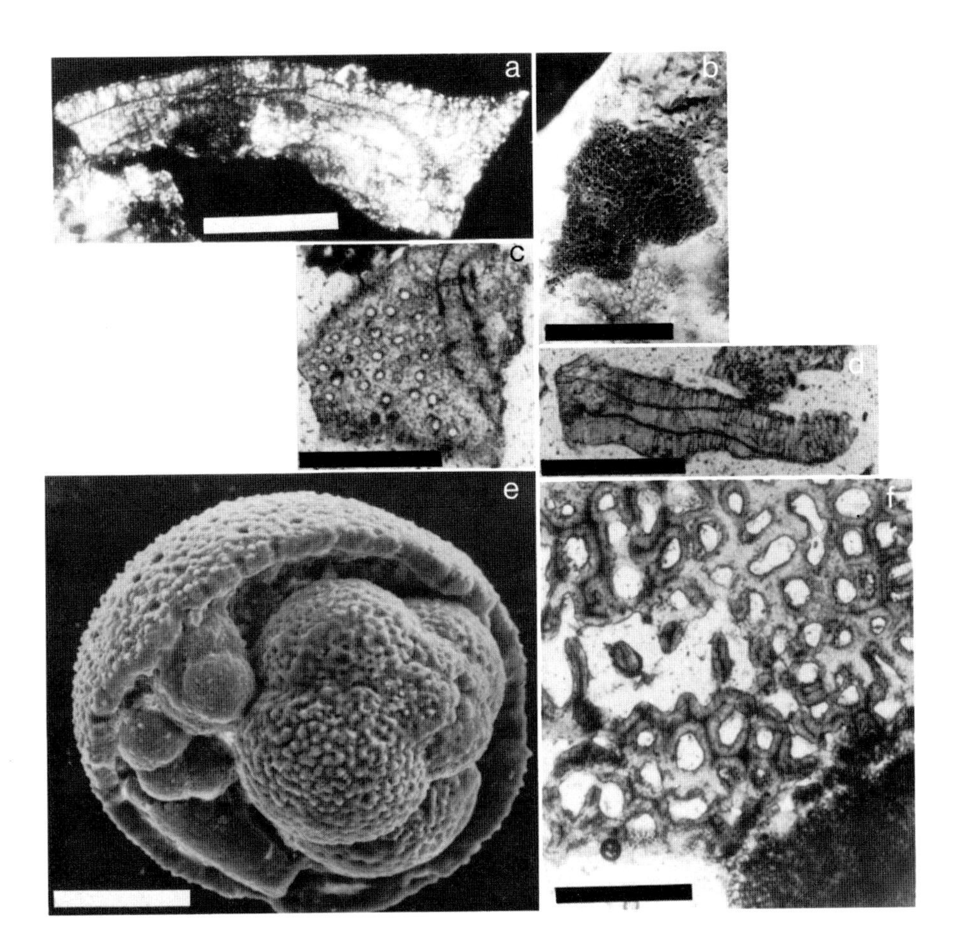

Plate XI. Foraminifera. (a) Hyaline foraminifera tests can be recognized, even in small fragments, by their diagnostic extinction patterns under polarized refracted light. Compare this photograph with Plate X g. Scale is 50 microns. (b) *Homotrema rubrum* is a common Caribbean reef flat foraminifera that encrusts available rubble and substrate. It is recognized by its pink to red color and its porous surface. Reflected light; scale is 1 mm. (c) and (d) Foraminifera fragments under refracted light. Scales are 100 and 50 microns, respectively. (f) Refracted light photomicrograph of *Homotrema rubrum*, coating a piece of coralline algae (lower right). Scale is 100 microns

Table 20. Magnesium carbonate content of various foraminiferia families. Data from BLACKMON and TODD (1959)

Family	Mean % $MgCO_3$	Range
Imperforate		
Miliolidae	13.6	(6–18)
Ophthalmidiidae	13.8	(7–18)
Peneroplidae	16.3	(15–18)
Alveolinellidae	17.7	(14–20)
Perforate		
High-Mg		
Camerinidae	12.7	(12–13)
Calcarinidae	> 19	(17–> 24)
Planorbulinidae	12.4	(8–15)
Homotremidae	15.7	(15–16)
Mixed		
Buliminidae	14 3	(1–15)
Rotaliidae	14 3	(1–20)
Discorbidae	10	(4–15)
Intermediate		
Amphisteginidae	4.6	(3–7)
Low		
Nonionidae	2	(2–4)
Elphidiidae	3	(2–4)
Lagenidae	< 1	(< 1)
Globigerinidae	1	(0–2) } planktonic
Globorotaliidae	tr	(0–1) } planktonic
Aragonite		
Robertinidae	Arag.	
Ceratobuliminidae	Arag.	

layer up to 50 microns thick in water depths between 120 and 700 meters[3]. Total time required for this secondary calcification may be close to one year (BÈ and ERICSON, 1963). PESSAGNO and MIYANO (1968) have found that the secondary outer layer is radially hyaline while the inner layer is granular, suggesting different modes of calcification and possibly different compositions.

Test size of most planktonic populations increases in lower latitudes (STONE, 1956). BERGER (1969) has suggested that this is related to the deeper thermoclines in tropical seas (thus a greater depth range for growth). Tests also become more porous with increasing water temperature (BÈ, 1968). The productivity of planktonic foraminifera can be estimated by knowing the number of living individuals within the plankton and their life span. BERGER (1969, 1970) has estimated that living

3 More recently, EMILIANI (1971) presented isotopic data which suggest that *Globorotalia menardii* and *G. truncatulinoides* do calcify over a considerable depth range. Other planktonic species, however, appear to calcify primarily at the surface and near surface.

populations generally range from 1 to 200 individuals per m^3 in surface waters, decreasing to less than 1 individual per 100 m^3 at depths greater than 1000 m. Empty (dead) shells generally are present in concentrations of 100 to 1000 per 10^5 m^3 and show little variation with depth (BERGER, 1970). Absolute productivity depends greatly upon the phytoplankton productivity in the area. Highest productivity of planktonic foraminifera occurs in the North Pacific, lowest in the Sargasso Sea. Although planktonic productivity decreases with increased distance from shore, the rapid decrease of other carbonate and non-carbonate components is so marked that the coarser fraction of most oceanic sediments in depths less than 4000 to 5000 meters is dominated by planktonic tests.

With advent of calcareous coccolithophorids and planktonic foraminifera in the early and mid-Mesozoic, the center of carbonate sedimentation quickly shifted from shallow water to the deep sea. The tests of these two planktonic protistids account for more than 80% of the modern carbonate deposited in the oceans (see p. 239 ff.). At present planktonic foraminifera are far more important than coccoliths, but this does not seem to have been the case in the early and mid-Tertiary (see above).

Composition

The composition of foraminifera tests is closely related to shell structure and generic factors. BLACKMON and TODD (1959) found that the five imperforate porcellaneous families are totally high magnesian calcite. The imperforate hyalines, on the other hand, can be aragonite (2 families within the superfamily Rotertinacea), magnesian calcite (7 families), calcite (10 families), or mixtures of magnesian calcite and calcite (3 families) (Table 20). Those perforates with consistently high magnesium content are shallow-water benthonics with radial fibrous extinction and unusually thick walls (for example, *Calcarina*).

CHAVE (1954) calculated that foraminifera increase their Mg content by about 1% for every 8 °C increase in water temperature (Fig. 43), but the absolute variation is stongly dependent upon generic factors (Fig. 23). Strontium values within calcitic tests are less than 0.2%, but the aragonitic *Höglundina* shows strontium concentrations greater than 0.7%. BOCK (1967) found relatively high concentrations of barium, boron and copper in the tests of *Archaias* and *Sorites* from Florida Bay (Table 21). Apparently these high values are not related to terrigenous contamination, since the foraminifera were collected alive from the blades of turtle grass in a predominantly carbonate-rich area. BOCK also reported definite changes in both barium and magnesium content throughout the year, but the exact cause of the fluctuations is not clear.

Early chemical analyses of planktonic foraminifera were performed on tests selected from deep-sea oozes. Because of contamination by terrigenous clays and authigenic minerals, these analyses were often erroneous (for example, see BRADY, 1884, cf. VINOGRADOV, 1953; CORRENS, 1941). EMILIANI (1955a) tried to eliminate the contamination effects by careful sample washing and by plotting regressional trends. He concluded that Ti, Al, Si, Fe, Mn and Mg concentrations in planktonic tests are caused by terrigenous contamination; only the Sr is totally biogenic.

Table 21. Elemental compositions of selected foraminifera genera. Data from CLARKE and WHEELER (1922), VINOGRADOV (1953), CHAVE (1954)[c], THOMPSON and CHOW (1955), ODUM (1957), BLACKMAN and TODD (1959)[b], KRINSELY (1960a), LIPPS and RIBBE (1967), BOCK (1967)[a], GOMBERG and BONATTI (1970) and G. THOMPSON and R. W. SHERWOOD (unpublished data). Range of values are in parentheses

	Percent organic matter	Mineral.	Percent			Parts per million							
			Ca	Mg	Sr	Fe	Mn	Ba	Cu	Ni	B	Si	Zn
Benthonic													
Archaias		MgC	35.3	1.35[a] (1.28–1.42) 3.35[b,c]	0.16 (0.15–0.18)		60 (40–80)	74 (12–210)	10 (6–22)		26 (20–40)		
Sorites		MgC	35.2	1.46[a] (1.30–1.57) 3.75[b,c]	0.15 (0.15–0.16)		50 (45–65)	115 (55–150)	16 (12–21)		60 (50–90)		
Amphistegina		MgC		1.25 (1.02–1.75)	0.19 (0.18–0.21)	110 (tr–310)	20	1	9 (tr–17)	<2	70 (65–80)	1000 (30–1500)	1 (tr–2)
Homotrema	3.44	MgC	34.0	3.20 (3.05–3.50)									
Orbitolites	1.30	MgC	35.4 (34.6–36.0)	3.01 (2.57–3.63)		1100 (tr–4750)						1350 (510–2700)	
Calcarina	6.72	MgC	31.6	3.75	0.19	10	2	2	2	<2	80	40	<2
Heterostegina		MgC		3.05	0.20	25	2	3	2	<2	45	150	<2
Planktonic													
Globigerinoides		Cal.		0.15 (0.12–0.20)	0.10 (0.10–0.11)	200	11 (9–13)		12 (8–15)	20			
Globorotalia		Cal.		0.11 (0.08–0.16)	0.16 (0.11–0.18)	140	12		9	20			
Globigerina		Cal.		0.18 (0.10–0.25)	0.13 (0.10–0.17)				8 (8–10)	20			

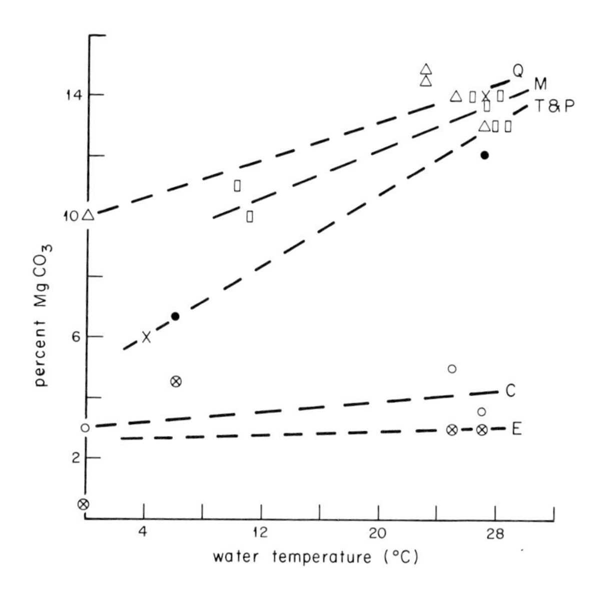

Fig. 23. Relation between magnesium content of various benthonic foraminifera and water temperature. Data from CHAVE (1954) and BLACKMON and TODD (1959). Q = *Quinqueloculina*; M = *Miliolinella*; T = *Triloculina*; P = *Pyrgo*; C = *Cibicides*; E = *Elphidium*

KRINSLEY (1960a) analyzed specimens collected in plankton nets (thereby avoiding terrigenous contamination) and found that Sr, Cu and Mg are biogenic, while Al, Mn and Ti are probably related to sediment contamination. WANGERSKY and JOENSUU (1964) obtained similar Sr and Mg values in bulk analyses of washed plankton species from deep-sea sediments, and electron probe analyses by LIPPS and RIBBE (1967) further substantiated these observations. Noteworthy are the exceedingly low magnesium and strontium contents (Table 21). Curiously, the concentrations of some trace metals (Fe, Mn, Cu, Ni) are as high or higher in planktonic tests than in shallow-water benthonic species (Table 21). Recent analyses have shown that deep-sea benthonic foraminifera have similar Sr and Mg concentrations as do planktonic tests (DUPLESSY and others, 1970). Fluorine content in planktonic foraminifera averages about 300 ppm (range from 75 to 505 ppm) (CARPENTER, 1969).

Foraminifera precipitate their tests in isotopic equilibrium with the ambient environment (UREY and others, 1951; EPSTEIN and others, 1953; EMILIANI, 1955b)[4]. Some isotopic values for benthonic foraminifera are listed in Table 22. Assuming that modern planktonic species precipitated their tests at roughly the same depths during the past, EMILIANI (1970 and references therein) has constructed Pleistocene paleotemperature curves which correspond closely (according to EMILIANI) with paleoecologic evidence. Other workers, however, have questioned both the paleontologic and paleoecologic validity of these isotopic curves. For further information on this debate, the reader is referred to articles by SHACKLETON (1967), DONN and SHAW (1967) and EMILIANI (1967).

4 Recent isotopic analyses of deep-sea benthic foraminifera show δO^{18} differences of more than 1% between various genera, suggesting that biologic activities also play an important role in isotopic composition (DUPLESSY and others, 1970).

Table 22. Stable isotopic values for various benthonic foraminifera. Data from EPSTEIN and LOWENSTAM (1953), GROSS (1964), GROSS and TRACEY (1966), SHACKLETON (1967), GOMBERG and BONATTI (1970) and DUPLESSY and others (1970)

	δO^{18}	δC^{13}
Shallow-water tropical genera		
Amphistegina	−2.0 (0.0 − −3.6)	+0.3 (+0.2 − +0.5)
Archaias	−0.6	+3.6
Calcarina	−3.3	−2.0
Marginopora	−2.7	+3.1
Homotrema	−1.6 (−0.2 − −2.5)	+2.5 (+2.4 − +2.6)
Deep-water genera		
Pyrgo	+3.2 − +5.0	
Planulina	+2.8 − +4.1	

Tintinnida

Tintinnids are loricate ciliated protozoa, possessing cup-shaped single cell tests. These organisms are present in modern sediments, especially in the deep ocean. Although Tintinnids possess hard carbonate tests, the tests consist of organic agglutinations of various small carbonate particles, such as coccoliths, rather than a precipitated shell (TAPPAN and LOEBLICH, 1968). At present there are no published data concerning the ability of Tintinnids to precipitate calcium carbonate.

Porifera

Sponges (Phylum Porifera) contain two distinct types of spicules (sclerites): 1. Megascleres, 3 to 30 microns in diameter and 100 to 500 microns in length, help to form a loose skeletal framework for the sponge; 2. Microscleres are usually less than 1 micron in diameter and 10 to 100 microns in length, and are found exclusively within the flesh (DE LAUBENFELS, 1955). The spicules can be composed of silica, calcite or spongin. In those sponges containing calcareous spicules (Class Calcispongea) most spicules are megascleres. Three basic calcareous spicule types are noted: monaxons, triradials, and quadriradials (Fig. 24). The monaxons generally protrude from the epidermis and serve as protection for the animal. Triradials and quadriradials provide the structural framework of the sponge; in extreme instances, such as the genus *Petrosoma*, these spicules (Fig. 24) merge and form a rigid framework. Spicules are secreted by one or more amoebic-like cells, called sclerocytes. Spicules become enclosed by a thin retractile membrane (sheath) that probably acts as a calcium (and possibly magnesium) pump (JONES, 1967, 1970).

Recently another type of calcareous sponge has been described by HARTMAN and GOREAU (1970a, b). These sponges, tentatively assigned to the class Sclerospongia, can precipitate massive skeletons composed of aragonite and siliceous

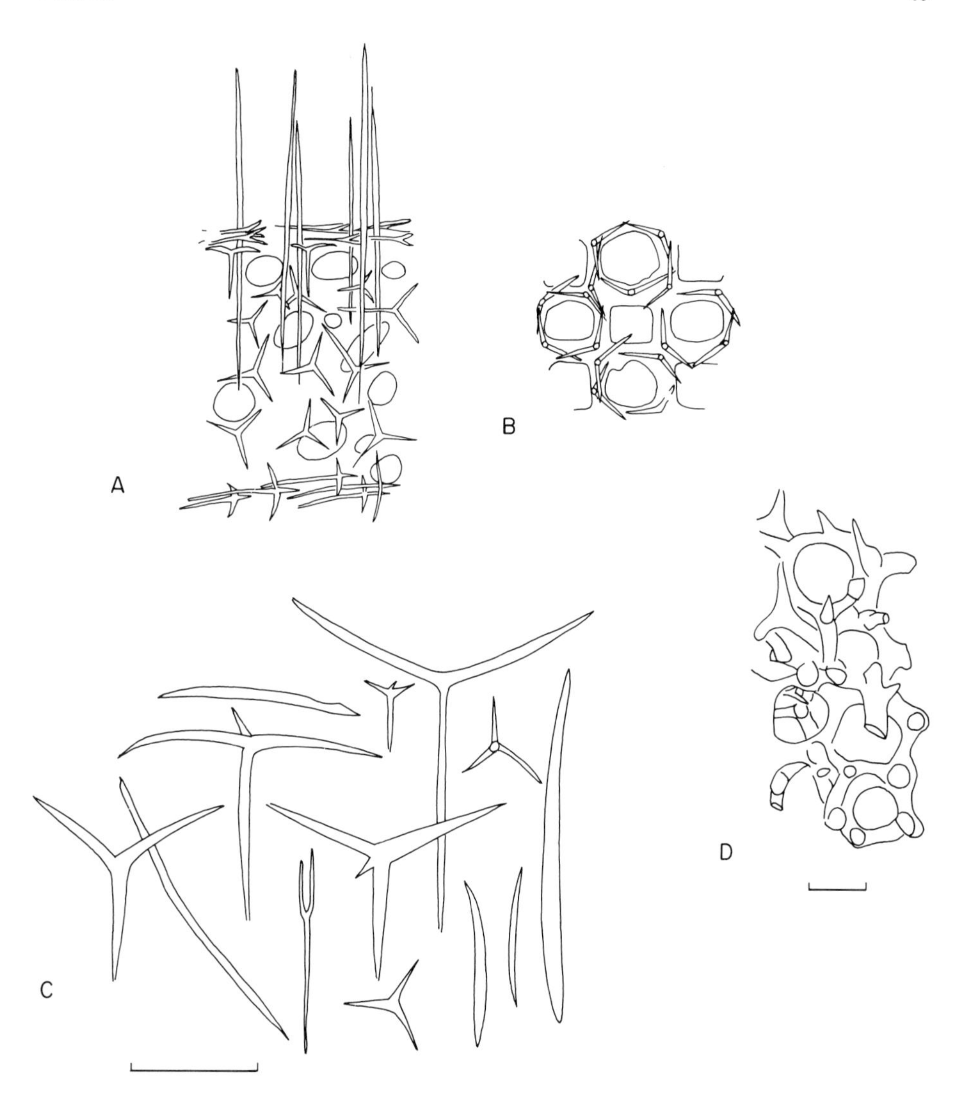

Fig. 24 A–D. Calcareous sponge spicules. A and B show the position of various spicules within the sponge. Triradiate spicules are used as a structural framework, and often surround radial canals (open spaces in Fig. B), as opposed to incurrent canals, which are unsupported. Monaxial spicules, which penetrate the epidermis, serve as protection (A and B after HYMAN, 1940). C illustrates various sponge spicules, monaxial, triradiate and quadriradiate. The tuning fork seen in this illustration is an example of an irregular triradiate (after DODERLEIN, 1898; MINCHIN, 1908). D the pharetrom, *Petrosoma*, in which the spicules have merged to form a rigid framework (after DODERLEIN, 1898). Bars are 100 microns

spicules and organic fibers. Calcification in these sclerosponges radiates from organic centers in a manner similar to that in scleractinian corals. These skeletons, marked by small pits, 0.5 mm across and 1 to 2 mm deep, closely resemble the enigmatic fossil stromatoporids (HARTMAN and GOREAU, 1970a). T.F. GOREAU considered these sponges to be potentially important elements in the formation and consolidation of the basal reef framework in Jamaica (W.D. HARTMAN, in YONGE, 1971), and recent studies on the deep reef fronts at Discovery Bay, Jamaica and at St. Croix, Virgin Islands show that below depths of about 80 m, the sclerosponge *Ceratoporella* is *the* dominant reef-forming organism (C.H. MOORE and H.G. MULTER, 1973, oral communication).

Most calcareous sponges live in water depths less than 100 m and are particularly abundant on hard bottoms with swift currents. Sclerosponges on the other hand, tend to be cryptic, living mainly in crevices and caves (HARTMAN and GOREAU, 1970a). Sponge spicules generally do not contribute large quantities of skeletal carbonate to marine sediments.

Sponge spicules are fairly easy to identify. Their smooth and rather simple geometrical shapes are contrasted to the relatively complex shapes of other carbonate spicules (Fig. 24).

Until recently few data were available concerning the composition of calcareous sponge spicules (Table 23). Apparently spicules from the grass Calcispongea are totally magnesian calcite, although JONES and JENKINS (1970) have found trace amounts of aragonite in the sponge *Sycon ciliatum*. The absolute amount of magnesium varies greatly with temperature (CHAVE, 1954), but also varies from genus to genus. JONES and JENKINS (1970) found $MgCO_3$ ranging from 5.2 to 12.0 mole % for various genera off England. Moreover, magnesium content varies with the size and shape of the spicule (Table 24); apparently each spicule has a uniform composition throughout (JONES and JAMES, 1969). The reason for the variation in composition with spicule size is not clear, although it may be that larger spicules precipitate faster and thus are unable to discriminate as well against Mg as the slower-forming spicules (JONES and JAMES, 1969). Strontium content is similar to that in other calcitic invertebrates, ranging from 0.20 to 0.26%. Sodium and sulfur are surprisingly high (Tables 23 and 24), but these high values may represent organic contamination rather than elemental concentrations within the carbonate itself.

Most sclerosponges have been reported to be aragonitic (LISTER, 1900; HARTMAN and GOREAU, 1970a), but some magnesium calcite species have also been found (HARTMAN and GOREAU, 1970b). Siliceous spicules and organic frameworks also are present, although in relatively small amounts. To date no analyses on the elemental composition of sclerosponge skeletons have been published.

Coelenterata

Colonial coelenterates are one of the major carbonate contributors in the modern oceans. Both hydrozoans and anthozoans deposit calcium carbonate. Among the anthozoans, octocorals and scleractinians are prominent carbonate formers.

Coelenterate skeletons have two distinct growth forms, the hard rigid calcareous stony corals (hydrocorals and scleractinians) and the "supple, highly deformable, but visco-elastic" octocorals (WAINWRIGHT and DILLON, 1969). The polyps of stony corals generally extend at night to capture plankton. Octocoral polyps,

Table 23. Chemical composition of sponge sclerites from the waters off England; range of values in parentheses. From data reported by JONES and JENKINS (1970)

	Mineral.	Percent				Parts per million									
		Mg	Sr	Na	S	Fe	Mn	K	Ba	Cu	Zn	Pb	P	Si	Al
Leuconia	MgC	2.58 (2.40–2.75)	0.22	0.57 (0.55–0.59)	0.28 (0.27–0.30)	< 100	10	< 100	10	< 10		< 10	< 15	100	100
Clathrina	MgC	3.15	0.20	0.49	0.32										
Leucosolenia	MgC	2.00	0.20	0.59	0.33										
Sycon	MgC	1.27 (1.25–1.30)	0.23 (0.22–0.25)	0.55 (0.54–0.56)	0.33 (0.30–0.37)										
Grantia	MgC	1.90	0.20	0.66	0.32										
Amphiute	MgC	2.10	0.22	0.55	0.33										

Table 24. Composition of various size sclerites within the same sponges. Data from JONES and JENKINS (1970)

Species	Type sclerite	Weight % of sponge sclerites	Mg	Sr	Na	S
Leuconia nivea	micromonoaxons	2.1	2.5			
	small tri- and tetr-racts	3.2	2.6			
	large triacts	19.2	2.65	0.21	0.58	
	giant triacts	75.5	2.80	0.21	0.53	0.30
Amphiute paulini	small tri- and tetr-racts	16.2	1.90	0.23		
	large oxea	83.7	2.20	0.21		

on the other hand, extend in the daytime, suggesting a greater nutritional dependence upon photosynthesis (see below) (WAINWRIGHT, 1967).

Ecology and Calcification

Hermatypic (reef-building) coelenterates are defined as having symbiotic dinoflagellates (zooxanthellae) in their epithelial tissues, whereas ahermatypic species do not. Most studies on the ecology of hermatypic and ahermatypic coelenterates have been confined to stony corals, and therefore much of the following discussion will concentrate on the results of these studies.

Hermatypic corals grow best in shallow-water (less than 20 meters) environments, with water temperatures between 25 and 30 °C and salinities between 34 and $37^0/_{00}$. However a much wider range of temperatures and salinities can be tolerated. Relatively healthy hermatypic corals have been found in waters as cold as 11° and as warm as 40 °C (MACINTYRE and PILKEY, 1969; KINSMAN, 1964b) and in salinities ranging from brackish to greater than $60^0/_{00}$ (SQUIRES, 1962; KINSMAN, 1964b). Ahermatypic species occur in deeper and colder waters, seldom shallower than 100 meters nor warmer than about 20 °C (JOUBIN, 1922; VAUGHAN and WELLS, 1943; TEICHERT, 1958).

Table 25. Calcium accretion compared to total growth rates for various scleractinians and hydrozoons. Data after GOREAU and GOREAU (1960a) and LEWIS and others (1968)

	μg Ca/mg N/hr	cm/yr
Acropora cervicornis	> 50	15–26
Acropora palmata	40–49	
Millepora camplata	40–49	
Porites porites	30–39	3–4
Millepora alcicornis	20–29	
Diploria labyrithiniformis	20–29	
Siderastrea	10–19	
Montastrea annularis	10–19	2–3
Porites astroides	0–9	
Madracis asperula		2–3

Hermatypic corals calcify at much faster rates than do ahermatypic corals. While reef corals can grow by as much as 15 to 26 cm per year (Table 25; see below), ahermatypic species probably grow at about 1 cm per year (PRATJE, 1924; TEICHERT, 1958). Laboratory experiments show that hermatypic corals calcify at faster rates during the day than at night (GOREAU, 1963) (resulting in daily growth bands in some species; WELLS, 1963; BARNES, 1970). These data suggest that the photosynthetic acticities of the zooxanthellae play a major role in calcification, but the exact role is not clear. Zooxanthellae do not appear to be a direct food source, since stony corals do not contain the proper enzymes required to digest plant material. During periods of starvation corals may actually expell the zooxanthellae (YONGE, 1958, 1963). More probably the zooxanthellae and corals exhibit a

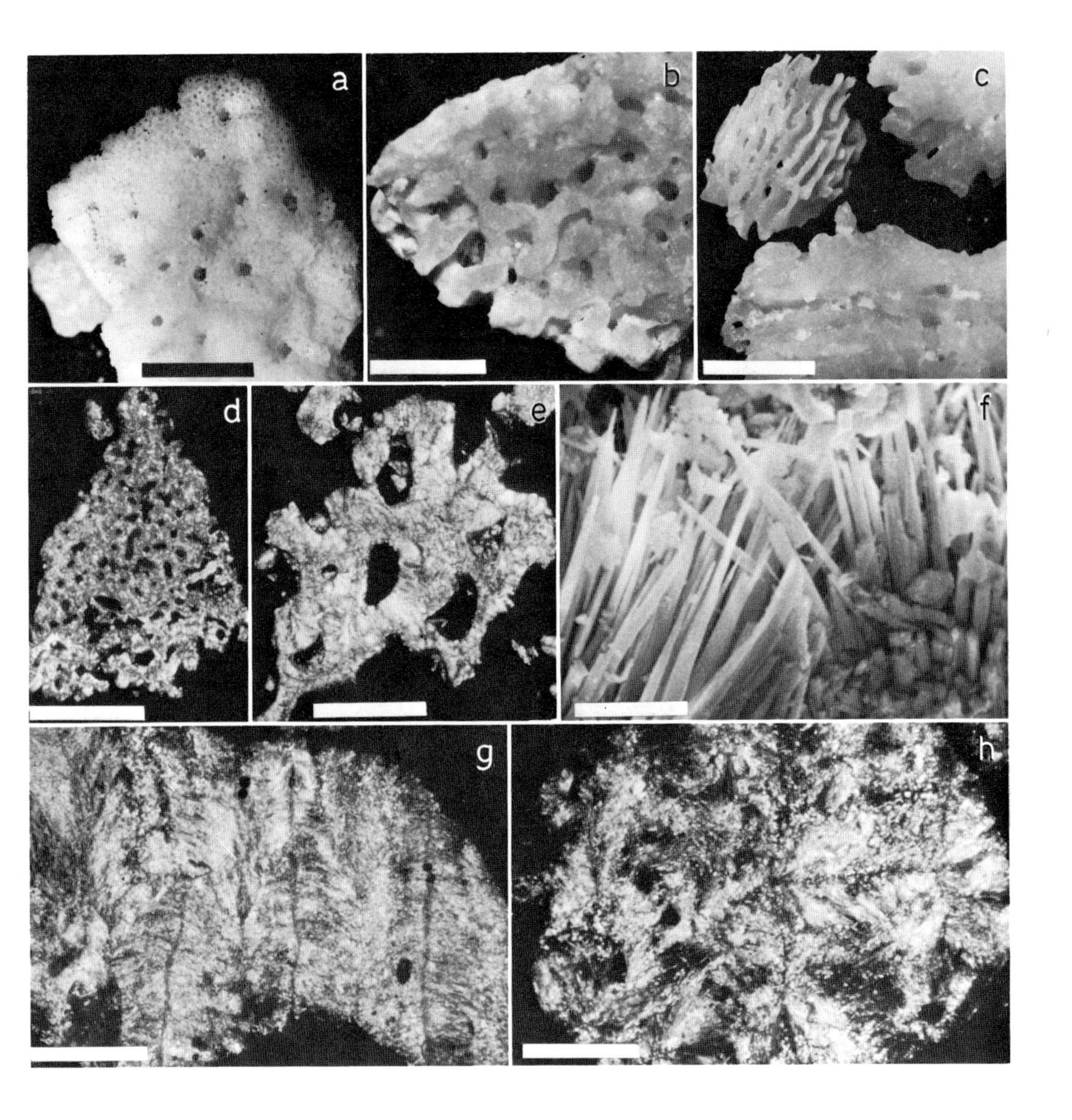

Plate XII. Coral. (a) *Millepora fragment*, with dactylopores (small pores) and gastropores (large pores). Reflected light; scale is 1 mm. (b) and (c) Scleractinian coral fragments. Reflected light; scales are 0.8 and 0.75 mm, respectively. (d) *Millepora* fragment under polarized refracted light. Scale is 250 microns. (e) *Porites* fragment under polarized refracted light. Scale is 250 microns. (f) SEM photomicrograph of scleractinian coral fragment showing possible sclerodermite structures. Scale is 25 microns. (g) and (h) Scleractinian fragments under refracted polarized light, showing the diagnostic petrographic characteristics. Scales are 250 microns

mutual symbiosis. The algae consume the corals' excretory products (GOREAU, 1961) while the coral polyps utilize the oxygen produced by the photosynthesizing algae. The gycerol liberated from the zooxanthellae also may be of nutritional value (MUSCATINE, 1967). Supporting this latter statement, PEARSE and MUSCATINE (1971) found that photosynthetic products tend to migrate towards those parts of the coral with the highest rates of calcification; impedence of this migration hinders calcification.

Other vitamins and hormones may also be used by corals (GOREAU and GOREAU, 1960b). Recent studies suggest that although reef corals are capable of ingesting large quantities of zooplankton (COLES, 1969), the zooplankton populations on most reefs are too small to account for significant nutrition of the average coral (JOHANNES and others, 1970).

There seems little doubt but that the rapid precipitation of calcium carbonate by hermatypic corals is related to the uptake of respiratory CO_2 by zooxanthellae:

$$Ca(HCO_3)_2 \rightarrow CaCO_{3\downarrow} + H_2CO_{3\uparrow}$$

(GOREAU, 1959, 1961) and in fact, metabolic CO_2 may be constitute a significant quantity of the $CO_3^=$ deposited in the coral skeleton (GOREAU, 1961; PEARSE, 1970). The presence of carbonic anhydrase is also important. Inhibition of this enzyme reduces calcification by about 50% (GOREAU, 1959b). WAINWRIGHT (1963) suggested that the amide part of chitin, the major organic compound in scleractinians, is critical for calcification, and that zooxanthellae contribute the glucose critical to the synthesis of chitin within corals.

The various groups of calcifying coelenterates exhibit different growth habits, petrographic characteristics and compositions. Therefore each of the three major groups will be discussed separately in the following paragraphs.

Hydrozoa

Two classes of hydrozoans are prominent carbonate contributors in modern seas, Stylasterina and Milleporina. The former live over a relatively great range of water depths and latitudes. In contrast, Milleporan corals, because of their association with zooxanthellae, are restricted to shallow depths (generally less than 30 m) in tropical latitudes (BOSCHMA, 1956). The genus *Millepora* is especially common in the reef front zones of Caribbean coral reefs, and locally can be a major contributor to both the reef rock and sediment. Its rate of growth compares with the faster growing scleractinians (Table 25).

The skeleton of *Millepora* is easily recognizable under reflected light by the presence of small pores (dactylopores) that surround slightly larger pores (gastropores) (Plate XII). These polyp openings, together with a shallow complex canal system, give the skeleton a porous structure that can be recognized under both reflected and refracted light. The linear arrangement of sclerodermites, so common in scleractinians, is absent. However, smaller, less well-developed bundles of aragonite crystals are common, often making it impossible to distinquish between milleporans and scleractinians in thin section.

Millepora skeletons aragonitic; they contain similar concentrations of strontium and magnesium to those found in scleractinians (Table 26). Stylasterinids are aragonitic in waters warmer than 3 °C; in colder environments the skeleton can be partially or totally calcitic (LOWENSTAM, 1964b). Aragonitic stylasterinids contain high amounts of strontium. The calcitic portion of these hydrocorals contains about 2% magnesium (LOWENSTAM, 1964b).

Octocorallia

Five of the eight octocorallian orders listed by BAYER (1956) contribute significant amounts of carbonate to modern sediments. These orders are Stolonifera, Coenothecalia, Pennatulacea, Alcyonacea and Gorgonacea. Coenothecalina is the only octocorallian to produce a stony skeleton; *Heliopora*, the best known of this group is a prominent reef builder in Indo-Pacific coral reefs. This coral has a blue color and large polyp cavities (up to 1 mm in diameter) separated by smaller tubules, facilitating recognition under reflected light.

Other octocorals contain numerous spicules (sclerites) secreted by the mesoglea, and concentrated in the base of the colony, in the body wall between the septa and in the axis (BAYER, 1956). The amount of axis calcification depends upon both species and the age of the individual (LOWENSTAM, 1964b). In most instances the spicules are loose, being held together by a horny protein, gorgonin. However in the Stoloniferan, *Tubipora*, the spicules, 0.4 to 1 micron in diameter, are fused into a hard skeleton (SPIRO, 1971). The red color and "organ-pipe" configuration of this Indo-Pacific stoloniferan makes its identification relatively easy.

Pennatulaceans (sea pens) are the only octocorals capable of growing in soft sediment; all others need hard substrate for attachment (BAYER, 1956). Sea pen spicules usually consist of long smooth rods or bars, although plates and disks are also common (Fig. 25).

Gorgonians and alcyonaceans are the major octocoral carbonate contributors in the modern seas. Gorgonians flourish in Caribbean coral reefs and are especially common in lower energy back reef areas (see below). Alcyonaceans are restricted mostly to Indo-Pacific reefs, and tend to be more massive and more heavily calcified than the Caribbean gorgonians (CARY, 1931; BAYER, 1956). Both gorgonian and alcyonacean spicules are easily recognized by their monaxial shape with protruding warts and spines. These spicules are commonly 100 to 300 microns in length and tens of microns in diameter (Fig. 25). Colors are characteristically purple, pink or white.

CARY (1918) calculated that the gorgonians at Dry Tortugas undergo a complete turnover every five years, and that the annual contribution of spicules is about one ton per acre. This estimate seems high, since it would mean an addition of some 25 grams (or about 10 to 12 cm) of spicules per cm^2 per 1 000 years. Still, the sedimentologic role of gorgonians and alcyonaceans should not be underestimated.

Heliopora is the only common aragonitic octocoral; its strontium content is somewhat lower and magnesium content higher than in the aragonitic hydrocorals and scleractinians (Table 26). The spicules of other octocorals are composed of

Table 26. Mineralogy and chemistry of soft and stony corals; range of values in parentheses. The first values are on an organic-free basis. Data from VINOGRADOV (1953), THOMPSON and CHOW (1955), WIAK (1962), HARRISS and ALMY (1964), KEITH and WEBER (1965), MATTHEWS (1966), MILLIMAN (1967), (unpublished

	Percent organic matter	Mineral.	Percent			
			Ca	Mg	Sr	Na
Hydrozoa						
Millepora	3.15	Arag.	38.8	0.16 (0.10–0.37)	0.83 (0.69–0.98)	
Stylaster	3.66	Arag.	38.8	0.07 (0.06–0.08)	0.83 (0.78–0.85)	
Octocorallia						
Heliopora	6.60	Arag.	39.6	0.26	0.69 (0.68–0.70)	
Tubipora	2.38	MgC	33.8	4.1 (3.5–5.0)	0.28 (0.22–0.35)	
Gorgonia		MgC	33.2	3.7 (2.7–3.9)		
Xiphogorgia		MgC	33.2	3.9 (3.7–4.0)		
Corallium		MgC	35.5	2.7	0.17	
Zoantharia (Order Scleractinia)						
Solenosmilia		Arag.	37.2 (35.2–39.4)		0.87 (0.82–0.90)	
Desmophyllum		Arag.	39.0 (38.9–39.1)		0.95 (0.90–1.00)	
Dendrophyllia		Arag.	37.6		0.88	
Madracis		Arag.	39.1 (37.5–41.0)		0.85 (0.83–0.88)	
Meandrina		Arag.	38.1 (37.6–38.5)		0.85 (0.83–0.87)	
Acropora	3.85	Arag.	38.8	0.12 (0.11–0.13)	0.81 (0.74–0.87)	0.44
Porites	5.38	Arag.	39.4	0.14 (0.09–0.18)	0.79 (0.71–0.87)	0.43
Siderastrea		Arag.		0.13 (0.06–0.29)	0.79 (0.76–0.81)	0.34
Montastrea		Arag.	39.0	0.11 (0.07–0.16)	0.68 (0.63–0.72)	0.40 (0.33–0.45)
Diploria		Arag.		0.32 (0.21–0.43)	0.80 (0.69–0.90)	

magnesian calcite and therefore are low in strontium. Iron, manganese and phosphorous concentrations are high, but the extent to which these elements may be organically complexed is not known. Deeper-water octocorals contain significantly less magnesium than shallow-water species (CHAVE, 1954).

three scleractinians are ahermatypic deep-sea corals; the rest are hermatypic reef corals. Elemental ODUM (1957), BOWEN (1956), LOWENSTAM (1963, 1964b), TATSUMATO and GOLDBERG (1959), WASKO-KINSMAN (1969b), MÜLLER (1970), SPIRO (1971), LIVINGSTON and THOMPSON (1971) and S. HUSSEINI data)

Parts per million												
Fe	Mn	K	Ba	Cu	Zn	Pb	P	B	U	Ni	Cr	Co
tr			8				tr					
tr			85				tr					
30							tr	5				
160	7			7	104		4800					
300	4			6	74		4800					
30	2		30 (16–60)	7 (2–14)	5 (3–9)	<2		75	3.8	<2	<2	<2
			26 (25–28)	10 (9–11)	3			70	4.3			
			30	7	35			100	4.6			
35 (5–95)	4 (2–5)		24 (11–38)	3 (2–5)	<2	<2		85 (80–95)	3.9 (3.5–4.5)	2	13 (tr–23)	0.1
13 (5–30)	2		42	5 (1–8)	<2	<2		78 (55–100)	3.9 (2.7–4.8)	2	<1	0.05
65 (15–120)	6 (2–10)	100		1		7	tr	21	1.9			
80 (25–170)	6 (3–14)	120	15	4	<2	<2	tr	21	2.1			
12 (5–20)	3 (<2–5)		17 (10–25)	6 (2–11)	<2	<2		57 (50–65)	2.6 (2.5–2.7)	2	5	0.1

LOWENSTAM (1963, 1964b) showed that in the gorgonian suborder Holaxonia the rind (body wall) spicules are magnesian calcite. Base and axis calcification is totally calcitic in temperatures below 15 °C, and totally aragonitic above about 22 °C; in intervening temperatures, the mineralogy is mixed. The strontium content

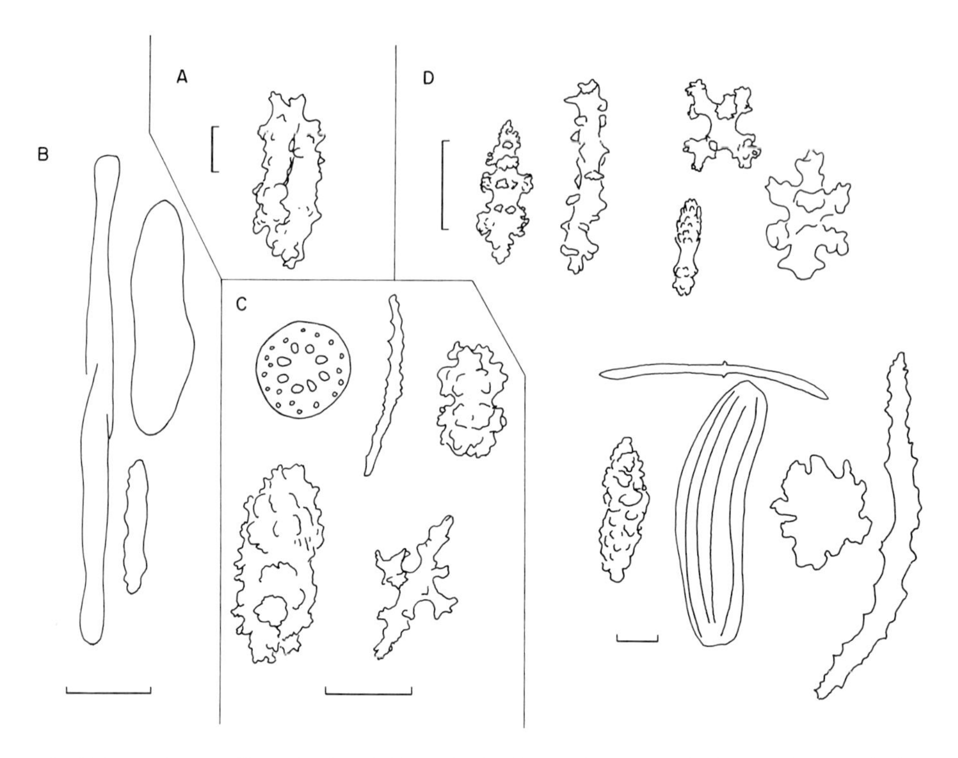

Fig. 25 A–D. Octocorallian spicules. A. Fused Stolonifera (*Tubipora* sp.); B. Pennatulacean spicules; C. Alcyonarian spicules; D. Gorgonian spicules. Bars and 100 microns. After BAYER (1956)

in the aragonitic phase is high (about 0.86%) and low in the magnesian calcite phase (0.27%). In contrast, the magnesium content in the calcitic phase is high (about 3.4%) and surprisingly high in the aragonitic phase (0.20%) (LOWENSTAM, 1964b). LOWENSTAM (1964b) further stated that the ratio of aragonite to calcite within *Plexaura flexusa* increases with increasing temperature, reaching about equal concentrations in the tropics.

Zoantharia (Scleractinia)

In terms of calcium carbonate produced, scleractinians (Madreporaria) are probably the most important coelenterate group in the oceans. Many of the environmental parameters required for growth have been mentioned earlier in this section and will be discussed in Chapter 6, but special mention should be made of scleractinian growth rates.

Scleractinian growth has been measured in two manners. The classic method involved measuring the change in length and mass of the coral with time. The other method, initiated by T.F. GOREAU, measures the uptake of Ca^{45} within the coral. A comparison of the results from these two methods is shown in Table 25. Fastest growers are the ramose (branching) species. For example, GOREAU and

GOREAU (1960a) found that *Acropora cervicornis* accretes at a rate greater than 50 µg of Ca per mg of N per hour, in comparison to about 10 to 19 µg of Ca for the massive *Montastrea*. Growth of a branching coral is not uniform along the skeleton; for instance, the tips of *A. cervicornis* grow about 3 times faster than the adjacent portions (GOREAU, 1959; GOREAU and GOREAU, 1960a).

In areas with relatively great annual temperature ranges, corals grow fastest in the warmer months. SHINN (1966) found that *A. cervicornis* grew $1\frac{1}{2}$ to 2 times faster in the summer than in the winter in Florida. In more tropical areas, where seasonal fluctuations are smaller, growth appears to be more constant throughout the year, resulting in larger annual growth increments; for example, *Acropora* grows about 10 cm/yr in Florida (SHINN, 1966) and 26 cm/yr in Jamaica (LEWIS and others, 1968). Branching *Porites* generally grows at about 2 to 4 cm/yr and the more massive corals, such as *Montastrea annularis*, at about 2 to 3 cm/yr (LEWIS and others, 1968, and references therein).

The scleractinian skeleton is composed of four elements, the basal plate, the septa, the epitheca and the dissepiments (BARNES, 1970). WELLS (1956) suggested that skeletal precipitation originates in the fold of the endodermal wall, but the exact mechanism of skeletogenesis is unknown. GOREAU (1961) found that calcium is transferred to the site of calcification which lies just outside the calcioblastic epidermis; here it is templated onto a polysaccharide. The $CO_3^=$ is obtained from metabolic processes.

The basic structural unit of the scleractinian coral is the sclerodermite, a cluster of aragonite fibers which fan out from a central growth surface, termed the "center of calcification" (WELLS, 1956; WAINWRIGHT, 1963; WISE, 1969; BARNES, 1970). All parts of coral skeletons that have been examined contain these three dimensional fans (BARNES, 1970). Under refracted light, the "feathery" aragonite fibers seem to radiate from the dark centers of calcification. Under polarized light these clusters exhibit radial extinction (Plate XII g, h).

Large scleractinian fragments are easily identified under reflected light. Septa, dentations, pali, mesentaries and numerous other morphological features often can be distinguished. The porous nature of *Porites* can be recognized both in reflected and refracted light (Plate XII). However, caution should be used, since *Porites* can be confused with *Millepora* in thin section.

All scleractinians are aragonitic. Strontium contents are high, generally between 0.75 and 0.90% (Table 26)[5]. LOWENSTAM (1964a) and KINSMAN (1969b) suggested that the higher amounts of strontium contained in deep-sea corals is related to the higher distribution coefficient of strontium in colder waters. Numerous analyses of deep-sea corals by THOMPSON and LIVINGSTON (1970), however, show that the difference in composition between hermatypic and ahermatypic corals is much less than previously reported, about 0.80% versus 0.88% THOMPSON and LIVINGSTON concluded that the higher strontium content in deep-sea corals is due to the lack of symbiotic algae rather than low temperatures.

Magnesium, iron, barium and manganese concentrations in scleractinian corals generally are lower than in the calcitic octocorals. The uranium content,

5 The strontium values reported by ODUM (1957), SIEGEL (1960) and HARRISS and ALMY (1964) are generally higher, averaging between 0.97 and 1.1%. All these workers used flame photometry, suggesting a possible systematic analytical error in the method.

however, is high (TATSUMOTO and GOLDBERG, 1960; THOMPSON and LIVINGSTON, 1970), as is fluorine (CARPENTER, 1969). WAINWRIGHT (1963) found that the skeleton of the reef coral *Pocillopora damicornis* contains only 0.01 to 0.1% organic matter. The relatively high organic contents reported by THOMPSON and CHOW (1955) (Table 26) may have resulted from impure samples, possibly polyp remains not washed from the skeletons.

Stable Isotopes

Available data suggest that hermatypic coelenterates are deficient in both C^{13} and O^{18} (Table 27; Fig. 26). The only exception seems to be the high C^{13} content

Table 27. Average stable isotope content in various corals. Most data are from WEBER and WOODHEAD (1970), with other data from KEITH and WEBER (1965) and GROSS and TRACEY (1966)

	δO^{18}	δC^{13}		δO^{18}	δC^{13}
Hermatypic Scleractinia			Ahermatypic Scleractinia		
Acropora	−3.20	−0.03	Dendrophyllidae	−3.69	−6.94
Porites	−3.95	−0.09	Hermatypic Hydrocorallia		
Pocillopora	−3.83	−2.38	*Millepora*	−1.33	+1.16
Fungia	−3.71	−0.47			
Goniastrea	−4.08	+1.09	Hermatypic Octocorallia		
Goniopora	−4.34	−3.02	*Heliopora*	−3.22	+3.76
Montipora	−3.36	−0.21	*Tubipora*	−4.78	−0.36

in *Heliopora*. KEITH and WEBER (1965) showed that reef corals of the suborder Faviina (families Favididae, Oculinidae, Meandrinidae and Mussidae) show a great range in δC^{13} values but remain relatively constant in δO^{18} content. In contrast, non-reef corals and those hermatypic corals belonging to the suborders Astrocoeniina and Fungiina show a positive correlation between δC^{13} and δO^{18}. KEITH and WEBER (1965) have suggested that this correlation may be due to the "mixing and utilization, at the calcification site, of both carbon and oxygen from

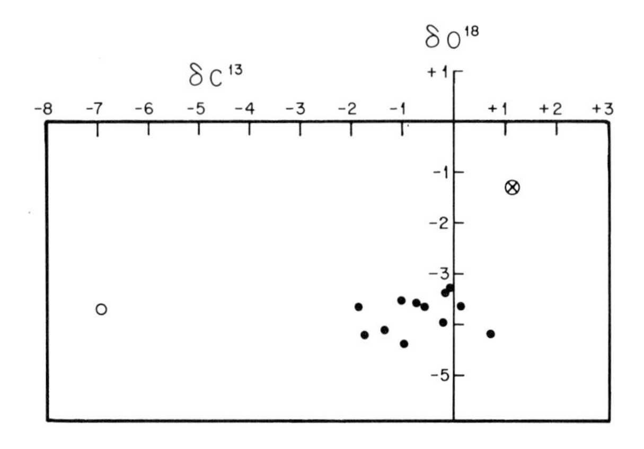

Fig. 26. Relation between oxygen and carbon isotopes in the various coral families. Hermatypic scleractinians (●); ahermatypic scleractinians (○); hermatypic hydrozoans (⊗). Data are from WEBER and WOODHEAD (1970)

different sources or metabolic pathways". This suggestion has been supported by recent data which show that ahermatypic corals have lower δC^{13} values than hermatypic corals (Fig. 26), inferring that respiratory C^{12} is preferentially utilized by zooxanthellae during photosynthesis (WEBER and WOODHEAD, 1970). This observation would further substantiate the opinion that the symbiotic algae are critical in the up-take of metabolic wastes produced by the coral polyps[6].

Bryozoa

Recent bryozoans (ectoprocts) have two major classes with mineralized skeletons, Cyclostomata and Cheilostomata. Cyclostomes generally have their tubes arranged parallel to one another; cheilostomes have short, box-like chambers. The greatest population densities of modern ectoprocts occur in shallow shelf waters, 20 to 80 m deep (RYLAND, 1967); locally these skeletons can constitute the dominant carbonate component in some shelf sediments. For instance, sands on the shelf off southern Australia contain as much as 85% bryozoan debris (WASS and others, 1970). Shallow-water species that encrust seaweeds also can be important sedimentary contributors KENDALL and SKIPWITH (1969) reported that *Thalamoporella gothica*, a prominent encruster of seaweed in the Persian Gulf, can form sedimentary lenses up to 20 cm thick.

Ecology and Petrology

Skeleton formation in the ectoprocts involves calcification within the epithelium of the zooids. The degree of calcification is more significant in advanced and complex superfamilies. For example, calcification in simple anascan cheilostomes is limited to the walls surrounding the zooid. In the complex anascans, a superficial skeleton, formed by the outfolding of epifrontal membranes, becomes prominent (BOARDMAN and CHEETHAM, 1969). Calcification also is strongly dependent upon the water temperature. Many ectoprocts grow extensively for periods of 3 to 6 months and increase in length at the rate of 1 to 6 mm/month (PEQUEGNAT and FREDERICKS, 1967; SMITH and HADERLIE, 1969). Turnover rates (the time required to replace an old population with an entirely new one) range from 2 to 36 months (SMITH, 1970). An extensive review of the skeletogenesis and colony formation in ectoprocts has been presented by BOARDMAN and CHEETHAM (1969).

At least seven distinctive crystalline ultrastructures have been recognized in bryozoan skeletons. Lamellar, massive, columnar cell-mosaic, parallel fibrous, crystal stacks and transverse fans are all calcitic, while transverse fibrous ultrastructures are aragonitic (SANDBERG, 1970).

6 More recent data indicate that δO^{18} content within any hermatypic genus is temperature dependent (WEBER and WOODHOUSE, 1972a) and that non-scleractinians deposit their $CaCO_3$ in isotopic equilibrium with sea water (WEBER and WOODHOUSE, 1972b). WEBER (1973) also has shown that ahermatypic corals have δO^{18} and δC^{13} values lower than predicted; partly this is in response to generic effects.

Petrographic recognition of the ectoproct skeleton is relatively easy, although smaller fragments may be confused with coral or possibly barnacle fragments. The characteristic cellular arrangement of the zooecia and their apertures is the principle characteristic used in reflected light identification (Plate XIII). Under transmitted light the distinctive features include the cellular arrangement of the zooecia and the granular texture of the carbonate walls.

Composition

Recent studies by SCHOPF and MANHEIM (1967), RUCKER and CARVER (1969) and CHEETHAM and others (1969) have helped to clarify the skeletal composition of ectoprocts. On the basis of composition, SCHOPF and MANHEIM (1967) divided bryozoan skeletons into three general groups. Phylactolaemata and Ctenostomata are poorly mineralized, containing more than 50% organic matter and water. Erect and bush-like Cheilostomates, such as *Bugula* and *Eucratia*, are better mineralized, but still contain 25 to 50% organic matter and water. These forms also contain relatively high concentrations of phosphorous and iron[7]. Heavily calcified ectoprocts (both Cheilostomata and Cyclostomata) form the third group, and generally contain less than 10 weight-% organics and water as well as lower amounts of phosphorous and iron. All known aragonitic forms occur in this third group.

The mineralogy (and therefore the chemistry) of the ectoprocts is highly variable, being related to phylogenic, ontogenetic and environmental factors. RUCKER and CARVER (1960) found that encrusting species of Anasca tend to be calcitic, while free forms are aragonitic or contain mixtures of aragonite and calcite. Many species within the suborder Ascophora also have mixed mineralogies. Aragonite usually occurs on the superficial skeleton during the adult stages of the organism (CHEETHAM and others, 1969; SCHOPF and ALLEN, 1970; SANDBERG, 1970)[8]. Mineralogy also can be dependent upon water temperature (LOWENSTAM, 1954b; RUCKER and CARVER, 1969); in Woods Hole aragonite constitutes 50 to 60% of the *Schizoporella* skeleton (SCHOPF and MANHEIM, 1967), 50 to 70% in Bermuda (LOWENSTAM, 1954b) and more than 80% in southern Florida (MILLIMAN, unpublished data). However, the aragonite content in other mixed species may not be temperature dependent (CHEETHAM and others, 1969).

Judging from the available data, most calcitic ectoprocts contain relatively low concentrations of magnesium; values rarely exceed 2% (8 mole % $MgCO_3$) (Table 28). The magnesium content apparently varies within the different morphological parts of some species. SAUDRAY and BOUFFANDEAU (1958), for instance, found that the magnesium content in *Bugula turbinata* is $2\frac{1}{2}$ times greater in the

---

7 SCHOPF and MANHEIM (1967) have suggested that the phorphorous may combine with calcium to form a sort of apatite structure, but no apatite has been detected by conventional X-ray diffraction techniques.

8 A recent study by POLUZZI and SARTORI (1973) shows that Cyclostomata ectoprocts contain no aragonite and have $MgCO_3$ concentrations between 2 and 9%. Of the more common types of Cheilostomata ectoprocts, most membraniforms are aragonitic, most reteporiforms, vinculariiforms and Adeoniforms are magnesian calcite (6 to 12% $MgCO_3$) and celleporiforms can be either aragonite or magnesian calcite.

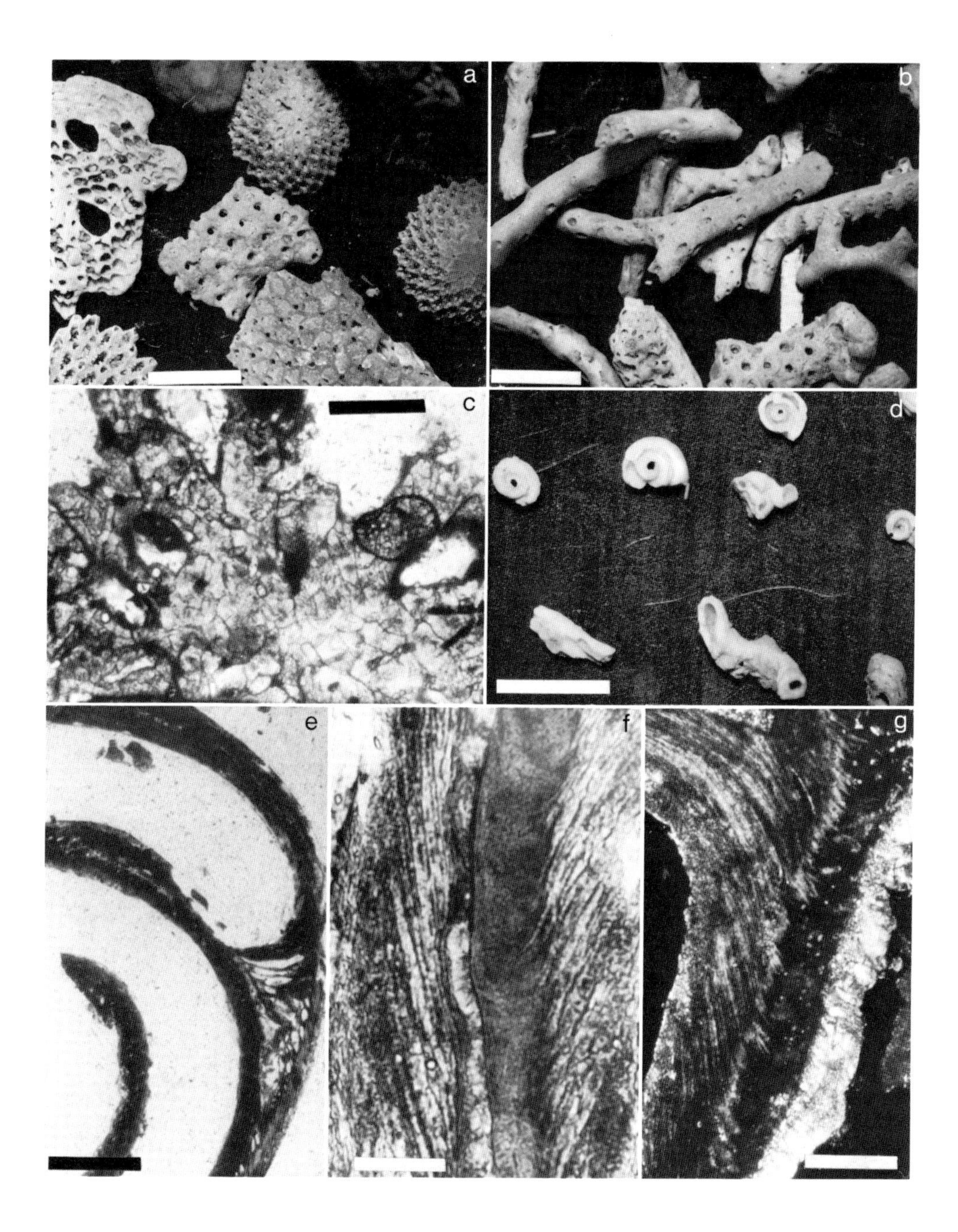

Plate XIII. Bryozoans and Serpulids. (a) Encrusting bryozoans. Reflected light; scale is 1 mm. (b) Branching bryozoans. Reflected light; scale is 1 mm. (c) Section of bryozoan. Note the characteristic skeletal fabric and zooecial openings. Refracted light; scale is 250 microns. (d) Serpulid tubes. Reflected light; scale is 1 mm. (e), (f) and (g) Serpulid tubes. Note the characteristically rough laminations and opaque nature of the center layer: (g) is under polarized light. Scales are 250, 40 and 100 microns, respectively

Table 28. Mineralogical and chemical composition of selected bryozoans, on an organic- and residue-free basis; range of values are in parentheses. Most data are from SCHOPF and MANHEIM (1967). Other data are from CLARKE and WHEELER (1917), THOMPSON and CHOW (1955), MAUCHLINE and TEMPLETON (1966), RUCKER and CARVER (1969) and SCHOPF and ALLAN (1970)

	Percent organic matter and H_2O	Mineral.	Percent					Parts per million	
			Ca	Mg	Sr	P	Fe	Ba	Zn
Cyclostomata									
Tubulipora	6.9+0.9	MgC	38.1 (37.6–38.6)	1.20 (1.12–1.28)	0.30	0.13 (0.12–0.14)	0.11 (0.06–0.17)	60 (31–89)	53
Cheilostomata									
Bugula	27.6+9.5	MgC	35.7 (34.4–36.3)	2.99 (2.49–3.83)	0.26 (0.16–0.32)	0.74 (0.59–0.83)	0.49 (0.29–0.75)	44 (32–55)	
Eucratia	32.5+5.6	MgC	35.9 (35.4–36.4)	2.67 (2.35–2.99)	0.19	0.80 (0.72–0.88)	1.30 (0.41–2.18)	98 (35–162)	
Membranipora	11.1+2.7	MgC	36.0 (35.4–36.6)	2.83 (2.46–3.19)	0.24 (0.23–0.25)	0.19 (0.09–0.29)	0.18 (0.11–0.24)	28 (17–39)	
Cryptosula	14.4+2.1	MgC+some Arag.	36.8 (36.3–37.7)	2.06 (1.71–2.41)	0.40 (0.26–0.55)	0.21 (0.12–0.45)	0.28 (0.21–0.37)	32 (25–44)	
Schizoporella	5.0+2.6	Arag.+MgC	38.7 (37.5–39.6)	1.03 (0.19–1.48)	0.54 (0.49–0.65)	0.16 (0.14–0.19)	0.10 (0.05–0.20)	26 (18–42)	37
Parasmittina	5.5+2.4	Arag.	39.2 (38.9–39.4)	0.16 (0.09–0.20)	0.87 (0.83–0.88)	0.09 (0.07–0.10)	0.10 (0.09–0.11)	21 (18–24)	

axis than in the branches. Similarly, phosphorous content is about 3 times greater in the axis. Preliminary microprobe analyses, however, show no significant variations in the calcium, magnesium or strontium across the skeletal walls of those species with constant mineralogy (SCHOPF and ALLEN, 1970). The aragonite in bryozoan skeletons is rich in strontium; species of the aragonitic genus *Parasmittina* contain an average of 0.87% Sr (Table 28). The phosphorous and iron contents within the ectoprocts (even well-calcified species) are far greater than in most other calcareous invertebrates.

Brachiopoda

Infaunal and epifaunal brachiopods generally live in water depths shallower than 300 meters. Although brachiopods are prominent in the fossil record, especially the Paleozoic, they have been considered relatively unimportant in modern carbonate sediments. Recently JACKSON and others (1971) pointed out that in cryptic shallow water tropical habitats, such as in reef crevices and underneath foliose corals, brachiopods (in association with sclerosponges) can occur in dense populations. For example, off Jamaica *Argyrotheca johnsoni* can reach concentrations of 50 per m^2 at depths of 60 m, while in shallower depths *Thecidellina barretti* can have populations as dense as 10000 per m^2 in crevices and caves. Clearly the potential importance of such communities merits further investigation.

Inarticulate brachiopods are composed of calcium phosphate and some are enriched with fluorapatite (MCCONNELL, 1963; VINOGRADOV, 1953). Articulate brachiopods contain a chitinous periostracum which covers a calcitic shell. The bilaminar shell is composed of a thin outer lamellar layer and a thick inner prismatic layer which is deposited by the mantle (JOHNSON, 1951). Three shell structures have been noted; a multilayered impunctate shell, the punctate shell structure which is penetrated by numerous small tubules, and the impunctate shell in which the inner portion is penetrated by structureless rods. The punctate structures aid in strengthening the shell. An excellent review on the shell morphology and structure of brachiopods has been given by WILLIAMS and ROWELL (1965). JOHNSON (1951) presented photographs showing cross-sections of several types of brachiopod shells.

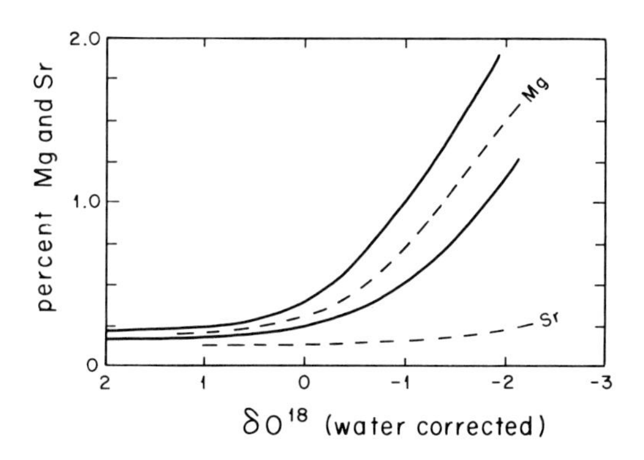

Fig. 27. Variation of Sr and Mg in recent articulate brachiopods with water temperature (as represented by δO^{18} values). Data from LOWENSTAM (1961)

Articulate brachiopod shells contain 1 to 5% organic carbon, mainly in the form of chitin (JOPE, 1965). The shells are believed to be low magnesium calcite, although high magnesium values in *Crania* have been reported by CLARKE and WHEELER (1922). LOWENSTAM (1961) showed that modern articulate brachiopods deposit their shells in isotopic equilibrium with the ambient water. Similarly, both the strontium and magnesium content vary directly with temperature (Fig. 27). DITTMAR and VOGEL (1968) reported very high manganese and vanadium values for *Waltonia* and *Gryphus* (Mn=0.002 to 0.143%; V=0.006 to 0.218%), and theorized that the ratio between these two elements is a function of the relative current energy in the depositional environment. Such high values also may be related more to the organic matrix rather than incorporation within the carbonate lattice.

Mollusca

Mollusks are divided into five major classes: Amphineura, Gastropoda, Pelecypoda, Scaphopoda, and Cephalopoda. A sixth class, Monoplacophora, has been proposed, but for purposes of this discussion can be ignored as a significant modern carbonate contributor. ALLEN (1963) has presented an extensive bibliography of recent papers on the ecology and morphology of various modern mollusks.

Typical molluscan shells contain three or more layers. The outermost layer is usually a chitinous periostracum. The inner layers (two or more) are calcareous, but can contain up to 9% organic matter (with an average content of 2 to 3%) (HARE and ABELSON, 1964; VINOGRADOV, 1953). Layers of organic matrix between the crystalline layers give the shells increased flexibility needed in environmental stresses (WAINWRIGHT, 1969).

BØGGILD (1930) recognized seven types of microstructures in mollusk shells: foliated, grained, homogenous, nacreous, prismatic, crossed-lamellar and complex. According to BØGGILD, foliated structures are always calcitic, nacreous and complex always aragonitic, prismatic and homogenous generally calcitic, crossed-lamellar generally aragonitic and grained structures are mixed. TAYLOR and others (1969) have distinquished only six types of microstructures within pelecypod shells: nacreous, crossed-lamellar, complex-crossed lamellar and homogenous structures are aragonitic, foliated is calcitic and prismatic can be either aragonite or calcite. Additional areas of calcification occur in the myostracal layer (which underlies areas of muscle attachment) and the inner layer of some ligaments; both types are aragonitic (TAYLOR and others, 1969; KENNEDY and TAYLOR). MAC CLINTOCK (1967) recognized four major microstructures in patelloid gastropod shells: prismatic, foliated, crossed and complex crossed; each of these four microstructures can be divided into distinct substructures, upon which the classification of patelloid gastropods is based. Excellent plates of various molluscan shell microstructures have been given by BØGGILD (1930), MACCLINTOCK (1967) and BATHURST (1971).

The concepts of mollusk shell calcification have been studied extensively. Excellent reviews have been presented by WILBUR (1964) and KENNEDY and

TAYLOR (1969). Shell formation occurs at the outer interface of the mantle in contact with the extra pallial fluid. Calcification probably involves a template phenomenon in which some charged proteins attract Ca^{++} and others attract $CO_3^=$ (HARE, 1963; TOWE and HAMILTON, 1968). DIGBY (1968) has suggested that this process is akin to an electrochemical reaction, the enzymes acting only as catalysts. JOHNSON and others (1964) have shown that calcium and carbonate ions have several possible sources, including oceanic waters and metabolic by-products. The mode of crystal growth depends upon the layer in which it occurs, but basically involves the merger of smaller crystals into larger ones by dissolution and re-precipitation (WILBUR, 1964). The specific mineralogy of the shell appears to reflect the composition of the proteins within the neighboring organic matrix; calcite shells have higher acid to base ratios than aragonitic shells (HARE, 1963; HARE and MEENAKSHI, 1968). In some instances the periostracum may actually influence the shell structure and thus the mineralogy (TAYLOR and KENNEDY, 1969).

BØGGILD (1930) and KOBAYASHI (1969) claimed that one could identify the genus or species of a mollusk on the basis of shell structure (see above). In most marine sediments, however, molluscan fragments generally are too small to permit such detailed identification. Viewed in thin section, the most common petrographic feature of a mollusk shell is its multilayered structure, often with nacreous or cross-laminated layers (Plate XIV). The shell's shape may allow one to determine class or even order. Under reflected light molluscan shell fragments generally can be distinquished from other carbonate components, although some pelecypod fragments may be confused with worn barnacle valves; micro-pelecypods can be mistaken for ostracod valves.

The following sections are devoted to the various molluscan classes. Pteropods are given special mention in view of their unique mode of living and their importance in some deep-sea sediments.

Amphineura

Amphineurans are divided into two orders. Aplacophorans are wormlike molluscans with tiny calcareous spicules usually 100 to 200 microns long. LOWENSTAM (1963) showed several pictures of these spicules and stated that the spicules are aragonitic. Their importance as sediment contributors is probably very small.

Polyplacophorans (chitons) occur throughout the oceans, but generally are most common in tidal and rocky subtidal environments. Because of their relatively slow turnover rates (SMITH, 1970), they are probably quantitatively unimportant as sediment contributors. Chiton shells are composed of 8 over-lapping articulated plates (valves). Viewed in transverse section, these valves exhibit a 4-layer structure: an outer periostracum, a relatively soft, opaque and porous tegmentum, and two inner crystalline layers, the articulamentum and the hypostracum (SMITH, 1960). The tegmentum (shown in BØGGILD, 1930, plate VIII, Figs. 4–5) is a particularly diagnostic petrographic feature.

Amphineurans are aragonitic. The few available chemical analyses indicate that chitons contain more strontium than other mollusks (average of 0.67%;

Table 29)[9]. This apparent inability to discriminate against strontium may be a result of the primitive nature of polyplacophorans compared to other mollusks (LOWENSTAM, 1964a). Magnesium content is high (0.11%), considering the aragonitic mineralogy. LOWENSTAM (1962) reported the presence of magnetite cappings on chiton teeth; in addition to representing a biochemical oddity, the coatings provide the means by which chitons can effectively erode carbonate substrate (see below).

Scaphopoda

Scaphopods have tapered curved tubes, open at both ends and generally less than 2 cm in length. They are found primarily in sublittoral depths, both in sandy and muddy substrata (LUDBROOK, 1960). They comprise an appreciable portion of the deep-sea molluscan fauna (CLARKE, 1962). Shells are characteristically aragonite, and contain low strontium and magnesium concentrations (Table 29).

Cephalopoda

Other than nautiloids, most modern cephalopods have meager or nonexistent shells. Most shells are aragonitic. A notable exception is the egg case of the octopod *Argonauta*, which is calcitic (BØGGILD, 1930); the average magnesium content of 1.22% (VINOGRADOV, 1953) suggests that this shell is in the lower range of magnesian calcite. In view of this mineralogy, it is surprising to find so much strontium (0.39%) (Table 29); the fact that the egg case is precipitated by the animal's arms and not its mantle (CHAVE, 1954) may explain this anomaly. Strontium content in *Nautilus* aragonite is also comparatively high (0.35%), although HALLAM and PRICE (1966) report considerably lower values (0.23 to 0.24%).

In both *Nautilus* and *Sepia* (a cuttlefish) different structural layers contain different elemental concentrations. PRICE and HALLAM (1967) report that the periostracum of *Sepia* contains 0.26% strontium and the phragmocone from 0.33 to 0.40%. In addition, strontium content for both species decreases with age (PRICE and HALLAM, 1967). If the strontium decrease were correlative with a migration to warmer waters, these data disagree with the findings of EICHLER and RISTEDT (1966), who concluded that changes in the δO^{18} and δC^{13} ratios in the early shell and septa of two *Nautilus* specimens indicate migrations to cooler (or deeper) waters. Perhaps the change in Sr concentrations is more related to biochemical changes during ontogeny than to environmental factors (PRICE and HALLAM, 1967).

Pteropoda

Pteropods are planktonic gastropods, belonging to the subclass Ophisthobranchia. The other planktonic gastropods, Heteropods (Prosobranchs) are less common,

9 BOWEN'S (1956) data for both calcium and strontium seem too low compared with other published data, and therefore have not been used in the compilation of the averages presented in Table 29.

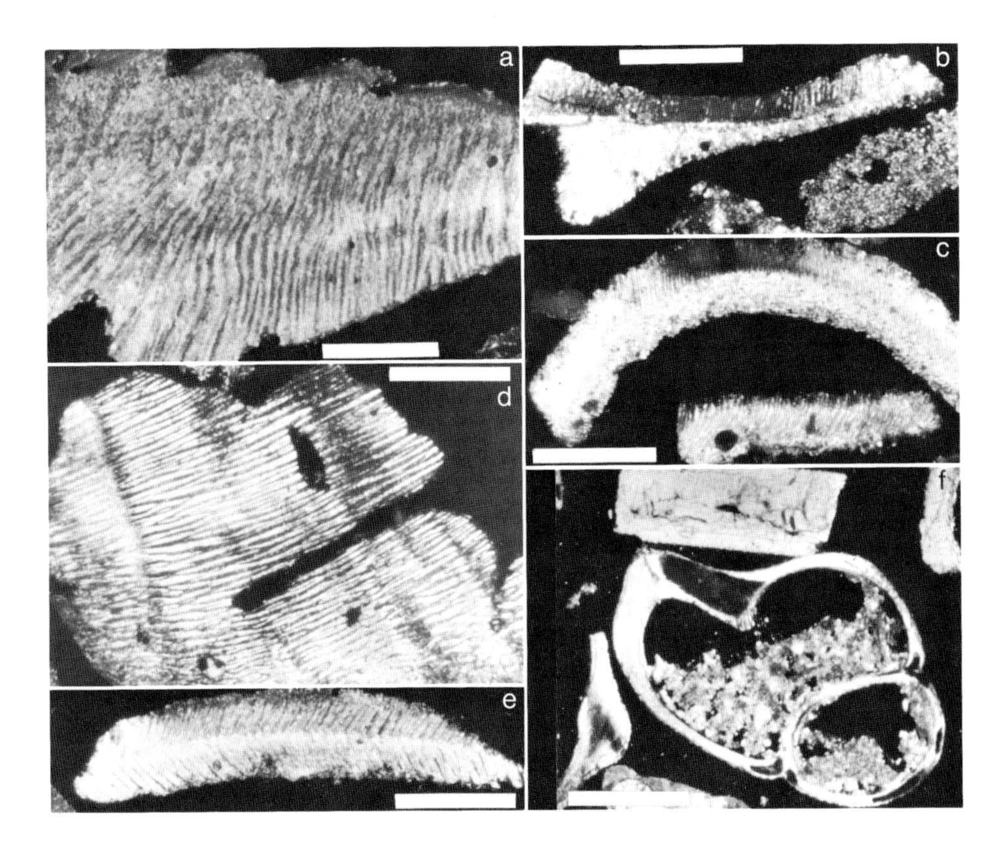

Plate XIV. Molluscan Fragments. All photomicrographs were taken with polarized refracted light. Scales in (a) and (c) are 100 microns; the others are 250 microns. The various diagnostic petrographic properties of molluscan skeletons can be seen in these photographs. Note the internal sediment within the gastropod shell in (f)

Table 29. Chemical and mineralogical data for Polyplacophora, Scaphopoda and Cephalopoda. Data from VINOGRADOV (1953), THOMPSON and CHOW (1955), BOWEN (1956), LOWENSTAM (1964a), HALLAM and PRICE (1966) and PRICE and HALLAM (1967)

	Percent organic matter	Mineral.	Percent			Parts per million						
			Ca	Mg	Sr	Ba	P	Fe	Mn	Pb	Ni	Cu
Polyplacophora	4.30 (3.01–6.10)	Arag.		0.11	0.70 (0.62–0.78)	10						
Scaphopoda	1.81	Arag.		0.05	0.19 (0.17–0.21)	7						
Cephalopoda												
Argonauta		MgC (?)		1.22 (0.95–1.72)	0.39		> 2500					
Sepia		Arag.		0.10 (0.02–0.20)	0.32 (0.26–0.40)			32 (20–45)	4 (2–6)	127 (24–230)	tr	2
Nautilus		Arag.	39.84	0.03 (0.30–0.42)	0.23 (0.18–0.32)							

Table 30. Mineralogic and chemical composition of pteropods. Data from KRINSLEY and BIERI (1959), NICHOLLS and others (1959) and PYLE and TIEH (1970)

	Mineral.	Percent		Parts per million													
		Mg	Sr	Mn	Al	Mo	Ni	Co	Pb	V	Cu	Zn	Cr	B	Ti	Sn
Styliola	Arag.	0.022 (0.011–0.036)	0.10 (0.09–0.11)	2[a] 14[b]	6 (3–8)						15 (8–20)					
Diacria			0.026							55 (50–60)		18 (12–24)				
Limacina						8	2	20	200	85	30		< 1	50	5	< 1
Miscellaneous Species	Arag.	0.015 (0.012–0.017)	0.10 (0.08–0.11)	3 (2–5)	5 (3–7)											

[a] Mediterranean and North Atlantic.
[b] Caribbean Sea.

and neither their mineralogy nor chemistry is well documented (although BØGGILD, 1930 stated that the heteropod shell is aragonitic).

Whole pteropod shells display characteristic shapes (Plate XXVIII). Their thin homogenous shells (BØGGILD, 1930) can facilitate their identification in thin section. Pteropod-rich sediments generally are limited to tropical and subtropical deep seas, in depths shallower than 2700 meters (see p. 234ff.). Pteropod shells contain an average of 0.1% strontium (about half that of other gastropods) and only about 0.2% magnesium (Table 30). KRINSELY and BIERI (1959) found that manganese content varies with location, being about 2 to 5 ppm in Atlantic and Mediterranean shells and 14 to 15 ppm in the Caribbean. Aluminum, manganese and magnesium concentrations are greater in dead tests than in living ones, no doubt reflecting the post-depositional absorption of clays on the sea floor (KRINSELY and BIERI, 1959). VINOGRADOV (1953) reported relatively large quantities of phosphorous in some pteropod shells, but how this phosphorous is incorporated into the shell is not known. Special note should be made of the large amount of lead in the shell of *Limacina*. The only other invertebrate shells containing similarly large lead concentrations are those of certain cephalopods (for example, *Sepia*).

Stable isotope concentrations in pteropods have been reported by several workers (DEUSER and DEGENS, 1969; VERGNAUD-GRASSINI and HERMAN-ROSENBERG, 1969), but the significance of these measurements is doubtful. Pteropods undergo considerable daily vertical migration; *Creseis acicula*, for example, migrates from the near surface at night to a depth of 500 m during the day (STUBBINGS, 1937). The range of temperatures and salinities to which the animal is exposed daily undoubtedly has a considerable effect upon the stable isotope composition of the shell.

Pelecypoda

Pelecypods (bivalved mollusks) inhabit nearly every marine environment. They are probably the major benthic carbonate contributor to modern shallow-marine sediments. Pelecypods tend to deposit their calcium carbonate shells more slowly than do gastropods, but absolute growth is strongly dependent upon the nature of the environment and substrate (SWAN, 1952). Most quiet-water pelecypods have thin shells, while those mollusks living in higher energy environments have thicker shells. Similarly, mollusks living in very cold waters usually have thin and chalky shells (NICOL, 1967). Perhaps the fastest measured rate of pelecypod carbonate deposition is that of *Tridacna gigas*, which deposits 23 grams of $CaCO_3$/100 cm^2/year (BONHAM, 1965). Oysters can precipitate about 1.5 grams/100 cm^2/year, but the rates of most other shallow-water pelecypods are slower. Turnover rates of most shallow-water pelecypods are probably about 3 or 4 years (SMITH, 1970). Deeper water forms exhibit even slower growth: the ocean quahog, *Arctica islandica*, for instance, can reach ages as old as 40 years and more (H. WAITE, unpublished manuscript).

Because calcification is dependent on many environmental and biologic fluctuations, shell deposition can be periodic. Growth lines on the outer layers of many pelecypod shells are thought to represent yearly, monthly, biweekly and daily growth increments[10]. Annual growth rings are defined by periods with little or no shell growth, probably during periods of spawning or marked temperature changes. During periods of extreme stress, the inner shell surface can actually dissolve (Crenshaw and Neff, 1969; Davies and Sayre, 1970). Although shell growth in many mollusks is greatest during warmer months (Epstein and Lowenstam, 1953; Lowenstam, 1954a), some species have their most rapid growth during the colder months (Orton, 1928; Hamai, 1935; Haskin, 1954; Zhirmunsky and others, 1967). Monthly and bi-monthly increments no doubt reflect tidal influences. The factors causing daily growth bands are not clear. Because most pelecypods lack photosynthetic algae, they do not necessarily deposit their calcium carbonate during light hours. For example, MacClintock and Pannella (1969) report that the clam *Mercenaria mercenaria* accomplishes most of its calcification during high tide at night.

Pelecypod shells are dominantly aragonite or contain a mixture of aragonite and calcite. Both shell structure and mineralogy ususally are constant within any superfamily (Kennedy and others, 1969). Most species with mixed mineralogy belong to the order Anisomyaria. At present only one genus (*Chama*) within the order Eulamellibranchiata has been found to have mixed mineralogy (Bøggild, 1930; Lowenstam, 1954a). The most common calcitic pelecypods arc the oysters *Ostrea* and *Crassostrea*.

Table 31. Comparison of the nacreous and prismatic layers in pelecypods with mixed mineralogies. Data from Waskowiak (1962)

		Mineralogy	Percent		Parts per million						
			Mg	Sr	Fe	Mn	Ba	B	Pb	Co	Ni
Mytilus	prismatic	Calcite	0.20	0.10	139	218	25	5	1	1	1
	nacreous	Aragonite	0.012	0.21	16	5	12	10	<1	1	—
Pinna	prismatic	Calcite	0.21	0.024		35		1	2		
	nacreous	Aragonite	0.006	0.10		18		3	1		

In those shells with mixed mineralogies, the outer layers (generally prismatic or foliated) are calcitic; the inner layers, generally nacreous or laminated, are aragonitic (Table 31). Dodd (1965) found that the pelecypod *Mytilus* contains an additional inner calcitic layer whose thickness varies inversely with water temperature. As a rule, aragonitic bivalve shells contain very little magnesium (generally less than 0.05%) and compared to other mollusks, relatively high concentrations of strontium (generally greater than 0.25%) (Table 32). Calcitic shells (or layers) contain more magnesium (greater than 0.20%), more iron and manganese and less strontium than do the aragonitic shells. The extent to which iron may be concentrated within the organic matrix is subject to question (Kessel, 1936; Waskowiak, 1962).

10 In some mollusks, however, periodic calcification may not reflect such readily definable time increments (Malone and Dodd, 1967).

Gastropoda

Most marine gastropods belong to the subclass Prosobranchia, pteropods (subclass Opisthobranchia) being the major exception. Like pelecypods, marine gastropods tend to be smaller and thin-shelled in colder, deeper waters. In contrast to the relatively slow rates of pelecypod calcification, however, gastropods can precipitate their shells rapidly. For instance, KEITH and others (1964) found a 23-cm specimen of *Strombus gigas* from the Florida Keys to be only two years old. Gastropods also seem to have shorter life spans and higher rates of population turnovers. COX (1960) states that most marine gastropods have an average life span of only 3 to 5 years with a maximum of about 10 years. The data of EPSTEIN and LOWENSTAM (1953) and KEITH and others (1964) suggest that the period of active gastropod calcification varies greatly with species; some species are active seasonally, others precipitate their shells uniformly throughout the year. This would explain the wide range of δO^{18} values found in the gastropod fauna at Bermuda (EPSTEIN and LOWENSTAM, 1953).

Most marine gastropod shells are aragonitic. LOWENSTAM (1954b) found only five genera (*Patella*, *Haliotis*, *Fissurella*, *Nerita* and *Littorina*) with mixed mineralogies, but four other occurrences (*Thais*, *Neptunea*, *Purpura* and *Tegula*) have been documented (WASKOWIAK, 1962; Table 33). Gastropods seem to have slightly lower Sr concentrations and higher magnesium values than do pelecypods (Table 33); the clear-cut differences cited by TUREKIAN and ARMSTRONG (1960) do not seem to exist. Iron, manganese and other trace element concentrations seem uniformly lower in gastropod shells than in pelecypod shells. The limpet *Patella* contains a notably different composition from other gastropods. WASKOWIAK (1962) found an average of 1.4% magnesium (from 7 analyses) in the shell of this gastropod. Since the analyzed shells contained more than 20% aragonite, it seems probable that the calcite fraction was high magnesian calcite. This is the only reported occurrence of magnesian calcite in any gastropod or pelecypod known to the writer.

Variations in Gastropod and Pelecypod Composition

Numerous exceptions exist in the rules governing the mineralogical and chemical data of gastropods and pelecypods. For example, some aragonitic mollusks, such as *Elliptio*, can heal shell wounds with calcite or even vaterite (WILBUR, 1964; ABOLINS-KROGIS, 1968), and the calcitic oyster *Crassostrea virginica* has an aragonitic residum and muscle scars (STENZEL, 1963). Although strontium concentration is generally related to aragonite content, its absolute concentration can show considerable variation. HALLAM and PRICE (1968) found that the average inner layer of the aragonitic *Cardium* contains half again the strontium contained in the outer layer (0.32% versus 0.21%). Such variations may reflect different metabolic influences in carbonate precipitation.

The composition of mollusks can change greatly with environment as well as in response to various organic phenomena, such as the composition of the organic matrix (WATABE and WILBUR, 1960; HARE, 1963). Generally in mollusks with

Table 32. Composition of selected pelecypod shells; range of values are in parentheses. The last three
DOV (1953), ODUM (1957), BOWEN (1956), THOMPSON and CHOW (1955), TUREKIAN and ARMSTRONG
BROECKER (1968), LOWENSTAM (1964b), LEHRMAN (1965)[e], NELSON (1965), BROOKS and

	Percent organic matter	Mineral.	Percent Ca	Mg	Sr	Na
Macoma	2.29	Arag.		0.023	0.22	
	(2.17–2.42)			(0.007–0.060)	(0.14–0.40)	
Mya	2.35	Arag.		0.024	0.21	
	(2.22–2.48)			(0.005–0.048)	(0.07–0.30)	
Cardium		Arag.		0.025[a]	0.31	
				(0.007–0.055)	(0.09–0.55)	
				0.012[b]	0.15	
				(0.008–0.019)	(0.12–0.18)	
					0.32-inner }—c	
					0.21-outer }	
			36.0[g]	0.046[g]	0.13[g]	0.56[g]
Mercenaria		Arag.	39.0		0.19[a]	
			33.0[g]	0.005[g]	0.017[g]	0.57[g]
Cyprina	1.93	Arag.		0.012	0.16	
				(0.006–0.020)	(0.13–0.22)	
Venus	2.38	Arag.		0.024	0.22	
				(0.011–0.045)	(0.07–0.40)	
Arca		Arag.		0.029[a]	0.25	
				(0.014–0.065)	(0.16–0.40)	
				0.009[b]	0.16	
				(0.007–0.014)	(0.12–0.25)	
Chama	2.47	Arag.		0.039	0.30	
				(0.010–0.090)	(0.12–0.55)	
Donax		Arag.		0.022	0.16	
				(0.014–0.038)	(0.12–0.19)	
Ostrea	1.37	Cal.		0.43	0.13	0.20
	(1.01–1.71)			(0.19–0.80)	(0.07–0.19)	
Crassostrea	2.10	Cal.	39.0	0.17[d]	0.10	0.24
	(1.79–2.34)			(0.11–0.27)	(0.07–0.16)	(0.16–0.33)
				0.36[e]	0.08	
				(0.27–0.40)	(0.08–0.09)	
Anomia		91% Cal.		0.37[a]	0.14	
		(5–17% Arag.)		(0.15–0.60)	(0.07–0.16)	
				0.26[f]	0.19	
				(0.21–0.31)	(0.15–0.24)	
Pecten	1.83	Cal.	39.2	0.24	0.14	
		(mostly)	(38.7–39.6)	(0.04–0.60)	(0.07–0.22)	
		Arag.		0.06	0.30	
Mytilus	3.35	59% Cal.		0.068	0.13	0.36
	(1.54–5.53)	(0–100%)		(0.01–0.30)	(0.03–0.30)	(0.30–0.43)
		Cal.		0.07	0.07	
				(0.06–0.07)	(0.03–0.12)	
		Arag.		0.03	0.21	
				(0.01–0.07)	(0.12–0.40)	
		(?)	39.0[g]	0.069[g]	0.001[g]	0.42[g]
Glycymeris		Arag. (?)	32.0	0.02	0.07	0.44
Modiolus		(?)	28.0	0.044	0.007	0.47
Chlamys		Cal. (?)	32.5	0.10	0.034	0.33
			(32–33)	(0.079–0.130)	(0.027–0.041)	(0.29–0.37)
Malletia	2.56	Arag.		0.034	0.17	
				(0.024–0.046)	(0.14–0.19)	
Abra	1.80	Arag.		0.017	0.13	
				(0.010–0.022)	(0.11–0.16)	
Cuspidaria	3.12	Arag.		0.088	0.24	
				(0.068–0.108)	(0.23–0.24)	

genera are from the deep sea. Most data are from WASKOWIAK (1962)[a]. Other data are from VINOGRA-(1960)[b], RUCKER and VALENTINE (1961), SMITH and WRIGHT (1962), PILKEY and GOODELL (1963, 1964)[f], RUMSBY (1965), PILKEY and HARRISS (1966)[d], HALLAM and PRICE (1966)[c] and KILHAM (1970)

Parts per million

Fe	Mn	K	Ba	B	Pb	Ni	Cu	V	U	Al	Zn	Co	Cd	P
348 (1–3000)	11 (tr–45)		14 (tr–55)	2 (tr–4)	2 (tr–10)	13 (tr–42)	1		0.11 (0.09–0.13)					
141 (5–420)	21 (3–100)		<10 (tr–70)	2 (tr–4)	1 (tr–4)	7 (tr–70)	2							
169 (5–750)	11 (1–150)		25 (tr–100)	4 (1–9)	2 (tr–4)	3 (tr–50)	1							
			10 (7–12)											
1600[g]	2.0[g]	160[g]			0.44[g]	0.11[g]	3.0			84[g]	6.8[g]	1.6[g]	0.34[g]	10[g]
1600[g]	1.5[g]	230[g]			0.43[g]	2.4[g]	1.7[g]			71[g]	3.4[g]	1.2[g]	0.80[g]	28[g]
90 (65–120)	3 (1–4)			4 (3–5)	<1	<1	<1							
81 (5–300)	14 (1–20)		8 (tr–40)	3 (1–7)	<1	<1	<1							
102 (23–300)	12 (1–34)		14 (tr–75)	5 (1–11)	10 (tr–80)	1 (tr–5)	1							
			16 (6–41)											
10 (1–34)	<1			2 (tr–3)	<1	<1	<1							
18 (9–34)	6 (3–8)			2 (1–5)	1 (tr–4)	<1	<1							
127 (7–750)	31 (3–200)			11 (2–30)	2 (tr–40)	1 (tr–15)	1	130						
267 (159–432)	33 (18–50)			5			3				2			
225 (6–540)	224 (45–300)			8 (1–12)	1	tr	2							
237 (38–1059)	140 (13–438)		14 (5–24)											
124 (30–340)	78 (36–130)		8 (6–12)	3 (1–5)	2 (tr–10)	8 (tr–40)	tr	49						
75	2			5	18(?)	2	>1							
78 (5–500)	27 (1–100)			4 (1–10)	3 (tr–75)	3 (tr–100)	2	110						
12 (4–15)	11 (2–26)			4 (2–7)	1									
51 (5–140)	16 (1–48)			5 (4–9)	1				0.03 (0.02–0.05)					
290[g]	3.6[g]	18[g]			0.40[g]	2.1[g]	2.0[g]			76[g]	0.04[g]	0.15[g]	0.95[g]	66[g]
47	1.5	43			0.60	0.16	0.09			420	160	0.20	0.03	36
15	20	88			1.9	0.20	1.9			150	6.2	0.30	0.03	630
92 (64–120)	18 (17–18)	70 (19–120)			1.2 (0.64–1.7)	0.43 (0.18–0.69)	0.54 (0.38–0.70)			365 (300–430)	9.1 (7.3–11)	0.24 (0.22–0.27)	0.08 (0.04–0.13)	630

Table 33. Composition of gastropod shells; range of values in parentheses. Most data are from WASKO-ODUM (1957), BOWEN (1956), KRINSLEY (1959, 1960b), TUREKIAN and ARMSTRONG (1960)[b], LOWENSTAM and others

	Percent organic matter	Mineral.	Percent Ca	Mg	Sr	Na
Strombus		Arag.	39.8 (39.7–39.9)	0.05 (0.05–0.06)	0.16 (0.12–0.21)	
Buccinum		Arag.	39.4 (39.2–39.6)	0.027[a] (0.007–0.095)	0.26[a] (0.07–0.40)	
			35.0[d]	0.022[d]	0.075[d]	0.55[d]
Vermetus		Arag.		0.027 (0.015–0.042)	0.21 (0.16–0.26)	
Conus		Arag.		0.07[a] (0.034–0.100)	0.37 (0.30–0.40)	
				0.015[b] (0.005–0.043)	0.12 (0.10–0.21)	
					0.15[c] (0.13–0.17)	
Natica	3.54	Arag.	39.7	0.020 (0.018–0.028)	0.17 (0.16–0.19)	
Cerithium		Arag.		0.029 (0.01–0.05)	0.12 (0.07–0.22)	0.21
Gibbula		Arag.		0.022 (0.012–0.055)	0.30 (0.16–0.55)	
Olivella	1.84	Arag.		0.012 (0.005–0.018)		
Busycon		Arag.		0.020 (0.013–0.026)	0.27 (0.26–0.28)	
Nucella		Arag.	37.0	0.057	0.12	0.53
Nassa	6.46	Arag.		0.033 (0.012–0.053)	0.25 (0.12–0.55)	
Neptuna		mixed		0.27 (0.20–0.30)	0.22 (0.12–0.30)	
Nerita		71% Arag.		0.32 (0.04–0.43)	0.14 (0.11–0.17)	
Littorina	2.58 (1.50–4.39)	mixed (Arag. <25%)		0.068 (0.025–0.250)	0.13 (0.07–0.30)	
Patella		mixed		1.40[a] (0.75–2.4)	0.18[a] (0.07–0.27)	
			38.0[d]	0.35[d]	0.057[d]	0.41[d]
Tegula	3.90 (3.64–4.06)	67% Arag. (54–76%)		0.044 (0.040–0.055)	0.12 (0.11–0.14)	
Crepidula	2.65 (2.36–3.22)	Arag. (Cal. tr–7%)		0.036[a] (0.017–0.205)	0.20[a] (0.14–0.34)	
			32.0[d]	0.040[d]	0.01[d]	0.52[d]

WIAK (1962)[a]; other data are from VINOGRADOV (1953), CHAVE (1954), THOMPSON and CHOW (1955), (1964a)[c], PILKEY and GOODELL (1963), BROECKER (1963), FRIEDMAN (1969), MÜLLER (1970) and SEGAR (1971)[d]

Parts per million

Fe	Mn	K	Ba	B	Pb	Ni	Cu	U	Al	Zn	Co	Cd	P
20	3		1	1	6		1	<0.01					
78[a]	18[a]			2[a]	1[a]	1[a]		0.48[a]					
(9–300)	(tr–100)			(tr–7)	(tr–20)	(tr–8)							
470[d]	1.2[d]	180[d]			0.45[d]	12[d]	9.1[d]		83[d]	13[d]	0.16[d]	0.048[d]	12[d]
165	10			1	20	1	1						
(30–300)					(1–40)								
65	1			2		tr	1						
(5–190)	(1–2)			(1–3)									
			7										
			(5–15)										
76	18			1	3	2	1						
(10–200)	(8–45)				(tr–100)	(tr–8)							
290	62		7	1	2	10	2						
(30–700)	(6–150)		(5–8)		(tr–3)	(tr–30)							
132	9			2	<1	1	<1						
(22–250)	(1–25)			(1–2)		(tr–7)							
65	3		3	2	<1	2							
(3–365)	(1–7)		(3–4)	(1–3)									
4	1.1	260			0.45	0.25	0.29[d]		170[d]	0.59	0.25	1.64	34
210	8			2	2	21	<2						
(45–600)	(1–34)			(1–6)	(tr–7)	(tr–100)							
110	2			1	1	5	<2						
(22–180)	(1–4)					(tr–10)							
65	8		5	2	20(?)								
			(tr–7)										
180	49		7	1	2	23	3	0.008					
(12–800)	(5–150)		(tr–9)	(tr–3)	(tr–10)	(tr–80)							
40[a]	34[a]			3[a]	5[a]	3[a]	1[a]						
(9–100)	(2–190)			(1–7)	(1–12)	(tr–20)							
430[d]	8.6[d]	140[d]			0.46[d]	0.54[d]	42[d]		470[d]	51[d]	0.80[d]	2.8[d]	
14	14			1		<1							
120[a]	16[a]		8[a]	2[a]	1[a]	2[a]	2[a]						
(5–659)	(1–73)		(3–26)	(1–3)		(tr–8)							
1600[d]	2[d]	130[d]	20[d]		0.4[d]	1.6[d]	1.8[d]		80[d]	1.1[d]	1.5[d]	2.4[d]	20[d]

mixed mineralogies, aragonite content increases with decreasing salinity (LOWENSTAM, 1954a; DODD, 1963, 1966). EISMA (1966), however, has found that *Mytilus edulis* (a species which both LOWENSTAM and DODD have studied) exhibits no mineral change with different salinities; genetic effects may explain this discrepancy (DODD, 1966).

Aragonite content also can increase with increasing water temperature. For example the shell of the limpet *Littorina* is entirely aragonitic in the Caribbean but contains significant amounts of calcite in the Gulf of Maine (LOWENSTAM 1954a, b). Because of such temperature influences, the exact mineralogy may depend upon the season during which the shell is precipitated. The exact temperature influence, however, is not always clear. PILKEY and GOODELL (1963) reported that the aragonite content in the pelecypod *Anomia* decreases with increasing temperature. DODD (1963) found a direct relationship between the aragonite in *Mytilus* and the water temperature, but in the same genus DAVIES (1965) found an inverse relationship, and WASKOWIAK (1962) found none. The fact that most deep-sea bivalves are aragonitic (KILHAM), would argue against a strict temperature control of mineralogy.

The most important factor influencing elemental concentrations is the mineralogy of the shell (or layer). This is most obvious in magnesium concentrations, but is also apparent in other elements, such as strontium, iron and manganese. This observation contradicts the conclusions of ODUM (1957), TUREKIAN and ARMSTRONG (1960) and HARRISS (1965), that crystal growth kinetics, perhaps combined with metabolic processes, is the most critical factor controlling elemental composition.

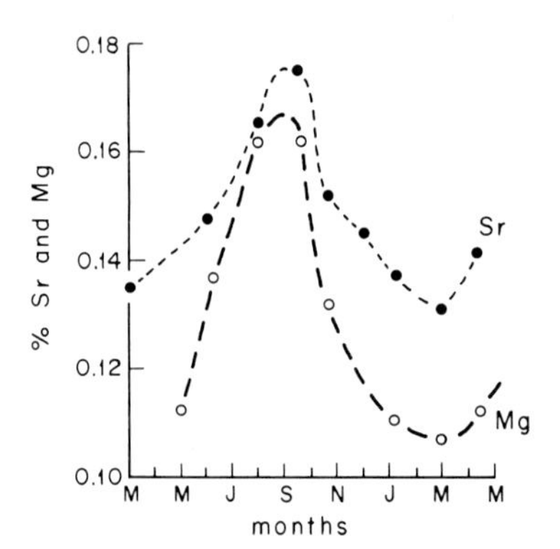

Fig. 28. Variation of Sr and Mg content in the last formed portion of the outer prismatic layer in the bivalve *Mytilus*. After DODD (1965)

The variation of magnesium with water temperature is not as well defined in mollusk shells as in other organisms (CHAVE, 1954). Both WASKOWIAK (1962) and DODD (1965) noted a positive correlation between magnesium and water temperature in *Mytilus* shells (Fig. 28). Of the seven mollusk species investigated by PILKEY and GOODELL (1963) two showed slightly inverse relations with temperature, the others showed no definite trend. MOBERLY (1968) found that the magnesium content in *Aequipecten* is twice as high in the wide growth bands (assumed to be precipitated during the summer months) than in the narrow bands. But in the oyster *Crassostrea* the thinner bands have higher magnesium concentrations (MOBERLY, 1968).

Only a few mollusk shells have been found to exhibit strontium-temperature relationships. In two of the three aragonitic forms, strontium content decreases with increasing temperature, while it increases in two calcitic shells (Fig. 29). Clearly the decrease in strontium in aragonite with increased temperature is more a function of physiologic controls than physiochemical reaction rates (KINSMAN, 1969 b). The relations of other trace element concentrations with temperature are poorly known; available published data show few significant trends.

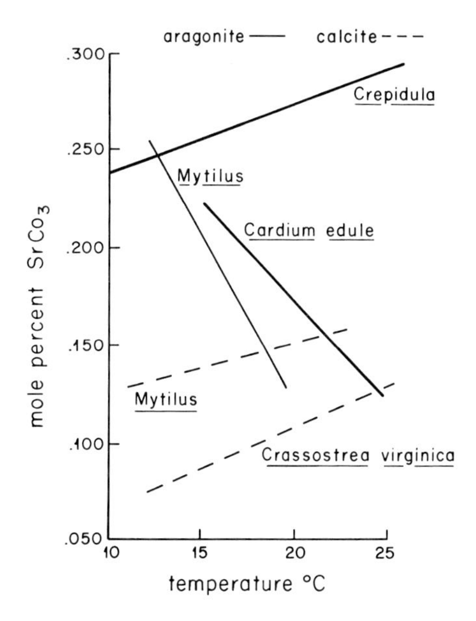

Fig. 29. Variation of Sr content in some calcitic and aragonitic mollusks with water temperature. Data from PILKEY and GOODELL (1963), DODD (1965), LEHRMAN (1965) and HALLAM and PRICE (1968)

Strontium also can be related to the rate of growth of the mollusk shell, but at present the exact trend is in question. PILKEY and GOODELL (1963) and DODD (1965) suggested that fast-growing shells contain higher strontium concentrations. SWAN (1956) offered the opposite view, that fast growth results in less strontium absorption. HALLAM and PRICE (1968) have invoked other physiologic factors to explain strontium variations within molluscan shells.

Some elements, notably magnesium and manganese, increase in concentration with decreasing salinity (RUCKER and VALENTINE, 1961; PILKEY and GOODELL, 1963, 1964; DODD, 1965; PILKEY and HARRISS, 1966). Concentrations of other elements, such as boron and sodium, increase with salinity (WASKOWIAK, 1962; RUCKER and VALENTINE, 1961; PILKEY and HARRISS, 1966), and some elements, such as strontium, can show either an inverse or direct relation to salinity (RUCKER and VALENTINE, 1961; CRISP, 1972) or show no apparent relation at all (DODD, 1965; HALLAM and PRICE, 1968) (Fig. 30).

LAHOUD and others (1966) showed that pelecypods contain highly variable uranium concentrations (up to two orders of magnitude) throughout the shell. Generally values are much higher on the outer surface of the valve than in the inner parts, suggesting that the shell surface absorbs uranium after calcification. Of special note are the relatively high vanadium concentrations (49–130 ppm) found in pelecypod (and pteropod) shells; NICHOLLS and others (1959) and BROOKS and RUMSBY (1965) have suggested that biogenically-fixed vanadium may have an important effect on the overall budget of this element in marine sediments.

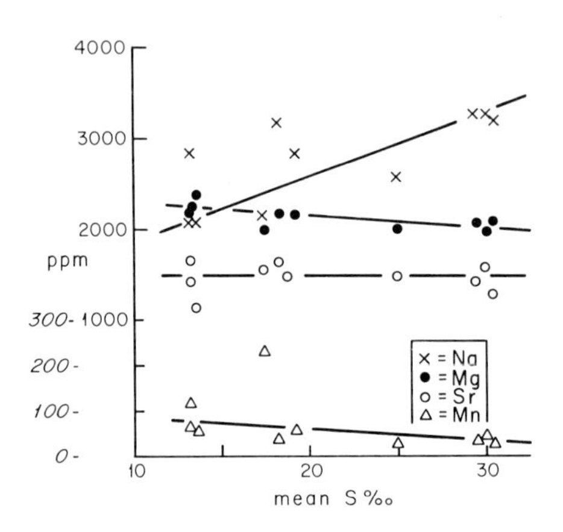

Fig. 30. Mg, Sr, Mg and Na content in the oyster *Crassostrea virginica* with varying salinity. Data from RUCKER and VALENTINE (1961)

Mollusks are believed to secrete their calcium carbonate in isotopic equilibrium with the ambient environment.[11] In normal sea water, the δO^{18} ratio in a mollusk shell will reflect the temperature and δC^{13} will average near zero (EPSTEIN and LOWENSTAM, 1953; LLOYD, 1964; KEITH and others, 1964). Normal δO^{18} values in pelecypods range from -4 to $+3^0/_{00}$, δC^{13} from -2 to $+2^0/_{00}$. Concentrations of δO^{18} in gastropods seem to be more restricted (-2 to $+2$), and the δC^{13} is mainly positive (-1 to $+4$) (KEITH and others, 1964). Differences in the average isotopic values of these two classes may reflect different seasons of calcification, as pelecypods calcify mainly in warmer months, while gastropods can calcify throughout the year (EPSTEIN and LOWENSTAM, 1953). Isolated or nearshore waters can exhibit anomalous δO^{18} and δC^{13} concentrations (see p. 31), and mollusks living within these waters can show similar variations (LLOYD, 1964; KEITH and PARKER, 1965).

As in trace elements, stable isotope content can vary within various parts of the shell. KEITH and others (1964) found that interior aragonitic layers of *Mytilus californianus* are enriched with δC^{13} and depressed in δO^{18} relative to the outer calcitic layers. This fractionation suggests different sources of the calcium carbonate utilized in shell deposition. Precipitation of the interior aragonite probably

---

11 TOURTELOT and RYE (1969) have found that the Cretaceous pelecypod *Inoceramus* was able to fractionate O^{18} relative to other mollusks, and fractionate C^{13} in the aragonitic portions of its shell relative to the calcitic portions. Possibly other mollusks also can do this.

utilizes more metabolic carbon dioxide while calcification of the outer layers may utilize more oceanic carbon and oxygen.

Annelida

Many polychaete annelids secrete calcium phosphate tubes, some of which contain small amounts of calcite (Table 34). Others form chitinous tubes, while still others, such as Sabellarians, form agglutinated tubes. The only polychaetes able to form calcareous tubes belong to the family Serpulidae. All serpulids are epibenthic, mostly encrusters. In contrast, many non-carbonate polychaetes are burrowers or borers. Whether this difference in life style affects tube composition is not known.

Relatively little is known about the calcification of serpulid tubes. Some of the physiological aspects of calcification have been covered in studies by ROBERTSON and PANTIN (1938), HANSON (1948), SWAN (1950), HEDLEY (1956a, b, 1958), QUIEVREUX (1963) and NEFF (1967, 1969). Calcification apparently can occur in two separate glands. NEFF (1967) noted that in two genera (*Eupomatus* and *Hydroides*) calcium-secreting glands precipitate aragonite, while the ventral shield epithelium may produce calcite. In *Pomatoceros* calcite was observed in the calcium-secreting glands. The mucus secreted by the carbonate-precipitating glands is rich in acidic amino acids and generally resembles the protein composition of other carbonate-producing invertebrates (MITTERER, 1971).

Under certain conditions serpulids exhibit high rates of carbonate production. *Hydroides* reefs in south Texas can accrete at rates of about 8 cm/yr (BEHRENS, 1968), although growth rates decrease with advancing age of the colony. ABE (1944) found that spirorbids in Japan reached an adult stage in 3 months, with an average growth of 0.023 mm/day. SMITH and HADERLIE (1969) estimated that spirorbids in southern California have an average life span of 2 months and can produce tubes 1 to 4 mm long. Many serpulids in temperate climates live from early spring to early winter, with annual growth of 5 to 10 cm (BORNHOLD and MILLIMAN, 1973).

Encrusting serpulid tubes occur in many sediments, and while not generally constituting a major sediment source, they can provide significant amounts of carbonate in some environments. For example, on the eastern United States continental shelf serpulids generally are rare, but in some locations, especially the outer shelf, they can contribute over 10% of the carbonate fraction (MILLIMAN, 1972). Under proper conditions, serpulids can construct large shallow-water reefs; apparently swift currents are required for maximum development (BEHRENS, 1968). Many reefs identified as serpulid (such as those in the Mediterranean and at Bermuda), however, are actually composed of vermetid gastropods (Fig. 31; see p. 200 ff.). In terms of the carbonate budget within the oceans, it is questionable whether the amount of carbonate contributed by serpulids can offset the amount eroded by other polychaetes that bore and burrow into the substrate (see Chapter 9). For instance, HARTMAN (1954) concluded that annelids in the Marshall Islands were more destructive with respect to the carbonate budget than they were constructive.

Serpulid tubes, with their chalky white, tubular appearance, can be identified relatively easily with reflected light. Spirorbid worms generally construct small, planispiral tubes (Plate XIII), while many other serpulids precipitate long, straight tubes. Under transmitted light longitudinal sections of tubes can be recognized as two parallel lines. Serpulid tubes tend to be laminar in cross section, but contrary to mollusks, they seldom possess smooth, well-defined laminations; rather, they have microcrystalline, subparallel laminae (Plate XIII) (Fig. 31).

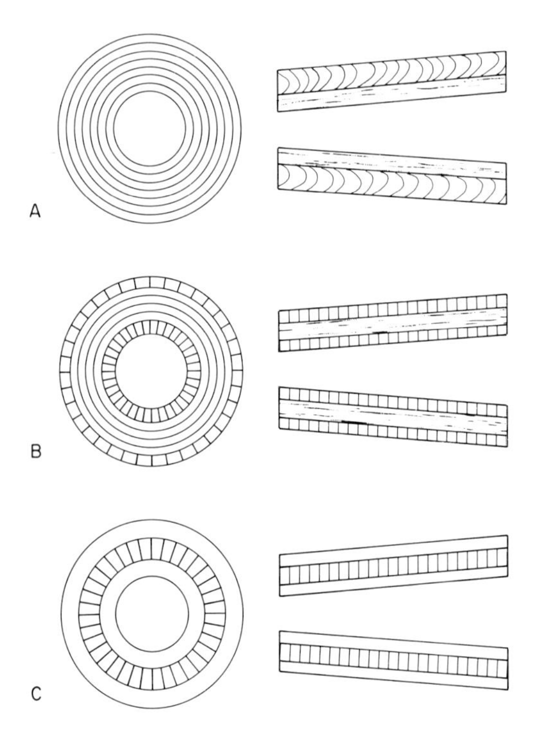

Fig. 31 A–C. Comparison of transverse and longitudinal sections of Serpulids (A), vermetid gastropods (B) and scaphopod tubes (C). The serpulid tube in A shows a cone-in-cone outer layer and a laminar inner layer. The vermetid in B has a prismatic inner and outer layer and a laminated inner layer. The scaphopod in C has a clear inner and outer layer and a prismatic inner layer. After Horowitz and Potter (1971)

Until recently the mineral and chemical composition of serpulid tubes was not well documented. Lowenstam (1954a, b) showed that serpulid tubes can be composed of calcite, aragonite, or mixtures of the two minerals. Lowenstam stated that the amount of aragonite within the tube increases with increasing water temperature, with polar species having calcitic tubes and tropical species having aragonitic tubes. Chave (1954) inferred that the $MgCO_3$ content within serpulids also increases with increasing water temperature. More recent studies (Bornhold and Milliman, 1973) however, have shown that the composition of a serpulid tube is more dependent upon genus than environment. Aragonitic tubes are not restricted to the tropics, but also can be found in temperate and polar waters (for example, *Filograna* and *Protula*). Similarly, calcitic tubes occur in the tropics as well as in colder waters (for example *Spirobranchus* and *Crucigera*) (Fig. 32).

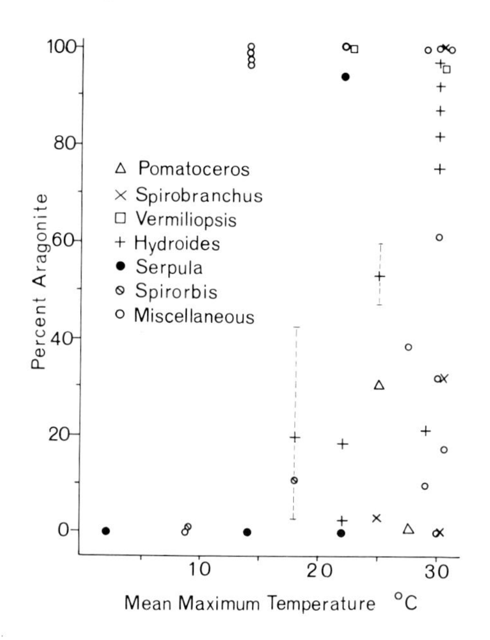

Fig. 32. Relation of aragonitic content within various serpulid genera with changing temperature. From BORNHOLD and MILLIMAN (1972)

In calcitic serpulid tubes, $MgCO_3$ concentration is directly related to water temperature (Fig. 33). But in those tubes with mixed mineralogy, magnesium content within the calcitic phase is directly related to aragonite content (Fig. 34). The rate of $MgCO_3$ increase within the calcite is such that between about 0 and 60% aragonite, the total amount of Mg within the tube remains more or less

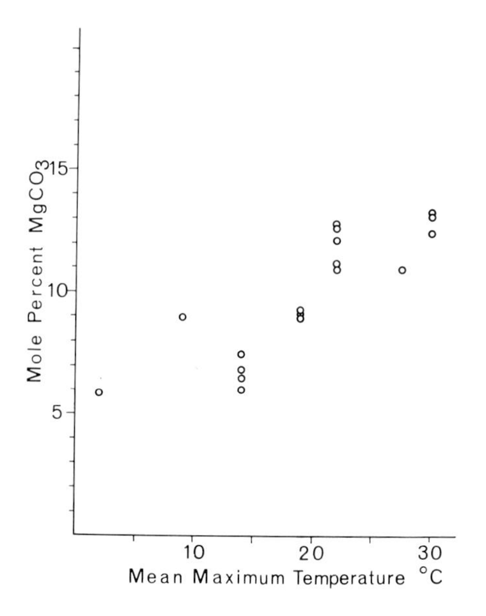

Fig. 33. Variation of $MgCO_3$ within calcitic serpulids with changing water temperature. From BORNHOLD and MILLIMAN (1972)

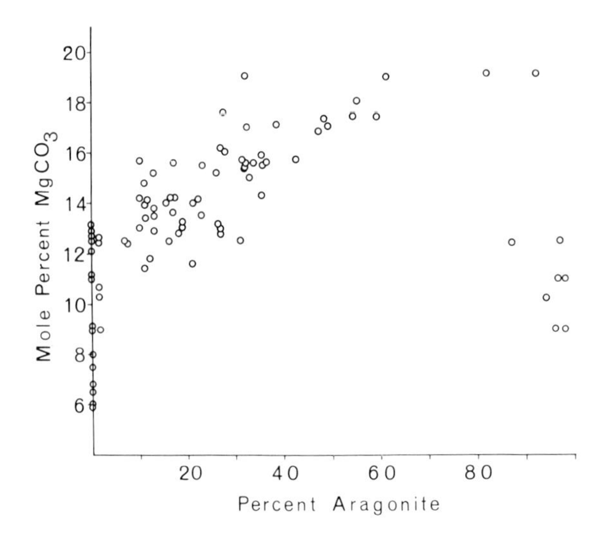

Fig. 34. Correlation between $MgCO_3$ content within the calcitic portion of serpulids and the amount of aragonite within those tubes. $MgCO_3$ increases with increasing aragonite (and thus decreasing calcite) so that the total $MgCO_3$ within the tube remains more or less constant. This trend stops suddenly at about 60% aragonite. From BORNHOLD and MILLIMAN (1972)

constant, between 2 and 3%. Above 60% aragonite, however, the amount of magnesium in the calcite decreases markedly and in tubes with more than 90% aragonite, the $MgCO_3$ content is significantly less than at lower aragonite values (Fig. 34).

WASKOWIAK (1962) presented a large number of chemical analyses on serpulid tubes. His magnesium analyses, however, appear to be in error; for example, he reported that Mg content in *Hydroides norvegicus* exceeds 10%, which is clearly higher than values reported by other workers (Table 34). The amount of strontium within serpulid tubes depends directly upon the aragonite content; at 100% aragonite, the Sr content is about 0.8%. In comparison with other more advanced invertebrates, the iron content in serpulid tubes is relatively high. Some serpulids may also contain minor amounts of calcium phosphate (VINOGRADOV, 1953).

Arthropoda

Most arthropods have soft-bodies and chitinous exoskeletons. Only in the superclass Crustacea do we find prominent examples of exoskeletons which are strengthened with calcium and phosphate deposits or are totally calcified. Three classes, Cirripedia, Ostracoda and Decapoda constitute the dominant calcifying crustaceans.

Cirripedia

Barnacles, within the order Thoracica, produce a hard calcareous skeleton. An outer circle of compartmental plates (usually numbering between 6 and 8) surround a series of inner opercular plates. In the most highly calcified barnacle

Table 34. Composition of some polychaete annelids. The top two genera represent annelids which have only partially calcified tubes. The other genera are from the carbonate-precipitating family Serpulidae. Data from VINOGRADOV (1953), THOMPSON and CHOW (1955), WASKOWIAK (1962), LOWENSTAM (1964b), and BORNHOLD and MILLIMAN (1973)

	Percent organic matter	Mineral.	Percent						Parts per million						
			Ca	Mg	Sr	P	Na	K	Fe	Mn	Ba	B	Pb	Ni	Cu
Onuphis	57.8 (53.1–62.5)	phos. +cal.	9.26 (2.41–16.11)	5.84		6.40 (3.30–9.49)									
Hyalinacia	61.6 (61.4–61.8)	phos. +cal.	3.76 (3.72–3.80)	5.02 (4.90–5.14)		9.03 (8.94–9.12)									
Serpulidae															
Spirorbis	12.5	MgC		1.64 (0.3–3.0)	0.22 (0.09–0.32)			0.02	467 (250–1000)	22 (tr–38)	26 (9–30)	5 (2–9)	2 (1–4)	40 (15–100)	2
Spirobranchus	3.3 (1.0–5.7)	MgC		2.98 (2.12–3.57)	.37		0.07 (0.06–0.08)		899 (238–1560)	tr					
	2.0	Arag.		0.10 (0.10–0.11)	1.0		0.03		214	tr					
	4.2	mixed (Arag. 32%)		2.32	0.59		0.03		759	43					
Serpula	6.2	MgC		1.35	0.17		0.02	0.01	230 (190–250)	65 (26–130)		8 (4–9)	10 (3–24)	10 (tr–30)	5 (1–7)
Hydroides		mixed (Arag. 22%) (9–35%)		2.81 (2.21–3.50)	0.36 (0.25–0.49)		0.25 (0.04–0.61)	0.18 (0.28–0.55)	325 (250–400)	48 (26–70)		8	6 (1–12)	32 (10–55)	2
Filograna	4.0	Arag.		0.19	0.70										
Pomatoceros	4.7	Arag.		0.12	0.93		0.04		924	57					
Mercierella	5.0	MgC		2.25	0.32										
Vermilopsis	1.7	Arag.		0.11	0.65		0.03		267	37					

suborder, Balanomorpha, the compartmental and opercular valves often rest upon a calcareous basal disc (although many species lack this disc). As a result, balanoid compartmental valves tend to remain attached to one another after death; the opercular plates, however, rarely remain fixed (NEWMAN and others, 1969). With sufficient postmortum agitation and abrasion, the compartmental valves also will separate.

Barnacles require a hard substrate for attachment and thrive best in high energy environments, where their filter-feeding mode of food-gathering can act with maximum efficiency. Although barnacles are commonly thought of as intertidal animals, they also occur in neretic depths. Barnacle growth is faster in warmer waters and in the younger stages of the individual development. WERNER (1967) found that *Balanus triganus* grows an average of 3.9 mm during its first two weeks (in Florida). COSTLOW and BOOKHOUT (1953, 1956) reported increases in the areal diameter of *B. amphitrite niveus* and *B. improvisus* of 12 mm^2 in 8 days and 16 mm^2 in 25 days, respectively. Similar species from different environments display widely variable growth rates, mostly due to differences in food supply (CRISP, 1964). Shell accretion rates range from 15 to 30 mm/year (PEQUEGNAT and FREDERICKS, 1967; SMITH and HAEDERLIE, 1969). Apparently individual barnacles are not long-lived; in Woods Hole and in Southern California they do not live more than about one year (GRAVE, 1933; SMITH and HAEDERLIE, 1969).

Because of this high turnover rate, barnacles can be important sediment contributors. On the eastern United States and Canadian continental shelf, barnacle plates probably are the second most abundant benthonic carbonate component (second only to mollusk shells) and in particularly rocky areas such as Onslow Bay (North Carolina) and on the Florida shelf they can contribute more than 50% of the carbonate sediments (MILLIMAN, 1972).

The most characteristic petrographic features of barnacle plates are the longitudinal tubes and ribs that give the valve a rather hollow appearance (CORNWALL, 1962; NEWMAN and others, 1967; 1969) (Plate XV c–g). Valves are mostly triagonal, although specific shape depends upon the valve and species. Under reflected light the raised sheath on the inner side of the plate is particularly diagnostic (Plate XV a, b). Traverse and longitudinal thin sections usually display the longitudinal tubes and ribs as well as the inner and outer laminae (Plate XV). Even smaller fragments are discernible by their non-laminar, grainy shell structure.

Numerous chemical analyses have been reported for the balanoid genus *Balanus*. Valves are calcitic and generally contain less than 0.30% magnesium and strontium; THOMPSON and CHOW's strontium values are slightly higher (Table 35). The high sodium content measured by PILKEY and HARRISS (1966) may reflect trapped sea salts rather than sodium within the calcite lattice, although GORDON and others (1970) reported similar sodium values but found only small amounts of chlorine. In brackish waters magnesium and manganese contents increase greatly, while sodium and strontium decrease (PILKEY and HARRISS, 1966) (Fig. 35).

LOWENSTAM (1963, 1964a) found that the basal disc of the balanoid *Tetraclita* is aragonitic while the opercular and compartmental valves are calcitic. The absolute percentages of aragonite within the total skeleton will depend upon the thickness of the basal disc, which increases in warmer climates. As might be

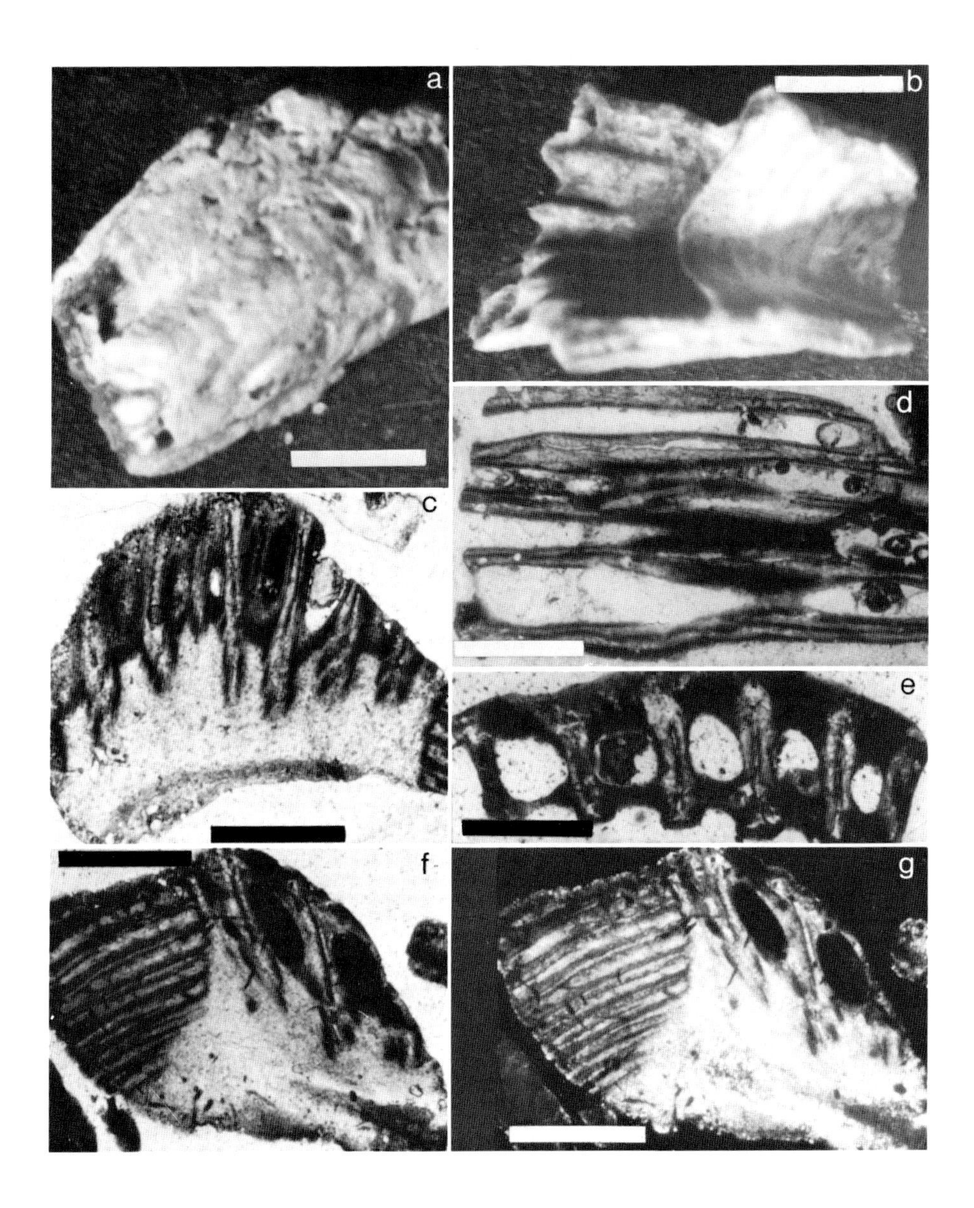

Plate XV. Barnacles. (a) and (b) Reflected light photomicrographs of barnacle plates, both convex and concave sides. Scales are 250 microns. (c), (f) and (g) Oblique sections of barnacle plates. Refracted light, (g) with polarized light. Scales are 250 microns. (d) Longitudinal section of barnacle plate. Note dominant longitudinal canals. Refracted light; scale is 250 microns. (e) Transverse section of barnacle plate. Refracted light; scale is 250 microns

Table 35. Composition of selected Arthopods. Elemental composition is on an organic-free basis. Data from VINOGRADOV (1953), THOMPSON and CHOW (1955)[a], LOWENSTAM (1964a, 1964b), WASKOWIAK (1962)[b], BROECKER (1963) and PILKEY and HARRISS (1966)[c]

	Percent organic matter	Mineral.	Percent					Parts per million							
			Ca	Mg	Sr	Na	P	Fe	Mn	Ba	B	Pb	Cu	Ni	U
Cirripedia															
Balanus	4.44 (1.80–12.3)	Cal.	38.7	0.29 (0.05–1.0)	0.35[a] (0.28–0.43)			170 (3–750)	60 (2–230)	39 (tr–130)	2	3 (tr–22)	3	16 (tr–100)	15 (7–21)
					0.29[b] (0.07–0.60)										
				marine[c]-0.28	0.24	0.22		130	200						
				brackish[c]-0.49	0.22	0.08		60	1900						
Tetraclita		side plates-Cal.		0.82 (0.77–0.87)	0.40 (0.39–0.42)										
		base plate-Arag.		0.08 (0.06–0.10)	0.83 (0.82–0.84)										
Decapoda															
Cancer	27.6 (20–32.7)	MgC		2.69 (2.40–3.39)	0.45 (0.44–0.49)		2.14	18	19		tr	3	tr		
Carcinus		MgC		2.0 (0.5–2.4)	0.39 (0.30–0.55)			100 (9–400)	40 (6–100)	6 (0–30)	1 (0–14)	3 (1–5)	1 (tr–1)	tr (0–1)	

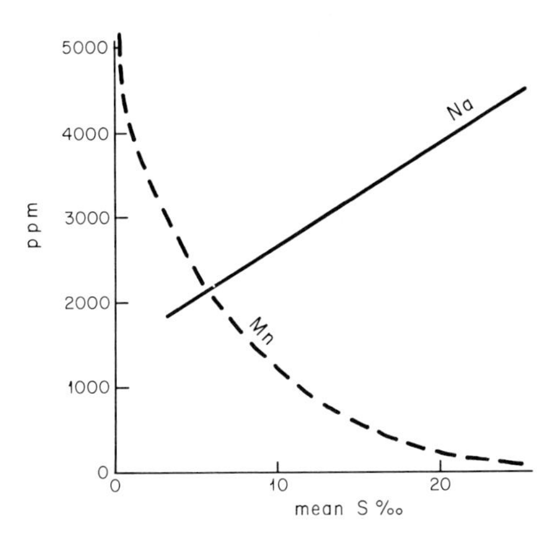

Fig. 35. Mn and Na content in barnacle valves with varying salinities. Data from GORDON and others (1970)

expected, the calcitic valves are relatively high in magnesium and low in strontium, while the aragonitic basal disc is low in magnesium and high in strontium. Strontium content in the basal disc is about 1.03% (LOWENSTAM, 1964a). This value seems higher than in other aragonitic skeletons, and may represent either a local anomaly or analytical error.

Ostracoda

Ostracods are benthonic bivalved crustaceans that can precipitate a calcified shell. Average shell length is 0.5 to 2.0 mm, but shell thickness depends upon the environment (SCOTT, 1961). As with most other arthropods, ostracod calcification is discontinuous, with periodic molting followed by precipitation of a new test. Although ostracods have stratigraphic and ecologic importance, they rarely comprise more than a small part of a carbonate sediment. SOHN and KORNICKER (1969) have pointed out the presence of calcareous nodules (composed of hydrated calcite) that are precipitated posthumously in the carapaces of myodocopid ostracods. The sedimentologic importance of these nodules, however, is suspected to be small. Most modern ostracod shells are composed of two layers, which may or may not be penetrated by numerous canals. A short discussion on the structure of ostracod shells was presented by JOHNSON (1951).

Relatively few chemical analyses of ostracod shells have been reported. Those shells that have been analyzed are mainly carbonate, although SOHN (1958) reported that between 5 and 20% of the test of *Chlamyolotheca* is protein and chitin. FOSTER and BENSON (1958) found that calcite is the only carbonate polymorph present in ostracods; chitin is usually relatively quickly decomposed after death and is replaced by calcite. According to SOHM's semi-quantitative analyses, potassium content is between 0.5 and 1.0%, mangesium, sodium, and silica between 0.1 and 0.5% and strontium between 0.05 and 0.1%. CHAVE (1954) noted a direct relation between magnesium content and water temperature, with ostracods in Florida containing 8 to 10 mole % $MgCO_3$.

More recent data, however, indicate that most ostracod shells contain between 1 and 5% $MgCO_3$ (CADOT and others, 1972). Although the $MgCO_3$ content within any genus depends partly upon water temperature, the distribution of magnesium within individual shells is more complex. Some shells have layers of Mg-rich and Mg-poor calcite (for example *Bairdia* sp. and *Macrocypris* sp.) while others (such as *Krithe*) contain uniform amounts of magnesium throughout the shell (CADOT and others, 1972).

Decapoda

The carapaces and appendages of some decapods, especially crabs and lobsters, can supply small quantities of carbonate to a sediment. Usually the fragments are rather large and can be easily recognized under reflected light.

The outer decapod shell (epicuticle) is chitinous. The inner (endocuticle) consists of a calcified layer, often sandwiched between chitinous (phosphatic) laminae. The relative amounts of calcite and phosphate depend both upon the species and age of the individual. For instance, the carapace of the crab *Cancer* contains about 2% phosphorous, while in the lobster *Homarus* phosphorus varies from 2.5% in small carapaces to more than 5% in large carapaces (STEIGER, in CLARKE and WHEELER, 1922; VINOGRADOV, 1953). The calcite phase is rich in magnesium, although concentrations do not exceed 3% (Table 35). STEIGER (in CLARKE and WHEELER, 1922) found a gradual increase of magnesium with increasing size of the carapace. Iron, manganese and barium concentrations are much lower than in cirripeds.

Other Arthropods

Both stomatopods and amphipods contain calcium carbonate within their carapaces, but their skeletons are primarily phosphatic. CLARKE and WHEELER (1922) reported 28.56% $CaCO_3$ and 15.99% $MgCO_3$ within the stomatopod *Chloridella*. This would correspond to a near protodolomite if all the magnesium were contained within the carbonate. According to GREZE (1967) the calcium carbonate in amphipod carapaces is calcite.

Echinodermata

The phyllum Echinodermata contains five major living groups: echinoids (sea urchins and sand dollars), asteroids (starfish), ophiuroids (brittle stars), holothurians (sea cucumbers) and crinoids. Although their calcareous tests and fragments seldom comprise more than about 15% of any carbonate sediment, echinoderms contribute to marine sediments in a wide spectrum of environments, from littoral reefs to the deep sea. An excellent compilation of papers concerning the general biology of echinoderms has been presented by BOOLOOTIAN (1966). Of particular interest are those papers by Raup (the endoskeleton) and Swan (skeletal growth).

Echinoderms are one of the few invertebrate groups that contain an endoskeleton. The skeleton is composed of two elements, the calcareous sterom and the organic stroma. In echinoids the sterom is continuous and forms a hard endoskeleton. In ophiuroids and asteroids the stroma is more prominent, and in holothurians the only hard parts are those in the calcareous ring and the spicules (ossicles).

Skeletal precipitation in echinoids usually begins with the formation of a dense triaxial spicule. An encompassing collagenous organic matrix acts as the substrate for growth (TRAVIS and others, 1967); as it expands the skeletal structure also expands (OKAZAKI, 1960; RAUP, 1966). Sterom growth occurs around the margin of individual plates (MELVILLE and DURHAM, in DURHAM and others, 1966), while the stroma provides the organic matrix by which the sterom can obtain nutrition for growth (BEAVER and others, 1966). The resulting spongy and porous skeleton serves as both a means for conserving calcium carbonate during skeletogenesis and offers spaces of attachment for connective tissue (NICHOLS, 1962).

The best documented studies on echinoderm growth rates have concerned asteroids that feed on commercially valuable shellfish. SWAN (1966) presents numerous growth rate estimates, most of which fall within the range of 20 to 100 mm/year. Maximum growth occurs in summer months. The ophiuroid *Ophiothrix* grows an average of 9 mm/year (PEQUEGNAT and FREDERICKS, 1967). Shallow-water sea urchins generally grow at rates of 10 to 20 mm/year, and turnover rates vary from 2 to 24 months (EBERT, 1968; NORTH, 1968). Both growth rate and calcification increase with increasing water temperature (DAVIES and others, 1971). Those urchins with more rapid growth rates tend to have more spongy skeletons (RAUP, 1966). Thus, in contrast to other invertebrates, echinoid tests can be thicker and denser in cold waters than in warm waters (RAUP, 1958).

Skeletal Elements

Many workers, beginning as early as SORBY (1879), referred to the reticulate or fenestrate echinoid sterom (excluding the teeth) as being single crystals (DONAY and PAWSON, 1969 and references therein). TOWE (1967) cites evidence that the outer margins may be polycrystalline aggregates. The orientation of the c-axis relative to the plate surface can be of considerable importance in defining the phylogenetic character of echinoderms (RAUP, 1965 and references therein). This unicrystalline nature results in unit-extinction of the entire fragment under polarized light, thus making identification of echinoid fragments extremely easy in thin section (Plate XVI). The dull surface luster and the porous, sponge-like skeletal structure are diagnostic features of the echinoid sterom (both body test and spines) (Plate XVI). Six-sided plates, spine knobs on the interambulacral plates and long spines are also common features. Echinoid spines can be either hollow (such as *Diadema*) or filled with spongy tissue (for example, *Echinus*). The degree of internal filling can help in taxonomic identification. Echinoids also contain internal spicules that resemble holothurian ossicles, but these spicules are quantitatively less important than the other skeletal elements.

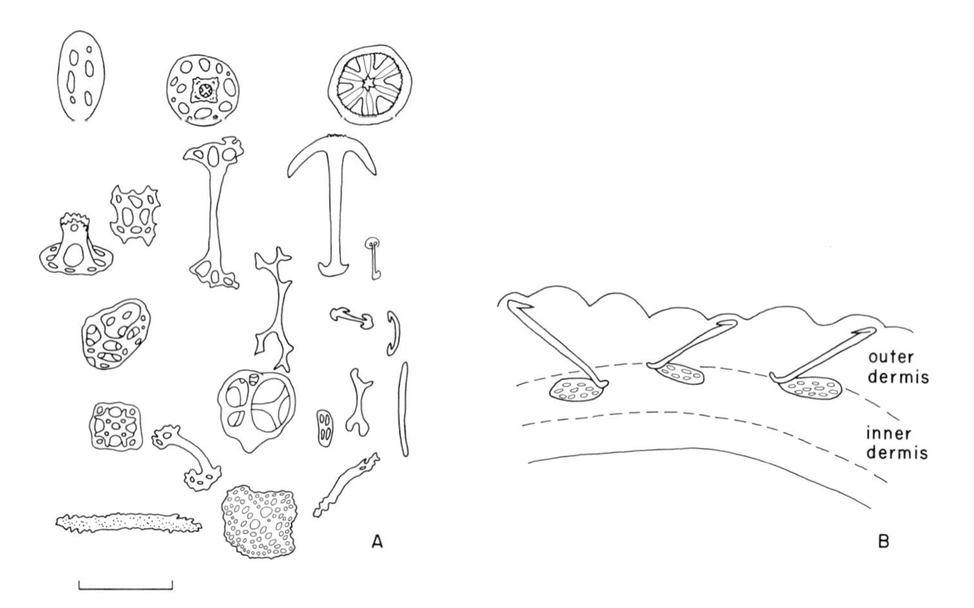

Fig. 36. A. Typical holothurian ossicles; scale is 100 microns (after ERWE, 1913). B. The relation of anchor and anchor plate ossicles within a holothurian. After HYMAN (1955)

Holothurian ossicles occur within both the dermal and inner layers and serve both for support and protection of the animal. These spicules are thought to represent an embryonic skeleton (HYMAN, 1955). The size and form of the ossicles are useful parameters in species identification. Names such as wheels, tables, anchors, buckles and hooks are often used to describe the widely divergent ossicle shapes (Fig. 36). Like other echinoderm skeletal parts, holothurian spicules appear as single crystals under polarized light. Sizes generally range from 75 to 125 microns. HAMPTON (1958) estimated that a single specimen of *Holothuria impatiens* contains more than 20000000 ossicles, although the ossicles constitute only about 10% of the total body weight. With such quantities of spicules produced, it is possible that holothurians could be of local sedimentary importance. Further reviews on holothurian ossicles can be found in CRONEIS and MCCORMACK (1932), CUENOT (1948) and FRIZZELL and EXLINE (1955).

The skeletal elements of ophiuroids and asteroids include aboral, lateral and oral arm plates, together with the buccal shield complex and arm spines. In addition, ophiuroids contain ossicles, about 2 mm in width, which act as internal supports within each arm. Little work has been done on either the petrography of these calcareous components or their sedimentary importance. The plates within crinoid stems can be recognized in thin section by their conspicuous recticulate pattern (JOHNSON, 1951).

As mentioned above, echinoderm fragments are common but seldom dominant in carbonate sediments. Echinoid plates and spines occur in tropical reef sediments and fragments of irregular echinoid tests contribute to lagoonal sands. In shelf

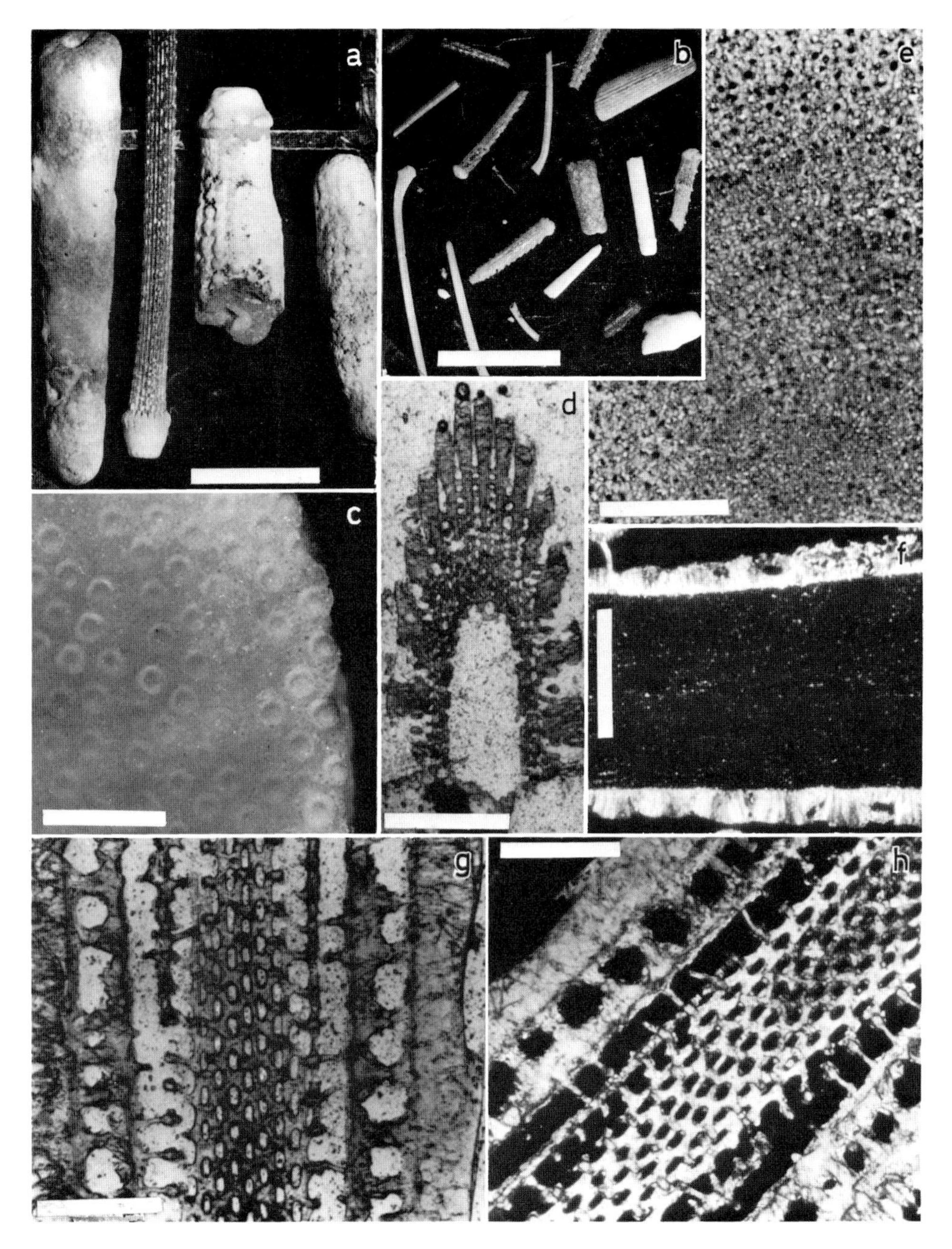

Plate XVI. Echinoids. (a) Echinoid spines in reef sediments tend to be large. Reflected light; scale is 1 mm. (b) In contrast, echinoid spines from continental slope deposits are finer grained. Reflected light; scale is 1 mm. (c) Echinoid plate. Note the pocked surface-texture. Reflected light; scale is 1 mm. (d) Oblique section of a hollow echinoid spine. Refracted light, scale is 250 microns. (e) Echinoid plate. Refracted light; scale is 100 microns. (f) Echinoid spine, with inner portion extinct under polarized light; scale is 250 microns. (g) and (h) Echinoid spine under plane and polarized light. Scale is 100 microns in each

sediments echinoids can be an important component, but usually only in low carbonate sands (see below). SCHÄFER (1962) tells of large accumulations of tests and spines within some areas of the North Sea. Echinoid spines are also common in upper continental slope sediments.

Other echinoderm fragments are usually less frequent in marine sediments. Crinoids and blastoids are present in many limestones, but crinoid stems seldom occur in modern sediments. Holothurian ossicles may be important in very fine sands and silts, but such sediments have not been studied in sufficient detail to warrant speculation as to their relative importance. In the English Channel skeletal fragments of the ophiuroid *Ophiothrix* can comprise more than 50% of the carbonate grains within certain biogenic sands (BOILLOT, 1964, 1965).

Composition

Echinoderm skeletal parts are calcitic, generally containing 2 to 4% magnesium (Table 36). PILKEY and HOWER (1960) and HARRISS and PILKEY (1966) found that echinoid tests in slightly brackish water contain less magnesium and sodium but similar amounts of iron, strontium and manganese compared with tests precipitated in normal ocean waters. Most of the sodium within echinoderm tests is

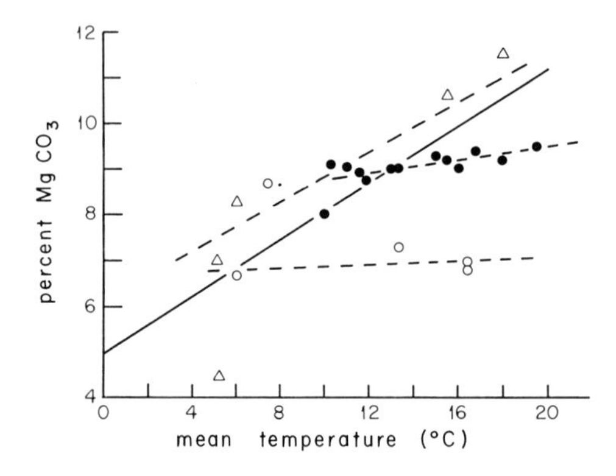

Fig. 37. Variation of Mg with water temperature within various echinoids. The solid line is the class trend as given by CHAVE (1954). The dotted lines represent generic trends (● = *Dendraster*; △ = *Encope*; ○ = *Strongylocentrotus*). Data from CHAVE (1954); PILKEY and HOWER (1960), and VINOGRADOV (1953)

present in a brine-like fluid within the living sterom, which crystallizes only upon death of the organism (DONAY and PAWSON, 1969). Average strontium values are between 0.2 and 0.3%.

Magnesium content within echinoid tests depends upon temperature (CLARKE and WHEELER, 1917, 1922; CHAVE, 1954), but generic trends may vary considerably from the general class trends (PILKEY and HOWER, 1960) (Fig. 37). Thus, different species within the same environment may have widely different magnesium contents. For instance, RAUP (1966) reports that echinoid tests from Eniwetok Atoll have $MgCO_3$ concentrations ranging from 8.4 to 14.9%. Sodium content within the echinoid *Dendraster excentricus* increases with water temperature, manganese and strontium content decrease with increasing temperature, and iron shows no significant change with relation to temperature (Fig. 38). Somewhat similar conclusions were reached by DAVIES and others (1971) in their study of echinoid spines.

Early data by CLARKE and WHEELER (1922; see VINOGRADOV, 1953) showed that the composition of individual echinoids is not uniform. Tests tend to contain higher magnesium concentrations than the spines, small spines have more magnesium than larger spines, and interambulacral plates contain more magnesium than ambulacral plates (Tables 36 and 37). More recent analyses have shown that

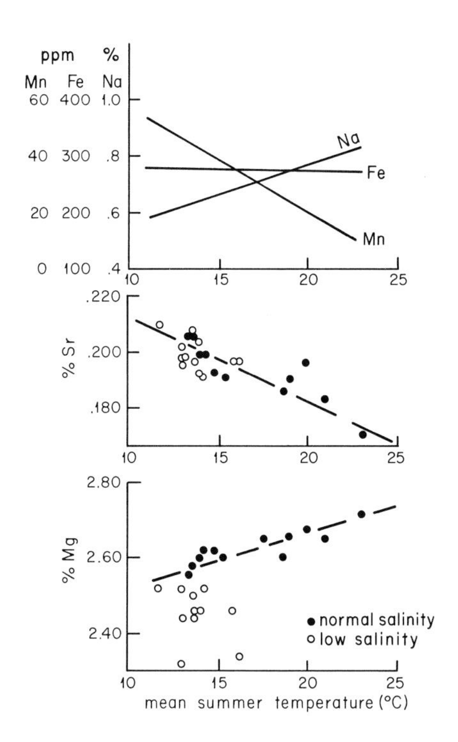

Fig. 38. Variation of Mg, Sr, Mn, Fe and Na in the test of the echinoid *Dendraster excentricus* with water temperature. After data from PILKEY and HOWER (1960) and HARRISS and PILKEY (1966)

magnesium content decreases in the test and increases in the spines towards the periproct (that is in newer skeletal elements) (FOWLER and DODD, 1969). FOWLER and DODD also found similar trends in the strontium content within the test and spines. Although teeth usually contain less than 10 mole % $MgCO_3$, the dense axial zone contains between 28.5 and 43.5%, with an average content of 41%, indicating a protodolomite composition (SCHROEDER and others, 1969).

The few studies on ophiuroids show that the body discs have slightly higher magnesium content than the rays (arms), suggesting similar trends with the echinoids (Table 36). No such trends have been found in the stalks and arms of the crinoids (CLARKE and WHEELER, 1922).

Table 36. Chemical composition of selected Echinoderms. Data from Phillips (1922), Vinogradov (1953), Chave (1954), Thompson and Chow (1955), Hampton (1958), Pilkey and Hower (1960), Waskowiak (1962), Harriss and Pilkey (1966) and Mauchline and Templeton (1966)

	Organic matter	Mineral.	Percent						Parts per million						
			Ca	Mg	Sr	Na	P	Si	Fe	Mn	Ba	B	Ni	Zn	Cu
Echinoidea															
Strongylocentrus															
test		MgC	36.7	2.09 (1.57–2.73)			0.13								
spines		MgC		1.08 (0.93–1.16)											
Dendraster		MgC		2.55 (2.17–2.96)	0.20 (0.16–0.22)	0.64 (0.38–1.22)			270 (140–620)	35 (12–94)					
Echinometra		MgC	33.5	3.81 (3.35–4.26)			0.81	0.06							
Mellita		MgC	34.0	3.45				0.07							
Encope		MgC	32.4	2.73 (1.30–4.00)	0.14		0.11	1.6							
Echinus		MgC		2.7 (2.4–3.0)	0.15 (0.07–0.22)			0.77	22 (9–34)	8 (6–10)	6	36 (12–60)	6		<1
Crinoidea															
Endocrinus															
stalk		MgC	35.2	3.37				1.0							
arms		MgC	35.1	3.46				1.36							
Asteroidea															
Asterias	35.8 (33.6–39.4)	MgC		2.61 (2.26–2.99)	0.24 (0.22–0.31)		0.22	0.31							
Astropecten		MgC		3.29 (2.93–3.83)	0.24			1.44 (0.11–4.07)	210	4				12	12

Ophiuroidea										
Gorgonocephalus	MgC		2.84 (2.44–3.48)	0.27	0.17	0.69				
Ophiocoma	MgC		3.09 (2.03–3.77)			0.02				
Ophiomyxa	MgC		4.45 (4.23–4.78)			0.19				
Holothuria										
Holothuria	MgC	32.5	3.60 (3.36–4.01)		0.20	0.17	90	6	24	6

Table 37. Distribution of major, minor and trace elements in the various hard and soft parts of the echinoid *Echinus esculentus*. Data from RILEY and SEGAR (1970)

		Oral shell	Aboral shell	Aristotle's lantern	Spines	Intestines	Gonads
Percent	Ca	30.0	30.0	30.0	32.0	0.4	0.22
	Mg	1.6	1.5	1.8	1.0	1.0	0.49
	Sr	0.09	0.08	0.09	0.08	0.01	0.002
	Na	0.72	1.1	0.81	1.0	5.2	2.7
Parts per million	K	360.0	460.0	330.0	650.0	9600.0	7000.0
	P	65.0	78.0	69.0	64.0	2600.0	8500.0
	Fe	6.9	2.5	1.6	16.0	22000.0	15.0
	Mn	56.0	57.0	88.0	4.5	11.0	0.52
	Ni	<0.23	<0.14	<0.13	1.6	<0.77	7.7
	Zn	110.0	12.0	22.0	28.0	550.0	110.0
	Cu	1.8	0.90	0.42	1.6	5.9	16.0
	Pb	<0.62	<0.62	<0.62	<0.58	<3.0	<0.68
	Co	<0.27	<0.16	<0.16	<0.17	<1.1	0.40
	Cd	0.67	0.30	0.20	0.66	8.9	0.65
	Ag	0.09	<0.05	0.53	<0.03	<0.75	<0.03
	Al	120.0	140.0	160.0	130.0	2100.0	<95.0

The partitioning of stable isotopes is also marked within the various morphological parts of the echinoid skeleton, but trends do not parallel the trace element distribution. Tests are enriched in C^{12} relative to the spines, but not nearly as enriched as various parts of the Aristotle's lantern (WEBER and RAUP, 1968) (Fig. 39). Irregular echinoids (sand dollars) have tests slightly enriched in C^{13}

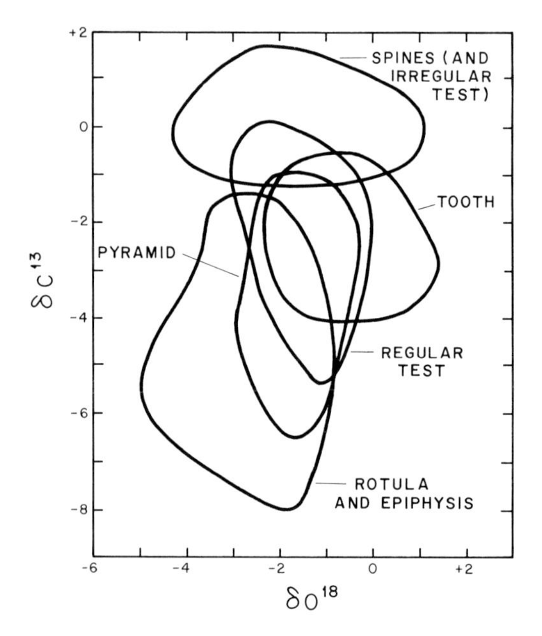

Fig. 39. C^{13} and O^{18} variations within the different morphologic parts of various echinoids. Data are from WEBER and RAUP (1966b)

relative to regular echinoids. Stable isotope ratios in crinoids and asteroids are similar to those in the pyramid, rotula, and epiphysis of echinoids (Fig. 39; Table 38). Ophiuroid calcite displays values closer to echinoid tests.

Table 38. Average δC^{13} and δO^{18} contents of selected families of Crinoids, asteroids and ophiuroids. Data from WEBER (1968)

	δC^{13}	δO^{18}
Crinoidea		
Antedon	−5.55	−2.51
Asteroidea		
Asterinidae	−5.64	−3.01
Astropectinidae	−6.10	−2.75
Ophiuroidea		
Gorgonocephalidae	−2.34	−1.36
Ophiocomidae	+0.66	−1.25
Ophiomyxidae	−0.81	−0.27

Chordata

Ascidians (tunicates) are epibenthic, filter-feeding colonial protochordates with soft bodies. Although superficially resembling sponges, the presence of a primitive backbone (notochord) within their larval stage signifies the high phylogenetic

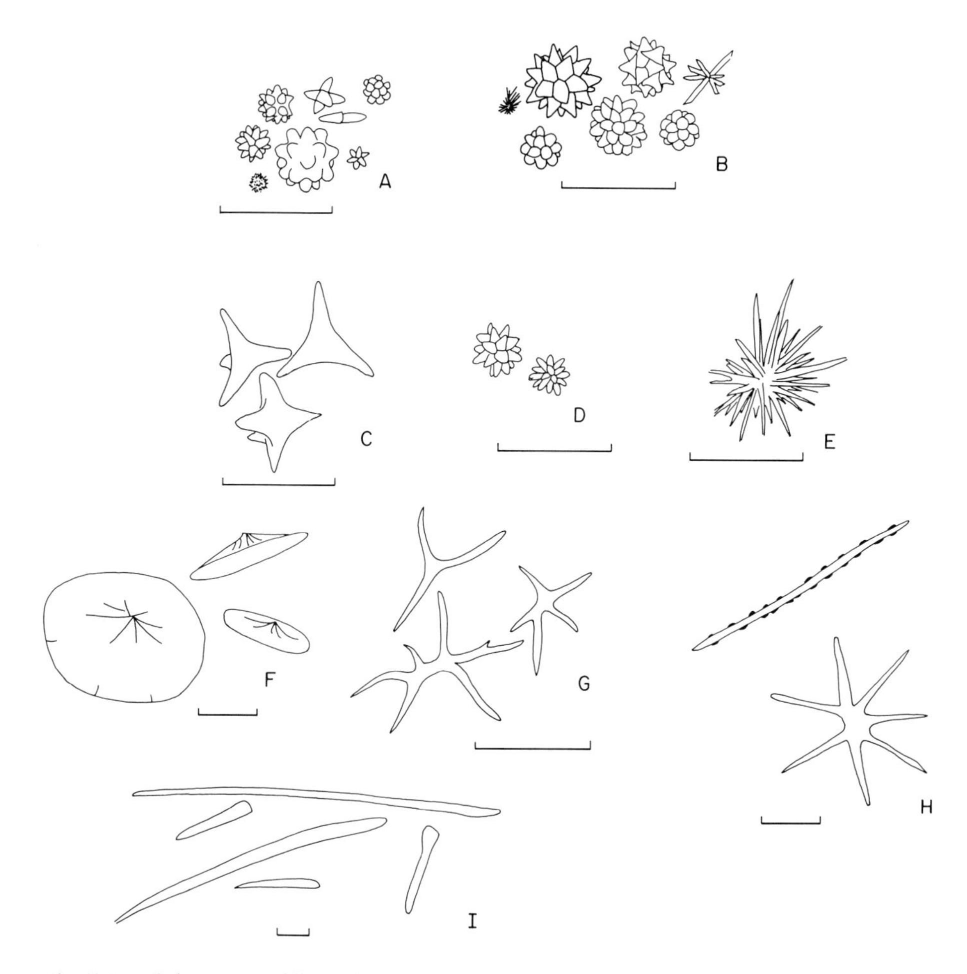

Fig. 40 A–I. Calcareous ascidian spicules. Groups A–F are from tests; G–I from brancial cavities. In each grouping the scale line is equal to 100 microns. A. *Didemnum;* B. *Trididemnum;* C. *Echinoclinum;* D. *Leptoclinides;* E. *Kukenthalia;* F. *Cystodites;* G and H. *Pyura;* I. *Herdmania.* After VAN NAME (1930, 1945)

development of the ascidians. Some families within this primitive chordate group contain carbonate spicules, making them the highest animal group to utilize calcium carbonate as a skeletal material.

Those ascidians containing large amounts of calcium carbonate spicules are limited mainly to the families Didemnidae and Pyuridae. As with sponges, body spicules serve both for protection and structural support. In the Didemnidae spicules are confined to the body wall. These spicules are generally 25 to 75 microns in diameter and spherical or stellate in shape. Common genera include *Didemnum*, *Trididemnum*, *Leptoclinides*, *Lissoclinum* and *Echinoclinum*. In contrast the Pyuridae contain spicules within the test, tentacles and branchial cavities. Some spicules are monaxial, but many are irregularly shaped and some display numerous long spines (Fig. 40). Branchial spines can be as long as 0.8 mm. *Pyura*, *Herdmania* and *Microcosmus* are representative genera. Several other families also contain spicules, and calcareous plates are found in the test wall of *Cystodites* (family Polycitoridae) (Fig. 40).

Although ascidians live throughout the ocean, most spicule-bearing species are restricted to depths shallower than about 200 m. Latitude does not seem to influence distribution, as spicule-bearing ascidians are present both in tropical and polar climates. Long ago HERDMAN (1885) suggested the possibility that tunicate spicules could be important sedimentary components. However, except for the Bahamas (where spicules are locally abundant; PURDY, 1963), the importance of these spicules in carbonate sediments has not been well documented. Few data concerning ascidian growth are available. PEQUEGNAT and FREDERICKS (1967) found that *Trididemnum* grows about 7 mm per year, but the annual increase in spicule concentration is not known.

Ascidian spicules are aragonitic (SCHMIDT, 1924; PRENANT, 1925). They are relatively rich in strontium (0.82%) and poor in magnesium (less than 0.15%) (LOWENSTAM, 1963, 1964a).

Vertebrata

Most vertebrates have phosphatic skeletons. Fish, however, also can precipitate calcium carbonate in their ears. These small carbonate grains, called otoliths, can be a visually obvious (but quantitatively unimportant) component of deep-sea carbonate sediments.

Fish form three types of otoliths. Lapillus and Asteriscus are small (100 to 300 microns) grains formed in the labyrinth and lagena respectively (Fig. 41) (CAMPBELL, 1929). The Sagita, the largest and most important otolith, is formed in the sacculus. These rather flat disks measure 2.5 to 10 mm in length and 1 to 6 mm in width. The disk has two distinct sides. The outer side is convex and the inner side is concave and usually has more ornamentation (Fig. 41). In thin section otoliths are seen to contain aragonitic crystals with the c-axis running parallel to the lenticular faces. Dark brown bands run approximately perpendicular to the aragonitic laths (DEGENS and others, 1969) and may represent daily growth lines (PANNELLA, 1971). The geological and taxonomic significance of otoliths has been discussed by CAMPBELL (1929) and SANZ-ECHEVERRIA (1949).

Otoliths are aragonitic and contain between 0.2 and 12% organic matter (VINOGRADOV, 1953; LOWENSTAM, 1963; DEGENS and others, 1969). LOWENSTAM (1963) reported that otoliths contain about 0.3% strontium and less than 0.01% magnesium. This magnesium content is the lowest amount found in any analyzed marine carbonate component. Iron and manganese are also exceedingly low

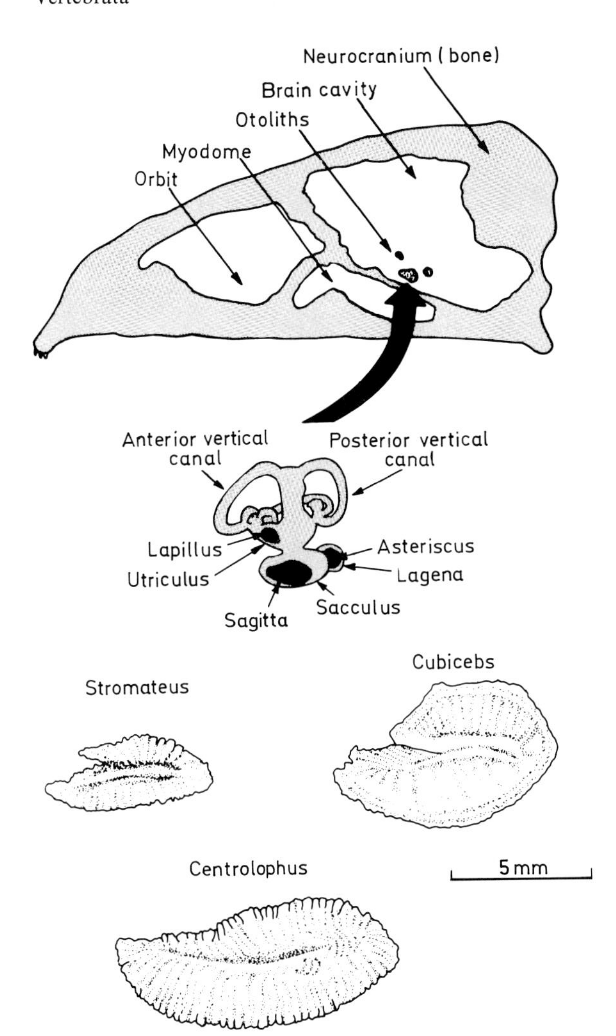

Fig. 41. Location of otoliths within the skull of bony fishes (above) and several representative otoliths (below). From DEGENS and others (1969)

(Table 39), but Na values can reach as much as 2 mole %; MORRIS and KITTLEMAN (1967) have suggested that the sodium may be present as either shortite ($Na_2Ca_2(CO_3)_2$) or pirssonite ($Na_2Ca_2(CO_3)_2 \cdot (H_2O)$). Oceanic otoliths generally have δC^{13} values of 0 to $-5^0/_{00}$ and δO^{18} values of $+2$ to $-1.5^0/_{00}$, suggesting that the source of the $CO_3^=$ is not metabolic but comes directly from ambient sea water (DEGENS and others, 1969).

Table 39. Composition of otoliths. Data from VINOGRADOV (1956) and LOWENSTAM (1963)

	Organic matter ($+H_2O$)	Mineral.	Percent				Parts per million				
			Ca	Mg	Sr	Na	Fe	Mn	K	S	P
Otoliths		Arag.	—	< 0.01	0.3						
Gadus	11.5	Arag.	38.75	0.04		0.33	20	54	750	76	310

Chapter 5. Summary of Carbonate Composition

This chapter presents a synthesis of the compositional data given in Chapters 3 and 4. Several problems, however, limit the thoroughness of such a discussion:

1. By now the uneven distribution of data concerning the composition of various carbonates should be apparent. Mollusks have been studied extensively, to the point where variations in composition can be related to various ecologic and physiologic factors. In spite of (or perhaps, because of) the many analyses, compositional trends within many pelecypod and gastropod shells are ill-defined. In contrast, the compositions of other carbonate contributors, such as serpulids and benthonic foraminifera, are poorly documented, and many further analyses are needed.

2. The distribution of strontium and magnesium in most carbonate components is well known (WASKOWIAK, 1962; LOWENSTAM, 1963, 1964a; DODD, 1967). Because of the small concentrations (and correspondingly more difficult analytical problems) trace element distribution is not as well documented.

3. Differences in analytical techniques (and accuracies) probably account for much of the spread in reported values. In an extreme example, WASKOWIAK (1962) found that the Sr content in the gastropod *Crepidula* averages 0.24%, but SEGAR and others (1971) reported an average of 0.01%. Other examples of analytical differences have been seen in the tables in Chapters 3 and 4.

Elemental Distributions in Carbonates

Calcium carbonate contains three major elements (calcium, carbon and oxygen), two minor elements (magnesium and strontium) and a considerable number of trace elements whose concentrations exceed 1 ppm. Generally the chemical composition of a carbonate reflects the incorporation of various cations and anions into the calcium carbonate lattice structure. However composition may also reflect: 1. incorporation of cations and anions into the organic matrix; 2. the presence of other mineral phases, such as brucite, dolomite, apatite, fluorite or complexed hydrates; 3. captured sea salts, such as NaCl or KCl; 4. adsorbed elements, such as iron and uranium, on the outer surface of shells (MCCAMMON and others, 1969; LAHOUD and others, 1966; SCHROEDER and others, 1970; LIVINGSTON and THOMPSON, 1971); or 5. captured foreign particles, such as grains incorporated by encrusting coralline algae or by accreting ooids. The presence of Al and Si is particularly diagnostic of foreign detrital grains.

Most trace elements are substantially more diffuse in shells than in the soft parts of the organism. For instance, only calcium, strontium and (perhaps)

Table 40. Distribution of major, minor and trace elements within the various hard and soft parts of the pelecypods, *Pecten maximus* and *Modiolus modiolus*. Data from SEGAR and others (1971)

		Pecten maximus						*Modiolus modiolus*				
		Muscle	Gut and digestive gland	Mantle and gills	Gonad	Upper shell	Lower shell	Muscle	Gut and digestive gland	Mantle and gills	Gonad	Shell
Percent	Ca	0.55	0.19	0.25	0.24	32.0	33.0	0.22	0.41	0.47	0.24	28.0
	Mg	0.20	0.20	0.40	0.30	0.077	0.059	0.40	0.53	0.57	0.59	0.014
	Sr	0.001	0.001	0.001	0.001	0.027	0.027	0.002	0.004	0.004	0.003	0.007
	Na	0.61	3.9	2.4	2.1	0.28	0.25	2.8	2.6	2.5	3.1	0.47
Parts per million	K	16000	3700	8800	10000	19	18	6400	4400	3900	5900	88
	P		3300	3000	8300	410	430	2300	4600	4200	8600	
	Fe	30	24.3	7.7	21.1	37	49	51	1000	220	130	15
	Mn	22	410	4.0	8.6	12	4.9	71	140	160	49	20
	Ni	1.7	0.96	0.82	1.5	1.2	2.4	2.0	0.59	3.2	1.2	0.20
	Zn	70	1100	420	180	4.1	6.1	68	190	180	140	6.2
	Cu	1.2	25	2.8	14	0.009	0.009	14	45	42	19	1.9
	Pb	17	1.7	1.5	0.40	0.62	0.60	7.5	22	18	25	1.9
	Co	0.34	0.68	0.58	0.18	0.20	0.20	0.30	1.1	0.30	0.72	0.30
	Cd	1.9	96	3.1	2.5	0.04	0.04	2.8	4.2	7.0	4.6	0.03
	Ag		8.9	0.36	0.059			0.03	0.81	0.24	0.07	
	Al	53	340	95	100	230	230	95	860	210	230	150

Table 41. Concentration factors of different cations present within various calcitic carbonates

	Coccoliths	Red algae	Benthonic Foraminifera	Planktonic Foraminifera	Porifera	Coelenterata	Bryozoa	Brachiopoda
Cations (in order of increasing ionic radius)								
B	7		5–15			1		
P							87–850	
Mg	0.8	20–70	0.5–40	0.1	20	30	15	2
Ni	2		<1	10				
Ca	3–5		3–5	3		2		
Fe		15–300	0.5–5	7–10	10	16–30	450	
Zn	0.3					5–7	2–3	
Mn	130	6–120	1–25	5	5	2–3		
V								
Na		0.2						
U								
Sr	0.2	0.3	0.2	0.1	0.3	0.2	0.3	0.2
Pb						300		
K		8						
Ba	6	0.01	1–7		0.7		1	

Table 42. Concentration factors of different cations present within various aragonitic carbonates

	Ooids	Aggregates	Coccoliths	Red algae	Brown algae	Green algae	Coelenterata	Bryozoa
Cations (in order of increasing ionic radius)								
B	12						4–20	
P	7						tr–465	87
Mg	0.4	4		6	3	0.6	1–2	1
Ni							1	
Cu	7	10					0.3	
Fe	10–30	30		20		16–360	1–17	100
Zn						3	0.4–7	
Mn	0.4	4		5		4–8	1–7	
V								
Na	0.3	0.3				0.2	0.4	
U	0.8					1	1	
Sr	1.3	1.2		1	1	1	1	1
Pb							20–330	
K	0.1	0.2						
Ba	0.6						0.3–3	0.7

magnesium are present in higher concentrations in mollusk shells than in corresponding soft parts; all other elements occur in lower concentrations (Table 40).

One way of documenting elemental distributions is by calculating their concentration in the carbonate and in sea water relative to the calcium concentrations in the carbonate and sea water:

Table 41. Continued

Mollusca						Serpulidae	Arthropoda			Echinodermata	Concentration factor
Polyplac.	Scaphopoda	Pelecypoda	Gastropoda	Pteropoda	Cephalopoda		Cirripedia	Decopoda	Ostracoda		
		1–2				1	0.4	0–3		7	$\times 10^{-3}$
		61						2100		100–800	$\times 1$
		2				10–40	2–3	20		10–40	$\times 10^{-3}$
		1–4			9	5–20	7	5		1	$\times 1$
		0.4				1–2	1	0.6		7	$\times 1$
		12–26				23–47	15	2–10		2–27	$\times 1$
		0.9								1–2	$\times 1$
		15–110				11–32	30–100	10–20		4–17	$\times 1$
		50									$\times 1$
		0.3				0.04	0.2			0.6	$\times 10^{-3}$
		0.1					0.05				$\times 1$
		0.2			0.5	0.3	0.3	0.5		0.3	$\times 1$
		30–600				70–330	100	30			$\times 1$
						0.3–0.6					$\times 10^{-3}$
		0.3			0.9	1	1	0.2		0.4	$\times 1$

Table 42. Continued

Mollusca						Serpulidae	Arthropoda	Chordata	Vertebrata	Concentration factor
Polyplac.	Scaphopoda	Pelecypoda	Gastropoda	Pteropoda	Cephalopoda		Cirripedia			
		0.4–1	0.2	10						$\times 10^{-3}$
		1–3	1–3		240				30	$\times 1$
0.8	0.4	0.2	0.5	1	0.4	1	0.6	1	0.3	$\times 10^{-3}$
		1–4	1–4	1						$\times 1$
		0.3	0.3	7	0.6					$\times 1$
		8–34	1–26		3	27–92			2	$\times 1$
		0.3–16	0.06–5							$\times 1$
		5–10	2–30		2	18–28			27	$\times 1$
		50		45						$\times 1$
		0.2	0.2			0.04				$\times 10^{-3}$
		0.1	0.1							$\times 1$
0.8	0.2	0.2–0.4	0.2–0.3	0.1	0.4	1	1	1	0.3	$\times 1$
		30	100	6700	4200					$\times 1$
									3	$\times 10^{-3}$
0.4	0.3	0.3	0.4							$\times 1$

$$\frac{(\text{Conc. element in carbonate/Conc. Ca in carbonate})}{(\text{Conc. element in seawater/Conc. Ca in seawater})}$$

If the dimensionless concentration factor of any particular element is greater than 1, it means that element is concentrated in the carbonate relative to calcium.

Similarly, concentration factors less than 1 indicate discrimination against that particular element. The concentration factors of the minor and trace elements within calcite and aragonite phases of the various inorganic and organic carbonate components are shown in Tables 41 and 42. From these data several generalizations can be made:

1. The concentration values for many elements, notably strontium, barium, uranium, copper and nickel, are near unity in one or both mineral phases; that is, partitioning of these elements relative to calcium appears to be minor. Other elements tend to be concentrated relative to calcium (manganese, iron and lead) and four elements (boron, magnesium, sodium and potassium) are notably excluded. Special mention should be made of the high concentrations of lead within carbonates. Most organisms concentrate lead over 100 times relative to calcium, but in pteropods and cephalopods the concentration factors are 7600 and 4200 respectively (Tables 41 and 42).

2. The range of values of some trace elements, such as iron and manganese, tends to be much greater than for the minor elements strontium and magnesium. For instance, strontium and magnesium content within shells of similar mineralogies seldom varies by more than a factor of 2 or 3 and usually much less. Iron and manganese concentrations, on the other hand, may vary by as much as two orders of magnitude, and one order of magnitude variation within a single genus is not uncommon.

3. The role of sodium in carbonates is not understood at present. Thermodynamic considerations suggest that Na^+ should occur in very minor amounts, but compositional analyses of various marine carbonates, and experimentally precipitated carbonates (KINSMAN, 1970) show Na values as great as 5000 ppm. The absence of Cl (as NaCl) has led KINSMAN (1970) to suggest that Na is present in a carbonate phase, but more recent data by G. MÜLLER (1970, oral communication) indicates that the Na probably occurs as a complexed hydrate within the carbonate lattice.

4. Sulfur and fluorine are mostly incorporated into compounds other than $CaCO_3$, presumably the sulfur is present as $SO_4^=$. Some fluorine may be present as sea salt, but the Fl/Cl enrichment factor is greater than 10^5 (CARPENTER, 1969) suggesting that most fluorine is incorporated into another form, perhaps fluorite (LOWENSTAM and MCCONNELL, 1968).

Variations in Composition with Mineralogy

Elemental composition depends strongly upon the mineralogy of the skeleton. Cations with large ionic radii, such as strontium and to a lesser extent barium, lead and uranium, tend to be more concentrated in aragonite than in calcite. Elements with small ionic radii (such as magnesium, manganese, iron, nickel and phosphorous) prefer calcite (Table 41). The relation between ionic radius and mineralogy seems logical in view of the lattice structures of aragonite and calcite (see above).

Strontium values are 3 to 5 times higher in most aragonites than in calcites. Values are greatest in oolite and grapestone (0.9 to 1.1 %) and somewhat lower in green algae, coral, bryozoans, cirriped basal plates and ascidian spicules. Only the mollusks (other than polyphacophorans) and vertebrates are able to discriminate effectively against strontium; their shells generally contain between 0.2 and 0.3% Sr. With the exception of cephalopods and cirripeds, strontium contents in calcites mostly are less than 0.3%.

Magnesium values within aragonites are uniformly low, never more than 0.5% by weight, and generally less than 0.25%. Magnesium concentrations in calcites can be either high or low, but seldom in between. Cocoliths, planktonic foraminifera, brachiopods and calcitic mollusks and arthropods (except decapods) generally contain less than 1 mole % $MgCO_3$. In contrast, red algae, many benthonic foraminifera, sponges, octocorals, bryozoans, decapods and echinoderms generally contain more than 8 mole % $MgCO_3$ (2% magnesium by weight). The only organisms that contain intermediate amounts of magnesium are the cephalopod *Argonauta* and some benthonic foraminifera.

Although mineralogy is the most important parameter in determining the elemental composition of a carbonate, other factors, such as phylogeny, the environment, mode of living, ontogeny, and biochemical and physiological considerations are also important.

Physiologic and Generic Factors

The mineralogy and elemental composition of any carbonate is partly dependent upon the mode of precipitation. The exact method of calcification is still debated. Most workers have supported an enzyme theory (the presence of various enzymes, such as carbonic anhydrase or phosphorylase, determines the site of calcification) or the template theory (the nucleation and form of the seed crystal is related to the composition of the organic template or matrix upon which it forms), but some workers have favored crystal poison or electrochemical models.

At present many data favor the template theory of calcification, although not all organisms necessarily utilize this technique (for example, *Halimeda* may precipitate physiochemically during photosynthesis; WILBUR and others, 1969). Those workers who support the template model point to the fact that all carbonates, even ooids (MITTERER, 1968), contain a matrix of protein and amino acids. According to this model, shell nucleation involves the coordination of Ca ions to free carboxyl groups of acidic amino acid residues in the matrix protein (DEGENS, 1967; DEGENS and others, 1967; MATHEJA and DEGENS, 1968; HARE and ABELSON, 1965; JACKSON and BISCHOFF, 1971). The spatial sequence of specific peptide chains and the nature of the organic matrix can determine the micro-architecture (HARE and ABELSON, 1965) as well as the mineralogy of the precipitated shell. MATHEJA and DEGENS (1968) have explained this latter phenomenon by the influence of oxygen absorption by various proteinaceous compounds upon the coordination polyhedra; $Ca^{+2}O_6$ will form calcite and $Ca^{+2}O_9$ will favor aragonite. The differences in the ratios of protein compounds from one species to another also may have phylogenetic implications (HARE and ABELSON, 1965; DEGENS and others, 1967; GHISELIN and others, 1967).

In many primitive phyla calcification is extracellular so that the concen-

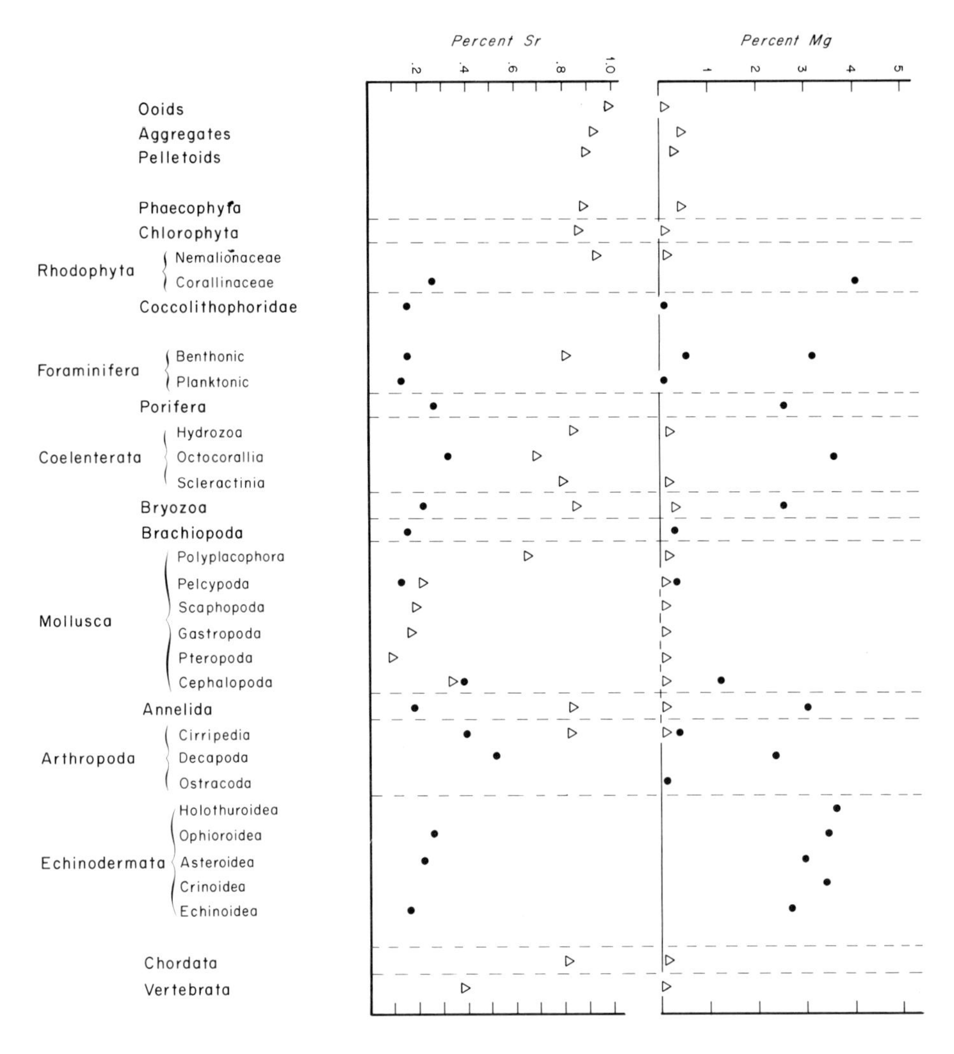

Fig. 42. Average Sr and Mg contents in the various marine carbonate components. Open triangles represent values from aragonitic carbonates; solid circles, calcite and magnesian calcite. Data are from the tables in Chapters 3 and 4

tration of trace elements is controlled partly by physiochemical processes (KINSMAN, 1969b). In contrast, advanced organisms, mollusks being the best example, calcify by passing cations through cellular tissue; this process can be very effective in partioning various trace elements.

The two elements most easily related to physiological processes are strontium and magnesium (Fig. 42), although other elements such as iron also may be dependent upon the complexity of calcification (Tables 41 and 42). LOWENSTAM (1963, 1964a, b) demonstrated that more phylogenetically advanced organisms contain less strontium, the strontium being preferentially excluded by more complex calcification processes. Excluding mollusks from this picture, one can see a slight decrease in strontium with increasing physiologic complexity (Fig. 42); mollusks appear to be totally unique in their ability to selectively discriminate against strontium (LIKINS and others, 1963). Within calcitic organisms, one can note a slight tendency towards *increased* strontium content with greater complexity (Fig. 42).

CHAVE (1954) showed a trend of decreasing magnesium content within magnesian calcites in more complex organisms. No such tendency is seen within the low magnesian calcites, but a slight trend of decreasing magnesium is noted in the aragonites (Fig. 42).

Stable isotope concentrations vary great within different skeletal groups (Fig. 19), and in part reflect different sources of oxygen and carbon utilized during calcification. Red algae, corals and echinoderms are out of isotopic equilibrium with their ambient environments, suggesting that these groups can utilize metabolic CO_2.

Mineralogical (and therefore chemical) content also can vary on order and class levels. For instance, octocorals are mostly calcitic, while hydrozoans and scleractinians are aragonitic. Porcellaneous foraminifera tests are high-magnesian calcite; hyalines can be low or high magnesian calcite, or even aragonite. Simple bryozoans are phosphatic and calcitic, while the more complex types contain mixtures of aragonite and calcite. Several other examples can be found in the previous sections.

Environmental Considerations

LOWENSTAM (1954a, b; 1964b) illustrated the dependence of mineralogy upon water temperatures. In species with mixed mineralogies, aragonite content increases with warmer waters, and calcite (either low or high magnesian) increases in colder waters. This relation holds for some species of gorgonians, bryozoans, pelecypods, gastropods, serpulids and cirripeds[1]. Furthermore, two of the common aragonite contributors, shallow water corals and codiacean algae, are restricted to tropical and subtropical waters. Since trace element composition depends partly upon mineralogy one would expect less Mg, Fe and Mn, and more Sr, Ba, U and Pb in carbonate sediments deposited in warmer waters. This assumption, however, is not totally valid, since the chemical composition of aragonite and calcite also will change with temperature. Thus, for instance, the increase magnesium in warm-water magnesian calcite may offset the decrease in calcite content within a sediment.

The temperature of the ambient environment exerts an effective control on the magnesium content in most calcitic organisms; higher temperatures mean higher magnesium content (CLARKE and WHEELER, 1917; CHAVE, 1954). The importance of temperature-related phenomena, such as growth rate, is not known (MOBERLY, 1968). A graphical presentation of magnesium content versus water

1 In many invertebrates, however, the absolute mineralogy is more generically dependent than environmentally controlled. Thus, although most tropical serpulids contain aragonite and most coldwater serpulids are calcitic, the absolute mineralogy at any locality depends greatly upon the genus (Fig. 32).

temperature was given by CHAVE (1954) (Fig. 43), but these trends must be re-evaluated in light of more recent work. For instance magnesium content in coralline algae increases at a much faster rate than does the modal $MgCO_3$ content (Fig. 21). In foraminifera, echinoids and serpulids generic factors play significant roles in determining the magnesium content at any given temperature (Figs. 23, 37 and 32). Similar generic factors probably apply in other invertebrate skeletons.

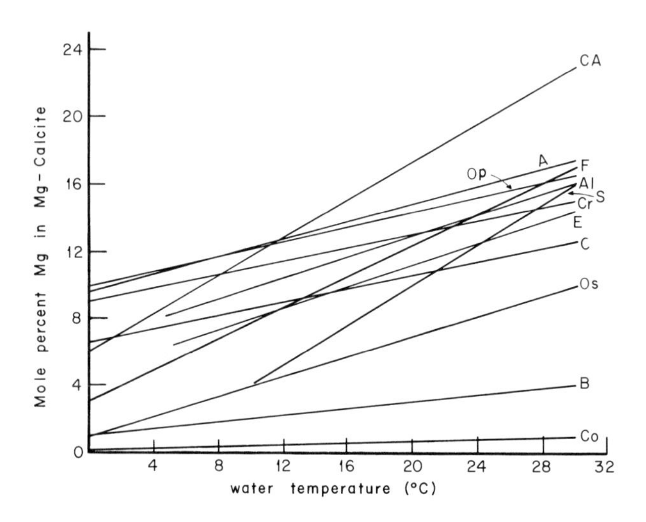

Fig. 43. Variation of Mg content in various invertebrate skeletons, according to CHAVE (1954). CA = coralline algae; Ast = asteroids; F = foraminifera; Op = ophiuroids; Al = alcyonarians; S = sponges; C = crinoids; E = echinoids; D = decapods; Os = ostracods; B = barnacles; Co = corals

Strontium decreases with increasing temperature in some aragonitic molluscan shells and increases with increasing temperature in some calcitic species (Fig. 29), but in most organic carbonates the strontium content remains more or less constant with temperature (THOMPSON and LIVINGSTON, 1970). The correlation between temperature and the concentration of other trace elements has not been studied in great enough detail to allow generalizations.

Salinity effects upon mineralogy are poorly documented, but aragonite content within some mollusks increases with decreasing salinity. In some mollusks Sr, Mg and Mn decrease with increasing salinity, but this relation must reverse itself at some point, since Sr and Mg concentrations in fresh-water shells tend to be lower than in saline waters. In echinoids Mg and Na decrease and Fe, Sr and Mn remain about constant with decreasing salinity; somewhat similar trends are seen in barnacle shells. According to FRIEDMAN (1969) Ba, Fe and Mn contents in mollusk shells are greater in fresh-water than marine. The increased concentrations of both Fe and Mn in reduced ionic states within stratified brackish or fresh-waters would explain this trend (F.T. MANHEIM, 1971, oral communication).

Ontogeny

The composition of some skeletons and shells changes with the development of the individual organism. The variation in magnesium content with the age of echinoid spines and tests (see p. 131 ff.) is a particularly good example. DODD (1965, 1967) has noted elemental variations within different age mollusks, but at present such trends are not clear.

Morphology

Mineralogy and composition in some organically-derived carbonates varies between different morphological parts, or even within a single part. The example of the aragonite distribution with cheilostome ectoprocts was cited in Chapter 4. In mollusks, certain shell structures are characterized by specific mineralogies. Even in organisms with a constant mineralogy, for instance pelecypod shells, elemental concentrations may change within the different layers.

In each instance we probably are witnessing the effect of skeletal precipitation by different organs or cells, each of which has a different effect on mineralogy or elemental partitioning. For example, bryozoan calcite walls are precipitated by the epithelium, while the superficial aragonite skeleton is formed by outfoldings of the epifrontal membranes (BOARDMAN and CHEETHAM, 1969).

Mode of Living

The most pure skeletal compositions among marine organisms occur in planktonic forms: coccolithophorids, foraminifera and pteropods. Cephalopods contain slightly higher elemental concentrations. On the other hand, lead concentrations within pteropods and cephalopods (concentrations within other planktonic skeletons have not been reported) are extremely high. Similarly, although planktonic foraminifera contain considerably less Mg and Sr than do benthomic forms, they do contain higher concentrations of Fe, Mn, Cu and Ni (see Table 21).

At present no comparison can be made between the composition of infauna and epifauna, but one could assume that these different life forms, with different sources of food and water, and different amounts of sediment ingestion, might show some compositional variations that reflect their different modes of living. The fact that serpulids are exclusively epifauna, may imply that the mode of living is critical in skeletal precipitation. PILKEY and HARRISS (1966) have suggested that the trace element concentration in barnacle shells can be influenced by their relative position within the intertidal zone.

Part III. Marine Carbonate Sedimentation

Carbonate sediments accumulate in nearly all parts of the world oceans. Their petrographic and compositional characteristics, however, depend upon the environment of deposition. Several environmental parameters control carbonate deposition: temperature, light, sedimentation, salinity, pressure and water depth. Of these the two most important parameters are temperature and water depth (which to a greater or lesser extent controls light, pressure and temperature).

Temperature not only defines the degree of saturation of calcium carbonate in sea water, but it also regulates the distribution of organisms which contribute calcium carbonate to the sediment. The distribution of mean surface temperatures throughout the world oceans is shown in Fig. 44 and the various ecologic zones are described in Table 43 and shown in Fig. 45.

Table 43. Temperature limits of various biogeographic zones (after VAUGHAN, 1940)

Temperature range	Biogeographic zone
>25 °C	Tropical
15–30 °C	Subtropical
10–25 °C	Temperate
5–10 °C	Subpolar
<5 °C	Polar

On the basis of water depth, the oceans can be divided into three zones:

1. Eulittoral-refers to shallow-water environments. SVERDRUP and others (1942) defined this zone as extending from the intertidal zone to a depth of about 50 m. However in discussing shallow-water carbonates, the maximum depth of vigorous hermatypic coral growth and non-skeletal sediment formation (about 15 to 20 meters) seems to be a more reasonable lower limit for this zone.

2. Sublittoral-represents a transition from the shallow-water environment to the deep sea. Decreases in light and temperature (and thus in photosynthesis) and increases in pressure characterize this zone. This depth range generally includes the continental self and upper continental slope.

3. Deep sea-refers to all depths greater than 200 meters. Photosynthesis is practically non-existent and carbonate-producing benthic communities are limited in their distribution. Deposition is predominantly pelagic.

Carbonate sedimentation within these three depth zones is discussed in the following three chapters.

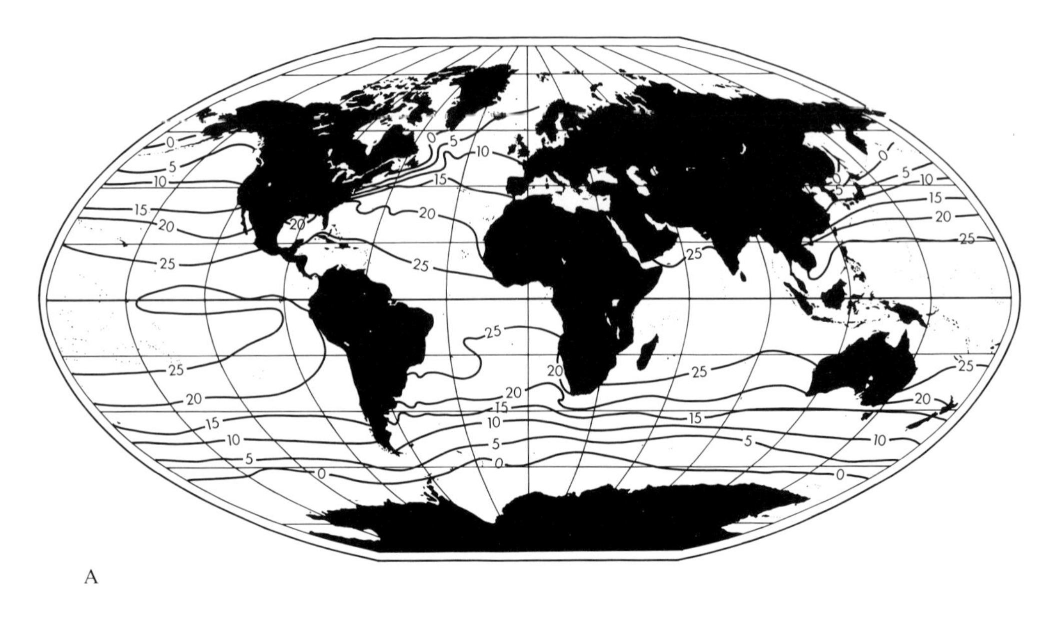

A

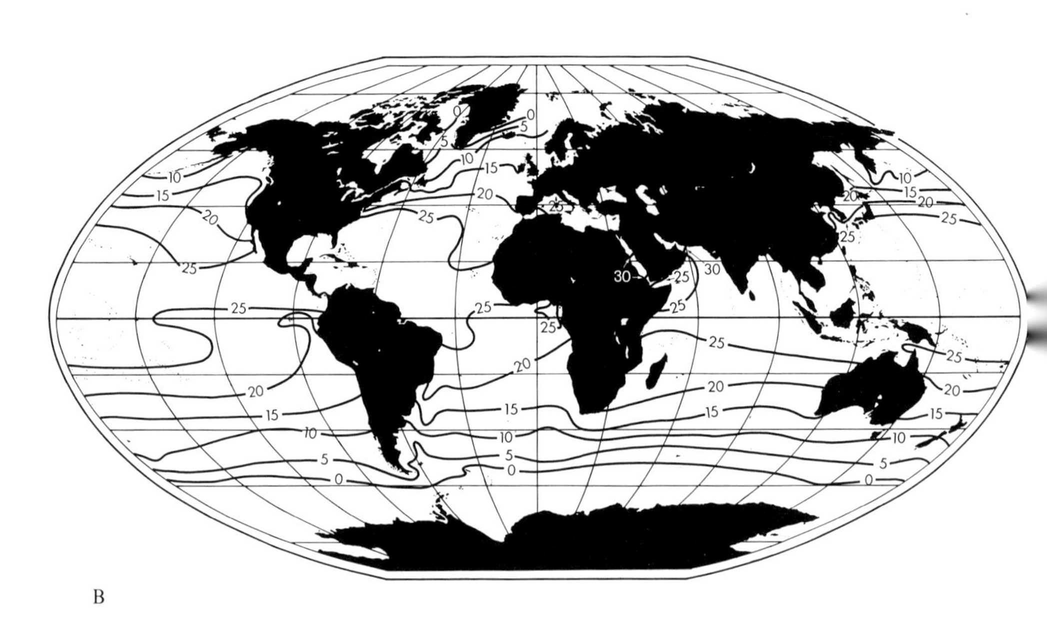

B

Fig. 44 A and B. Surface temperatures (in °C) throughout the world oceans in February (A) and August (B). After SVERDRUP and others (1942)

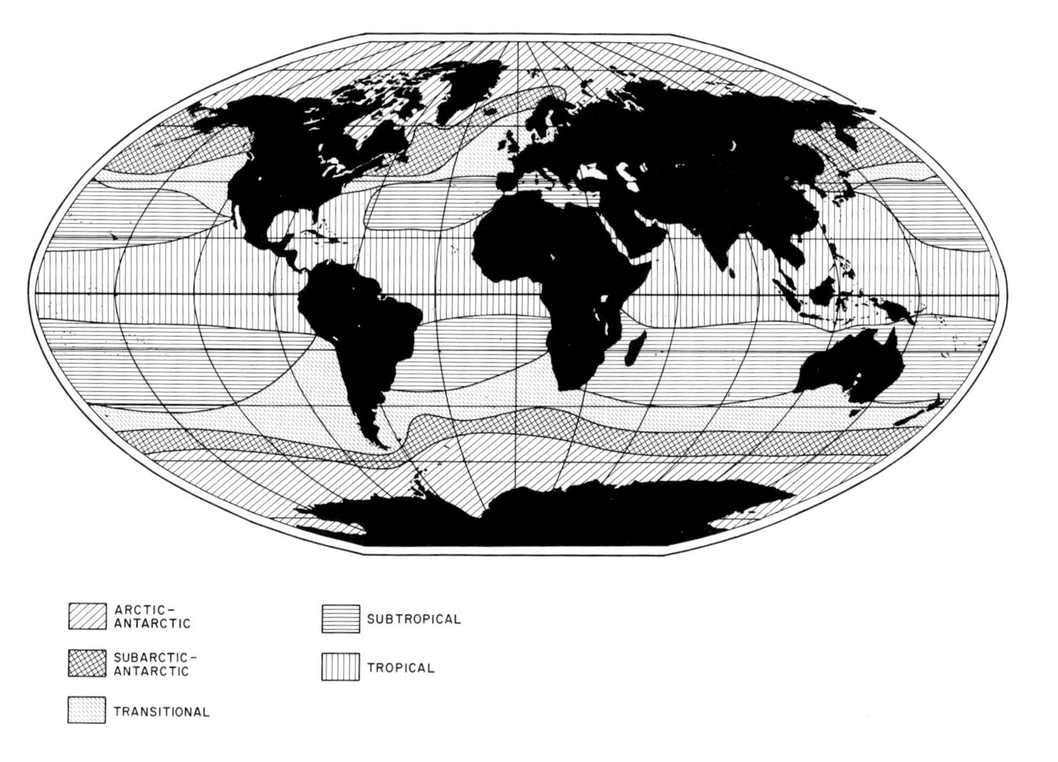

Fig. 45. Distribution of the temperature-related biologic zonations in the world oceans. After BÉ and TOLDERLUND (1971)

Chapter 6. Eulittoral (Shallow-Water) Sedimentation

Rapid shallow-water carbonate sedimentation generally is restricted to tropical and subtropical climates. Some of the major carbonate contributors, such as hermatypic corals and calcareous green algae are limited to warmer climates. In addition, warmer waters are more favorable for non-skeletal precipitation. This is not to say that carbonate-rich shallow-water sediments are limited to tropical climates. Serpulid and vermetid reefs, as well as high latitude eulittoral organisms, can create carbonate-rich deposits in colder climates. Some examples will be given in the latter part of this chapter.

The Coral Reef Environment

Of all the sites of modern carbonate deposition, the best known is the tropical coral reef environment. Within this discussion, the coral reef environment is viewed within rather broad limits. It includes not only coral reefs but also the lagoons adjacent to them, as well as many bank areas upon which oolite, aggregates and lime muds can accumulate. Some other reasons for viewing reef and non-skeletal sediments as a quasi-continuum have been given in Chapter 3, and others will be discussed in this chapter.

The extensive amount of coral reef research has been related to several factors. First, for the geologist, modern coral reefs represent probable analogs with ancient reefs, many of which are reservoirs for petroleum. Second, and of historical interest, coral reefs were strategically important in the Pacific Ocean during the Second World War, and were the sites of nuclear tests from the late 1940's to the 1960's. Third, the availability of coral reefs and their classroom and textbook examples of organism-sediment interrelationships provide unique opportunities for scientists and students. Fourth, but perhaps not least important, coral reefs happen to be located in particularly pleasing areas, such as Tahiti, the Marshall Islands and the Bahamas. It is questionable whether these reefs would be nearly so well described if they were restricted to less pleasant and romantic environments.

Early reef research in the mid and late 19th century was mainly descriptive; the studies of DARWIN, DANA, WOOD-JONES, WALTHER, GARDINER and CROSSLAND are particularly noteworthy, as they described modern reefs in such detail and with such insight that their reports are still used by many reef workers. Later work, especially that at the Tortugas Laboratory in the Florida Keys, at Palau in the equatorial Pacific and on the Great Barrier Reef, emphasized the biology of various reef organisms and tried to relate biological, chemical and geological parameters to an integrated understanding of the reef itself.

Within the past 20 years the study of tropical carbonate sedimentology has advanced rapidly, especially in the Caribbean. In particular, the studies of reef disintegration and cementation and delineation of deep fore-reef facies have revolutionized many of our concepts about reef formation and diagenesis. These significant advances, particularly within the past 5 years, have been the result of both technological advances in equipment used in underwater observations and collections and the improved technologies used in describing and analyzing the rock and sediment samples (such as stable isotopes, X-ray diffraction, carbon-14, sediment impregnation for subsequent thin-sectioning and scanning electron microscopy).

The literature concerning coral reefs and their lagoons exceeds all other carbonate literature combined. Excellent reviews have been given by numerous authors (DAVIS, 1928; GARDINER, 1931; VAUGHAN, 1933; KEUNEN, 1950; WELLS, 1957a; LADD, 1961; YONGE, 1940, 1963; STODDART, 1969; and NEWELL, 1971) and a two-volume text on the geology and biology of coral reefs is forthcoming (ENDEAN and JONES, 1973). Bibliographies by WELLS (1957b), RANSON (1958) and MILLIMAN (1965) have incorporated most of the pertinent literature to 1964. Particularly excellent reviews of the carbonate environments and sediments in Florida and the Bahamas have been given by BAARS (1963), GINSBURG (1964), PURDY and IMBRIE (1964), MULTER (1969) and BATHURST (1971). Much of the biologic, geologic and oceanographic information from Pacific reefs has been presented by WIENS (1962). A recent symposium on Indian Ocean coral reefs (STODDART and YONGE, 1971) has closed the gap in knowledge between this relatively undocumented ocean and the Atlantic and Pacific.

Environmental Factors Governing Coral Reef Distribution

Although corals often are not the dominant carbonate components within a coral reef (coralline algae usually are more common), they are the critical organism in reef formation, since their skeletons form the basis of the reef framework. The limitations of coral growth with respect to water depth, temperature and salinity, therefore, determine the distribution of coral reefs:

1. Temperature. The ideal temperature range for optimum coral growth is about 23 to 27 °C. At higher and lower temperatures the corals begin to lose their ability to capture food (MAYER, 1915; EDMONDSON, 1929). In general, reef corals do not grow actively in areas where winter temperatures fall appreciably below 18 °C (VAUGHAN, 1919a; WELLS, 1957a). Some species can survive far lower temperatures; for example, *Solenastrea* occurs on the North Carolina shelf, where water temperatures can be as low as 10 °C (MACINTYRE and PILKEY, 1969), but growth is restricted to individual colonies rather than anastomosing reefs. Because waters are warmer on the western sides of ocean basins, most coral reefs are confined to these areas (Fig. 46). The upper temperature limit is about 30 °C, although some genera, such as *Porites*, can survive considerably higher temperatures (KINSMAN, 1964b).

2. Water Depth. Because hermatypic corals rely upon symbiotic zooxanthellae for rapid calcification (Chapter 4), reef corals generally are limited to

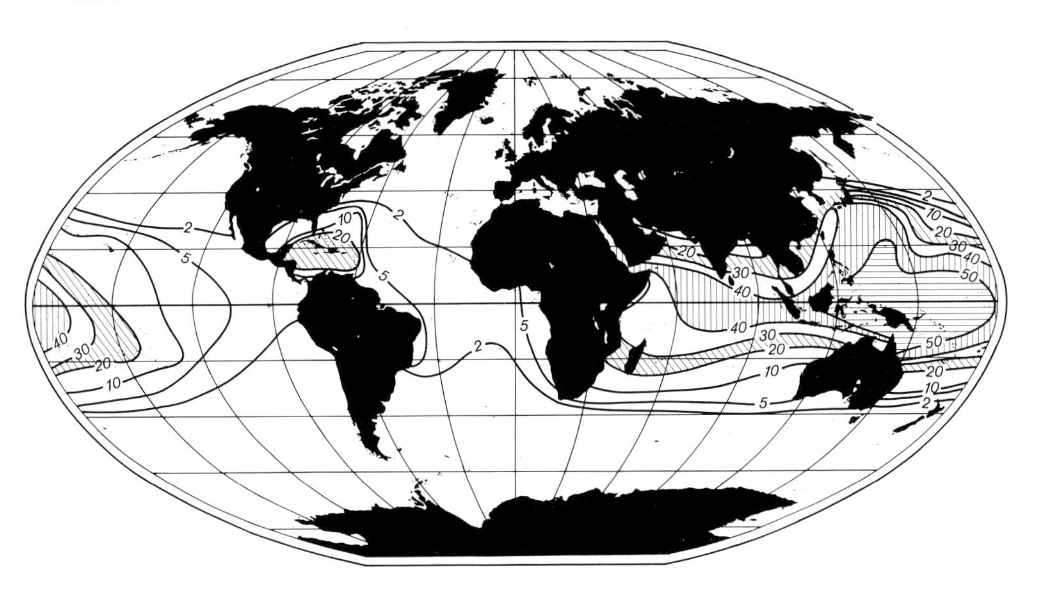

Fig. 46. Distribution of number of hermatypic coral genera in the world oceans (after STODDART, 1969). The large number of genera in the western parts of the tropical Atlantic and Pacific oceans and throughout the equatorial Indian Ocean coincide with the location of the most prominent modern coral reefs

those shallow depths at which zooxanthellae can photosynthesize. WELLS (1957a) reported that the number of coral species at Bikini Atoll decreases markedly with increasing depth and decreasing light intensity. More recent work in the Caribbean by the late T.F. GOREAU and in the Red Sea by Israeli scientists, however, shows that the total number of hermatypic species may not decrease until depths of at least 30 to 40 m. In fact, within this depth range the total number of species may even increase (Y. LOYA, 1971, oral communication). Perhaps the rapid decrease of species with depth described by WELLS was an artifact of sampling methods rather than actual distribution.

Although species diversity and abundance may not decrease markedly with depth (at least in the upper 30 or 40 meters), the rigidity of the corals and the rates of growth are reduced. Deep-water hermatypic corals tend to be flattened (thus utilizing the available light more efficiently) and therefore are more susceptible to bio-erosion and degradation (GOREAU and HARTMAN, 1963; GLYNN, 1973).

3. Salinity. Corals normally grow in salinities between 30 and $40^0/_{00}$. Severe evaporation (such as in the Persian Gulf, KINSMAN, 1964b) or exposure to torrential rains or river outflow (SQUIRES, 1962) may affect species distribution or even the entire reef development. Torrential rains can kill coral reefs (SLACK-SMITH, 1959) both by lowering salinities and by increasing terrigenous sedimentation near high-standing islands.

4. Turbidity and Wave Action. Corals thrive in areas of high wave action, because of the increased supply of both food and oxygen. Similarly, the crashing waves prevent silt accumulation which could suffocate the corals. Corals can

use two other mechanisms to combat heavy sedimentation: 1. branching types generally do not accumulate sediment simply by virtue of their shape; 2. some corals, such as *Porites*, use ciliary action to remove physically the sediment grains from the polyps (MANTON and STEPHENSON, 1935). Even the most hardy corals, however, can be killed by heavy sedimentation resulting from hurricanes or floods.

Reef Types and Their Evolution

Coral reefs commonly are classed into fringing reefs, barrier reefs and atolls (Fig. 47). Fringing reefs lie adjacent to land, with only shallow water separating them from the shore. Barrier reefs lie further offshore, with intervening lagoon

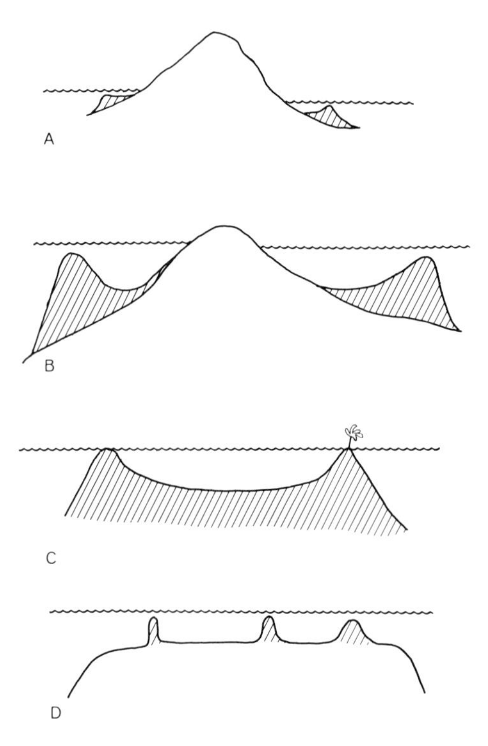

Fig. 47 A–D. Comparison between fringing reefs (A), barrier reefs (B), atolls (C), bank reefs (D)

depths generally greater than ten meters. Atolls have lagoons enclosed by reefs, with only low-lying carbonate cays exposed. This simplistic classification, however, does not take into account special reef forms, such as bank reefs, which form on bank tops. MAXWELL (1968a) has proposed that the terms atoll, barrier reef and fringing reef be restricted to oceanic environments, and be differentiated from reefs near high-standing land, where physiographic conditions (such as increased runoff and terrigenous sedimentation) are critical (TAYLOR, 1968). MAXWELL (1968a, p. 101) has classified the reefs within the Great Barrier Reef Complex of Australia into 14 reef types.

CHARLES DARWIN (1851) theorized that atolls represented the final stage of reef formation on a subsiding volcanic foundation; the first stage involves the formation of fringing reefs, then with increased subsidence, a barrier reef, and finally an atoll. Although many alternative theories were offered (summarized by DAVIS, 1928), the only theory which offered a serious alternative to DARWIN was developed by REGINALD DALY (1910) who concluded that atolls were formed by the transgression of post-glacial seas over wave-eroded platforms. Thick vertical sequences of reef rock present on many Pacific atolls, plus the finding of volcanic rock at the bottom of test holes at Eniwetok, Mururoa and Midway atolls (LADD and SCHLANGER, 1960; LADD and others, 1967), however, has substantiated DARWIN'S basic postulate, although the subsiding basement need not be volcanic (HOFFMEISTER and LADD, 1935, 1944). In fact, in most nearshore areas, the subsiding basement is non-volcanic, for example the structural control of the British Honduras atolls (STODDART, 1962), the terrigenous basement of the Great Barrier Reef (MAXWELL, 1968a) and the 6–10 km of carbonates underlying the present-day Bahamas (UCHUPI and others, 1971).

On the other hand, there is no question but that modern reefs possess many remnants of pre-existing low stands of sea level. MACNEIL (1954) and EMERY and others (1954) suggested that lagoons may actually represent karst topography formed by meteoric waters during glacially-lowered sea level. Recent echo soundings in some Caroline Island lagoons (SHEPARD, 1970) show substantial evidence of karst-like features. Furthermore, many modern bank and barrier reefs probably formed on pre-existing topographic foundations since the last rise of sea level. For instance, many of the modern reefs along the Florida Keys and in the Bahamas may be shallow cappings on top of Pleistocene bedrock (NEWELL, 1959).

Indo-Pacific versus Caribbean Reefs

In terms of general environment and component species, the reefs in the Indian and Pacific Oceans are roughly similar and quite distinct from those found in the Atlantic (Caribbean). Indo-Pacific reef corals total more than 80 genera, while only about 20 genera are present in the Caribbean (VAUGHAN, 1919b; WELLS, 1957a; GOREAU and WELLS, 1967) (Fig. 46). The total number of hermatypic scleractinian species in the Caribbean is about 50 compared with more than 700 in the Indo-Pacific. The Caribbean contains only three species of *Acropora* and four species of *Porites*, as compared to 150 and 80 species of these corals in the Indo-Pacific (WELLS, 1957a). Many other prominent reef organisms, such as the octocorals *Heliopora* and *Tubipora*, mollusks *Tridacna* and *Hippopus*, and some coralline algae (notably *Porolithon onkodes*) are also lacking in the Caribbean (EKMAN, 1953; WELLS, 1957a). Similarly, there are about 300 atolls in the Indo-Pacific, as compared to only about 10 in the Caribbean (BRYAN, 1953; MILLIMAN, 1973).

The differences in reef populations and reef zonations (see below), plus the assumed absence of algal ridges in the Caribbean, have led many workers (especially those who have concentrated on Indo-Pacific studies) to conclude that Caribbean reefs are poor cousins in terms of both morphology and growth (for example, see WELLS, 1957a; NEWELL, 1959; HELFRICH and TOWNSLEY, 1963). Poor reef development, however, seems to be limited to the northern Caribbean (Florida and the Bahamas are the most well-studied areas and the areas which have been quoted by many workers, such as WELLS and NEWELL, as possessing typical Caribbean reefs). Reefs in the warmer, southern Caribbean show development and growth that are comparable with Indo-Pacific counterparts (GOREAU and WELLS, 1967; MILLIMAN, 1969a; GLYNN, 1973).

The difference in species distribution between the Indo-Pacific and the Caribbean can be explained by both Tertiary history and size of the two biogeographical areas. During the Mesozoic and early Tertiary the Pacific and Atlantic were connected by the Tethyan Sea, which stretched across what is now Central America. Reef genera in these two oceanic areas were closely related. Since the closing of the Isthmus of Panama in the Miocene, however, organisms in the two areas have evolved separately. Indo-Pacific corals evolved from the eastern Tethys genera, while Atlantic corals represent western Tethys genera (NEWELL, 1971). The total area within the tropical Indo-Pacific is considerably greater than in the tropical Atlantic, thus favoring increased speciation. In addition, the tropical Atlantic, being closer to continental land masses, experienced greater climatic fluctuations during the late Tertiary and Quaternary than did the more oceanic Indo-Pacific (EMILIANI and FLINT, 1963). All of these factors have favored the increased speciation in the Pacific and a decrease of species in the Atlantic. Thus many scleractinian genera, such as *Stylophora*, *Pocillopora*, *Goniastrea*, *Goniopora*, *Pavona* and *Seriatopora*, have vanished from the Caribbean but remain important members of the Indo-Pacific reefs (WELLS, 1957a).

Morphology of the Coral Reef Habitat

Any particular environment within the coral reef ecosystem is a function of bottom topography, wave energy, sedimentation rate, oxygen supply, diurnal temperature range and many other factors. As these factors change, so do the environment, the component organisms and the sediments. The atoll, being an ideal self-contained biotope (WELLS, 1957a), is an excellent example of the coral reef habitat (Plate XVII), although most fringing and barrier reefs display similar zonations. The zonations of both West Indian and Indo-Pacific reefs are discussed in the following paragraphs. For the sake of this discussion, the coral reef habitat has been subdivided into two distinct environments: reefs (generally hard substrate) and lagoons (generally sandy to muddy substrate).

Reef Environments

Morphology and Ecology

Reef Front. The reef front (fore reef) extends from the windward edge of the reef flat to the lower limits of coral growth (Fig. 48). As mentioned in earlier paragraphs,

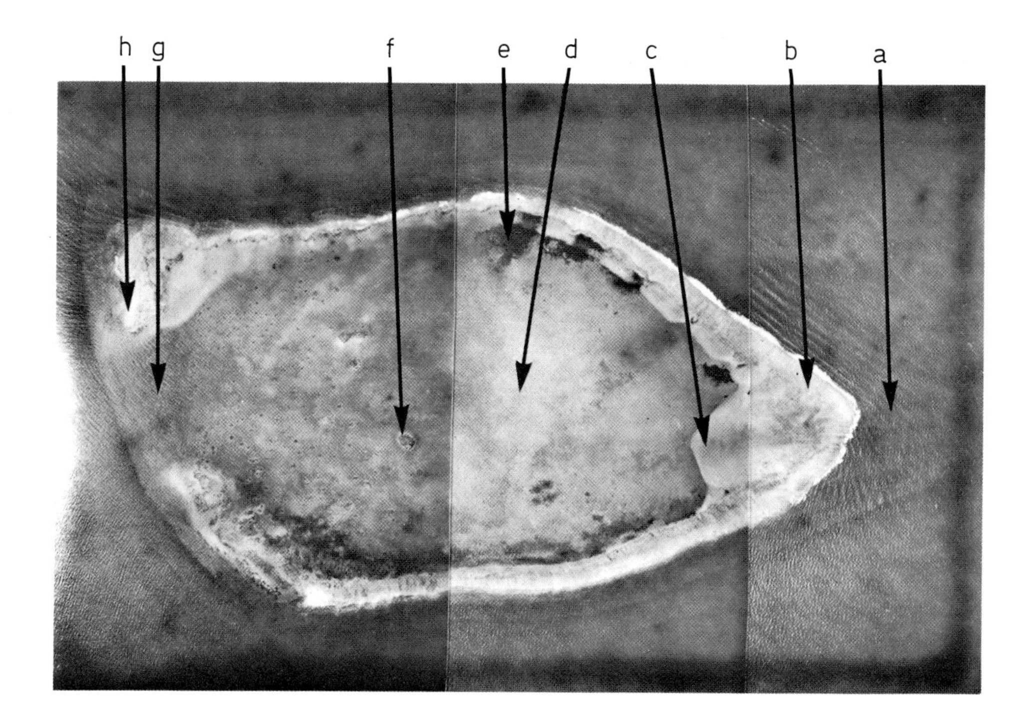

Plate XVII. Physiographic Zones of a Coral Atoll. This aerial photomosaic of Hogsty Reef, in the southeastern Bahamas, provides a good example of the various physiographic zones within a coral atoll: (a) fore reef; (b) peripheral reef flat; (c) inner reef flat, with fans of reef sediment grading into the lagoon; (d) lagoon; (e) grass bed within lagoon; (f) lagoonal patch reefs; (g) lagoon entrance; (h) sand cays on peripheral reef flat

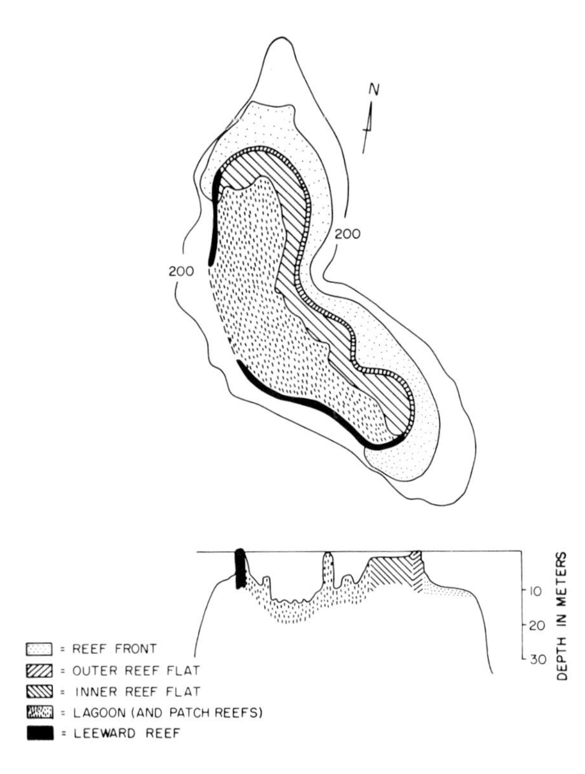

Fig. 48. The distribution of reef zonations at Courtown Cays, southwestern Caribbean. These zonations are found in most Caribbean and Indo-Pacific atolls. After MILLIMAN (1969a)

this lower depth limit probably exceeds 40 meters. Rapid horizontal accretion in the deeper reef-front zones at Jamaica has been documented by LAND and GOREAU (1970)[1]. The reef front generally drops quite steeply to a depth of about 5 to 15 meters and then gently to about 18 m; below this depth the bottom gradient is much steeper, and in some localities it may almost be vertical. The −18 m terrace occurs at many atolls and is judged to be a eustatic feature formed during a lower stand of Pleistocene sea level (NEWELL, 1961; STANLEY and SWIFT, 1967).

Studies of Pacific reef fronts have been limited by the difficulties in crossing the shallow outer reef flat, either by small boat or by skin diving. Caribbean reef fronts, generally having smaller surf and less hungry sharks, have been investigated more thoroughly. LOGAN (1969) defined 5 major reef communities on the reef front at Campeche Bank (Mexico). The shallower communities are dominated by branching corals (*Acropora*) as well as encrusting *Millepora* and coralline algae, while the deeper communities contain mostly hemispheroid and foliose corals (Plate XVIII a). With increasing depth coralline algae become dominant. Similar trends have been described in the Pacific (WELLS, 1957a).

An important feature of the reef front is the spur and groove (buttress) system. This feature consists of closely-spaced, steep-sided linear highs and lows that extend perpendicularly seaward from the reef flat (Plate XX). Some grooves in

1 In a recent article, GINSBURG and JAMES (1973) reported that horizontal reef accretion on the reef front of the British Honduras Barrier Reef can exceed 30 cm per thousand years at water depths greater than 100 m. Reef talus constitutes a major sedimentary component of these limestones.

Plate XVIII. Fore reef and Reef Flat of Caribbean Atolls. (a) The fore reef often is characterized by large quantities of heads of *Montastrea annularis*, here shown in both columnar (left) and platey (right) growth forms. (b) The outer reef flat on most Caribbean contains oriented *Acropora palamta* colonies and *Millepora* zones that extend up to low tide level (see MILLIMAN, 1972) but on some reefs in the southeastern Caribbean (this particular photograph was taken at low tide at Roncador Bank), the coralgal zone is emergent, thus somewhat resembling the Pacific algal ridge. (c) In this zone the dominant components are *Millepora*, soft corals and coralline algae. (d) Inner reef flat sediments often are dominated by coral rubble washed in from the outer reef; these fragments are mostly coated with encrusting red algae. (e) Inner reef flats generally contain more quiet-water growth forms of corals, such as this patch of *Acropora cervicornis*. (f) The transition from the outer reef flat to the inner reef flat often is marked by a marked increase in depth, decrease in living coral cover and increase in rubble and reef debris

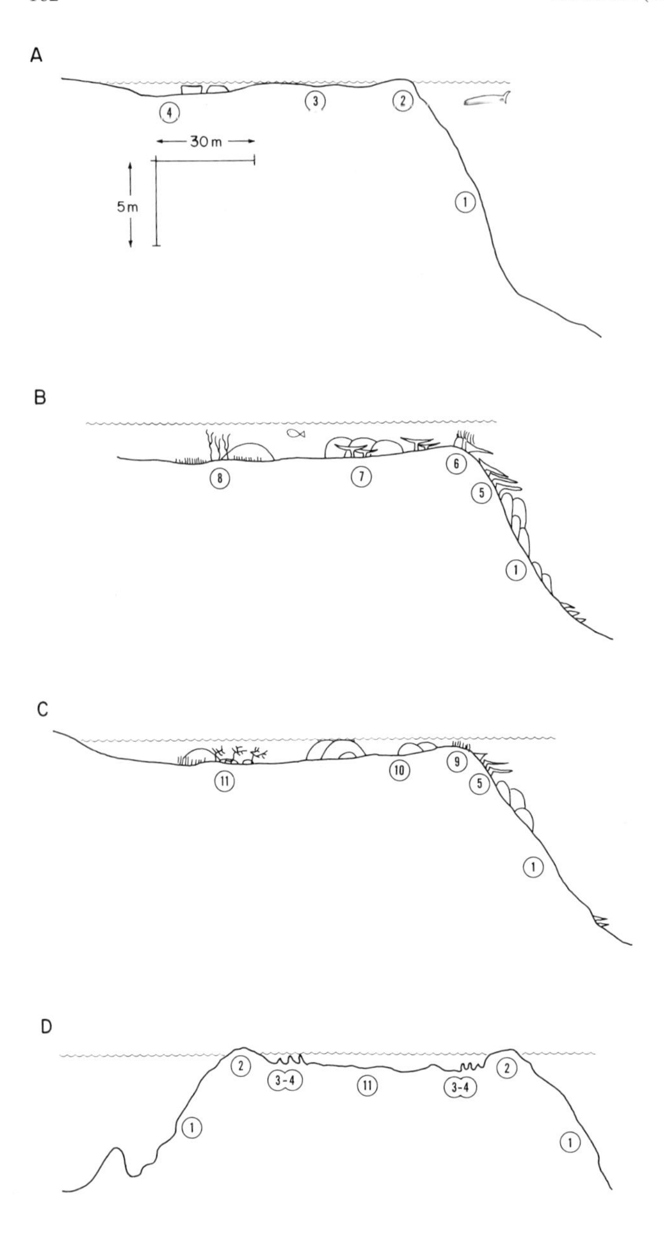

Fig. 49 A–D. Four windward reef flats. A portays a "typical" Indo-Pacific reef flat at Bikini Atoll (according to WELLS, 1957a). B is a typical Bahaman-Floridan type reef flat (after NEWELL, 1959). C is a southern Caribbean reef flat (after MILLIMAN, 1969a), and D is a platform reef from the Great Barrier Reef (after MAXWELL, 1968a). The zones listed numerically are: 1. Fore reef; 2. algal ridge; 3. *Acropora digitifera* (coral-algal) zone; 4. *A. palifera* and *Heliopora* zone; 5. *A. palmata-Montastrea* zone; 6. *Millepora* zone; 7. hemispheric coral zone; 8. inner reef flat, characterized by alcyonaceans and gorgonians; 9. *Millepor-red* algae—*Playthoa* zone (ecologic equivalent of 2 and 6); 10. *Diploria* zone (ecologic equivalent of 7 and 4); 11. inner reef flat, characterized by sand and gravel, some grasses and corals, few gorgonians

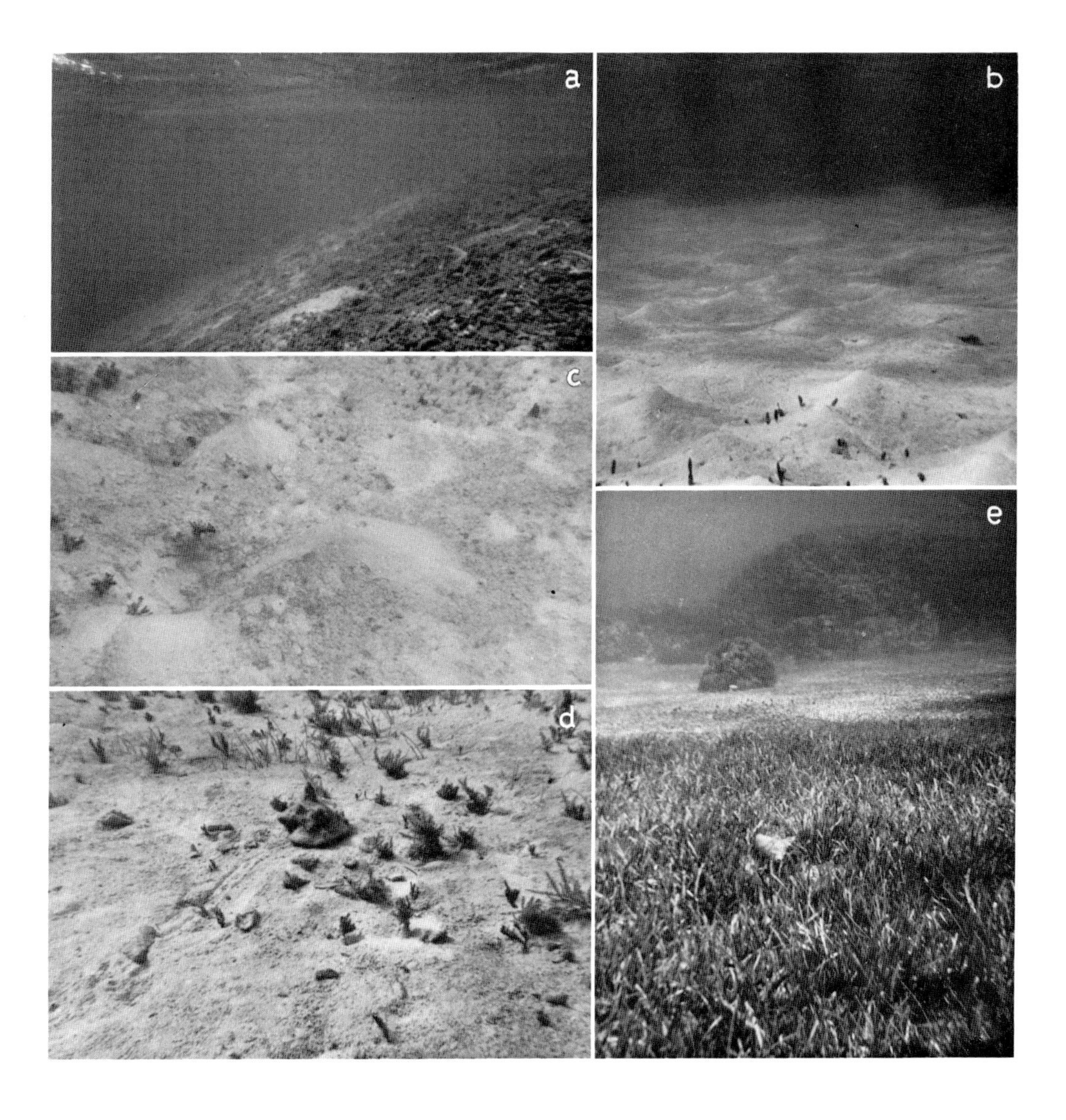

Plate XIX. Lagoonal Environments in Caribbean Atolls. (a) The lagoon slope (sand cliff) connects many reef flats to interior lagoons. Sediments often are coarse at the top and become progressively finer towards the bottom. (b) Lagoon floors often are covered with the mounds of burrowing organisms. These particular mounds are nearly 0.5 meters across and probably were produced by the burrowing shrimp *Callianasa*. (c) Close-up of mounds, notice the opening at the top of the central mound. (d) Browsing gastropods (in this instance, the conch *Strombus*) and various green algae (here mostly *Halimeda*) are common on lagoon floors. (e) Grass beds can help in the vertical accretion of lagoonal sediments. The grass bed in the foreground is some 0.5 meters higher than the floor surrounding the patch reef in the background

Pacific atolls actually extend onto the reef flat in the form of surge channels (Plate XX b–d). The relief of the spur and groove system can exceed 5 meters, and the system can extend to depths as great as 20 meters. This system apparently can help in dispersing the enormous incoming wave energy that continually is expended upon the reef face (MUNK and SARGENT, 1954).

Some debate has arisen as to whether spurs and grooves are constructional or erosional features. Early workers in the Indo-Pacific (DAVID and SWEET, 1904; KUENEN, 1933) recognized the importance of both constructional and erosional processes. Coral growth is most active on the spurs. Outwash of both water and sediment along the groove axis prevents coral growth and may also erode the substrate. These two processes together can accentuate the relief of the spur and groove system. Similar processes were recognized in the Marshall Islands (EMERY and others, 1954) and on Guam (TRACEY and others, 1964). In the Caribbean, GOREAU (1959) suggested that the spur and groove system off Jamaica was the result of the growth of *Montastrea annularis*, while SHINN (1963) showed that *Acropora palmata* is the major coral component in the spurs off Florida. Several other workers, however, have suggested that grooves are erosional in origin (CLOUD, 1959; WIENS, 1962) and this conclusion also was reached by NEWELL (NEWELL and others, 1951), who found that the spurs and grooves off Andros Island (Bahamas) are cut into Pleistocene bedrock.

Windward Reef Flat. The outermost portion of the windward reef flat is generally the shallowest part of the coral reef ecosystem. Corals and algae in this zone can grow above low tide level and yet be moistened sufficiently by the crashing surf to withstand long periods of subaerial exposure.

Many Pacific outer reef flats are characterized by an algal ridge system (sometimes erroneously termed a "Lithothamnion Ridge"), which is a mound of crustose coralline algae that parallels the reef front and which can grow as much as one meter above low tide level (Fig. 49; Plate XX). The dominant algal genus in most ridges is *Porolithon* (*P. onkodes* and *P. gardineri*), although other genera, such as *Lithophyllum*, *Lithothamnium* and *Neogoniolithon*, also are common. Although algal ridges occur on many Pacific reefs (WELLS, 1957a) they are absent from most Indian Ocean reefs, as well as those Pacific reefs exposed to monsoon conditions (GUILCHER, 1965; STODDART, 1969; STODDART and YONGE, 1971). Apparently the seasonal shift in wind direction does not allow any section of the reef margin to receive the constant surf necessary for the construction of such ridges.

→

Plate XX. Pacific Coral Reefs. (a) Aerial photograph of the eastern reef flat and reef front at Bikini Atoll, showing the spurs and grooves (lower part of picture), the reef flat proper, dune-like bodies of sand on the leeward portion of the reef flat, and the lagoon in the background. (b) The algal ridge at Bikini Atoll is emergent during all but high tides. This particular ridge is composed predominantly of the coralline algae, *Porolithon*. (c) and (d) The beneficial effect of wave splash upon coralline algae is shown by this sequence of two photographs, in which the surf crashes into a surge channel, thereby moistening the emergent reef-flat organisms. (e) Aerial photograph of the leeward portions of Bikini Atoll, showing the deep lagoon passes and the reef flats (with islands) that define the leeward margin. Photographs courtesy of J.I. TRACEY, JR.

Plate XX.

Until recently no counterpart of the algal ridge system had been documented in the Caribbean. The lack of continually large surf (due to short fetch in many areas) and the cold winter air temperatures would account for this condition in the northern Caribbean. In the southern Caribbean, however, fetch is sufficiently long and seasonal temperatures sufficiently warm, that counterparts of the algal ridge system do occur. MILLIMAN (1969a) reported a *Millepora* zone (common on the outer reef flat of most Caribbean and Pacific reefs) that contains large quantities of coralline algae (*Neogoniolithon*?) and soft corals (*Palythoa*). This zone parallels the outer reef flat and extends as much as 30 cm above low tide level (Plate XVIII), thus occupying the same niche and somewhat resembling the Indo-Pacific algal ridge. More recently, GLYNN (1973) found an algal ridge system in the coral reefs off Panama that apparently is very similar to those in the Indo-Pacific; moreover, the dominant crustose algal species, *Porolithon pachyderma*, is very similar to *P. onkodes* in the Pacific (ADEY and MACINTYRE, 1973).

The general ecologic zonations on the reef flats in the Caribbean and Indo-Pacific display both distinct and subtle differences from one another (Table 44). WELLS (1954, 1957a) divided the windward reef flats in the Marshall Islands into several major zones. Proceeding lagoonward, these zones include the coral-algal zone (including the algal ridge system), which is covered by only a few cm of water and contains relatively few species of corals (*Acropora*, *Pocillopora* and *Montipora*); the *Acropora palifera* zone (microatolls) which contains scattered low microatolls (generally less than 3 m across and 1 m high); and the *Heliopora* zone, in which microatolls of *Heliopora coerula* are found (Fig. 49). In some atolls this latter zone is considerably shallower and contains only a few hardy forms of corals (such as *Porites lutea*). These zonations, it should be noted, apply specifically to the Marshall Islands, and many deviations can be found on other Indo-Pacific reef flats (STODDART, 1969).

The role of corals on most Indo-Pacific reef flats is not nearly as important as that of encrusting coralling algae. Loose sediment is rapidly cemented into the solid bottom by coralline algae, especially *Neogoniolithon* (WORMSLEY and BAILEY, 1969). *Halimeda* is not common, perhaps because of the shallow depths and relatively poor circulation on most Indo-Pacific reef flats (MAXWELL, 1968a). Gastropods tend to be common on hard reef substrate, while in unconsolidated sediments bivalves prevail (ODUM and ODUM, 1955; TAYLOR, 1968). Benthonic foraminifera can grow profusely on the inner reef flat, especially in protected environments, where sufficient algae are available for grazing (EMERY and others, 1954; JELL and others, 1965).

Coral growth assumes a greater importance in Caribbean reef flats. In contrast to the cohesive shallow reef flats in the Indo-Pacific many Caribbean reef flats are irregular, being composed mostly of individual coral heads (Fig. 49). Reef flat depths generally are greater than those in the Indo-Pacific, varying from a few centimeters to more than one meter. Dominant corals in the outer reef flat include massive and encrusting species, such as *Montastrea annularis*. *Porites astroides* and various species of *Diploria*. Coralline algae encrust loose reef debris but are not nearly as important as in the Indo-Pacific (GOREAU, 1959a; STODDART, 1962; MILLIMAN, 1969a). Green algae (especially *Halimeda*), however, are present in large quantities. The inner reef flats of Caribbean windward reefs generally have

Table 44. Comparison of Indo-Pacific and (southern) Caribbean coral reefs. Partly derived from FOSBERG (1962)

Indo-Pacific	Caribbean
1. About 700 coral species.	1. About 50 coral species.
2. Foliose corals often found in low energy conditions; encrusting corals in higher energy environment and branching corals in lagoonal environments.	2. Foliose and encrusting corals often found in low energy conditions, while branching and massive corals found in higher energy conditions.
3. Many corals can survive prolonged subaerial exposures, as long as they are wetted occasionally by splash.	3. Many corals can survive prolonged subaerial exposures, as long as they are wetted occasionally by splash.
4. Algal ridge present on many outer windward reef flats.	4. Emergent *Millepora* zones present on some reefs; algal ridges on some others.
5. Reef flat often paved with coralline algae, resulting in a relatively smooth and shallow reef flat.	5. Coralline algae often not so prevalent and thus the reef flats often are not as solid.
6. Reef flat sediments dominated by coralline algae and benthonic foraminifera.	6. Reef flat sediments dominated by coral, coralline algae and *Halimeda* fragments.
7. Sand cays mostly found on windward reef flats.	7. Cays occur on slightly more leeward reef flats.
8. Lagoon depth often related to atoll diameter; large atolls can have lagoons deeper than 50 m.	8. Lagoons generally shallower than 15 m.
9. Most lagoonal sediments biogenic.	9. Bank and lagoonal sediments can be non-skeletal.
10. Large lagoons can have distinctive depth-related sedimentary facies.	10. Lagoonal facies usually similar to those found on peripheral reefs or related to patch reef sedimentation.
11. Leeward reefs often well-developed.	11. Leeward reefs usually poorly developed, resulting in rather open lagoons.
12. More than 300 atolls and extensive barrier reefs.	12. About 10 atolls and only 2 easily definable barrier reefs.

water depths varying from 0.5 to 1.5 meters. The bottom is covered mostly with sand and gravel debris derived from the outer reefs (Plate XVIII) but heads of some corals, notably *Siderastrea*, *Porites* and *Montastrea* are prominant locally. Some inner reef flats also contain large patches of green algae and various marine grasses (*Thalassia*, *Halodule* and *Syringodium*). Gorgonians are common in the inner reef flats in the northern Caribbean, but decrease in importance to the south.

Leeward Reefs. Because the wind direction on many atolls is more or less constant throughout the year, windward reefs receive the bulk of the heavy surf and waves, and therefore tend to be better developed than the leeward reefs. Reefs on the leeward sides of atolls resemble weaker and less well developed versions of the windward sides. Zonations on the leeward sides of many Pacific reefs are similar to those on the windward sides, but the algal ridge is less well-developed and often is entirely absent (WELLS, 1957a). Spur and groove features often are missing from leeward reefs (for example, Mahe in the Seychelles; LEWIS, 1968). In the Caribbean reefs the entire reef flat structure is generally less well developed, often being composed of anastomosing patch reefs rather than a continuous solid reef flat (see MILLIMAN, 1969a).

Reef Sedimentation

In coral reefs carbonate is deposited in two distinct manners: as reef framework and as unconsolidated sediment. The relative importance of the various skeletal carbonate components to modern reef accretion has been documented largely on the basis of limited reef borings and from the dynamiting of reef faces (STODDART, 1969). Comparisons of elevated Pleistocene reefs with neighboring modern reefs in such areas as Florida (HOFFMEISTER and others, 1967; HOFFMEISTER and MULTER, 1968), Barbados (MESOLELLA, 1967), the Red Sea (SHÄFER, 1967), SAIPAN (CLOUD and others, 1956; JOHNSON, 1957) and GUAM (TRACEY and others, 1964; SCHLANGER, 1964) also have given additional insights into the construction of modern coral reefs. The rigid reef framework is produced mainly by hermatypic corals and is cemented by coralline algae. Although other organisms, such as *Halimeda*, mollusks, bryozoans, foraminifera, echinoderms, brachiopods and sponges may supplement the reef framework, their importance generally is minor compared to that of coralline algae and coral.

One of the most significant developments in understanding coral reef dynamics has been the recognition of the extent of biodegradation of the reef rock and its subsequent infilling and cementation. In their study of freshly exposed reef surfaces (obtained by dynamiting the reef), GINSBURG and others (1971) noted an interesting sequence of events in the Bermudan "boiler reefs". Burrowing organisms destroy much of the original reef structure. The resulting secondary voids, ranging in diameter from microns to decimeters, are coated by epibenthic organisms (such as *Homotrema rubrum*) and filled with internal mud. Subsequent lithification produces a rock that is markedly different from the original coralline algae-coral-vermetid reef. The importance of the process has been substantiated by similar observations in Jamaica (LAND and GOREAU, 1970) and the Red Sea (FRIEDMAN, 1972). Clearly this concept of continual alteration and cementation of the reef structure represents an important new direction in understanding reef accretion and sedimentation. Further studies of this process, both in modern and ancient reefs, should be accelerated.

In terms of total carbonate production, the unconsolidated sediments produced by reefs appear to be far more important than the carbonate incorporated into the reef itself. STODDART (1969) estimates that 4 to 5 times more loose sediment is produced than is incorporated as reef framework. Because of the greater ease of collection and subsequent analysis, the nature and composition of reef sediments are understood much more completely than is the nature of the reef itself.

Plate XXI. Reef Flat Sediments. (a) Reef flat sand from the windward reef of Courtown Cays, southwestern Caribbean. The sediment is composed mostly of fragments of coralline algae and coral, with lesser amounts of mollusks, *Halimeda* and echinoid spines. Scale is 1 cm. (b) *Halimeda*-rich sand from the lagoonward portion of the windward reef flat at Courtown Cays. Scale is 1 cm. (c) Reef flat sand from Kayangel Atoll, Pacific. Note the abundance of worn benthonic foraminifera, which is typical of Indo-Pacific reef flat sands. Scale is 1 cm. (d) Lagoon sand from Albuquerque Cays, southwestern Caribbean. Larger particles are benthonic foraminifera, mollusks and *Halimeda;* the finer sediment includes coral and coralline algae derived from the neighboring reef flats. Scale is 5 mm

Because of the generally high wave energies, unconsolidated reef sediments are mostly composed of sand and gravel; silt and clay usually constitute less than 1 to 2% of the total carbonate (Plate XXI). In fact, many reef flats contain little loose sediment: what is not washed into the lagoon or cemented onto the reef flat is found in local depressions or leeward of large coral heads. Five components contribute the bulk of most reef flat sediments: corals, coralline algae, *Halimeda*, foraminifera and mollusks (Plate XXI). Miscellaneous constituents, such as bryozoans and echinoids, can be important locally, but generally comprise less than 5% of most reef sediments.

The relative concentration of any carbonate component depends upon the depositional environment and the populations and productivities of the various component organisms. Coral fragments generally constitute 15 to 30% of a coral reef sediment. The total number of coral species, however, does not necessarily dictate the amount of coral in the sediment. For instance, tropical Pacific reefs contain more than 500 species of coral (WELLS, 1954) while Caribbean atolls generally have less than 50 and Midway and Kure Atolls (central north Pacific) have only 5 prominent species (DANA, 1971). Yet the proportion of coral fragments found in the sediments of these various reefs is surprisingly constant (Tables 45 and 46).

Table 45. Average composition of peripheral reef sediments from various Caribbean reefs and atolls

<table>
<tr><th></th><th>Florida[a]</th><th>Alacran Reef[b]</th><th>Andros, Bahamas[c]</th><th>Abaco, Bahamas[d]</th><th>Ragged Islands, Bahamas[e]</th><th>Hogsty Reef[f]</th><th>Courtown Cays[g]</th><th>Albuquerque Cays[g]</th><th>Roncador Bank[g]</th><th>Serrana Bank[g]</th></tr>
<tr><td>Skeletal fragments</td><td></td><td></td><td></td><td></td><td></td><td></td><td></td><td></td><td></td><td></td></tr>
<tr><td>Coral</td><td>20</td><td>26</td><td>24</td><td>27</td><td>12</td><td>27</td><td>35</td><td>30</td><td>25</td><td>33</td></tr>
<tr><td>Mollusks</td><td>12</td><td>7</td><td>6</td><td>11</td><td>18</td><td>22</td><td>10</td><td>9</td><td>15</td><td>5</td></tr>
<tr><td>Foraminifera</td><td>6</td><td>8</td><td>12</td><td>11</td><td>13</td><td>5</td><td>3</td><td>2</td><td>3</td><td>3</td></tr>
<tr><td>Coralline algae</td><td>10</td><td>11</td><td>33</td><td>10</td><td rowspan="2">39</td><td>19</td><td>21</td><td>21</td><td>24</td><td>24</td></tr>
<tr><td>Halimeda</td><td>30</td><td>40</td><td>17</td><td>16</td><td>2</td><td>28</td><td>32</td><td>17</td><td>17</td></tr>
<tr><td>Misc. skeletons</td><td>7</td><td>1</td><td>6</td><td>5</td><td>4</td><td>1</td><td></td><td>1</td><td>1</td><td>1</td></tr>
<tr><td>Non-skeletal fragments</td><td rowspan="6">10</td><td></td><td></td><td rowspan="5">4</td><td rowspan="6">?</td><td></td><td></td><td></td><td></td><td></td></tr>
<tr><td>Oolite</td><td></td><td></td><td></td><td></td><td></td><td></td><td></td></tr>
<tr><td>Pelletoids</td><td>2</td><td></td><td>1</td><td></td><td></td><td></td><td></td></tr>
<tr><td>Aggregates</td><td>4</td><td></td><td>tr</td><td></td><td></td><td></td><td></td></tr>
<tr><td>Cryptocrystalline lumps</td><td></td><td></td><td>17</td><td></td><td>2</td><td></td><td>3</td></tr>
<tr><td>Misc. non-skel.</td><td></td><td>2</td><td>15</td><td></td><td></td><td></td><td>11</td><td>10</td></tr>
<tr><td>Unknown</td><td>6</td><td></td><td></td><td></td><td></td><td>5</td><td></td><td></td><td></td><td></td></tr>
</table>

[a] GINSBURG, 1956; [b] HOSKIN, 1963; [c] GOLDMAN, 1926; [d] STORR, 1964; [e] ILLING, 1954; [f] MILLIMAN, 1967a; [g] MILLIMAN, 1969b.

Table 46. Composition of peripheral reef flat sediments from some Pacific atolls

	Ifaluk[a]	Midway[b]	Kure[b]
Coral	36	25	15
Mollusk		12	12
Foraminifera	23	14	10
Coralline algae	23	35	52
Halimeda	10	6	2
Miscellaneous skeletons	9	8	9

[a] TRACEY et al., 1961; [b] GROSS et al., 1969.

Mollusk fragments appear to be more common in reef sediments than one would intuitively expect. Population estimates (see ODUM and ODUM, 1955; KORNICKER and BOYD, 1962) suggest that living mollusks are relatively sparse compared to other carbonate producers. Moreover, the predicted rate of carbonate production is lower than in foraminifera, corals or algae (see Chapter 4). However, mollusks generally constitute 10 to 20% of a reef flat sediment; this may be the result of their greater ability to resist diminution by abrasion and corrosion (GROSS and others, 1969) or it may indicate that estimates of mollusk populations and productivities are too low.

The most important biogenic component in most Caribbean reef flat sediments is *Halimeda* (Plate XXI a, b). Coralline algae, although important, are not as plentiful as in Indo-Pacific sediments (Tables 45 and 46). With the exception of the encrusting *Homotrema rubrum* (MACKENZIE and others, 1965), benthonic foraminifera are relatively unimportant on Caribbean reefs. In contrast, benthonic foraminifera (such as *Amphistegina madagascariensis*, *Marginopora vertebralis* and *Calcarina spengleri*) are major contributors to Indo-Pacific reef flat sediments. Many Indo-Pacific sediments also contain large quantities of coralline algae; for instance, GOREAU and others (1972) estimated that 90% of the reef flat sediment at Saipan is algal.

Reef flat carbonate production is relatively high, and most unconsolidated reef flat sediments that have been age-dated by carbon-14 analysis are less than 2 thousand years old. In contrast, sediments on the deeper parts of the fore reef usually represent admixtures of recent and relict sediments, the latter increasing with increasing water depth (MACINTYRE, 1967 a). This observation conflicts with the theoretical calculations by CHAVE and others (1972), who concluded that the upper and lower slopes have the highest gross productivities within the general coral reef environment.

Lagoonal Environments

Types of Lagoons

In terms of carbonate sedimentology, lagoons are defined as shallow marine environments which are bounded on at least one side by a topographic high. Although many tropical lagoons are bordered by coral reefs, some lagoons are delineated by land masses (for instance Great Bahama Bank, Florida Bay and Batabano Bay). In terms of hydrography, lagoons can be divided into three classes normal, brackish and hypersaline. Normal lagoon waters display temperatures and salinities that are more or less similar to those in the surficial waters of the adjacent ocean. In brackish lagoons, fresh water influx from runoff and rainfall exceeds evaporation; the opposite is true in hypersaline lagoons. Hydrography and circulation can exert a major influence upon the sedimentological characteristics of a lagoon. Therefore, each of these three types of lagoons will be discussed separately.

Normal Lagoons. Most coral atoll lagoons have a relatively rapid exchange of water with the adjacent ocean. Oceanic waters primarily flow into lagoons over the windward reefs and exit across leeward reefs and through lagoon channels. In open lagoons (lagoons with numerous lagoon channels whose depths are about equal to the average lagoon depth, thus allowing free exchange with the ocean) currents generally are unidirectional throughout the water column, flowing in a direction coincident with the wind (see MILLIMAN, 1967, 1969a) (Fig. 50); residence time of these lagoon waters generally is less than a few days. Many large atolls have lagoon entrances which are shallower than the lagoon itself. This tends to restrict circulation, and in fact, may aid in the formation of bottom currents that flow in the opposite direction to the surface currents (Fig. 50). The ultimate direction and magnitude, however, is still wind dependent (BERTHOIS and others, 1963).

In many atolls the lagoon depth is related to the diameter of the atoll; small atolls generally have shallow lagoons while large atolls, such as Eniwetok in the Marshall Islands, can have depths as great as 70 meters. This generalization, however, does not apply to Caribbean atolls where depths range from 5 to 15 meters, regardless of the size of the atoll. Most atoll lagoons contain three major morphologic features: lagoon slopes, lagoon floors and patch reefs. Each of these exhibit distinct fauna and flora, as well as distinct sediment types.

Lagoon slopes connect the peripheral reef flats to the lagoon proper. Some slopes are gentle but many are steep, sometimes approaching the angle of sediment repose. The steep lagoonal slopes in the southwestern Caribbean atolls have been compared to sand cliffs (MILLIMAN, 1969a), and the analogy is apt. One literally can jump from some reef flats into lagoonal depths greather than 5 meters. Carrying the analogy further, sediment continually moves down the steep gradient into the lagoon (Plate XIXa). Such sediment movement, of course, can influence the development of benthonic communities. Patch reefs, however, can grow on some lagoon slopes, and on the deeper and more gradual lagoonal slopes in the Pacific, luxurient *Halimeda* thickets occur.

The lagoon floor is defined as that part of the lagoon not directly covered by patch reefs. This environment generally contains large populations of holothurians, echinoids, browsing gastropods, clams, benthonic foraminifera, and green algae. The sedimentologic and ecologic characteristics of the lagoon floor, however, are strongly influenced by the current regime, which in turn is related to the morphology of the atoll (see preceding paragraphs). Where currents are slow, sediments tend to be fine grained and infaunal filter feeders are common. In areas exposed to greater current velocities, sediments tend to be

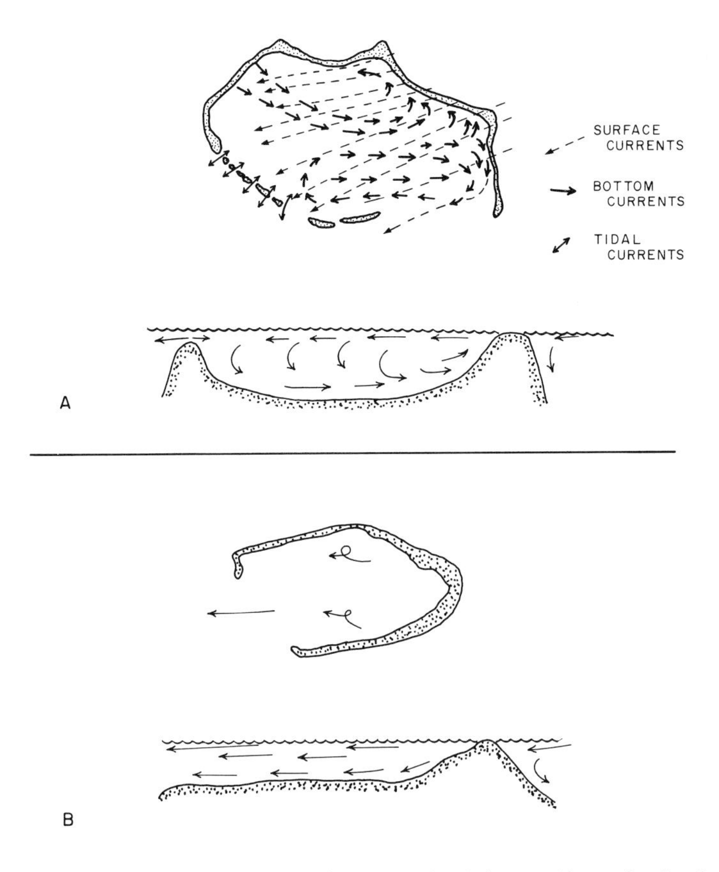

Fig. 50 A and B. Two extremes of lagoonal circulation. A. Shows the distribution of currents within Bikini lagoon. Because the lagoon is relatively deep and enclosed, its circulation is characterized by a prominent bottom countercurrent. Residence time of the lagoon water is on the order of weeks to months. After von ARX (1954). B. Hogsty Reef lagoon is not only an order of magnitude smaller than Bikini (both in vertical and horizontal dimensions), but it also is relatively open. Currents are unidirectional and coincide almost directly with wind direction (except for periodic tidal reversals). Residence time is in terms of hours and days. After MILLIMAN (1967)

coarser and the organisms often are epifaunal and more capable of filter feeding (RHOADS and YOUNG) (Plate XIX b–d).

Patch reefs are coral reefs that rise above the lagoon floor. These reefs can be small and low lying (coral knolls) or can be represented by very large reefs that extend to the water surface (patch reefs proper). Some patch reefs in the Marshall Islands have dimensions approaching those of small atolls (EMERY and others, 1954). The distribution of patch reefs within a lagoon is highly variable. In some atolls (for example, Kapingamarangi) patch reefs are relatively sparse; in contrast, the windward lagoon at Serrana Bank is covered by abundant patch reef growth (Fig. 51).

Although some patch reefs can contain reef zonations similar to those found on windward reefs (MILLIMAN, 1969 a), most such reefs lack high concentrations

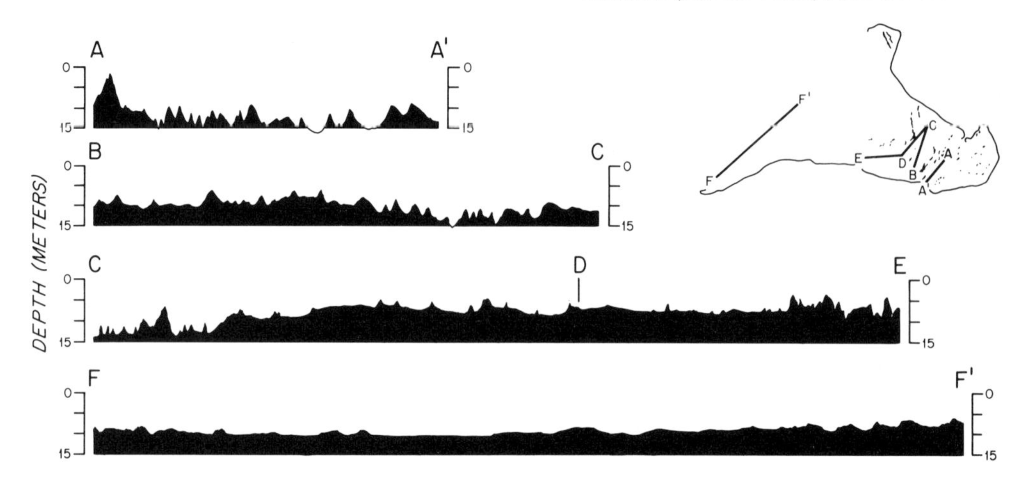

Fig. 51. Bathymetric profiles across the lagoon at Serrana Bank, southwestern Caribbean. The atoll is approximately 20 km long. Note the abundance of patch reefs in the first two profiles and the relatively smooth lagoon bottom on the last profile. It is in this more subdued and unproductive lagoon bottom portrayed in profile $F-F'$ that non-skeletal carbonates are abundant. From MILLIMAN (1969a)

of coralline algae. The absence in most lagoons of continuous agitation by crashing surf may explain this difference (WELLS, 1957a). The character of a patch reef and its biota is also dependent upon the hydrographic and sedimentologic character of the lagoon water. For example, in Fanning Atoll patch reef corals in turbid waters tend to be branching, while in clear waters they are massive (ROY and SMITH, 1971).

Patch reefs influence lagoonal sedimentation in several ways. First, they contribute sediment that is different from normal lagoon floor sediments and remarkably similar to that found on peripheral reefs. Depending upon their abundance within a lagoon, patch reefs can be an important source for lagoonal sediments. Second, patch reefs can act as sedimentary baffles. Large numbers of reefs can be as effective in impeding bottom currents and trapping sediments as shallow sills at lagoon entrances.

Brackish-water Lagoons. The prime example of a lagoon that is strongly influenced by the influx of fresh water from adjacent land is the inner part of Florida Bay. This area receives large dilutions of fresh water during the rainy months, when salinities in the interior part of the bay can drop to 10–15$^0/_{00}$. During dry months these same areas can have salinities in excess of 50$^0/_{00}$ (MCCALLUM and STOCKMAN, in GINSBURG, 1964; SCHOLL, 1966). Evidently these gross fluctuations in salinity, together with the rather quiescent conditions caused by the presence of the Florida Keys (which protect the bay from waves and currents from the Straits of Florida) have restricted the development of reef organisms and thereby influenced carbonate sedimentation. Corals, coralline algae and *Halimeda* are restricted to areas directly influenced by tidal currents (JINDRICH, 1969). Carbonate-depositing organisms in Florida Bay are predominantly mollusks and foraminifera. Calcifying plants include the green algae

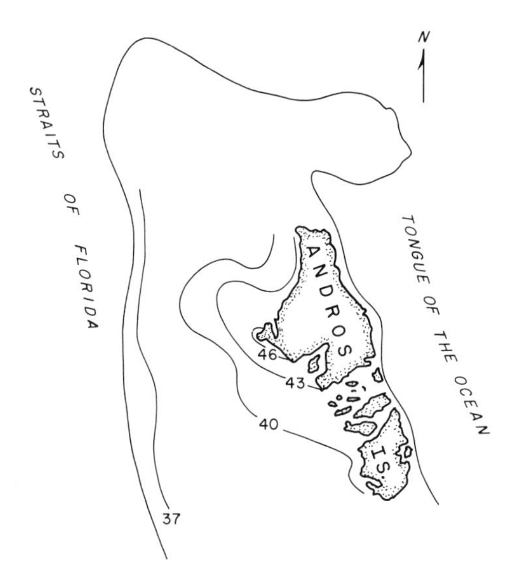

Fig. 52. Surface salinities ($^0/_{00}$) on Great Bahama Bank, MAY, 1955. After CLOUD (1962a)

Penicillus, *Udotea*, *Rhipochephalus* and *Acetabularia* (STOCKMAN and others, 1967). Grass banks can host many epibenthic plants (such as *Melobesia*) and animals (such as foraminifera and serpulids) which upon the decay of the grass, contribute to the ambient sediment.

Hypersaline Lagoons. Hypersaline lagoons include such areas as Great Bahama Bank (NEWELL and others, 1951, 1959; CLOUD, 1962a), the nearshore areas of the Persian Gulf (EVANS and others, 1969; KINSMAN, 1964a; KENDALL and SKIPWITH, 1969), Laguna Madre (RUSNAK, 1960; DALRYMPLE, 1964), and Shark Bay, Australia (LOGAN and CEBLUSKI, 1970). Hypersalinity is usually the result of restricted or sluggish circulation combined with high rates of net evaporation. On Great Bahama Bank, waters become increasingly saline towards the inner part of the bank west of Andros Island (SMITH, 1940; CLOUD, 1962a; BROECKER and TAKAHASHI, 1966; TRAGANZA, 1967) (Fig. 52). Salinities can exceed $45^0/_{00}$; water temperatures exceed 30 °C in summer months and generally remain above 24 °C even in winter. LOGAN and CEBULSKI (1970) report that salinities in the inner extremities of Shark Bay exceed $70^0/_{00}$ and similar salinities have been reported from the inshore areas of the Persian Gulf (KINSMAN, 1964a). Many of the organisms and biologic communities in hypersaline lagoons are similar to those found in normal and brackish lagoons. NEWELL and others (1951, 1959) reported six dominant biologic facies on Great Bahama Bank, each of which is associated with a distinct sediment type (compare Figs. 53 and 59): mollusks, bryozoans, foraminifera (especially peneroplids; TODD and LOW, 1971) and green algae are the most abundant carbonate-producing organisms. The relative paucity of skeletal-producing organisms in some hypersaline bank and lagoon deposits suggests that organic populations and productivities are insufficient to offset large-scale non-skeletal precipitation; in fact, low populations and productivities *may* enhance non-skeletal deposition) see Chapter 3 and p. 194 ff.).

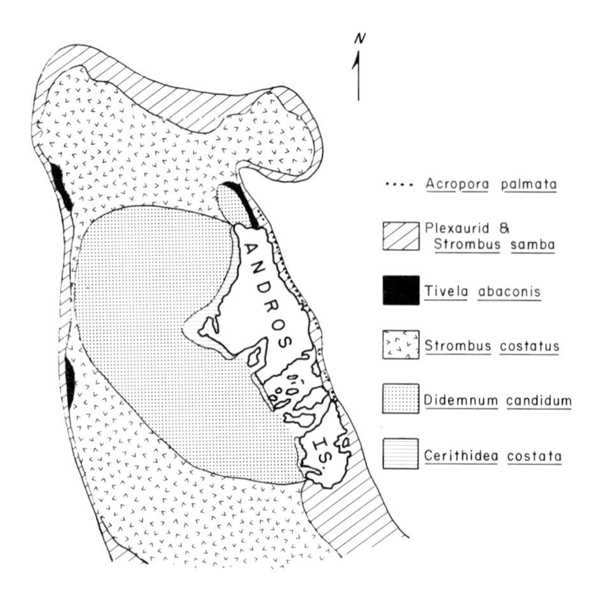

Fig. 53. Distribution of biologic facies on Great Bahama Bank (after NEWELL and others, 1959)

Grass Banks and Algal Mats

Both grass banks and subtidal algal mats play important roles in the ecology and sedimentology of lagoonal sediments. Grass banks, usually composed of the sea grasses *Thalassia* and *Cymodocea* (*Posidonia* can be important in some areas; MOLINIER and PICARD, 1952; DAVIES, 1970a), occur in shallow lagoons and banks (generally less than 10 m deep). Grass blades trap fine particles by slowing currents; in addition, the roots (rhyzomes) can extend as deep as 30 m into the sediment, bind the sediment and thus stabilize the bottom (GINSBURG and LOWENSTAM, 1958). Because of the trapping and binding ability of grass blades and roots, sea grass beds generally rise above the non-grass covered bottoms (Plate XIX) and contain appreciably finer sediments (Fig. 54). Neighboring tidal channels can accentuate the relief of such banks even further (JINDRICH, 1969; DAVIES, 1970a). The most impressive example of the accreting ability of grass banks is the large bank in Shark Bay, which covers an area greater than 1000 sq. km, and reaches a thickness of more than 7 m (DAVIES, 1970a). MOLINIER and PICARD (1953) reported that *Posidonia* banks in the Mediterranean are capable of vertical accretion in excess of 1 meter per hundred years.

Grass beds house numerous benthic and epiphytic invertebrates and calcareous plants, many of which provide most of the sediment deposited within the banks. Not only do the benthic calcareous algae contribute calcium carbonate to the sediments, but they also are capable of binding the sediment, both by trapping (*Batophora*) and binding with holdfasts (*Halimeda*, *Penicillus*, *Udotea* and *Rhipocephalus*) (SCOFFIN, 1970). Echinoderms, gastropods (such as *Strombus*) and many bivalves also are common in grass banks (NEWELL and RIGBY, 1957; TAYLOR and LEWIS, 1970). The grass acts both as a protective substrate and as a food source. In addition, decaying grass can serve as an important diagenetic agent, as well as a possible catalyst in the inorganic precipitation of calcium carbonate (see the discussion on p. 185 ff.).

One special grass community includes branching coralline algae (*Neogoniolithon strictum* is the most common species) and branching coral (*Porites* and *Pavona* are the common types). On the seaward side of Rodriguez Key in the Florida Keys this community is particularly well-developed (MULTER, 1969). Similar communities have been reported from elsewhere in the Florida Keys (JINDRICH, 1969), the Bahamas (NEWELL and RIGBY, 1957; NEWELL and others, 1959), Puerto Rico (GLYNN, 1963) and Shark Bay (DAVIES, 1970a).

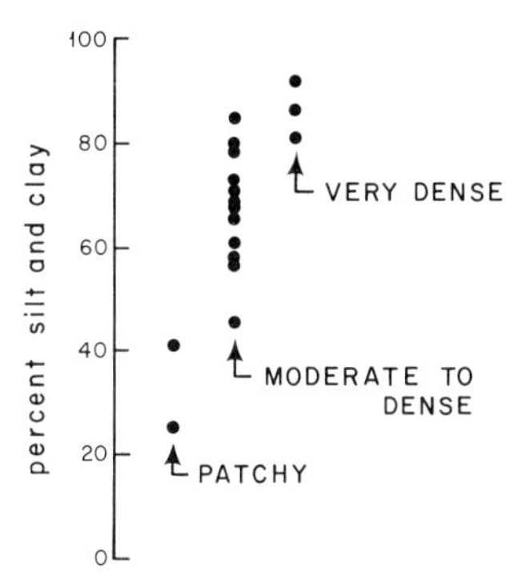

Fig. 54. Variation of sediment size (as expressed in percent silt and clay) with density of *Thalassia* grass cover in Buttonwood Sound, Florida Bay. The greater the extent of grass cover, the greater the amount of fine particles that are trapped. Data from LYNTS (1966)

Grass blades also serve as a substrate upon which epibionts attach and grow. The most common organisms include the encrusting coralline algae *Melobesia* and *Fosliella* (HUMM, 1964), serpulids (NAGLE, 1968), foraminifera (BOCK, 1967) and bryozoans. HUMM (1964) suggested that these carbonate-producing organisms may provide an important source of calcium carbonate, and LAND (1970) estimated that between 40 and 180 gm of $CaCO_3$ can be produced per m^2 of grass bed per year in Jamaica[2]. Apparently most of the epiphytic organisms decompose soon after death and contribute mainly to the finer sediment fraction. Sand-size grains in grass bank sediments generally reflect the benthic animals living in the bank (JINDRICH, 1969) or material transported to the bank from neighboring areas rather than those species growing on the surrounding grass blades (BOCK, 1967).

The role of subtidal algal mats was discussed in a previous section (p. 51 ff.), and will only be summarized here. Algal mats serve three important purposes in subtidal lagoonal sedimentation. First, their sticky filaments tend to bind the sediment, and thereby prevent erosion (DOTY and others, 1954; BATHURST, 1967c; SCOFFIN, 1970; NEUMANN and others, 1970). Subtidal algal mats in the Bahamas bind both mud- and sand-size sediments into cohesive layers that are capable of withstanding currents 5 to 9 times faster than the velocity which would erode these sediments if they were not bound (SCOFFIN, 1970; NEUMANN and others, 1970). SCOFFIN (1970) recognized four types of subtidal algal mat communities in

2 More recent estimates by PATRIQUIN (1972) in Barbados indicate average carbonate production of 2.8 kg/year. Part of this difference with LAND's values is related to differences in analytical technique, but part is real.

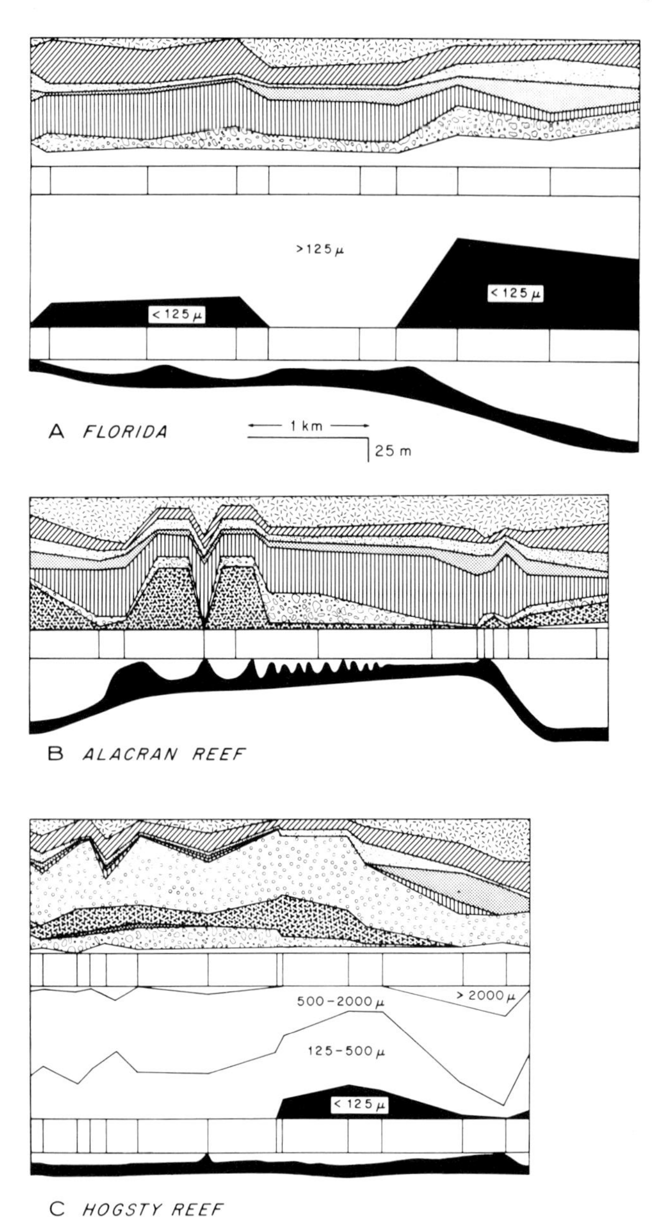

Fig. 55 A–F. Distribution of size grades and composition of the gravel and sand size components across six Caribbean coral reefs and lagoons. From MILLIMAN (1973)

the Bahamas: the green algae *Enteromorpha* and *Lyngbya*, the red *Laurencia* and the blue-green *Schizothrix*.

Second, algal mats provide an excellent source of food for sediment scavenging animals in lagoonal areas (BATHURST, 1967c). Third, the metabolic processes which

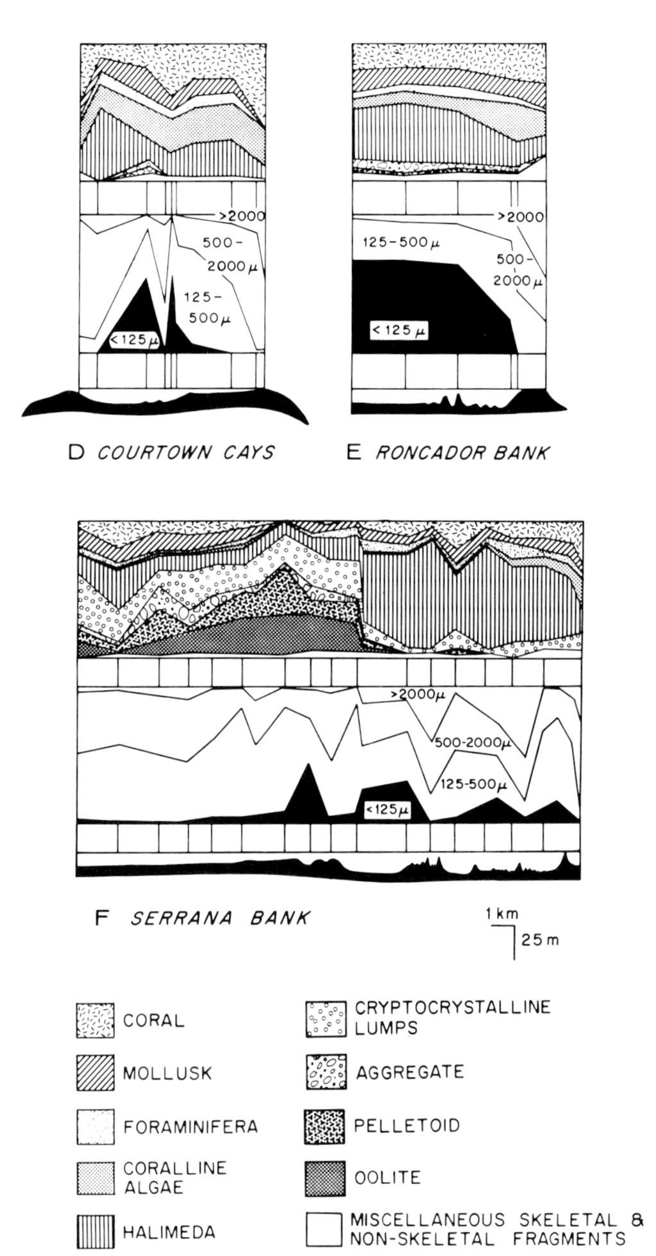

Fig. 55. Continued

occur in both living and decaying algae, together with the fine-grained nature of the filament-sediment mixture, provide numerous micro-environments that are ideal for diagenetic alteration and cementation of carbonates (BATHURST, 1967c; see Chapters 9 and 10). Endolithic algae may be important factors in the alteration of water chemistry in some lagoonal sediments. The effect of these algae upon the diagenesis of carbonate sediments is discussed in Chapter 9.

Burrowing Organisms

The formation of sedimentary burrows by invertebrate organisms can modify the bottom sediment considerably: 1. As the burrow is formed, sediment is expelled by the organism and fine grained sediments are preferentially removed by currents. Thus the mounds around burrows often contain coarser sediment than in non-burrowed sediments (SHINN, 1968 c). However, abandoned burrows tend to act as sediment traps, and can fill with material that is considerably finer than the surrounding sediments. 2. The organisms that form burrows tend to rework the ambient sediment and form fecal pellets, thereby influencing the diagenesis of fine-grained sediments. 3. The presence of such deposit feeders also can adversely affect the distribution of filter-feeding organisms by burying larvae and thus preventing attachment to the bottom (RHOADS and YOUNG, 1970).

SHINN (1968 c) found that five invertebrates are responsible for most of the intertidal and subtidal burrows in south Florida. The shrimps *Alpheus* and *Callianassa* and the anemone *Phyllactis* are the common subtidal types. *Callianassa* is by far the most obvious, producing volvano-like mounds that extend up to 30 cm above the bottom (Plate XIX). Intertidal burrows are formed mostly by the land crab *Cardisoma* and the fiddler crab *Uca*. FARROW (1971) found that the dominant burrowers at Aldabra Atoll (Indian Ocean) also are crustaceans. Each major environment within the sediment-covered bottom at Aldabra contains diagnostic burrowing crustaceans. The only environment in which crustaceans are not dominant is the lagoonal sand flats, where the polychaete *Arenicola* predominates (FARROW, 1971).

Lagoonal Sediments

Lagoonal carbonate sediments generally display greater textural and compositional variations than do most reef sediments. In part these variations reflect the wide range of environmental conditions within lagoons. Specifically, the texture and composition reflect the biologic, hydrographic and chemical regime within the depositional environment. These parameters, in turn, are influenced by the climate, current system and physical dimensions (such as depth and area) of the lagoon (or bank) and the proximity to peripheral or patch reefs.

Carbonate texture depends upon the component organisms (or non-skeletal fragments) that compose it. To a lesser extent, texture can reflect the circulation patterns within the lagoon. Open lagoons (see above) generally contain larger amounts of sand and gravel than do lagoons with restricted circulation (Fig. 55). This simplistic model, however, does not take into account the effect of blue-green algal mats and grass banks which tend to stabelize the bottom regardless of the current regime.

The sedimentary influence of peripheral reefs decreases as the area of the atoll (or bank) increases and correspondingly, the area covered by peripheral reefs relative to lagoonal area decreases. For example, the ratio of peripheral reef flat area to lagoonal area at Courtown Cays (24.7 km^2) is 0.69, while at Serrana Bank (191 km^2) the ratio is 0.089, and at Eniwetok Atoll (943 km^2) the ratio is less than 0.06. Thus smaller atoll lagoons receive proportionately greater quantities of peripheral reef flat sediments than do large atolls (Fig. 55). The relative ease with which peripheral reef sediments move into lagoons, of course, depends upon the circulation within the lagoon. Small open lagoons tend to contain sediments that are quite similar in composition to those found on the peripheral reef flats. This is especially true of the finer grained fractions which are transported more readily than the coarser-grained components (Fig. 56).

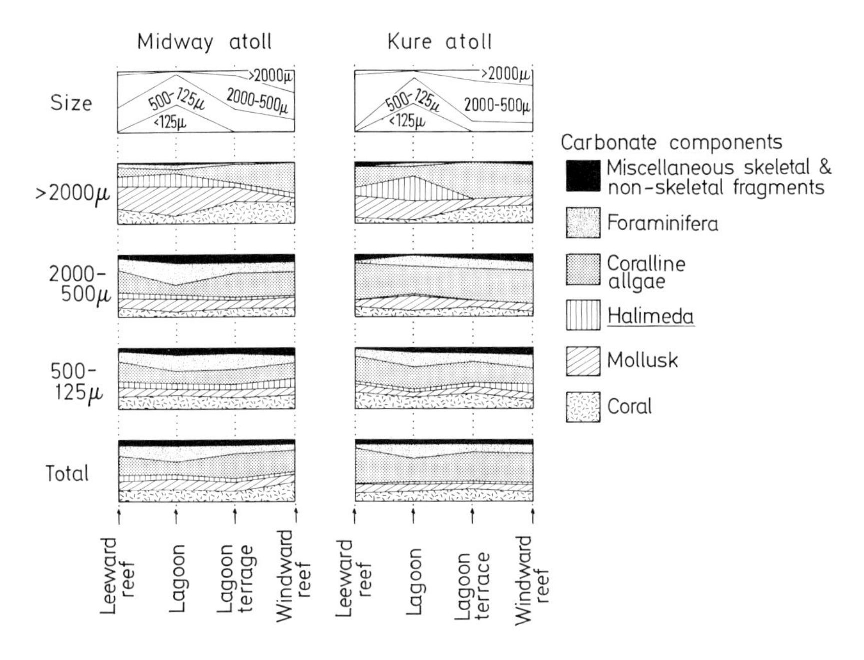

Fig. 56. Size and composition of sand and gravel within various depositional environments at Kure and Midway Islands (from GROSS and others, 1969). The coarser particles are derived from sources nearby to where they are deposited. For instance the greater than 2000 micron fraction in the lagoonal sediments contains considerably more *Halimeda* and mollusks and less foraminifera, coralline algae and coral than do the peripheral reefs. The fraction finer than 500 microns, on the other hand, shows a more or less even distribution of components across the atoll, suggesting a leeward transport of these finer components from the windward peripheral reefs

Large and deep lagoons can display endemic sediments and sedimentary zonations. In the Caribbean, for instance, fecal pellets, mollusks, foraminifera (mostly peneroplids; BOCK and MOORE, 1967; WANTLAND, 1967) and aragonitic muds are common (for instance, Alacran Reef; HOSKIN, 1963) (Plate XXII a, b). Large Pacific atoll lagoons tend to contain large quantities of benthonic foraminifera and *Halimeda*. The distribution of components in some deep Pacific lagoons also can be depth related, resulting in a series of roughly concentric

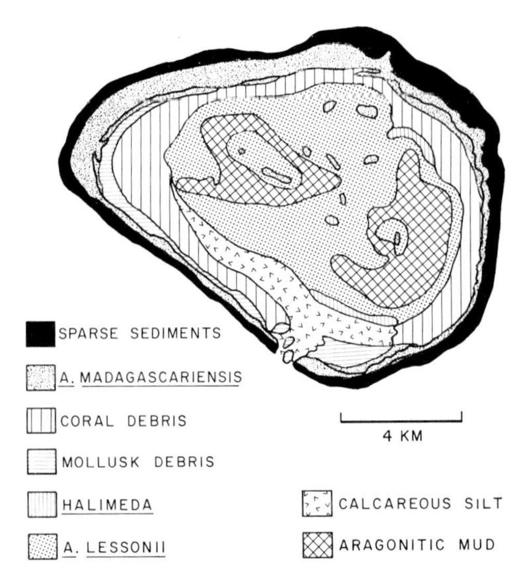

Fig. 57. Distribution of carbonate facies at Kapingamarangi Atoll. After McKEE and others (1959)

sedimentary bands within the lagoon (Fig. 57). It is interesting to note the importance of *Halimeda* within many Pacific lagoon sediments (Table 47), as opposed to the dearth of this component in most peripheral reef flat sediments (see above); this is quite different from the Caribbean (Table 48). *Halimeda*, however, is not uniformly important in all Pacific lagoons. It is nearly absent at Raroia (NEWELL, 1956), Fanning (ROY and SMITH, 1971) and Midway and Kure (GROSS and others, 1969).

As mentioned in a previous section, patch reefs serve two important functions within a lagoon, they can restrict both water and sediment movement, and they can be an important sediment source. While *in situ* lagoon floor sediments are dominated by mollusks, green algae and foraminifera, patch reefs contribute material rich in coral and coralline algae (although the amount of coralline algae is somewhat less than what one would expect from a peripheral reef sediment).

Table 47. Composition of the sands in some Pacific Ocean lagoonal sediments

	Funafuti[a]	Bikini[b]	Rongelap[b]	Rongerik[b]	Eniwetok[b]	Johnson Isl.[c]	Cocos Lagoon (Guam)[d]	Raroia[e]	Ifaluk[f]	Midway[g]	Kure[g]	Pearl and Hermes Reef[h]
Coral	tr	17	12	20	42	24	45	43	15	18	18	17
Mollusk	tr	9	12	8	8	12	18	tr	—	15	10	18
Foraminifera	21	15	21	38	14	2	2	28	29	19	18	6
Coralline algae	tr	—	—	—	—	61	18	tr	6	31	42	48
Halimeda	77	57	55	40	35	1	17	tr	32	7	4	
Misc. skeletons	tr	1	—	—	—	—	—	—	15	10	8	—

[a] JUDD, 1904. [b] EMERY et al., 1954. [c] EMERY, 1956. [d] EMERY, 1962. [e] BYRNE, *in* NEWELL, 1956. [f] TRACEY et al., 1961. [g] GROSS et al., 1969. [h] THORP, 1936b.

Table 48. Average composition of the sand-size components of lagoonal sediments from various Caribbean reefs and atolls

	Florida[a]	Alacran Reef[b]	Hogsty Reef[c]	Courtown Cays[d]	Albuquerque Cays[d]	Roncador Bank[d]	Serrana Bank (east)[d]	Serrana Bank (west)[d]
Skeletal fragments								
Coral	7	15	4	28	20	22	9	8
Mollusks	18	7	10	12	12	12	7	10
Foraminifera	7	7	4	6	6	5	5	3
Coralline algae	3	5	1	21	14	12	4	1
Halimeda	38	23	1	28	31	37	61	13
Misc. skeletons	7	2	1	1	3		1	1
Non-skeletal fragments								
Oolite			1			tr	1	15
Pelletoids	12	33	16				1	15
Aggregates		9	13					8
Cryptocrystalline lumps			46	2	12	6	4	21
Misc. non-skel.						1		
Unknown	7		1	2	2		5	4

[a] GINSBURG, 1956. [b] HOSKIN, 1963. [c] MILLIMAN, 1967. [d] MILLIMAN, 1969 b.

Patch reefs also can form suitable substrates for *Halimeda* growth. Generally, the influence of patch reef sedimentation is limited to the areas surrounding the individual reefs (EMERY and others, 1954; MCKEE and others, 1959; MORELOCK and KOENING, 1967; HOSKIN, 1966; JORDAN, 1971) (Fig. 58). At Bermuda patch

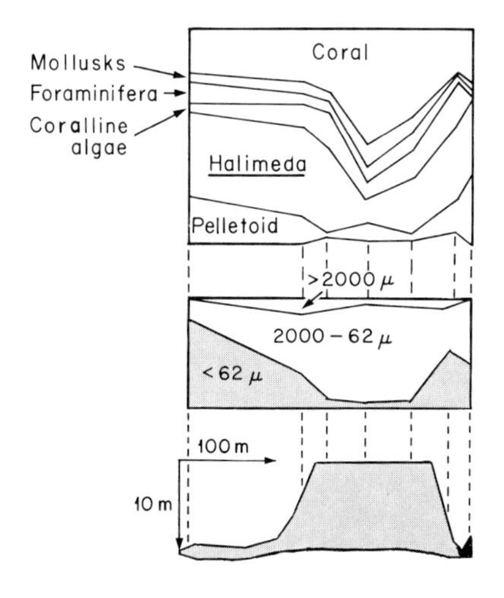

Fig. 58. The influence of patch reef sedimentation in Alacran Reef lagoon is limited to the immediate vicinity of the reef. Within a distance of 100 meters, the sediment displays the textural and compositional characteristics of normal lagoon floor sediments. Data from HOSKIN (1963, 1966)

Table 49. Average composition of the sands from various Caribbean bank areas

	Florida Bay[a]	Batabano Bay, Cuba[b]	Ragged Keys, Bahamas[c]	Great Bahama Bank[d]			
				Oolite facies	Grape-stone facies	Pellet-mud	Mud
Skeletal fragments							
Coral		<5	tr	tr	tr	—	—
Mollusk	80	10	6	1	4	5	12
Foraminifera	11	<5	4	1	3	7	14
Coralline algae	tr	<5	1	tr	tr	—	—
Halimeda	tr	0–10		2	3	2	9
Misc. skeletons	3		1	tr	1	1	1
Non-skeletal fragments							
Oolite				70	16	12	4
Pelletoids	3	60	6	7	5	57	33
Aggregates			27	8	36	10	12
Cryptocrystalline lumps			55	7	28	3	5
Misc. non-skel.						1	2
Unknown	1			2	2	5	7

[a] GINSBURG, 1956. [b] DAETWYLER and KIDWELL, 1959. [c] ILLING, 1954. [d] PURDY, 1963.

reef sedimentation extends no more than 10 m away from individual reefs (GARRETT and others, 1971). In lagoons in which they are abundant, however, patch reefs can be a major sediment source. For instance, patch reefs at Roncador Bank occupy more than 30% of the total lagoon area (MILLIMAN, 1969a); the composition of the coarse fraction within this lagoon is very similar to that found on the peripheral reefs (Tables 45, 48; Fig. 55).

In some respects sedimentation in brackish and hypersaline lagoons can resemble that in normal lagoons. If salinities are not extreme, the types of carbonate-producing organisms can be quite similar to those found in more normal lagoons. Brackish and hypersaline lagoons, however, can also contain distinct sedimentary facies. Florida Bay, for example, is remarkable for its almost complete lack of sand-size skeletal organisms other than mollusks and foraminifera (Table 49). This restricted skeletal composition is probably more a function of the turbid waters and lack of available hard substrate upon which coral, coralline algae and *Halimeda* can attach, than the fluctuating salinities. The fine fraction, which is dominated by codiacean algal remains (see below), appears similar to that found in many other normal and hypersaline lagoons.

High salinities can facilitate inorganic precipitation of calcium carbonate. However, as has been shown in Chapter 3, the deposition of non-skeletal grains (ooids, aggregates, hardened pelletoids and cryptocrystalline lumps) is more dependent upon the lack of skeletal deposition than upon hypersalinity: non-skeletal grains can form in normal lagoons (for example, Hogsty Reef and Serrana Bank; Table 48) as well as in hypersaline lagoons (Table 49). Similarly, the exact character of the non-skeletal grain (that is, whether it is an ooid, aggregate or lump) may be more a function of water depth and current regime than of hydrographic conditions (see Chapter 3; Fig. 59). Extreme environmental conditions

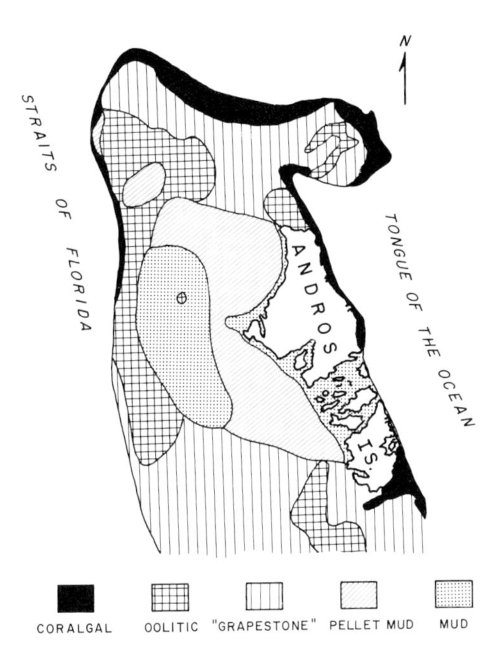

Fig. 59. Distribution of carbonate facies on Great Bahama Bank (as defined by the 200-meter isobath). After PURDY (1963)

(such as the Coorong, where salinities exceed $70^0/_{00}$) may limit the distribution of many carbonate-producing organisms and thus increase the dominance of non-skeletal sedimentation. Not only does the lack of biogenic calcification mean greater supersaturation of calcium carbonate within the waters, but the lack of skeletal deposition also results in the greater relative abundance of non-skeletal grains as well as a longer duration to which grains are exposed to supersaturated conditions (see Chapter 3).

Genesis of Shallow-Water Carbonate Muds

Accurate petrographic identification of most carbonate components is limited to grains coarser than 62 to 125 microns (see Table 8). The nature and origin of silt and clay-size particles, therefore, has been restricted primarily to the interpretation of mineralogical and chemical data. Perhaps it is precisely the dearth of definitive petrographic studies that has caused the continued debate concerning the origin of fine-grained lagoonal carbonates. The following paragraphs will discuss the most studied area, Great Bahama Bank, but these theories also can be applied to other areas, which will be discussed later.

Great Bahama Bank—The shallow (1 to 3 m) hypersaline tropical waters on Great Bahama Bank leeward of Andros Island (Fig. 52) present an ideal environment for the deposition of fine-grained carbonates. Sediments in this area are mostly silt and clay. The predominant mineral is aragonite, the clay fraction containing more than 85% aragonite. The rapid deposition of this carbonate is evidenced by the losses of calcium, pCO_2 and alkalinity in the waters across the bank (SMITH, 1940; CLOUD, 1962a; BROECKER and TAKAHASHI, 1966; TRAGANZA, 1967), but the mode of deposition, although widely debated, is not known. Excellent summaries of the various theories have been presented by CLOUD (1962a, 1962b) and BATHURST (1971).

AGASSIZ (1894) and NEWELL and others (1951) suggested that an appreciable amount of bank mud is transported from Andros Island. Such an origin, however, would not explain the losses in Ca^{+2} and CO_2 from the bank waters. Moreover, the general lack of terrestrial fossils in bank muds suggests that transport (such as during hurricanes) is limited to the immediate area leeward of Andros Island (CLOUD, 1962a).

Most theories concerning the origin of Andros muds have invoked the chemical precipitation of calcium carbonate, either through direct chemical precipitation or by biochemical processes (bacteria). Standard arguments include:

1. The aragonite needles are almost identical to those precipitated inorganically in laboratory experiments (JOHNSON and WILLIAMS, 1916; CLOUD, 1962a) (Plate XXII).

2. The extreme degree of supersaturation of calcium carbonate in the bank waters infers that precipitation *should* occur.

3. The rate of carbonate precipitation is directly proportional to the degree of carbonate supersaturation within the waters (BROECKER and TAKAHASHI, 1966).

4. By summing all the possible biologic sources, CLOUD (1962a) could account for no more than 25% of the total carbonate sedimentation.

The actual mode of precipitation, however, has not been defined. Those workers favoring direct chemical precipitation have pointed out the occurrence of "whitings", milky suspensions of carbonate that periodically cover large areas of shallow bank waters (CLOUD, 1962a). Once formed, whitings seem to grow and spread rapidly, suggesting that the suspensions represent an instantaneous inorganic precipitation, perhaps triggered by some biochemical mechanism. Carbon-14 data, however, suggest that the suspended material in some whitings is predominantly resuspended bottom material (BROECKER and TAKAHASHI, 1966). BROECKER and TAKAHASHI also have shown that the slow CO_2 exchange rates between water and the atmosphere are not conducive to such rapid precipitation. In addition, the lack of a measurable pH differential between the waters inside and outside such whitings infers that precipitation has not occurred. For these reasons, as well as others mentioned in following paragraphs, whitings are thought to represent resuspended bottom sediment, probably stirred up by fish or currents, rather than evidence of rapid chemical precipitation.

Most workers have invoked biochemical mechanisms for the precipitation, although the exact organisms and reactions are not known. DREW (1911) was the

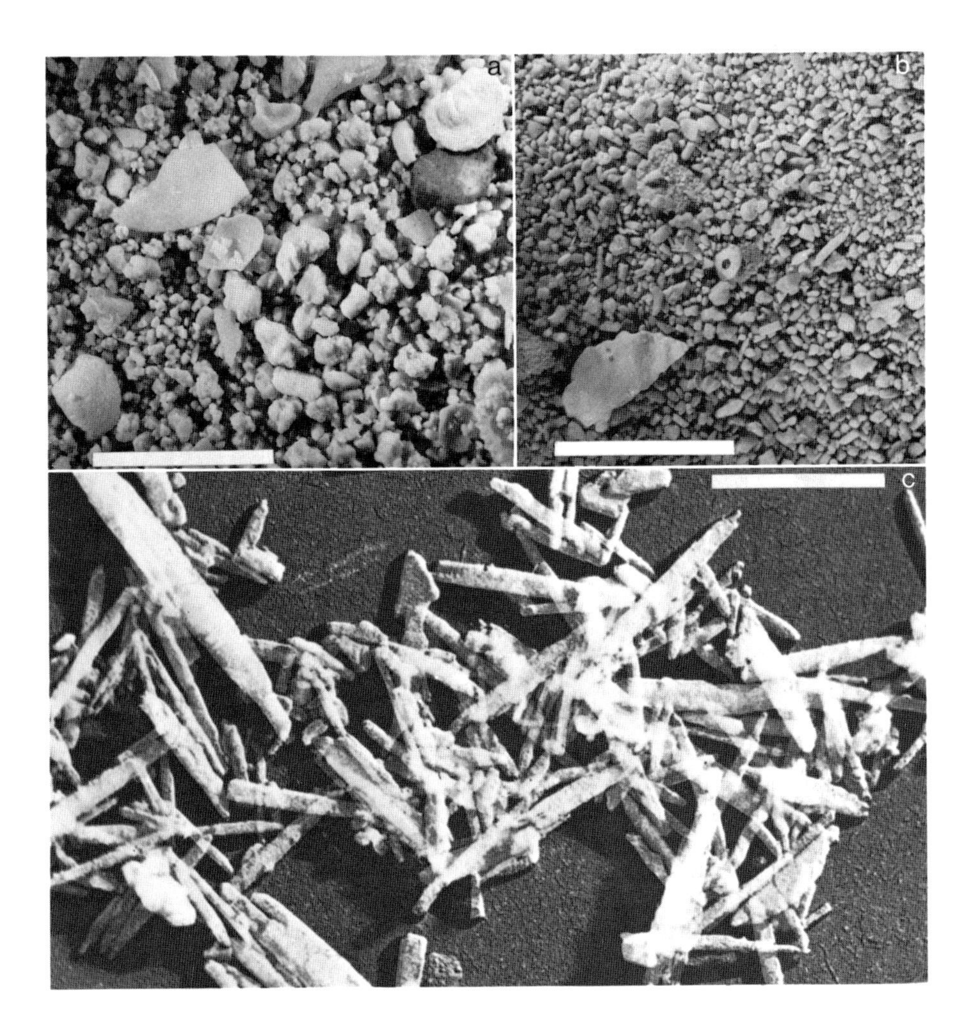

Plate XXII. Lagoon Sediments. (a) Grapestone sediment from Hogsty Reef lagoon, southeastern Bahamas; lagoon depth is 6 meters. Reflected light; scale is 5 mm. (b) Pelletal sand from Serrana Bank lagoon; water depth is 7 meters. Reflected light; scale is 5 mm. (c) Aragonitic needles from the Great Bahama Bank mud facies. TEM; scale is 1 micron

first to suggest that denitrifying bacteria could precipitate calcium carbonate directly. Further studies supported this theory (KELLERMAN and SMITH, 1914; VAUGHAN, 1914; SMITH, 1926) and others suggested that sulfate-reducers also could precipitate fine-grained carbonate (SMITH, 1926; BAVENDAMM, 1932; BLACK, 1933a). Three dominant reactions were envisioned:

$$Ca(NO_3)_2 + H_2 + C \rightarrow CaCO_3 + N_2O(OH)_2 + H_2O$$ (DREW, 1911)

$$Ca(HCO_3)_2 + 2\,NH_4OH \rightarrow CaCO_3 + (NH_4)_2CO_3$$ (KELLERMAN and SMITH, 1916)

$$(NH_4)_2CO_3 + CaSO_3 \rightarrow CaCO_3 + (NH_4)_2SO_4$$

(MURRAY and IRVINE, 1890; BLACK, 1933a).

Subsequent studies have strongly supported the role of sediment-inhabiting denitrifying bacteria as carbonate precipitators, producing either calcite (LALOU, 1957a, 1957b) or aragonite (MONAGHAN and LYTLE, 1956; OPPENHEIMER, 1960; GREENFIELD, 1963; CARROLL and others, 1965). Moreover, GREENFIELD (1963) has shown that dead bacteria cells can serve as excellent nuclei around which carbonate can precipitate.

Other workers, however, have questioned the role of bacteria in carbonate precipitation. LIPMAN (1924) speculated whether bacterial populations were sufficiently dense to produce the vast quantities of aragonite mud found in the Bahamas (see also GEE, 1934). Sulfate-reducing bacteria, for example, are restricted mostly to mangrove swamps around Andros Island and thus probably do not contribute much to the aragonite deposition leeward of Andros. Other workers (such as CLOUD, 1962a) have dismissed bacteria as insignificant agents on the basis of their apparent inability to alter any measurable parameter of pore water composition.

In recent years some workers have favored a purely biologic origin for the Bahama aragonitic muds. Because of their close resemblance to the aragonitic needles secreted by codiacean green algae, LOWENSTAM (1955) suggested that the Bahama Bank aragonitic muds were derived directly from algal disintegration. Subsequently, LOWENSTAM and EPSTEIN (1957) showed that the δO^{18} and δC^{13} contents of Bahama aragonitic muds fall closer to codiacean algal aragonite than ooid aragonite. If these carbonates were precipitated in equilibrium with the ambient environment, the temperature of formation (based on δO^{18} values) would be 22.8 to 39.8 °C for the codiacean algae, 27.6 to 31.7 °C for the aragonitic muds and 24 to 25.7 °C for the ooids. Although the latter values agree closely with the mean annual temperature in this area (24–27 °C), the algal and mud values are considerably higher. LOWENSTAM and EPSTEIN concluded that the muds were more similar to the algae than to the ooids (which presumably are "inorganic"). SEIBOLD (1962) also concluded that the aragonitic muds near Bimini Island (on the western side on Great Bahama Bank) were derived from codiacean algae. NEUMANN and LAND (1969) found that green algae in the Bight of Abaco (Little Bahama Bank) can produce 6 to 9 crops a year, which is more than enough to account for all the aragonitic mud produced in that area.

CLOUD (1962a) presented several counter-arguments which should be considered before one supports the attractive algal-origin of Bahama aragonitic muds. First, he calculated that the standing crops and productivities of the codiaceans leeward of Andros Island are not nearly large enough to produce such large amounts of mud. Second, CLOUD pointed out that carbonate precipitation occurs mostly during the summer months, a fact supported by hydrographic data (SMITH, 1940; TRAGANZA, 1967). Summer temperatures are sufficiently high for inorganically precipitated muds to contain the δO^{18} values observed by LOWENSTAM and EPSTEIN.

One factor that has not been investigated in sufficient detail is the chemical composition of the aragonitic muds. Codiacean algae commonly contain between 0.7 and 0.9% Sr, while other potential organic aragonite contributors (gastropods and pelecypods) contain significantly less Sr. If the mud were inorganic in origin, one might expect higher Sr values, such as the 0.95 to 1.0% found in oolite and other non-skeletal carbonates. A preliminary analysis of 5 mud samples from the bank leeward of Andros Island (Table 50) shows an average aragonite content of

Table 50. Mineralogy and elemental composition of five Bahamian mud samples taken from the bank leeward of Andros Island

Sample	Mineralogy			Trace elements				
	Arag.	Mg Calcite	Calcite	%			ppm	
				Mg	Sr	Na	Fe	Mn
1	84	12	4	0.66	0.82	0.43	340	10
2	84	12	4	0.96	0.90	0.50	540	11
3	84	12	4	0.65	0.95	0.33	973	17
4	89	10	1	0.48	0.88	0.40	180	20
5	84	13	3	0.69	0.91	0.62	325	7

85%, and a Sr content within the aragonite (assuming about 0.25% Sr in the calcite) of about 1.0%. These data, while only preliminary, would suggest a possible inorganic precipitation.

Other Areas—The composition and origin of carbonate muds in several other shallow-water areas have been studied. Some of these are discussed in the following paragraphs.

1. British Honduras. By use of petrographic analysis with immersion oils together with mineralogical and elemental data, MATTHEWS (1966) found that the 20 to 62 micron fraction in the southern British Honduras coastal lagoon (5 to 64 m deep) is produced mostly by the *in situ* decomposition of lagoonal organisms. Most prominent contributors are mollusks and foraminifera. Sediments derived from the windward reefs contribute only to the area immediately adjacent to the reefs, not the inner lagoon. *Halimeda* is not prominent, and judging from the rather low Sr values within the lagoonal muds, other codiacean algae are not major contributors. Recently SCHOLLE and KLING (1972) reported that coccoliths constitute an important part of the carbonate fraction finer than 20 microns, as much as 20% in some samples.

2. *Florida Bay.* The muds in Florida Bay are primarily aragonite (FLEECE, 1962; TAFT and HARBAUGH, 1964; PILKEY, 1964; MÜLLER and MÜLLER, 1967). Most of this aragonite is algal in origin. STOCKMAN and others (1967) found that the populations and growth rates of the codiaceans *Penicillus, Rhipocephalus* and *Udotea*, as well as *Acetabularia*, can account for all the aragonitic mud found in Florida Bay and the adjacent inner reef flats. Some inorganic precipitation may occur on or near photosynthesizing or decaying grass beds (JINDRICH, 1969), but at best this is probably of minor importance. Calcitic mud, which is less common, may come from the red alga *Melobesia* sp., which grows profusely on *Thalassia* blades (STOCKMAN and others, 1967).

3. *Alacran Reef.* The lagoon muds at Alacran Reef are composed of comminuted skeletal fragments and fecal pellets. On the basis of mineralogy and elemental analyses, HOSKIN (1966, 1968) concluded that *Halimeda*, corals and coralline algae are the main sedimentary contributors to the mud fraction, suggesting derivation from peripheral and patch reef sediments. The sand-size particles, on the other hand, are derived mostly from lagoon-floor organisms (Table 48).

4. *Persian Gulf.* Many shallow areas in the southern Persian Gulf are covered with highly aragonitic muds (EVANS, 1966; WELLS and ILLING, 1963; KINSMAN, 1964a). The high Sr values (about 0.94% Sr; KINSMAN, 1969b) are a bit too high for organic aragonite. Moreover, codiacean algae, the most reasonable organic source for aragonitic muds, are totally absent in the Persian Gulf (HILLIS, 1959). The lack of any reasonable organic source plus the high Sr values of the aragonite and the hypersaline warm waters suggest that the aragonite has been precipitated inorganically. The mode of precipitation, however, remains a mystery. WELLS and ILLING (1963) suggested that the aragonite is instantaneously precipitated as whitings. These writers speculated that photosynthetic activity by diatoms may trigger such a precipitation. DE GROOT (1965) found that the sediment suspended within one Persian Gulf whiting contained aragonite, magnesian calcite and calcite in almost exactly the same proportion as the bottom sediments. The opinion that whitings represent resuspended bottom material rather than direct inorganically precipitated carbonate agrees with the general conclusions reached in the Bahamas.

5. *Coorong, Southern Australia.* Both aragonitic and magnesian calcite muds occur in the evaporitic Coorong marine lagoons of southern Australia (VON DER BORCH, 1965a). While few mineralogical or chemical data are available, the extreme temperatures and salinities in this environment, plus the presence of other inorganically precipitated sediments (such as dolomite and magnesite; see Chapter 11), suggest that these muds are inorganically precipitated. The large-scale precipitation of magnesian calcite has not been reported from other shallow-water environments, but this mineral is dominant in many deep-water carbonates (see Chapter 8).

6. *Bimini Lagoon, Great Bahama Bank.* STEIGLITZ (1972), using a scanning electron microscope, found that most grains between 16 and 62 microns in

Bimini Lagoon are skeletally-derived. Foraminifera, codiacean algae and aggregate grains are the most abundant components.

Summary—Shallow-water carbonate muds can have a number of possible sources: 1. comminuted skeletons; 2. disintegrated codiacean green algae; 3. precipitated calcium carbonate; and 4. bacterially precipitated carbonate. Most lagoonal muds are biogenic. The source of high-Sr aragonitic muds in many lagoons (such as Florida Bay) is codiacean algae; abraded reef corals also may be prominent in areas immediately adjacent to peripheral or patch reefs (Alacran Reef). In those lagoons in which codiaceans are not common, abraded mollusks and foraminifera may be major sediment sources (British Honduras). Calcium carbonate also can be deposited by inorganic precipitation in hypersaline environments (Persian Gulf, Coorong). Some workers suggest that the precipitation is direct, but others invoke bacterial catalysis; as of yet the exact mechanism of precipitation is not understood.

Carbonate Cays

Carbonate islands are represented by two end members, modern cays consisting of unconsolidated sediments, and islands composed of pre-recent limestone. In keeping with the theme of this book, the latter type will not be discussed further. Modern carbonate cays include both mud and sand cays, and each represents unique environmental and sedimentological characteristics.

Mud Cays and Tidal Flats

Most mud cays and muddy tidal flats form in quiet shallow waters. The specific sedimentary features of these cays and flats depend greatly on the degree of exposure to subaerial and subaqueous conditions (Table 51), but many structures

Table 51. The distribution of primary sedimentary structures across a Bahama tidal flat in relation to the time during which the various structures are exposed to subaerial conditions. After GINSBURG and others (1970)

	Percent of time exposed to subaerial conditions
Polygon mud cracks (1–3 cm)	95–100
Birdseyes	90–100
Blistered and curled algal mat	75–100
Thin (mm) laminations	75–100
Knobby algal stromatolites	75–100
Algal stomatolites with palisade structure	50–95
Finger burrows (*Uca*)	50–90
Wide, shallow polygon mud cracks (5–15 cm)	50–75
Deep prism polygon mud cracks (20–30 cm)	50–75
Antler burrows (*Alpheus* sp., *Callianassa major*)	0–50
Spaghetti burrows (polychaete worms)	0–50

and sediment types also reflect the climate and depositional environment. For example evaporitic minerals (dolomite, gypsum, magnesite, halite, and so forth) and polygonal mud cracks are diagnostic of arid climates. Studies on mud flats and cays have been made in the Persian Gulf, the Coorong, the Bahamas, Florida Bay, British Honduras and Shark Bay, Australia. Many of these studies are discussed in connection with shallow-water dolomites in Chapter 11. For the purposes of this discussion, three types of carbonate mud flats will be treated briefly: the sabkhas along the Trucial Coast of the Persian Gulf, Andros Island, and Florida Bay.

The Trucial Coast on the southern Persian Gulf is characterized by broad (10 to 15 km) salt-encrusted coastal flats, called sabkhas. The surficial sabkha sediments, which may be fine or coarse depending upon the waves and winds, are supratidal in origin; underlying sediments are intertidal and subtidal, indicating a gradual seaward progradation of the coast line (KINSMAN, 1969a). Sabkhas, almost by definition, are arid. The influx of less saline waters is limited to occasional flooding during high tides and storms and periodic rainstorms. In order to compensate for the high rates of evaporation of surface waters, pore waters migrate upwards through the sediments by capillary action. The result is a dense subsurface brine from which evaporitic minerals can form; the upper crust often consists of one or more halite layers, each representing a period of evaporation following a sabkha flooding (Plate XXIII d). A further description of sabkha sediments can be found in Chapter 11.

The leeward portion of Andros Island has a semi-arid climate. SHINN and others (1969) have recognized three distinct depositional zones within this area: 1. a subtidal zone, including tidal channels and adjacent ponds; 2. an intertidal zone; and 3. a supratidal zone in which sediments usually are deposited only during storms or extreme high tides; beach ridges, marshes and levees occur within this zone (Plate XXIII). The various environmental zones and their sediments are listed in Table 52.

The mud cays in Florida Bay are located on shallow banks that rise 1 to 2 m above the average bottom depth (GORSLINE, 1963). The individual cays rarely stand more than $^1/_2$ m above high tide. Sediment may contain discrete sand layers, but the dominant sediment is silt and clay-size carbonates (FLEECE, 1962). Cay accretion is dependent upon the ability of mangrove trees to trap sediments and prevent erosion (DAVIS, 1940; GORSLINE, 1963; SCOFFIN, 1970)[3].

Equally important are the blue-green algal mats which have been discussed in Chapter 4 and earlier in this chapter (Plate XXIII d). Sediment from nearby banks is transported by storms and by abnormally high tides. BALL and others (1967) found an average of 5 cm of carbonate mud deposited on the inner Florida Bay cays by Hurricane Donna. Although such marked effects may not occur during every storm (PRAY, 1966; PERKINS and ENOS, 1968), catastrophic deposition can provide an major amount of mud cay sediment.

---

3 According to BASAN (1973), mud bank formation in Florida Bay is controlled by bedrock configuration. Soft sediment tends to accumulate in sink holes. Subsequently this mud is colonized by baffle-forming algae and grasses, which inturn increase the ability of the bank to accrete vertically.

Plate XXIII. (a) Intertidal and supratidal area off NW Andros Island showing the position of tidal channels as well as two areas in which recent supratidal dolomite occurs (designated as D). (b) Beach ridge and intertidal pellet mud, near Wide Opening, SW coast of Andros Island, Bahamas, at very low tide. Note erosion of seaward dipping beach ridge sediments. The thinly bedded nature of these beds is visible, especially at left of picture. Note the abundance of lithified and unlithified clasts on the intertidal pellet mud surface. (c) Area of lithified mud crack polygons bordering an interior (and now abandoned) tidal channel levee, SW Andros Island. Lithification of the polygons is by aragonite, magnesian calcite and (rarely) dolomite. (d) Soil pit in Cross Bank, Florida Bay, showing the stromatolitic structures which result from periodic storm deposition followed by long periods of supratidal exposure. Pencil serves as a scale. (e) Sabkha surface near Jidda, Saudi Arabia. The crust is composed of a thin algal mat, underlain and "cemented" by halite crystals. Photographs (a) and (d), courtesy of H. G. MULTER, (b) and (c), courtesy of C. D. GEBELEIN

Table 52. Characteristic of sedimentary environments on the tidal flat at Andros Island. After SHINN and others (1969)

	Sediment composition	Color and structure	Dolomite	Fauna and flora
Subtidal zone				
Adjacent marine	soft pellets, arag. muds	laminations absent, burrows present, gray color, H_2S	no	cerithid gastropods *Codakia and Chione*, forams, *Callianassa*
Channels	extremely variable, generally soft, foram tests mixed with lumps, dolomitic crust	inclined and cross bedding		*Peneroplis, Batillaria*
Ponds	soft and muddy, few pellets	few structures	no	*Peneroplis, Batillaria*
Intertidal flats	pellets, 40–80 m	laminations few, root structure present	no	*Batillaria, Uca, Rhizophora*, algal mats
Supratidal zone				
Beach Ridge	size decreases landward, pellets and fine skeletons, many gastropods, mud chips and pebbles	cross bedded, laminated and graded	high concentration on broad landward flanks	*Batillaria, Uca, Cardisoma, Avincinnia, Casurinia*
Levees	pellets	laminations <3 mm thick; burrows increasing away from channel	high concentration on broad landward flanks	*Uca, Batillaria*
Marsh	pelleted mud, organic rich	tan to white with dark laminae	present, but $<20\%$	*Cardisoma, Cerion*, algal mats

Sand Cays

Most sand cays occur in high energy environments, such as reef flats. Cay heights generally range from 1 to 5 meters, lengths and widths from tens to hundreds of meters. The long axis usually parallels the reef front, and lengths are 3 to 9 times longer than widths. Sand and gravel beaches form the intertidal and supratidal zones. Sometimes lithified beach sands (beachrock) occur in the intertidal zone, usually on the seaward side of cays (see Chapter 10). Many beaches merge landward with slightly elevated coarse deposits, called rubble (rounded fragments) or shingle (flattened fragments) ramparts. These ramparts often form ridges that line the island's periphery. Some ramparts stand only a few 10's of cm high (resembling a storm berm), but more commonly (especially on Pacific atolls), they stand 3 to 6 m high and represent the highest elevations on the island.

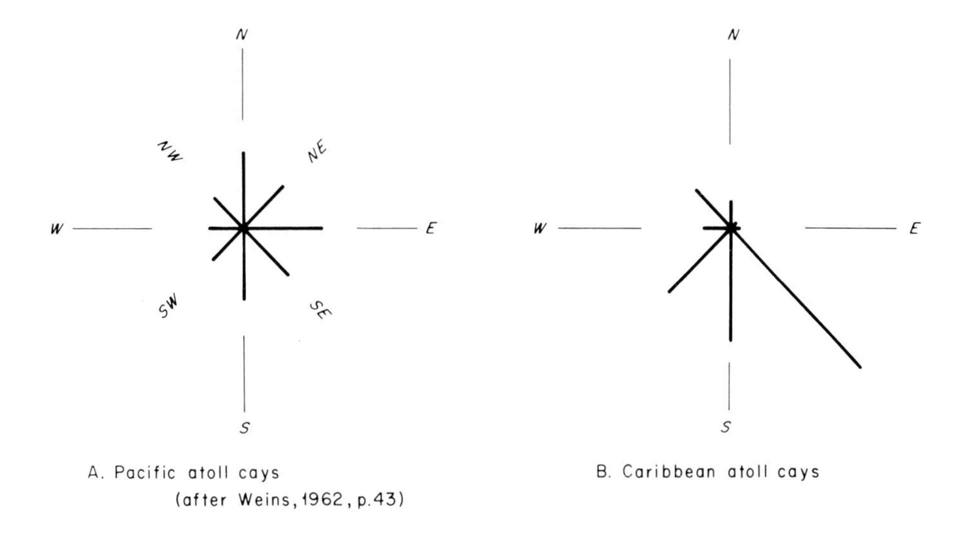

Fig. 60. Distribution of sand cays on the reef flats of Pacific and Caribbean atolls

Most Pacific atoll sand cays are located on windward reefs, while many Caribbean cays occur on more leeward portions of the reef flat (Fig. 60). Apparently the cays on many Indian Ocean atolls also occur on the more protected sides of the reefs (Kuenen, 1950). Since most Indian Ocean and Caribbean reef flats lack algal ridges, they probably are exposed to a greater amount of incoming surf than the more protected Pacific reef flats (most of which have algal ridges), thus explaining the more leeward location of sand cays.

Attempts to classify cays have resulted in recognition of 6 sand cay types: 1. submerged sand bar, possibly only emergent at low tide; 2. the simple low-lying sand or shingle cay; 3. the low-lying cay with pioneer standline plant communities, such as creepers and shrubs; 4. higher standing cays with a more complex and better-developed plant cover, often including trees; 5. sand cays with mangrove growth and sometimes mangrove swamps; 6. cays with exposed platforms or interiors of older rock (SPENDER, 1930; FAIRBRIDGE, 1950; FOLK, 1967).

A sand cay can be considered to be a wave of sediment migrating across a reef flat. Because of specific current patterns (STODDART, 1962), relict morphology (ZANS, 1958; EBANKS, 1967), as well as fortuitous circumstances, the migrating sand may accumulate into a low bank or emergent cay. If left as such, the cay will be ephemeral, as it is not likely to survive cataclysmic storms. The ability of a cay to survive depends upon its chances of attaining increased stability. Although rubble ramparts and beachrock may assist in forming a protective "exoskeleton" (WELLS, 1951), these features probably do not offer much protection during major storms (STODDART, 1963). Stabilization appears to be primarily dependent upon the degree and type of terrestrial vegetation cover. Plants aid both by anchoring existing soils within their root systems and by accreting new land through the trapping of windblown sand. Rarely will an unvegetated cay stand more than a few meters high, whereas vegetated cays can exceed 4 meters in height.

Vegetation cover slows but does not stop the eventual fate of cay migration into the lagoon. Beachrock, which is exposed by eroding shorelines, characterizes many seaward beaches; in some atolls it occurs in successive rows across the reef flat (Plate XXIV), each row indicating a pre-existing shoreline. Migration is not always gradual; typhoons and hurricanes can obliterate entire cays within the span of a few hours (see below).

Sand cay sediments generally are well-sorted and contain only small quantities of silt and clay. Being exposed to greater wind and wave energies, the seaward beaches usually contain coarser sediments than do lagoon beaches. FOSBERG and CARROLL (1965) found that the seaward beach sands in the northern Marshall Islands average 1.3 mm in diameter, while lagoon sands average 0.9 mm. STODDART (1964) reported a size range of 4 meters to 1 mm on the seaward side of Half Moon Cay, British Honduras; the lagoon sands generally are finer than 1 mm.

The composition of a beach sand is partly a function of sediment size and partly a function of source. Material coarser than a few millimeters in diameter is usually coral, mollusk or algal-coated debris. Finer sands generally are rich in foraminifera and *Halimeda*. Thus, the coarser seaward beaches may have distinctly different compositions than the finer lagoon sands, even though the sediment source may be the same. For the most part, however, the composition of a beach sand reflects its source, the surrounding reef flat. The compositions sometimes are so similar that only textural characteristics can differentiate certain reef flat and sand cay sediments (compare Tables 45, 46, 53 and 54; and Fig. 61).

The differences noted between Caribbean and Indo-Pacific reef flat sediments are also seen in the composition of their beach sands. Excluding the Bahamas, Caribbean beach sediments generally contain more than 30% *Halimeda* (Table 53) and lesser amounts of coralline algae and coral (see also STEERS and others, 1940; FOLK and ROBLES, 1964). Benthonic foraminifera are quantitatively unimportant, although EBANKS (1967) found that miliolids and peneroplids can contribute as much as 30% of the beach sand at Ambergris Cay, British Honduras. In contrast, benthonic foraminifera and coralline algae dominate Pacific beach sands, while *Halimeda* seldom constitutes more than a few percent (Table 54) (see also DAVID

Plate XXIV. Sand Cays. (a) Many sand cays begin as boulder piles on the windward sides of peripheral reefs. (b) With increased sand accumulation, a sand cay may form, but usually both vertical and horizontal growth are limited without vegetation. This unvegetated cay at Hogsty Reef is only about 0.5 meters high. (c) and (d) The effect of stabilizing role of cay vegetation can be seen at this cay at Hogsty Reef. The vegetated portion is more than one meter higher than the unvegetated portions of the cay. Note the beach rock in the foreground in (c). (e) The leeward migration of sand cays is evidenced by the occurrence of bands of beach rock windward of the cay, each band presumably representing a previous position of the windward beach. The cracks to which the man is pointing probably formed through the diurnal heating and cooling of the rock

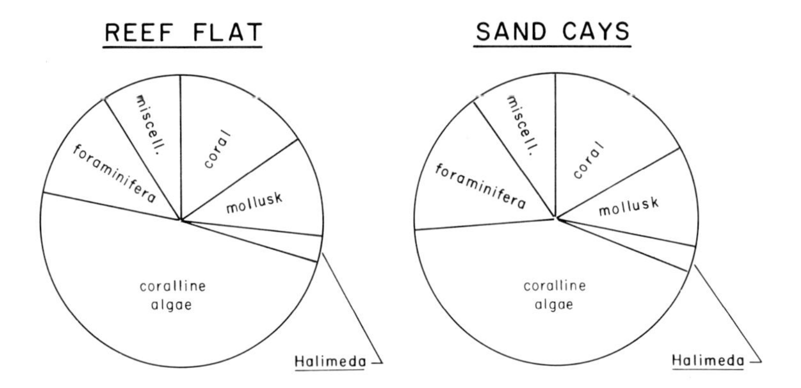

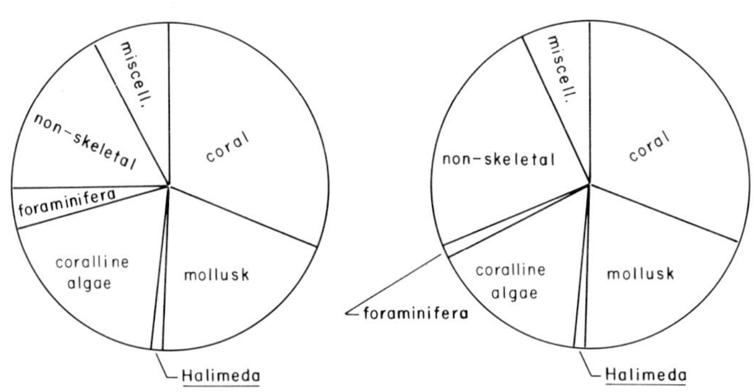

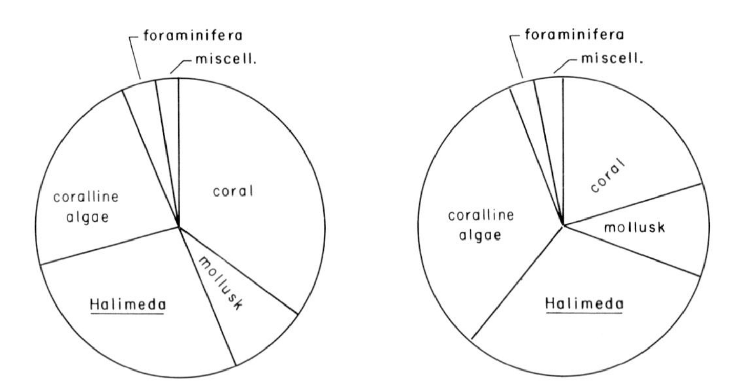

Fig. 61. Comparison between the composition of reef flat and sand cay sediment compositions at Kure Atoll, Hogsty Reef and Courtown Cays

Table 53. Composition of sand cay sediments from some Caribbean reefs

	Windward beaches Bahamas[a]	Leeward beaches Bahamas[a]	Hogsty Reef[b]	Isla Perez, Alacran Reef[c]	Windward Ambergris Cay[d]	Windward cays, Courtown[e]	Leeward cays, Courtown[e]	Albuquerque[e]
Skeletal fragments								
Coral	4		31	25	9	20	22	19
Mollusk	14	2	20		18	10	6	6
Foraminifera	18	2	1	15	7	3	3	3
Coralline algae	27	tr	16		15	33	10	22
Halimeda	5	1	1	60	35	31	54	39
Misc. skeletons	3	tr	1		3	tr	tr	1
Non-skeletal fragments								
Oolite	16	38	tr					
Pelletoids	tr	3	5					
Aggregates	6	49	1					
Cryptocrystalline lumps			18		6			6
Cay rock					4			5
Misc. non-skel.	6	6			2			

[a] ILLING, 1954. [b] MILLIMAN, 1967. [c] FOLK and ROBLES, 1964. [d] EBANKS, 1967. [e] MILLIMAN, 1969b.

Table 54. Composition of gravel and sand-size sediments on sand cays of some Indo-Pacific reefs

	Windward, Bikini Isl.[a]	Leeward, Bikini Isl.[a]	Hawaii[b]	Ifaluk[c]	Midway[d]	Windward, beaches, Kure[d]	Leeward, beaches, Kure[d]	Kayangel[e]
Skeletal								
Coral	20	45	9	26	19	19	15	10
Mollusk			33	10	9	13	10	6
Foraminifera	30	20	21	19	16	19	19	51
Coralline algae	35	25	32	47	47	38	47	9
Halimeda	tr	2		4	5	3	4	14
Misc. skeletons	10[f]	10[f]	5	3	2	2	3	3

[a] EMERY et al., 1954. [b] EMERY and COX, 1956. [c] TRACEY et al., 1961. [d] GROSS et al., 1969. [e] MILLIMAN, unpublished data. [f] Misc. includes mollusks.

and SWEET, 1904; NEWELL, 1956; MCKEE and others, 1959; SCHLANGER and BROOKHART, 1955; FOSBERG and CARROLL, 1965).

Effect of Cataclysmic Events (Storms) on Coral Reefs

Tropical storms (especially major storms such as hurricanes and typhoons) are both major constructive and destructive sedimentary agents in tropical eulittoral carbonate areas. Storms can deposit large quantities of silt on animals; they also can uproot corals, branching corals (such as *Acropora*) being the most easily destroyed (BLUMENSTOCK, 1961; STODDART, 1963). Lower osmotic pressures also can force corals to expell their zooxanthellae; although this will not kill the corals, it probably will reduce their ability to calcify (GOREAU, 1964). Islands can be destroyed by major storms, although the ability of an island to survive in large part depends upon the type and degree of vegetation cover (see above). Eroded sand cays often "migrate" into the leeward lagoons or sometimes seaward onto the reef front. Storms can bisect a reef flat with new channels that drain into the lagoon. Generally these destructive phases occur during the early stages of a hurricane; the constructure (depositional) phases occur mostly during the waning stages (VERMEER, 1963). The speed and degree of recovery of a community depends upon the specific organisms involved. For instance, grass beds usually recover within a short time because of their rapid regenerative powers (THOMAS and others, 1961). The repopulation of the reefs may take longer, but a large amount of reconstruction can occur within a few years (BLUMENSTOCK and others, 1961).

Sedimentary remnants of storms can remain as permanent deposits. Large amounts of rubble can be deposited both on islands (BLUMENSTOCK, 1961; MCKEE, 1959) and on reef flats (BALL and others, 1967). Incongruent layers of sand in muddy areas and mud in sandy areas, as well as the formation of spill-over lobes on the leeward side of breached islands are common (BALL and others, 1967). But probably the most permanent deposition occurs in supratidal environments, where sediments are inaccessible to subsequent reworking by normal marine processes (BALL and others, 1967; HIGH, 1969).

Not all hurricanes are capable of extensive erosion and deposition. Five years after Hurricane Donna struck southern Florida and the Bahamas (BALL and others, 1967) Hurricane Betsy arrived, but the geological effects were substantially less. The only significant reef damage was to the more fragile organisms. Both intertidal and supratidal sedimentation was extremely limited (PRAY, 1966; PERKINS and ENOS, 1968). Apparently these areas had not yet reached the pre-Donna conditions when Betsy arrived. Even so major storms play a much more important role in both sedimentation and erosion of reefs and lagoons than their short time span would suggest.

Other Tropical and Subtropical Carbonates

Although coral reefs, lagoons and non-skeletal banks have monopolized warm-water carbonate investigations, several other types of tropical and subtropical carbonate assemblages should be mentioned. Most prominent are the reefs formed by vermetids, serpulids and oysters.

Vermetid Reefs

One of the most common but least studied carbonate-producing organisms in the tropics and subtropics is the tube-building gastropod, belonging to the family

Vermetidae. These gastropods form prominent encrustations on many rocky intertidal zones in the Caribbean (LEWIS, 1960), Brazil (KEMPF and LABOREL, 1968), Florida (SHIER, 1965, 1969), Bermuda (STEPHENSON and STEPHENSON, 1954; GINSBURG and others, 1971) and the Mediterranean (MOLINIER, 1955; PÉRÈS, 1967b). KEEN (1961) has reclassified the entire family Vermetidae and has discussed ways to distinguish them from serpulids, with which vermetids often are confused.

Vermetids commonly flourish in warm marine waters; temperatures should not fall below freezing nor should the waters become too brackish (SHIER, 1969). Most vermetids also require clear, agitated waters, although at least one species, *Vermetus nigricans*, can thrive in turbid water (SHIER, 1969). All vermetids require a hard substrate on which to settle. In rocky intertidal zones vermetids commonly occupy the lower to middle part of the zone. They also thrive on hard substrate that lie just below the lower tide level.

Periods of rapid reef growth apparently alternate with long intervals of non-growth. For example SHIER (1969) reported that during the Holocene transgression of sea level, vermetids were able to form long barrier reefs, 3 meters thick, along the southwestern Florida coast. These reefs at present, however, are devoid of living vermetids and are being eroded at a relatively rapid rate. The reason for the mass mortality is not known, but apparently such periods of growth and death have occurred in the past (SHIER, 1969).

Another type of vermetid growth occurs in form of "Boiler Reefs" that line on the edges of the Bermuda Platform. These reefs are composed primarily of vermetids, milleporan corals and coralline algae. Internal cavities are filled with other skeletal components, such as *Homotrema rubrum*, and internal cement. Contrary to the conclusions of some workers (NEWELL, 1959; STANLEY and SWIFT, 1967) these reefs do not represent thin veneers over Pleistocene aeolian limestone, but rather are sites of rapid accretion of Holocene limestone. GINSBURG and others (1971) report that rock $3^1/_2$ m below the living surface of a Bermudan cemented reef is 2980 ± 160 years old, implying a growth rate of more than 1 m per 1000 years.

Vermetid accretion occurs on many rocky shorelines in the Mediterranean and can accentuate the inter-tidal nip (MOLINIER, 1955; PÉRÈS, 1967b) (Plate XXVa). This zone, which is cemented by large amounts of coralline algae (often coralline algae is dominant or even exclusively present), is called the Trottoir zone (PÉRÈS, 1967a).

Serpulid Reefs

Although serpulids encrust many substrates, very few serpulid reefs have been investigated. Most of the reported occurrences actually are mis-identified vermetid reefs. The best studied reefs occur along the southeastern Texas Gulf Coast, expecially in Baffin Bay. In this hypersaline lagoon, reefs form crusts up to $\frac{1}{2}$ m thick over a Pleistocene coquina (ANDREWS, 1964; DALRYMPLE, 1964). Unlike many vermetid reefs, these reefs are subtidal, the average water depth being about 0.6 to 1 meter (DALRYMPLE, 1964). The Baffin Bay serpulid reefs show signs of cyclic growth; laminar bands of barnacles occur within the reef, suggesting that serpulid growth was interrupted periodically and followed by barnacle growth (Dalrymple, 1964). In addition to barnacles, these reefs also can be associated with oysters and bryozoans (Shier, 1969).

Although growth rates are not well documented, serpulid reefs are capable of rapid accumulation. Behrens (1968) has calculated that serpulids growing in water tunnels off southeastern Texas grow at a rate of about 8 cm per year. Dalrymple (1964) has estimated that some of the Baffin Bay reefs have accreted more than 3 cm per year over the past 40 years.

Oyster Reefs

Oyster reefs are probably the most widely distributed carbonate reefs in the modern oceans. They occur in latitudes ranging from 64° N to 44° S (Galtsoff, 1964). Unlike other carbonate producers, oysters possess considerable nutritional and economic value, and therefore their biology has been studied extensively (Galtsoff, 1964). Oyster reefs also provide excellent habitats for other carbonate-producing organisms. Large populations of gastropods, bivalves, barnacles, crustaceans and serpulids owe their presence to the reef substrate. The boring sponge *Cliona* can destroy much of the shell material and thus seriously weaken the reef structure. In spite of the many studies on modern oyster reefs (as well as their Tertiary and Pleistocene counterparts, few studies have concentrated on either the sedimentary aspects of an oyster reef or the ecologic interactions between the various other hard-shelled invertebrates. Two comprehensive studies of modern oyster reefs and their fossil counterparts have been made by Ladd and others (1957) and Wiedemann (1972).

Oyster reefs generally are confined to shallow depths, between the intertidal zone and 30 meters. Most oyster reefs grow in slightly brackish waters, such as estuaries, but different genera exhibit different degrees of tolerance to salinity depression. *Crassostrea* can stand salinities as low as 5 to $10^{0}/_{00}$ for brief intervals, but the genus *Ostrea* is limited to more oceanic conditions (Galtsoff, 1964). Parker (1960) reported that *O. equestris* tends to replace *C. virginica* in Texas Gulf Coast oyster reefs when salinities become relatively high. Oysters cannot tolerate freezing temperatures for more than a few hours, thus limiting higher latitude reefs to subtidal depths (Galtsoff, 1964). Colder water forms often are inactive and will not feed or grow during winter months.

Oyster reefs also require firm substrate on which to attach; this includes rocks, old shells and firm mud. Loose mud and sand are not conducive for oyster development. Rapid deposition of fine-grained sediments can limit or kill off oyster reefs. Galtsoff (1964) mentions that "silting" by the Colorado River destroyed more than 6000 acres of oyster beds in Matagorda Bay (Texas) in a span of 36 years.

Optimal oyster development is restricted to estuaries, where the oysters are exposed to the slightly saline waters and the currents necessary to retard sedimentation. In the Texas Gulf Coast, oyster reefs tend to build perpendicular to the circulation patterns (Parker, 1960), probably in order to maximize the current effect. Reefs often resemble a low mound with a high center; living oysters and

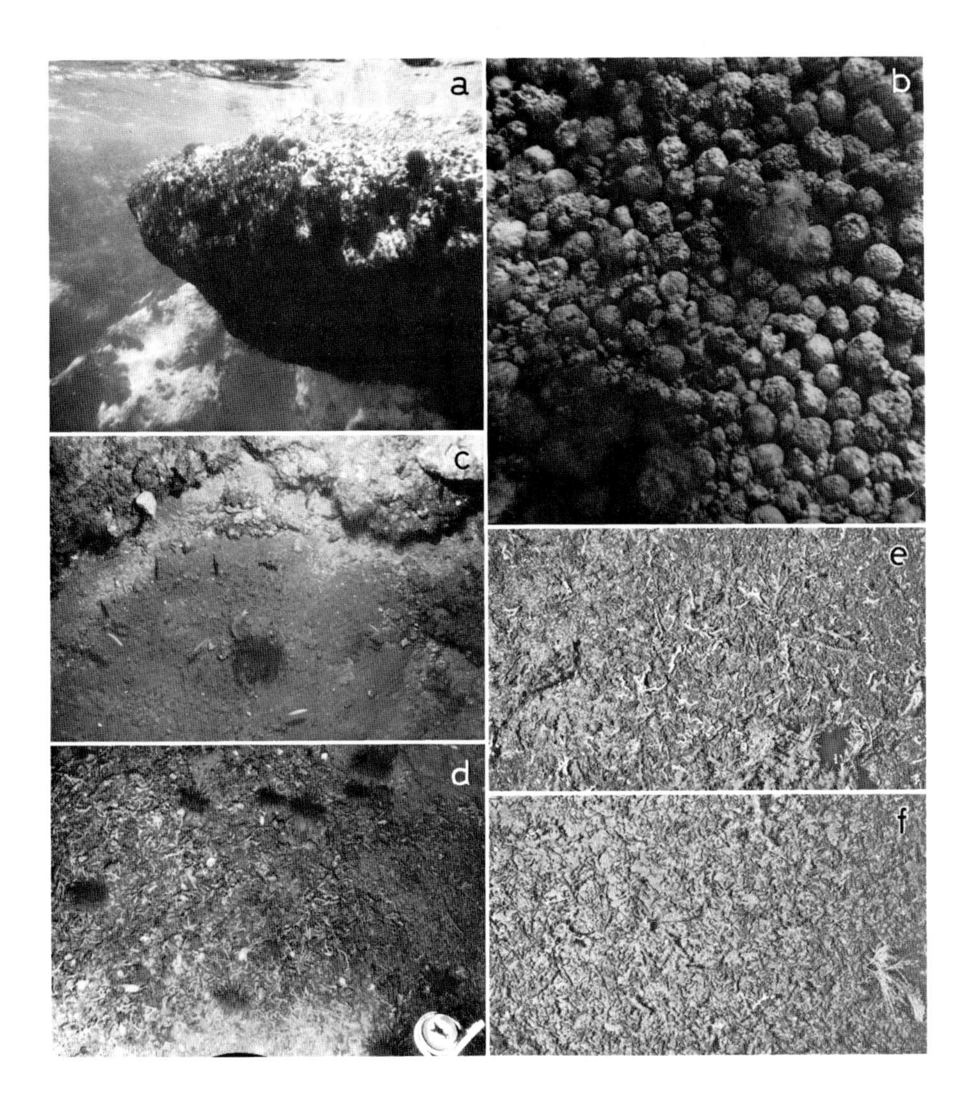

Plate XXV. Miscellaneous Carbonate Environments. (a) "Trottoir" zone, composed of coralline algae and serpulids; intertidal zone, Rhodes, Greece. (b) Algal balls, continental shelf off Canary Islands. Picture courtesy of J. MÜLLER. (c) Algal ridge outcrop off North Carolina. Water depth is 50 m. Photograph courtesy of I. G. MACINTYRE. (d) Oolitic ridge at shelf break off eastern Florida. The ridge is topped with growths of the coral *Oculina recta* and abundant molluscan debris. Water depth is 66 m. Photograph courtesy of I. G. MACINTYRE. (e) and (f) Deep-sea coral reefs on the Blake Plateau. Some corals in the upper photograph are living (white color), but the lower photograph shows mostly dead debris. Water depth is about 500 m

associated organisms are concentrated on the shoulders of the reef (LADD and others, 1957).

Oyster reefs seem susceptible to the same sudden mass mortality seen in serpulid and vermetid reefs. Estuaries are ephemeral features and thus oyster reefs can not survive for long periods of time. Either they build themselves out of existence (after they reach sea level, they must extend laterally) or they are buried by the shoaling estuary. Emergent reefs can serve as mud traps and subsequently be settled by intertidal and supratidal grass beds, while the living reef continues to accrete laterally around the periphery (WIEDEMANN, 1972).

Non-Tropical Carbonates

Many geologists and oceanographers view the temperate and polar regions as areas of detrital sedimentation; carbonate sedimentation is considered unimportant. In large measure this assessment is true. For instance, beach sands along the eastern United States north of Cape Hatteras average about 1% calcium carbonate, and the beaches as far south as Palm Beach (south-central Florida) contain no more than a few percent. However, given the proper conditions carbonate sedimentation can be important in higher latitudes. As in the tropics, the importance of carbonate sedimentation is dependent upon both the rates of detrital and carbonate sediment accumulation. In areas where the substrate is available for attachment, where biologic productivity is sufficiently high, and where detrital deposition is low, carbonates can constitute an important shallow-water component. In western Ireland, for example, KEARY (1967) found that in those areas in which detrital influx was small (such as Galloway Bay), carbonate content in the beach sediment can exceed 50%. Nearby areas with large influxes of detritus, on the other hand, have considerably smaller carbonate contents.

Many carbonate-rich areas occur on higher latitude continental shelves, and these are discussed in Chapter 7, but there are also excellent examples in the eulittoral environment (Table 55). In most instances high-latitude shallow-water carbonates are concentrated in coarse sands and gravels. Generally corals and green algae are absent, and echinoids, bryozoans and foraminifera are rare; BLANC (1968), however, reports that in the *Posidonia* and *Cymodocea* banks ("herbiers") of the northern Mediterranean, foraminifera (miliolids and nonionids) are common, as are ostracods and bryozoans. Mollusks, barnacles and coralline algae dominate most non-tropical carbonates, although the relative amount of each component depends upon the specific area and type of bottom. In the Gulf of Maine, the dominant carbonate contributors in sandy areas are mollusks (RAYMOND and STETSON, 1932), but on hard or rocky bottoms, encrusting red algae and barnacles also are important. Similarly, the rocky areas off Brittany provide ideal substrate on which the coralline alga *Lithothamnium calcareum* can attach, and this one species supplies the bulk of the carbonate to these carbonate-rich nearshore and beach sands (GUILCHER, 1964; VANNEY, 1965).

One would expect that cold waters, being undersaturated with respect to calcium carbonate, would limit the distribution of shallow-water carbonates in higher latitudes, and to some extent this must be true. For example, in many

Table 55. Some carbonate-rich sediments from high latitude intertidal and shallow-water environments

Area	Latitude	Water depth	Average carbonate	Major components	Worker
Gulf of Maine	44° N	Beach and nearshore	26–67%	pelecypods, barnacles gastropods	RAYMOND and STETSEN (1932)
John O'Groats, Scotland	59° N	Beach	>90%	pelecypods	RAYMOND and HUTCHINS (1932)
Brittany (France)	49° N	Beach to 10 m	Locally as great as 75%	coralline algae	VANNEY (1965) GUILCHER (1964)
Western Ireland	53° N	Beach	Mostly greater than 50% locally 100%	*Lithothamnium* barnacles mollusks	KEARY (1967)
Sitka Sound, Alaska	57° N	Beach and inner bay	66%	barnacles (74%) mollusks (12%)	HOSKIN (1971)

rocky areas around New England the mussel *Mytilus* and the red alga *Amphiroa* sp. are prominent organisms. With their expected high productivity and the lack of terrigenous influx in this area, one would expect high carbonate content in the nearby sediments, but seldom do the values exceed a few percent. This suggests that to some extent the carbonate must be dissolved, probably during the colder winter months. ALEXANDERSSON (1972a, c) has shown that biological boring of carbonates can hasten greatly the dissolution of skeletal fragments in higher latitude sediments. Another possible explanation is that these sediments are reduced mechanically, either by mechanical abrasion or through the ingestion and chewing by browsing animals; in their reduced size they are more easily dissolved during cold periods. RAYMOND and STETSON (1932), however, concluded that the coarse nature of the carbonate sediments along the Maine coast is not the result of preferential dissolution, but rather the preferential removal and transport of the fines by waves and longshore currents.

HOSKIN and NELSON (1969) have suggested that shallow-water carbonate sediments in higher latitudes are generally richer in calcite and poorer in aragonite and magnesian calcite than their tropical counterparts. This is probably true with respect to aragonite. Not only are corals and green algae (both of which are aragonitic) absent, but organisms with mixed mineralogies tend to have lower amounts of aragonite in colder climates (see Chapters 4 and 5). However, abundance of coralline algae, bryozoans and foraminifera in some deposits suggests that magnesian calcite can be an important mineral. The amount of magnesium within the cold-water calcite, however, should be considerably below levels found in the tropics (see Chapters 4 and 5).

Chapter 7. Transitional (Sublittoral) Carbonate Sedimentation

The sublittoral environment occupies water depths between 20 and 200 meters (see above). This interval coincides with the depth range of most continental shelves and some upper slopes. Carbonate sediments within this habitat generally represent a transition between shallow-water and deep-sea facies. Many sublittoral carbonates consist of material deposited during the last rise in sea level, 5 to 15 thousand years ago (MILLIMAN and EMERY, 1968), admixed with present-day sublittoral carbonates and planktonic components. Much of the discussion in this chapter is taken directly from a paper presented during an AGI (American Geological Institute) lecture (MILLIMAN, 1971).

Distribution of Calcium Carbonate on Continental Shelves

Because of the dependency of calcium carbonate solubility on water temperature, EMERY (1968) suggested that biogenic carbonate sedimentation increases in lower latitude shelves and correspondingly, is more important on the warmer western sides of ocean basins than on the colder eastern sides (Fig. 62). The increased carbonate content in shelf sediments south of Cape Hatteras (Fig. 63),

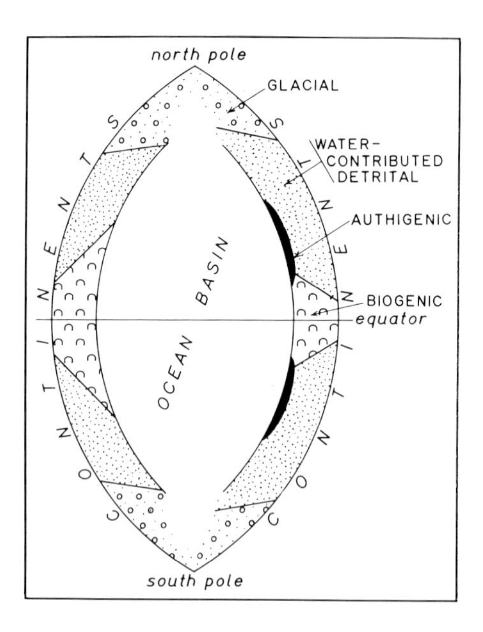

Fig. 62. Idealized distribution of biogenic sediments on continental shelves. From EMERY (1968)

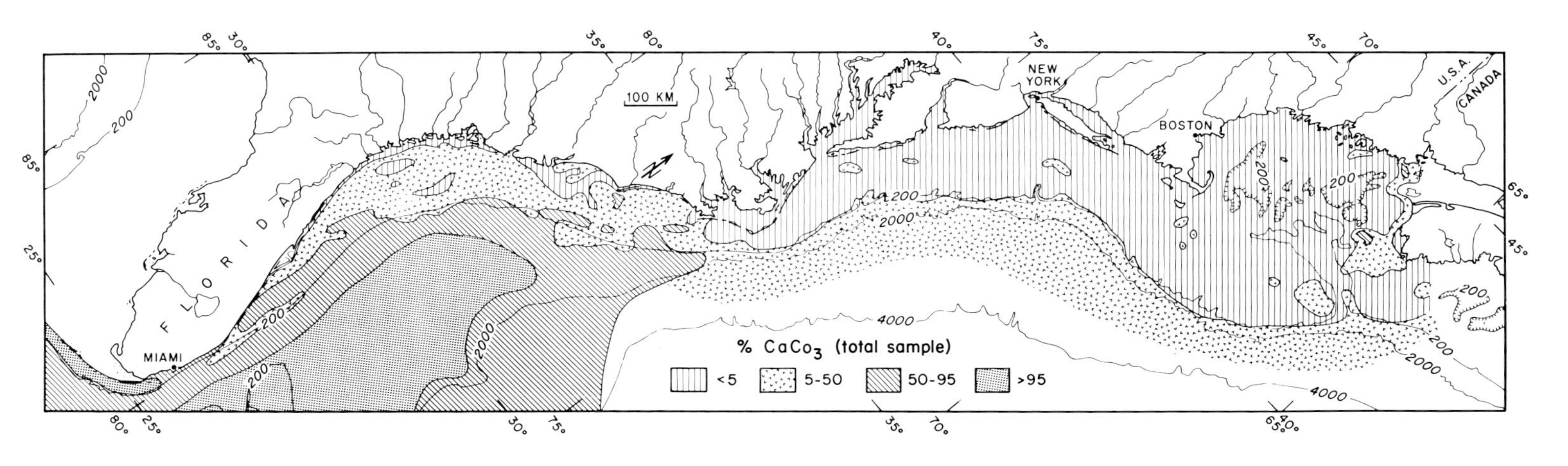

Fig. 63. Calcium carbonate content in suficial sediments of the eastern U.S. continental shelf and slope. From MILLIMAN and others (1972a)

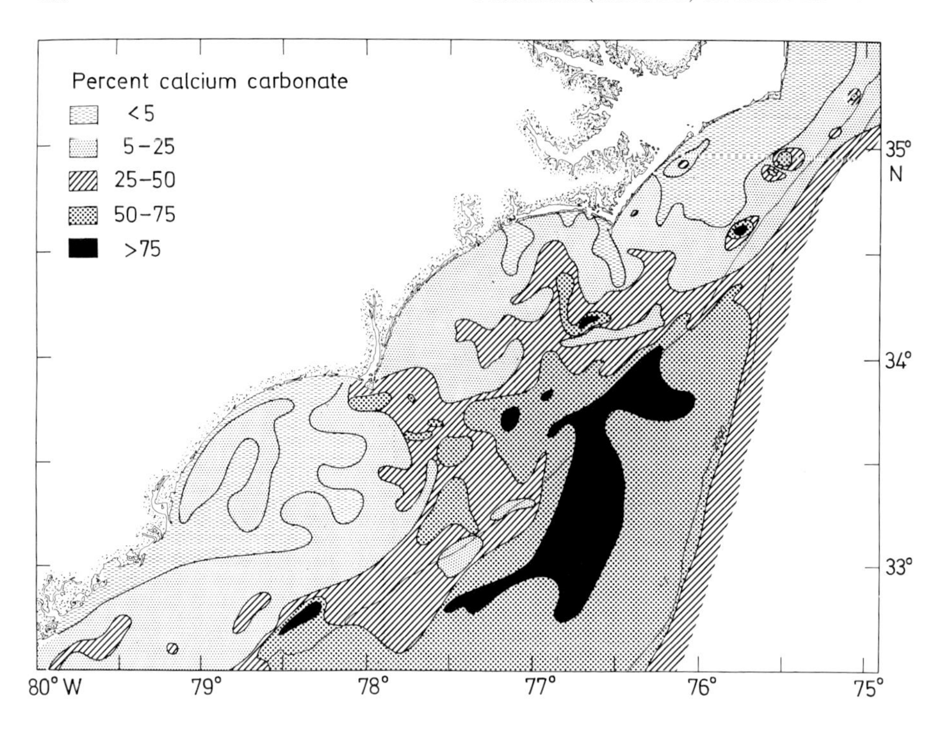

Fig. 64. Distribution of calcium carbonate in three "bays" on the continental shelf off North Carolina (U.S.A.); from north of south the bays are Raleigh Bay, Onslow Bay and Long Bay. From MILLIMAN and others (1968)

for instance, corresponds to a marked increase in water temperature (CERAME-VIVAS and GRAY, 1966; MILLIMAN and others, 1968). Similarly, sediments off eastern Asia display increased carbonate content with decreasing latitude (NIINO and EMERY, 1966).

Unfortunately, the correlation between carbonate content and water temperature breaks down when one takes into account the role of terrigenous sedimentation. The influx of terrigenous sediment not only dilutes carbonate grains, but also can bury many carbonate-producing organisms. Thus, in areas where terrigenous influx is small or where currents prevent deposition, carbonate sediments will predominate. For example, carbonate content is very high on the shelf adjacent to the Sahara Desert (northwestern Africa), but decreases markedly to the south, where tropical rivers contribute large quantities of terrigenous sediments (MCMASTER and others, 1969). Similarly, sediments on tropical shelves adjacent to large rivers may contain far less carbonate than sediments on polar shelves (such as Labrador and Alaska) where terrigenous influx may be small (MULLER and MILLIMAN, 1973; HOSKIN and NELSON, 1969). On a smaller scale, low piedmont river runoff into Onslow Bay (North Carolina) results in relatively carbonate-rich sediments, averaging about 35% $CaCO_3$ (Fig. 64). In contrast, the neighboring areas of Raleigh Bay and Long Bay receive considerably more terrigenous sediment from the Pamlico, Cape Fear and Pee Dee Rivers, and as a result contain far less carbonate than does Onslow Bay.

Because terrigenous sedimentation usually is greater in nearshore areas, the carbonate content of shelf sediments generally increases offshore. Maximum carbonate concentrations usually occur on the outer shelf or upper slope.

In summary, the carbonate content of a shelf sediment is a first-order function of sediment supply and a second-order function of water temperature and corresponding carbonate productivity. When dealing with the composition of a carbonate assemblage, however, the importance of these two factors often is reversed.

Carbonate Sediments on Continental Shelves

Most carbonates on continental shelves are sand-size. However, such components as mollusk shells and limestone fragments can contribute significant quantities of gravel in some areas. For instance, most of the gravel occurring on the shelf and upper slope south of Cape Hatteras is carbonate (MILLIMAN, 1972). Silt and clay-size carbonates are derived from bio-mechanical degradation of larger particles (MOLNIA and PILKEY, 1972) and to a lesser extent, from microconstituents, such as spicules and coccoliths. Aragonitic mud (interbedded with ooids) that occurs on the outer shelf off western India may be a relict of shallow-

Table 56. Carbonate assemblages within the surficial sediments of the Atlantic continental margin, eastern United States. After MILLIMAN and others (1972)

Carbonate assemblage	Average percent $CaCO_3$ (range in parentheses)	Diagnostic components (range of values in parentheses)
1. Mollusk		
North of Cape Hatteras	3 (tr–45)	Mollusk (60–100)
South of Cape Hatteras	12 (2–60)	Mollusk (45–95+)
2. Echinoid		
North of Cape Hatteras	2 (tr–5)	Echinoid (20–100); Mollusk (0–80)
South of Cape Hatteras	5 (2–15)	Echinoid (20–50); Mollusk (40–75)
3. Barnacle	55 (5–95)	Barnacle (20–55); Mollusk (25–70)
4. Barnacle—Coralline algae	61 (25–95)	Barnacle (5–55); Coralline algae (10–45)
5. Oolite	55 (10–95)	Ooids (10–55); Calcareous pellets (5–55)
6. Coral reef	95 (80–99)	Coral (10–20); *Halimeda* (10–35); Coralline algae (15–35)
7. Benthonic foraminifera	2 (1–5)	Benthonic foraminifera (20–70); Mollusk (15–70)
8. Planktonic foraminifera	(35–95)	Planktonic foraminifera (60–80)
9. Planktonic foraminifera-Coral	90 (85–99)	Planktonic foraminifera (10–80); Coral (15–90)

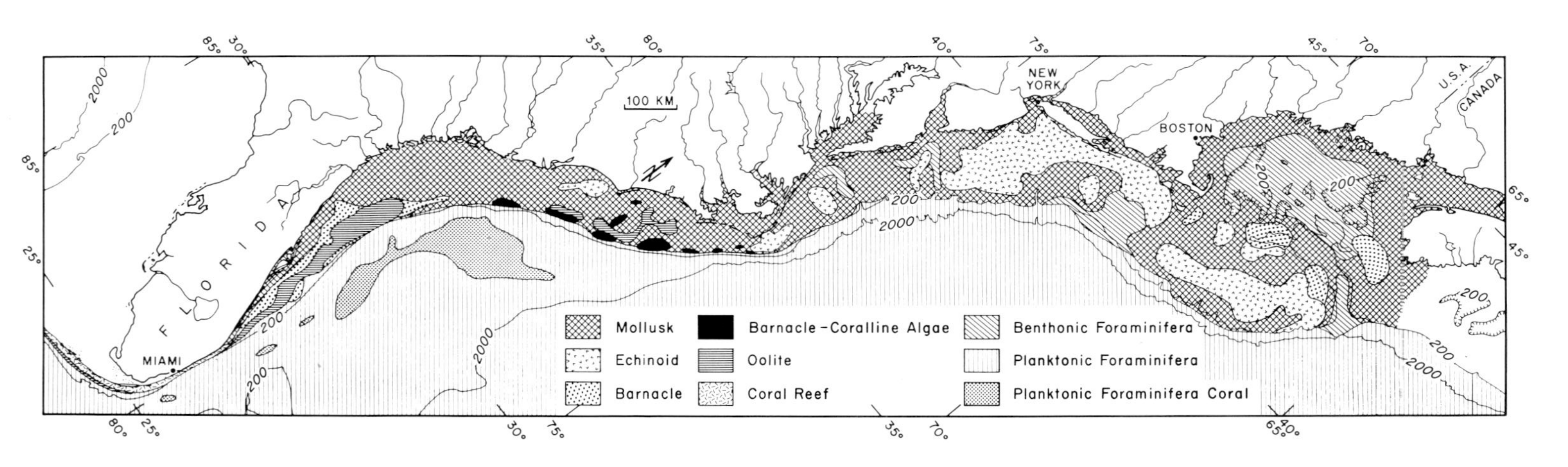

Fig. 65. Carbonate assemblages off the eastern United States. From MILLIMAN and others (1972a)

water deposition in an environment similar to the present-day Bahamas (VON STACKELBERG, 1970). Most other fine-grained shelf carbonates probably reflect modern deposition.

Most shelf sediments contain relatively few major carbonate components; 4 to 6 components make up more than 90% of the carbonate fraction in most sediments. On the basis of the relative abundance of major components plus the carbonate content, continental shelf sediments along the eastern U. S. have been catagorized into seven carbonate assemblages; two others occur on the continental slope and Blake Plateau (Fig. 65; Table 56).

When carbonate components and assemblages from various shelves throughout the world are compared, many similarities are found. These similarities will be discussed in the following paragraphs.

Inner Shelf Sediments

Inner shelf environments usually are exposed to greater sediment influx (from neighboring rivers) and greater siltation (because of generally lower current activities) than are outer shelves. As a result, many inner shelves are covered with relatively fine-grained, low-carbonate sediments. Carbonate-producing organisms found on the inner shelf are predominately infaunal filter-feeders (such as bivalves) and bulk-detritus feeders (such as echinoids) (see EMERY and others, 1965). In sandy areas, mollusks and echinoid fragments constitute most of the carbonate; in muddier sediments, benthonic foraminifera also are important (Figs. 63 and 71) (see also GAMULIN-BRIDA, 1967; KORNICKER and BRYANT, 1969; SARNTHEIM, 1971).

In nearshore areas in which surface and bottom currents are sufficiently fast to prevent terrigenous sedimentation or in which outcrops stand above the surrounding bottom, more elaborate carbonate assemblages can be found. These restricted assemblages resemble those occurring on the outer shelf, and contain such organisms as branching coralline algae, bryozoans and ophiuroids (PEARSE and WILLIAMS, 1951; GUILCHER, 1964; BOILLOT, 1964, 1965; OELTZSCHNER and SIGL, 1970) (Figs. 66, 67 and 68).

Outer Shelf Sediments

In those areas exposed to extensive terrigenous sedimentation, outer shelves may contain similar carbonate concentrations and carbonate assemblages as those on the inner shelf (Figs. 65, 66, 69). The major difference between these inner and outer shelf assemblages may lie in the relative abundance of benthonic (inner shelf) and planktonic (outer shelf) foraminifera.

Because of the greater current activity and the larger distances from rivers, many outer continental shelves contain sediments that are significantly richer in carbonate than sediments on the adjacent inner shelf. Moreover, carbonate components tend to be more diversified than those on the inner shelf. Off the eastern United States, for instance, mollusks and echinoids seldom constitute

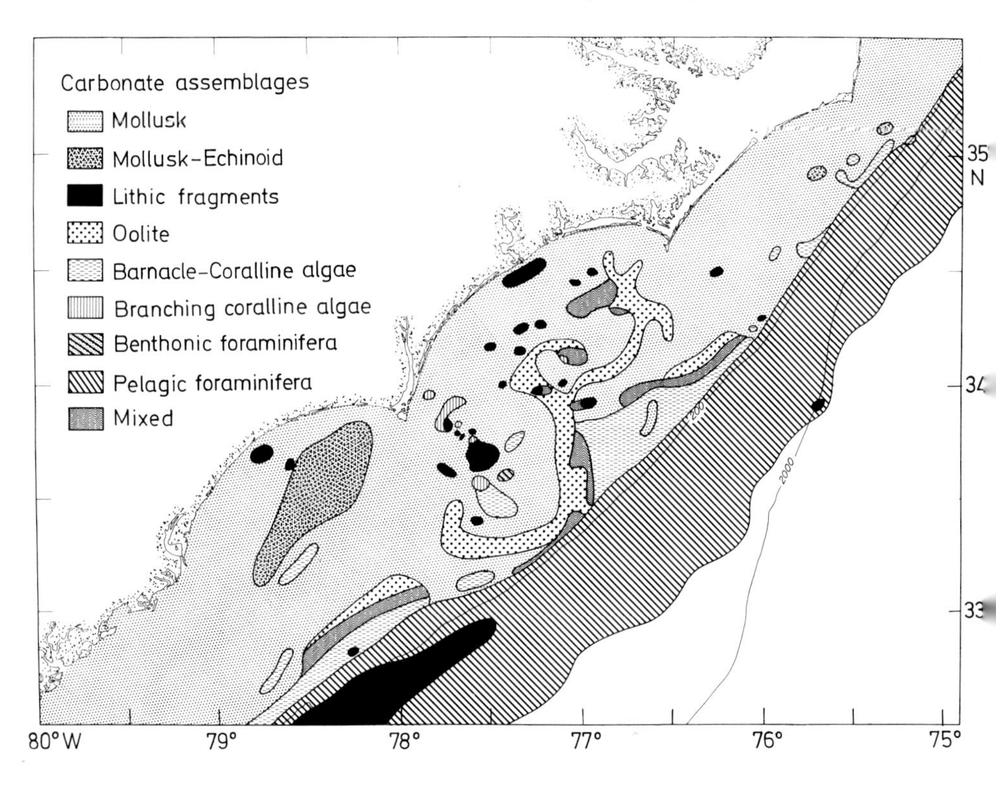

Fig. 66. Carbonate assemblages on the continental shelf off North Carolina. From MILLIMAN and others (1968)

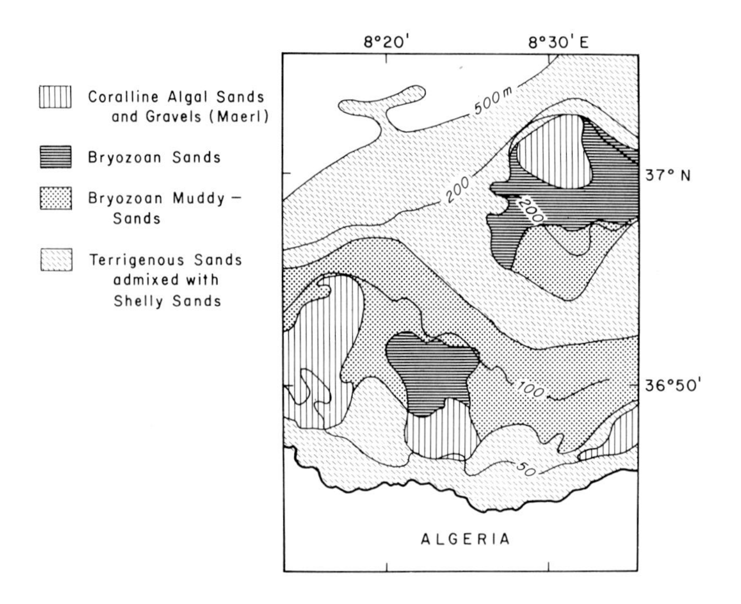

Fig. 67. Carbonate assemblages on the Algerian continental margin. After CAULET (1971)

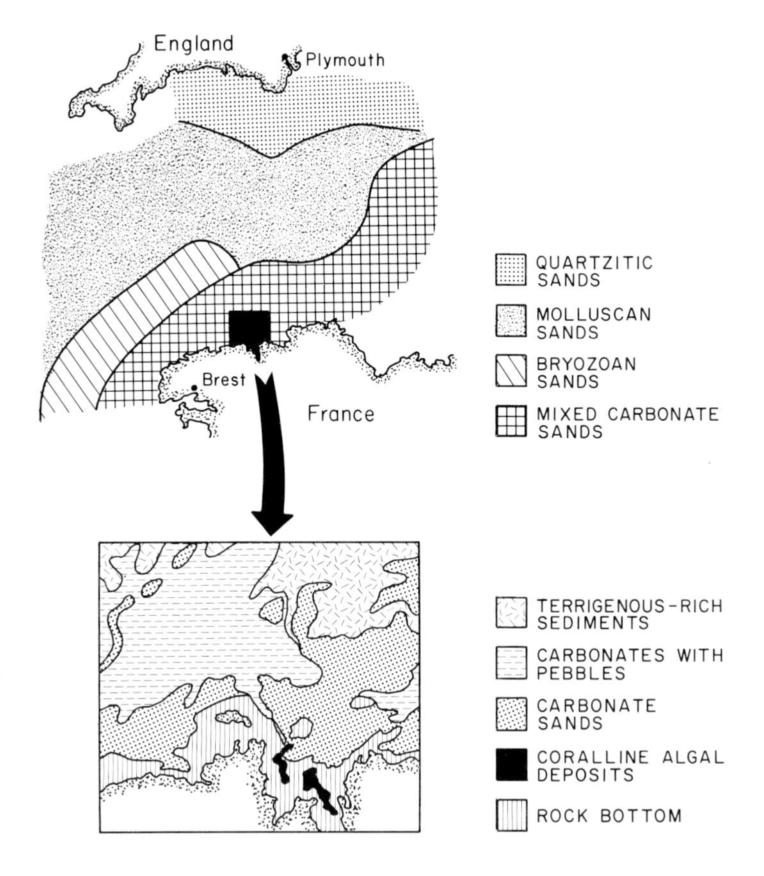

Fig. 68. Distribution of sediment types in the English Channel off northern France. Because of the swift currents and lack of sedimentation, sediments in the southern English Channel are rich in carbonate content. After BOILLOT (1965)

more than about 40% of the total carbonate fraction on the carbonate-rich outer shelves, whereas they rarely comprise less than 60 to 90% on the inner shelf. Prominent components on the outer shelf include coralline algae, epibenthic filter feeders (such as bryozoans, barnacles, sponges, tunicates and corals), benthonic and planktonic foraminifera, and oolite deposits (Plate XXVI and XXVII). Other components, such as serpulids, *Halimeda*, echinoids and ophiuroids, can be locally prominent.

Coralline algae and epibenthic filter feeders usually require hard substrate, such as rock outcrops or shells, upon which to grow. Branching coralline algae, however, also can grow in sand or gravel. In the Mediterranean Sea, algal sands (*Lithothamnium calcareum* and *L. solutum*) form a prominent shelf carbonate facies, called "maerl" (HUVE, 1956; PÉRÈS and PICARD, 1958; JACQUOTTE, 1962) (Plate XXVII b); a similar assemblage occurs off northeastern Brazil (ZEMBRUSCKI, 1968; COUTINHO and MORAIS, 1971) as well as off northern Africa (SUMMERHAYES, 1970; MILLIMAN, in preparation). In some warm shelves such as Brazil, the

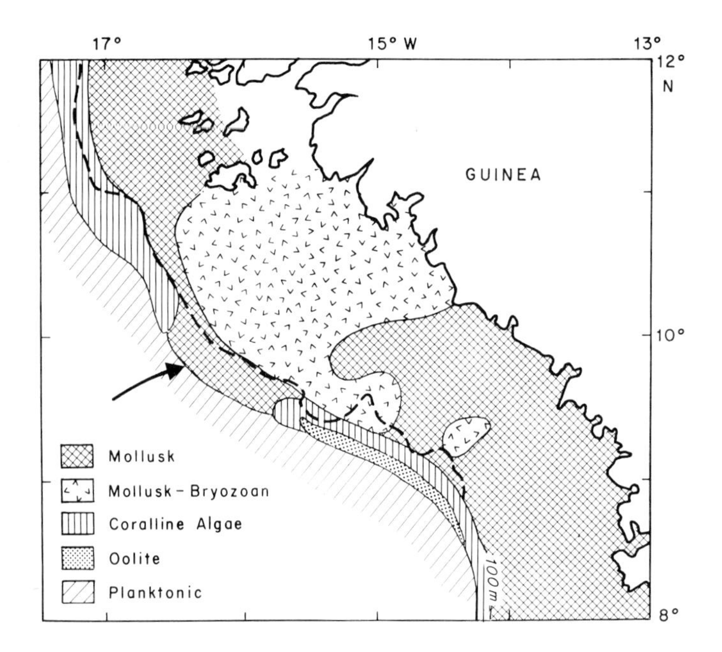

Fig. 69. Carbonate assemblages off Guinea and Sierra Leone. Note that the outer shelf and upper slope sediments off the Orango Delta (arrow) contain distinctly different assemblages than the sediments on either side. Presumably this reflects the lower carbonate content and the higher rate of terrigenous sedimentation in the area adjacent to the Orango Delta. After McMASTER and others (1971)

Mediterranean and the southeastern United States, *Halimeda* also can be an abundant component (MABESOONE and TINOCO, 1965; KEMPF, 1970; PÉRÈS, 1967a; MILLIMAN, 1972).

Encrusting coralline algae, together with corals, bryozoans and barnacles, form a prominent facies along many outer shelves (Figs. 65, 66, 67, 70, 71 and 72) (Plate XXVI). The French refer to such hard substrate communities as the "coralligenous facies" (PÉRÈS and PICARD, 1964; PÉRÈS, 1967b). Along the outer edges of many world shelves, coralline algae play an important role in the formation of the "algal ridge" system, which will be discussed in a following section.

Coralline algae also occur as algal nodules and balls. Generally these sediments are restricted to outer shelf and upper slope depths (MENZIES and others, 1966; PÉRÈS and PICARD, 1958; LABORAL and others, 1961; McMASTER and CONOVER, 1966; MÜLLER, 1969; MILLIMAN and others, 1973). They apparently are capable of accreting in water depths exceeding 100 meters (MILLIMAN and others, 1972b).

Epibenthic filter feeders usually grow wherever there is a suitable substrate upon which to attach. On the shelf off the eastern United States and Canada, the major epibenthic carbonate producers are barnacles (Plate XXVI c, d). Most barnacle fragments occur near the algal ridges off the southeastern United States and in rocky areas off Georges Bank, Nova Scotia and the Grand Banks (Fig. 65). Strangely, however, barnacle-rich sediments are rare on most other shelves,

Plate XXVI. Continental Shelf Carbonate Sands. (a) Relict coral reef sand from the Florida shelf. Dominant components are coralline algae, coral and *Halimeda*. Scale is 1 cm. (b) Bryozoan-rich sand from the north African shelf. Other components include coralline algae and mollusks. These sediments contain both relict and modern components. Scale is 1 cm. (c) Mollusk-barnacle sand from the central Florida Shelf, eastern United States. The mollusks are primarily relict, while the barnacles, judging from their fresh appearance, are modern. Scale is 5 mm. (d) Mollusk-barnacle sand from Grand Bank, Labrador. This mechanically and biologically degraded sediment has been dated at 19000 years B. P., and probably was deposited as an intertidal deposit during the last lower stand of sea level (MÜLLER and MILLIMAN, in press). Scale is 5 mm

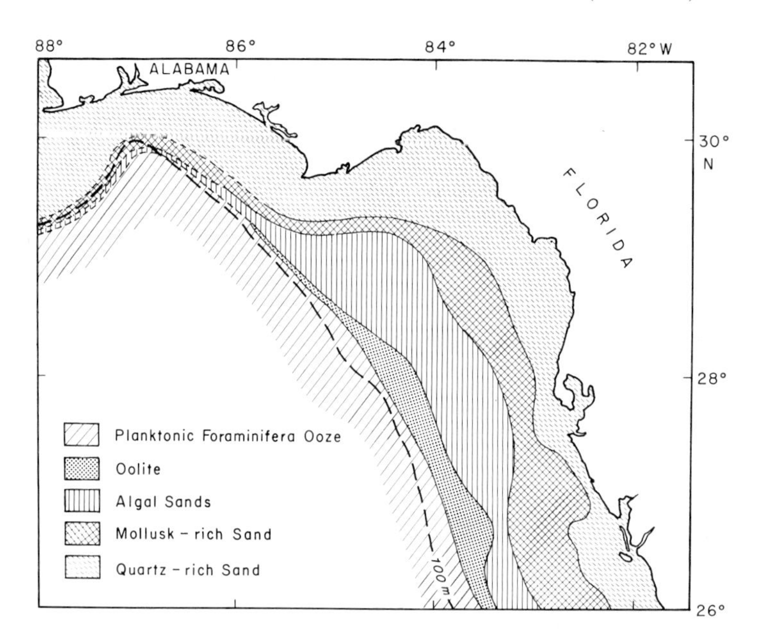

Fig. 70. Carbonate assemblages on the northwestern Gulf of Mexico continental shelf and slope. Note the decrease in carbonate-rich assemblages towards the Mississippi Delta (upper left). After GOULD and STEWART (1955), LUDWICK and WALTON (1964) and UPSHAW and others (1966)

although some deposits have been reported off the coast of Ireland (HERDMAN and LOMAS, 1898; KEARY, 1967; LEES and others, 1969; LEES and BUTLER, 1971), Alaska (HOSKIN and NELSON, 1969) and off parts of eastern Africa (STUBBINGS, 1939). Barnacles apparently are not common on tropical shelves; this may be the result of grazing fish that continually (and effectively) scrape the substrate, not allowing the settlement of barnacle larvae (NEWMAN, 1960).

→

Plate XXVII. Algal-rich Sediments from Continental Shelves. (a) Algal nodules from north Africa; water depth is 86 mm. Carbon-14 dates show that this material is contemporary in age. Reflected light; scale is 2.5 cm. (b) Coralline algal "maerl" from the north African shelf. Most of the coralline algae is braching and is mixed with lesser amounts of bryozoans and molluscan debris. Reflected light; scale is 2.5 cm. (c) Algal nodules (one is sectioned) from the outer shelf of the southeastern United States. In contrast to the nodules in (a), these nodules were deposited during the last lower stand of sea level, some 12000 years ago. Scale is 2.5 cm. (d) Coralline algae-barnacle sand from the outer shelf of the southeastern United States. As shown in Figure 65 this assemblage is found on much of the outer shelf along the southeastern United States, and probably was derived (at least in part) from the algal ridge system that lines this area. Reflected light; scale is 5 mm. (e) Algal limestone from the southeastern United States. This porous limestone is comprised of coralline algae that coat a serpulid-rich nucleus. The serpulid tubes have been filled with secondary magnesian calcite. The age of this limestone is about 11000 years and was deposited during the last lower stand of sea level. Scale is 2.5 cm. (f) Recrystallized algal limestone from the southeastern United States. In contrast to the limestone in (e), this rock is considerably less porous, the result of extensive filling by secondary magnesian calcite. This rock dates at 27000 years old, but since this partly reflects the date of recrystallization, it probably formed prior to that date. Reflected light; scale is 1 cm

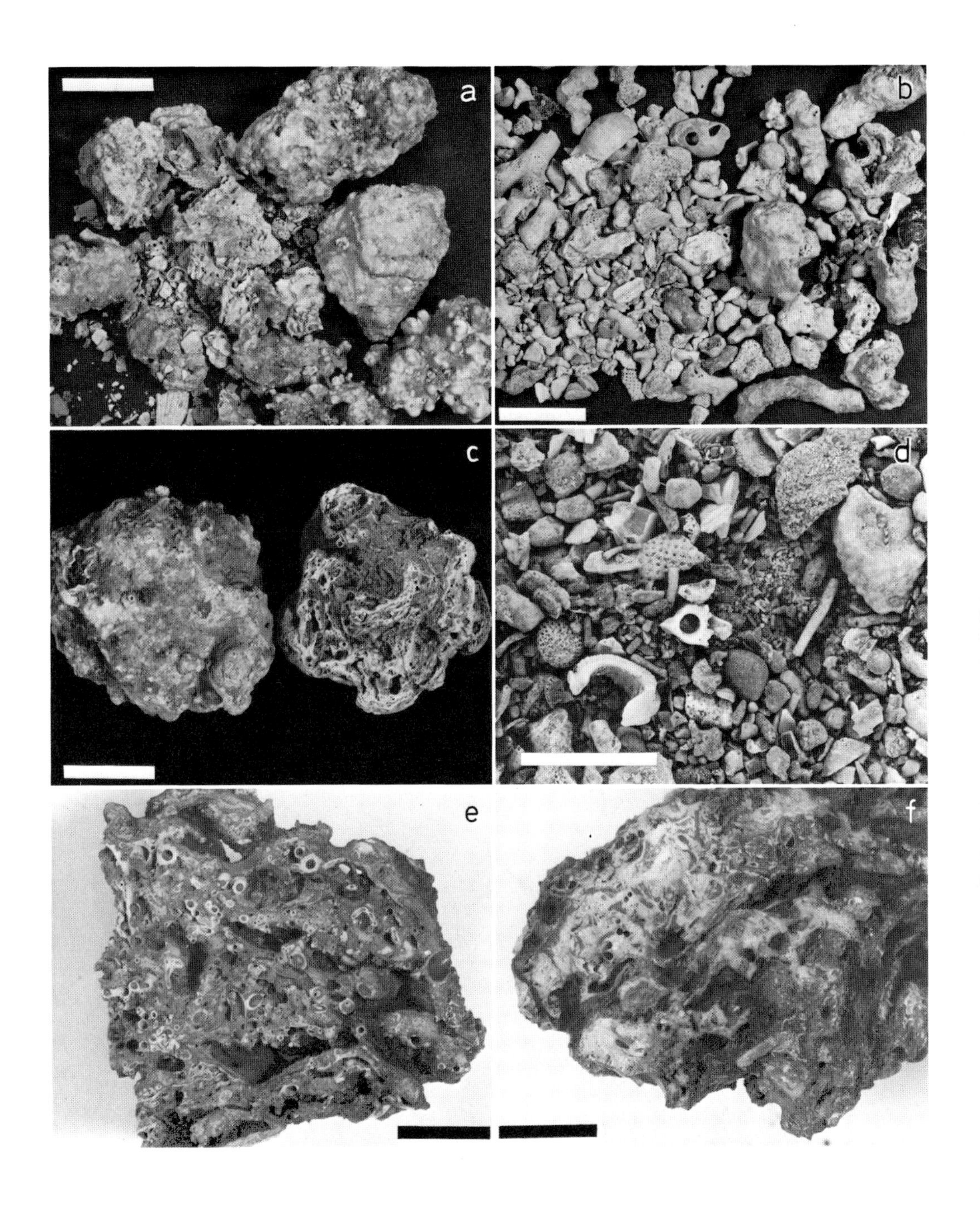

Plate XXVII.

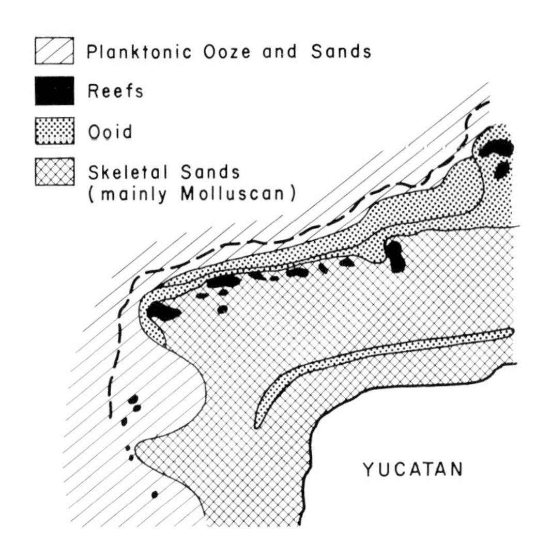

Fig. 71. Carbonate assemblages on Campeche Bank. After LOGAN and others (1969)

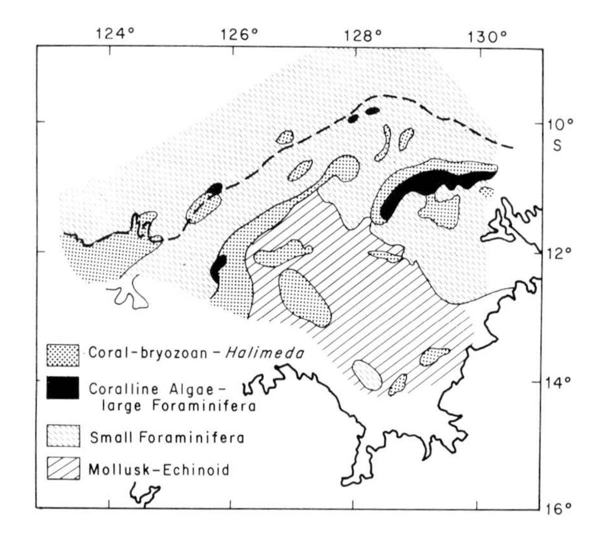

Fig. 72. Carbonate assemblages on the northwestern Australian continental margin. After van ANDEL and VEEVERS (1967)

The dominant epibenthic filter feeders in most other shelf areas are bryozoans. In some places bryozoan remains constitute the dominant carbonate component (Plate XXVI b). For example, WAAS and others (1970) report that bryozoans comprise up to 85% of the sediment off southern Australia. Other bryozoan-rich sediments are found off Venezuela (RUCKER, 1967), Brazil (MABESOONE and TINOCO, 1965), Alaska (HOSKIN and NELSON, 1969), western France (BOILLOT, 1964, 1965), northwestern Africa (MÜLLER, 1969; MCMASTER and others, 1971), northwestern Australia (BASSETT-SMITH, 1899; CARRIGY and FAIRBRIDGE, 1954) and much of the Mediterranean (BOURCART, 1957; BLANC, 1968; CAULET, 1972; MILLIMAN and others, 1973). In constrast to barnacles, which are mainly encrusters, bryozoans can, with the aid of coralline algae, construct prominent bioherms. Why bryozoans are dominant on some shelves and barnacles on others is not known.

The Algal Ridge System

Many outer shelves and upper slopes are characterized by small topographic features, rising 1 to 5 meters above the bottom and covered with various carbonate-producing organisms and sediments. Water depths vary from about 40 to 180 meters. Dominant carbonate components include coralline algae, large benthonic foraminifera (such as amphisteginids and miliolids), bryozoans and corals. Many such features serve as substrate for modern colonies of deep-water corals, gorgonians, bryozoans, barnacles and serpulids. These "reefs" tend to parallel the shelf break, but are discontinuous, thus more resembling scattered mounds than a ridge system. Among the areas in which such features have been observed are the Gulf of Mexico (JORDAN, 1952; STETSON, 1953; GOULD and STEWART, 1955; PARKER and CURRAY, 1956; LUDWICK and WALTON, 1957; JORDAN and STEWART, 1959; MATTHEWS, 1963; LOGAN, 1969), the southeastern United States (MENZIES and others, 1966; MACINTYRE and MILLIMAN, 1970; DUANE and MEISBURGER, 1969), the Guiana-Venezuela shelf (NOTA, 1958; KOLDEWIJN, 1958; SQUIRES, 1963; RUCKER, 1967), eastern Central America (KORNICKER and BRYANT, 1969), the Caribbean (MACINTYRE, 1967b, 1972), the west African shelf (ALLEN and WELLS, 1962; MCMASTER and others, 1970), Australia (CARRIGY and FAIRBRIDGE, 1954; VAN ANDEL and VEEVERS, 1967; MAXWELL, 1968b; DILL, 1969; VEEH and VEEVERS, 1970), Baja California (DILL, 1969), and western India (R.R. NAIR, 1970, written communication). No doubt similar features are present on many shelves that as of yet are unexplored.

The reef-like substrate plus the dominance of coralline algae in many of the reef rocks and associated sediments indicate that these features are algal reefs. The shallow-water nature of the components suggests that the reefs are relict features, having grown during lower stands of sea level; preliminary carbon-14 dates (NOTA, 1958; KOLDEWIJN, 1958; MENZIES and others, 1966; ALLEN and WELLS, 1962; MACINTYRE and MILLIMAN, 1970) support this belief. Why the reefs have formed mainly at or near the shelf break is not understood; it may be related to increased biologic productivity at the shelf break due to upwelled ocean water.

MACINTYRE and MILLIMAN (1970) found that the ridge system off southeastern United States is considerably more complex than previously thought. Four basic types off "reef rock" occur: 1. algal limestone similar to that described in the preceding paragraphs; 2. coquina-like limestone, strongly resembling beachrock; 3. oolitic rock; and 4. coral reef rock (Plate XXV c, d). Similar types of limestones have been reported from the Brazilian continental shelf (MABESOONE, 1971). Carbon-14 dates generally range from 10 to 15 thousand years B. P., although some are considerably older. Comparison with eustatic sea level curves suggests that most of these limestones were deposited during the Holocene transgression of sea level (MILLIMAN and EMERY, 1968; MACINTYRE and MILLIMAN, 1970). Algal and coral growth apparently was too slow to keep pace with the relatively rapid rise in sea level and the reefs were drowned[1]. Beach rock was deposited and cemented in the intertidal environment. The oolitic ridge off central and southern Florida may well represent an early Holocene analog of the modern Bahamas and the Pleistocene Miami Limestone (HOFFMEISTER and others, 1967; MACINTYRE and MILLIMAN, 1970).

Shelf Oolite

Oolite deposits are present on many shelves and upper slopes, generally seaward of the algal ridge system in depths as great as 163 m (Figs. 65, 69, 70 and 71). Locations include the southeastern United States (STETSON, 1938; PILKEY and others, 1966; TERLECKY, 1967; MILLIMAN and others, 1968), the Gulf of Mexico (GOULD and STEWART, 1955; LOGAN and others, 1969), the Mediterranean Sea (FABRICIUS and others, 1970a, 1970b), the Persian Gulf (EMERY, 1956; HOUBOLT, 1957), western Africa (MCMASTER and others, 1971), Australia (VAN ANDEL and VEEVERS, 1967), India (SUBBA RAO, 1958, 1964; NAIR and PYLE, 1969; VON STACKELBERG, 1970), and the Amazon shelf.

Modern ooids tend to form in agitated warm waters, generally in depths less than 2 meters (NEWELL and others, 1960), thus suggesting that shelf ooids are relicts of lower sea levels. The relict nature is also suggested by the fact that many shelf ooids appear reddish brown rather than the polished cream color of modern ooids (see below). Carbon-14 dates show that most shelf ooids range from 10 to 15 thousand years old (TERLECKY, 1967; MILLIMAN and others, 1968; NAIDU, 1969; LOGAN and others, 1969; FABRICIUS and others, 1970a), and thus probably were formed during the last transgression of the Holocene sea level.

One exception to this generalization is the oolite deposit off North Carolina (Fig. 66). Instead of occurring seaward of the algal ridge, it lies landward, in the central part of Onslow Bay. Carbon-14 dates are from 24 to 27 thousand years B. P., suggesting that the ooids were deposited during the last Würm regression (TERLECKY, 1967; MILLIMAN and others, 1968).

Relict oolites on continental shelves have a wider distribution than modern shallow-water oolites. Considering that early Holocene temperatures were cooler than at present, one can speculate on what environmental conditions facilitated such a wide-spread deposition of ooids. The stable isotope content of the shelf oolite off the southeastern United States suggests that deposition occurred in a supersaline environment, perhaps one similar to Laguna Madre on the southern Texas shelf (RUSNAK, 1960; MILLIMAN and others, 1968). This does not explain, however, why transgressional oolite generally occurs on the outer shelf edge nor why the egressional oolite off North Carolina occurs on the mid shelf.

Age of Shelf Carbonates

Carbonate components in shelf sediments display a wide range of ages. At one end of the spectrum are plants and animals presently living on the shelf. At the

1 The shallow coral reefs on the outer part of Campeche Bank (Fig. 71), may have accreted at rates which kept pace with the Holocene transgression (LOGAN, 1969).

other end are fossils derived from underlying strata; for example the Miocene fossils on Georges Bank (STANLEY and others, 1967) and in Onslow Bay (ROBERTS and PIERCE, 1967). Generally most shelf carbonate components are Holocene in age. Ooids, shallow-water foraminifera and shallow-water mollusks (such as oysters; MERRILL and others, 1965) probably were deposited during the transgression of Holocene sea level. Relict sediments often are characterized by their reworked appearance. Many components are bored and fragmented so that positive identification may be impossible without study under refracted light. The mechanisms of alteration are discussed in Chapter 9. Glauconite and phosphorite can form within both primary and secondary voids. Pyrite formation in both relict ooids and shells is well documented and probably is the result of anoxic post-depositional environments (HOUBOLT, 1957; TERLECKY, 1967; DOYLE, 1967; MILLIMAN and others, 1968; NAIR, 1969).

Many mollusk and echinoid fragments on the inner shelf were deposited by organisms living in present-day shelf conditions. Similarly, many of the barnacles and bryozoans that encrust the various substrates are modern (ROSS and others, 1964; ZULLO, 1966; BASSETT-SMITH, 1899; PÉRÈS and PICARD, 1964; LAGAAIJ and GAUTIER, 1965; BOILLOT, 1965; MAXWELL, 1968 b) as are most of the planktonic foraminifera in outer shelf sediments.

Coralline algae in continental shelf sediments may be relict but others are modern, since some forms are capable of living at considerable depths. For instance, in the Mediterranean living coralline algae have been dredged from depths greater than 100 meters (JACQUOTTE, 1962) although most growth is restricted to somewhat shallower depths (MILLIMAN and others, 1973). Similarly, some coralligenous and reef-like deposits may be modern. BLANC (1968), for example, reports that reefs seaward of Marseille (in water depths of 40 m) are accreting at a rate of 1.5 meters per thousand years.

Summary

At present our knowledge of shelf carbonate sedimentation lags far behind that of shallow-water carbonate sedimentation. However several basic points can be made.

1. The abundance of calcum carbonate in a shelf sediment is dependent primarily upon the rate of terrigenous sedimentation and secondarily upon water temperature. Thus on the basis of only carbonate content, one can gain some insight into the environment of deposition.

2. The type of carbonate components within a shelf sediment depends upon both the environment and the age of deposition. The general similarity of carbonate assemblages throughout the world shelves (Figs. 65–72) suggests that similar environmental conditions must occur on many shelf areas. Carbonate content generally is low and infaunal organisms common in inner shelf areas where sediment is sandy and muddy. Epifaunal filter feeders and coralline algae require hard substrate and reasonably strong currents in which to grow, and therefore generally are restricted to outer shelves. Some outer shelf carbonates, such as hermatypic corals and *Halimeda*, are restricted to warm waters. Fragments of

other components, such as coralline algae, barnacles, bryozoans and mollusks, can occur at nearly any latitude, although the species depends upon water temperature.

3. Modern shelf carbonates usually reflect a mixture of modern (sublittoral, including both benthonic and planktonic) and late Pleistocene-early Holocene (intertidal) components. Some components, such as ooids and shallow-water foraminifera, clearly represent relict shallow-water environments. Similarly planktonic foraminifera and deeper water benthonic fauna probably were deposited at modern depths. Differentiating between recent and relict coralline algae, barnacles and bryozoans, however, presents a more difficult problem, since these organisms can live over a rather wide depth range. This mixture of environments and ages (and thus degree of diagenetic alteration) can help one differentiate between shelf and shallow-water sediments.

4. At present terrigenous sedimentation on many shelves is restricted to estuaries and nearshore areas. Sediment which escapes the nearshore environment often by-passes the middle and outer shelf and is deposited in deeper water. Thus, on many shelves carbonate deposition may represent a major portion of modern sedimentation. If present-day shelf conditions were to continue for a sufficiently long time, the surficial sediments probably would be represented as a mixture of shallow-water and sublittoral carbonates displaying a wide range of post-depositional reworking and alteration (MILLIMAN and others, 1972).

Chapter 8. Deep-Sea Carbonates

Factors Influencing Distribution

The distribution and composition of deep-sea carbonates depend upon three major factors: carbonate productivity, terrigenous sedimentation, and the degree of carbonate saturation within the ambient waters. Carbonate productivity is greatest in the surficial waters of those oceanic areas having high biologic productivity. Such locations include areas of upwelling (for example, equatorial upwelling) and those areas peripheral to the major gyres of oceanic circulation (such as the Gulf Stream). Lowest productivity occurs within central oceanic gyres, such as the Sargasso Sea.

The rate of non-carbonate (primarily terrigenous) deposition is also important. For example, areas adjacent to large rivers may have substantially higher biologic productivity than arid areas, but increased carbonate productivity can be masked by rapid terrigenous sedimentation. Similarly, carbonate deposition near areas of equatorial upwelling is comparatively great, but so is the deposition of siliceous radiolarians and diatoms; as a result the average carbonate content of the bottom sediments may be no greater or even less than in normal oceanic areas.

The most critical factor controlling carbonate content in deep-sea sediments is the degree of carbonate saturation in the ambient and overlying waters. Only the surficial layers of the world oceans are supersaturated with respect to both aragonite and calcite (WATTENBERG and TIMMERMAN, 1936; BERNER, 1965; PYTKOWICZ, 1965; PETERSON, 1966; BERGER, 1967; LI and others, 1969, and references therein). The underlying waters are either saturated or undersaturated. At some depth in the water column, termed the "compensation depth", the rate of carbonate dissolution equals the rate of carbonate deposition (PYTOKOWICZ, 1970). Below this depth carbonate sediments normally do not accumulate.

The increased solubility of calcium carbonate with water depth is related to increased hydrostatic pressure, decreasing temperature and increasing CO_2 content within the ambient waters (see Chapter 1), as well as to the mineralogy of the specific carbonate. Pressure increases the solubility of $CaCO_3$ (OWEN and BRINKLEY, 1941) and also affects the dissociation of both carbonic acid and boric acid in sea water (BUCH and GRIPENBERG, 1932; CULBERTSON and PYTKOWICZ, 1968); for example, PYTKOWICZ and CONNORS (1964) calculated that at 1000 atmospheres (a depth of 10 kilometers) aragonite is approximately three times more soluble than at atmospheric pressure. Low temperatures also increase the solubilities of both $CaCO_3$ and CO_2. In addition, the carbon dioxide content of deep waters increases with the oxidation of organic matter:

$$(CH_2O)_{106}(NH_3)_{16}H_3PO_4 + 138(O_2) - 106(CO_2) + 122(H_2O) + 16(HNO_3) + H_3PO_4$$

(RICHARDS, 1965). In a closed system the production of CO_2 will decrease the pH. PARK (1968) found that this apparent oxygen utilization is a prime factor in regulating pH in the deep-sea.

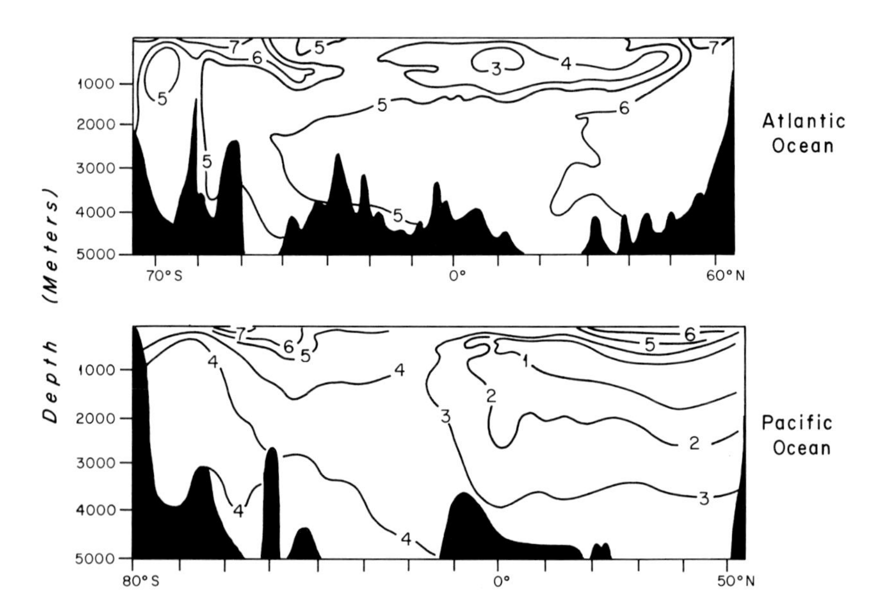

Fig. 73. Oxygen concentrations (ml/l) along latitudinal transects in the Atlantic and Pacific Oceans. After SVERDRUP and others (1942)

The depths of carbonate saturation and carbonate compensation vary greatly throughout the world oceans, suggesting that CO_2 and temperature are more critical in regulating carbonate dissolution than is pressure (REVELLE, 1934). In Arctic and Antarctic waters, CO_2 content is high because of cold temperatures, frequent mixing (and further hydration of CO_2) by storms and the inhibition of photosynthesis by thick ice cover. The resulting low pH's allow dissolution to occur at depths of less than 400 m (KENNETT, 1966; CHEN, 1966). Calcium carbonate contents in Antarctic sediments generally average less than 1%. Values in the Arctic are somewhat higher, but usually less than 30%. Some abnormally high concentrations occur in the Baltic Sea and Hudson Bay, probably due in part to the influx of terrestial limestone detritus.

Deep-water circulation in the Pacific Ocean is greatly influenced by the shallow sill depth of the Bering Straits, which effectively excludes Arctic deep water from the north Pacific. Thus deep circulation is driven mainly by the Antarctic bottom water, with the result that the circulation in the north Pacific is particularly sluggish. Oxygen is utilized progressively as the Antarctic bottom water flows north (Fig. 73) and thus the CO_2 content increases (Fig. 74). As a result, pH values in the north Pacific are distinctly lower than those in the south Pacific (Fig. 75), and saturation and compensation depths are shallower (Figs. 76 and 77). Carbonate values in north Pacific deep-sea sediments seldom exceed a few percent (Fig. 83).

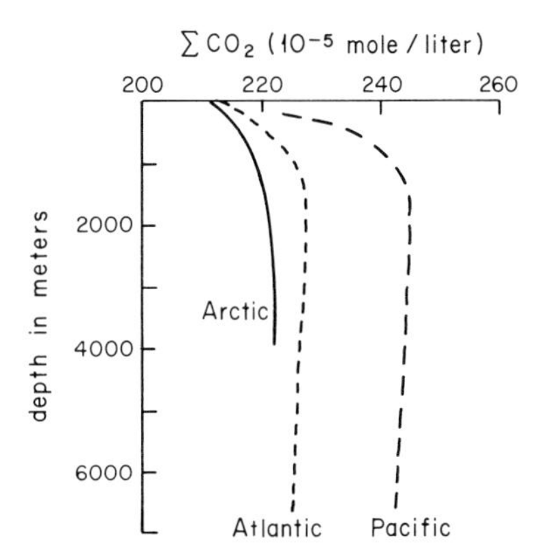

Fig. 74. Variation of total CO_2 content with water depth in the Pacific and Atlantic Oceans. After LI and others (1969)

PETERSON (1966) and BERGER (1967) found that the central Pacific waters are undersaturated with respect to calcite below depths of about 500 m. Dissolution rates, however, are slow down to about 3700 m; below this depth dissolution

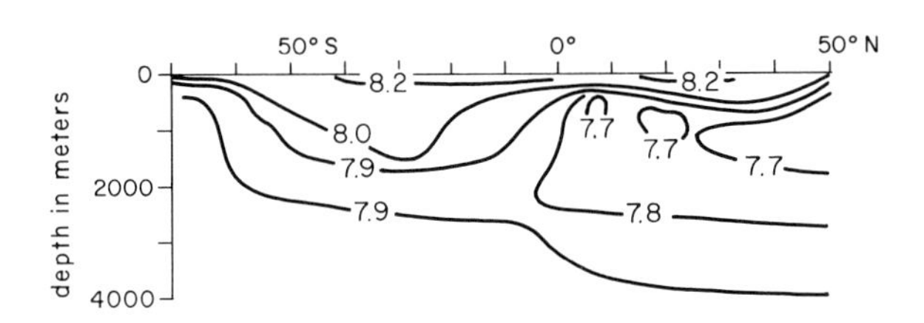

Fig. 75. Variations in pH along a tatitudinal transect in the Pacific Ocean. Note the low pH's in the north Pacific and the close correlation between them and oxygen concentrations shown in Fig. 73. After IVANENKOV (1966)

increases rapidly (Fig. 78). These observations agree with theoretical calculations (LI and others, 1969) (Fig. 79) and with the measurements of the carbonate content of sediments in the southern and central Pacific (Fig. 77).

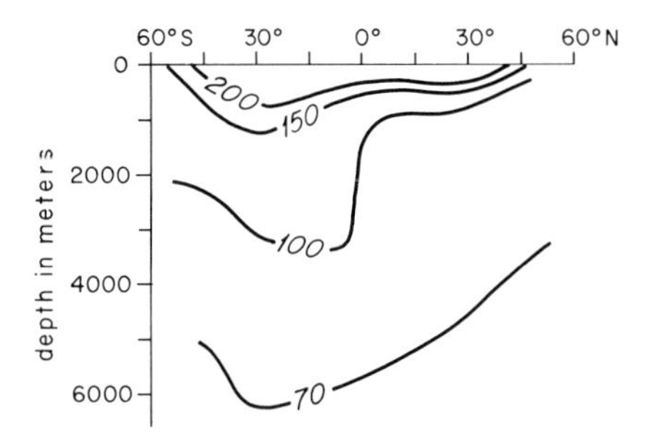

Fig. 76. Degree of calcite saturation relative to depth and latitude in the Pacific Ocean (along transect at 170° W). After HAWLEY and PYTKOWICZ (1969)

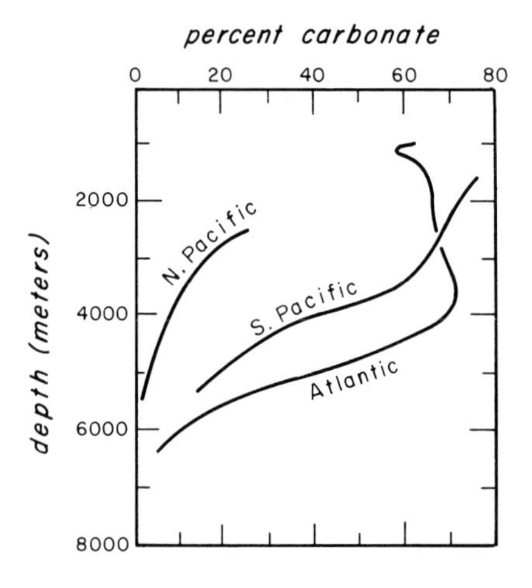

Fig. 77. Variation of carbonate content within deep-sea sediments with water depth. After SVERDRUP and others (1942)

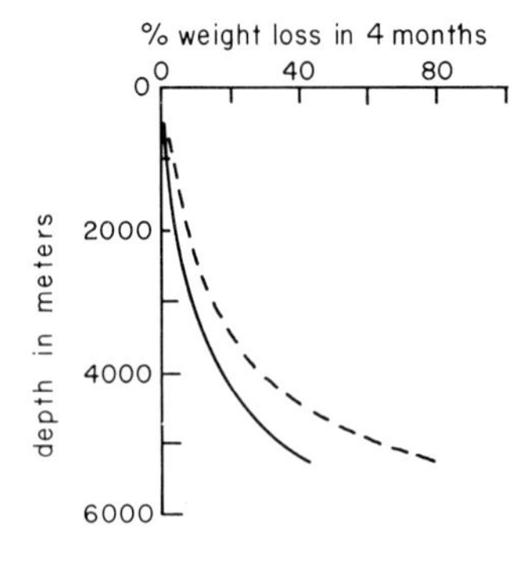

Fig. 78. Weight loss of calcium carbonate in foraminifera tests during emplacement at various depths in the central Pacific for a four month period. The solid line represents weight loss for untreated foraminifera. Dashed lines represent foraminifera treated with H_2O_2. Note the marked increase in dissolution at about 3700 m, which corresponds to decreased calcite saturation seen in Fig. 79. After BERGER (1967)

In the Atlantic Ocean, Arctic deep water plays a critical role in the deep circulation of the entire ocean. The increased circulation (and thus smaller residence times) of Atlantic deep waters means a smaller degree of oxygen utilization and thus lower total CO_2 content (Figs. 73 and 74). Calculations by LI and others (1969) suggest that the upper 5000 m of the Atlantic Ocean are saturated or just

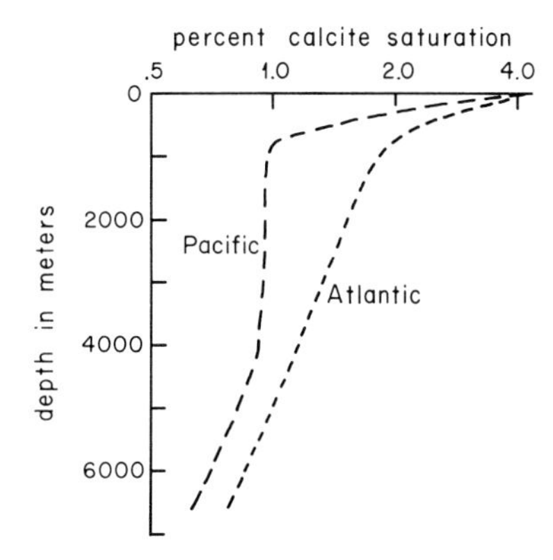

Fig. 79. Variations in calcite saturation with water depth in the Atlantic and Pacific Oceans. After LI and others (1969)

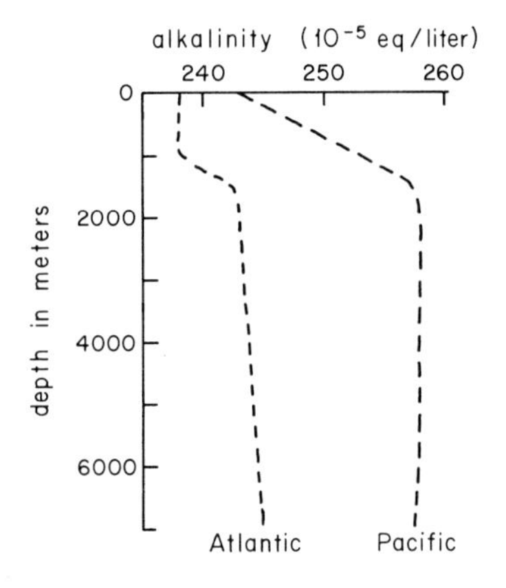

Fig. 80. Variation in alkalinity in Pacific and Atlantic Ocean waters with water depth. After LI and others (1969)

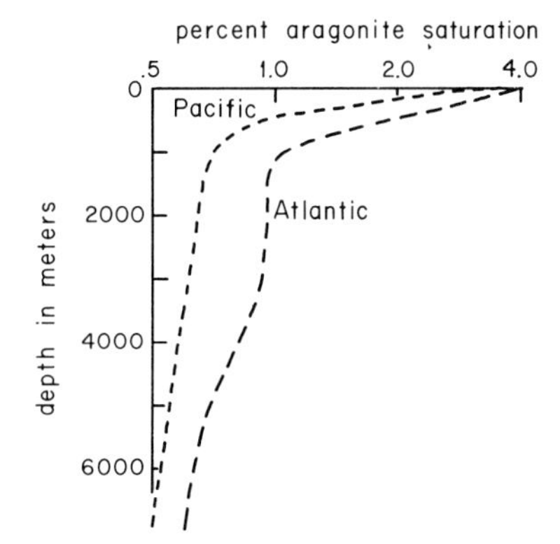

Fig. 81. The degree of aragonite saturation with water depth in the Pacific and Atlantic Oceans. After LI and others (1969)

slightly undersaturated with respect to calcite (Fig. 79). The fact that the compensation depth is deeper than 5000 m (Fig. 77) confirms this estimate. As a result of the smaller amount of dissolution, deep-sea alkalinities are distinctly lower than in the Pacific (KOCZY, 1956; LI and others, 1969) (Fig. 80).

Aragonite and magnesian calcite are both metastable with respect to calcite (CHAVE and others, 1962; CHAVE and SCHMALZ, 1966) and therefore both tend to dissolve at shallower oceanic depths than does calcite. Apparently, as will be discussed in the next chapter, magnesian calcite is slightly less soluble than aragonite in the deep sea. Pacific waters become undersaturated with respect to aragonite at depths shallower than 500 m; in the Atlantic undersaturation is reached at about 1000 m (Fig. 81). Therefore aragonite and magnesian calcite generally contribute only minor amounts of carbonate sediment in water depths greater than 1000 to 3000 m (PILKEY and BLACKWELDER, 1968) (Fig. 82). In tur-

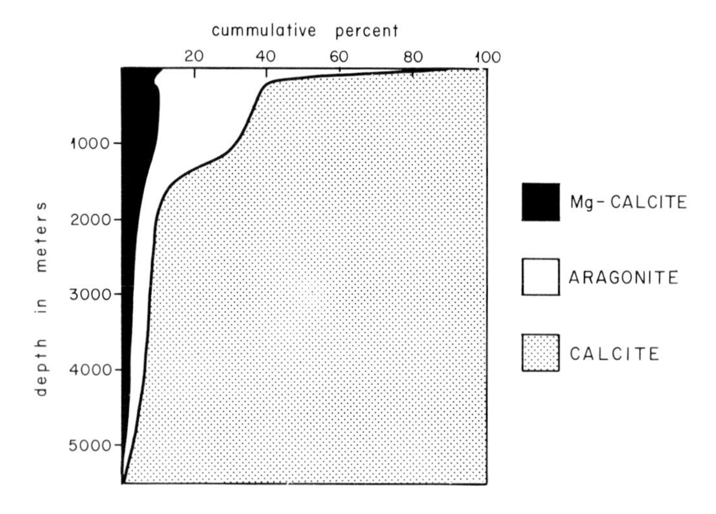

Fig. 82. Distribution of the various calcium carbonate polymorphs in oceanic sediments with varying water depth. After PILKEY and BLACKWELDER (1968). The deeper-water examples are taken mostly from the Atlantic Ocean. Pacific Ocean sediments show shallower distributions, and sediments from semi-enclosed basins, such as the Red Sea and the eastern Mediterranean Sea, have much different distributions of carbonate mineralogy

bidite sequences, however, both of these metastable phases may persist, although finer grains usually alter to calcite (FRIEDMAN, 1964, 1965; PILKEY and BLACKWELDER, 1966; GOMBERG and BONATTI, 1970).

Recent studies have shown that the correlation between water depth and the distribution of calcium carbonate within oceanic sediments is not as clearly defined as previously thought (SMITH and others, 1969). The presence of organic coatings on carbonate grains may retard dissolution (CHAVE, 1965; WANGERSKY, 1969; CHAVE and SUESS, 1970; SUESS, 1970). BERGER (1967) found that foraminifera treated with H_2O_2 (to remove organic coatings) exhibit dissolution rates at 5000 m which are twice as high as the rates of untreated foraminifera (Fig. 78). In addition, the rate of dissolution depends upon particle size (CHAVE and SCHMALZ, 1966); the smaller the particle, the more rapid the dissolution. Specific examples will be given in following sections.

Distribution of Calcium Carbonate in Deep-Sea Sediments

The distribution of calcium carbonate in deep-sea sediments is shown in Fig. 83. Carbonate contents are greatest in shallow-water areas that are isolated from terrigenous sedimentation, the best example being mid-ocean ridges. Because of their comparatively deep compensation depths, carbonate contents in the south Pacific and north Atlantic are relatively great, although the north Atlantic, by virtue of its generally shallower depths, has a greater average carbonate content.

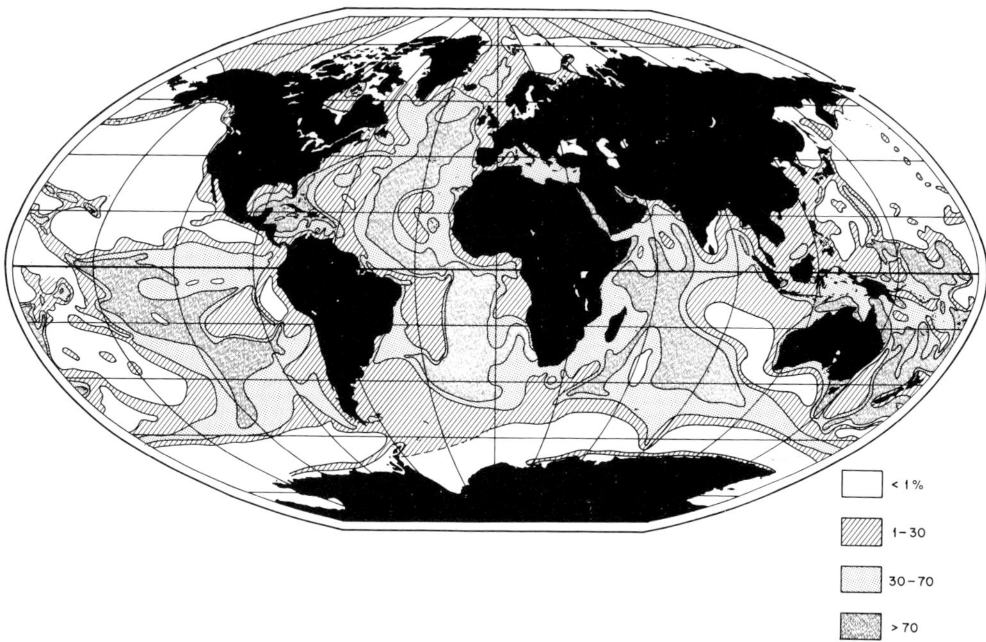

Fig. 83. World-wide distribution of calcium carbonate in ocean sediments. Atlantic and Pacific Ocean values are taken from the Chief Office of GEODESY and CARTOGRAPHY (USSR) (1969) and LISITZIN and PETELIN (1970). Indian Ocean carbonate distributions are less well documented, and the values shown on this chart have been derived from LISITZIN (1971), together with data from MURRAY and PHILIPPI (1908), KUENEN (1950), OLAUSSON (1960), LISITZIN (1960), NIINO and EMERY (1961), BEZRUKOV (1964), MÜLLER (1966) and SIDDIQUIE (1967)

The north Pacific has both a shallow compensation depth and great water depths; for the most part carbonate concentrations are less than 1%. The Indian Ocean deep-sea sediments contain carbonate concentrations that are roughly similar to those in the south Pacific although water depths in the former area are considerably shallower. Carbonate values in the polar regions, as mentioned in the previous section, are generally less than 1%.

Oozes

Deep-sea carbonates are composed primarily of planktonic tests. Echinoid spines, benthonic foraminifera and mollusks can be prominent in slope sediments but decrease rapidly with increasing water depth. Deeper water carbonates are almost completely planktonic; one notable exception is the occurrence of deep-water coral reefs (see below).

An ooze is defined as a deep-sea sediment containing more than 30% of any particular biogenic constituent. An excellent but short discussion of the various deep-sea biogenic components has been given by RIEDEL (1963). Five major oozes are recognized in the world oceans; planktonic foraminifera (commonly called *Globigerina* ooze), coccolith, pteropod, radiolarian, and diatom. The first three types are carbonate oozes, the latter two are siliceous. Carbonate oozes, represented in Fig. 83 as those areas containing more than 30% calcium carbonate, cover 47% of the sea floor, generally between the latitudes of 45° N and 45° S (SAITO, 1971).

Planktonic Foraminifera (Globigerina) Ooze

The term *Globigerina* ooze is applied to sediments in which more than 30% of the sediment is composed of planktonic foraminifera; benthonic foraminifera generally contribute less than 1% of the carbonate in deep-sea sediments (PARKER, 1971). "*Globigerina* ooze", however, is a misnomer, since component foraminifera often are dominated by other genera. The writer prefers the term "Planktonic Foraminifera Ooze", although convention and the frequency of usage probably will dictate the continued usage of "*Globigerina* ooze".

Whole tests of planktonic foraminifera generally comprise a major portion of the sand in most deep-sea sediments. The finer fractions of foraminiferal oozes are dominated by both broken foraminifera (generally silt size) and coccoliths

Table 57. Distribution of major planktonic foraminifera species in oceanic surface waters. Italicized zones refer to deep-sea areas in which that particular species constitutes an important sedimentary component. Arctic and Subarctic can be equated with Antarctic and Subantarctic. Data from BÉ and TOLDERLUND (1971)

Species	World distribution
Globigerina pachyderma	*Arctic-subarctic*-transition
Globigerina quinqueloba	Arctic-*subarctic*-transition
Globigerina bulloides	Arctic-*subarctic*-transition-subtropical
Globigerinita bradyi	Arctic-subarctic
Globorotalia inflata	Subarctic-*transition*-subtropical
Globigerinita glutinata	Subarctic-tropical
Globorotalia truncatulinoides	Subarctic-*transition-subtropical*-tropical
Globiqerina dutertrei	Transition-tropical
Orbulina universa	Transition-tropical
Globigerinella aequilateralis	Transition-*subtropical*-tropical
Hastigerina pelagica	Subtropical
Globigerinoides ruber	Transition-*subtropical-tropical*
Globigerinoides conglobatus	Transition-*subtropical*-tropical
Globorotalia menardii	Subtropical-tropical
Pulleniatina obliquiloculata	Subtropical-tropical
Globigerinoides sacculifer	Transition-*subtropical-tropical*

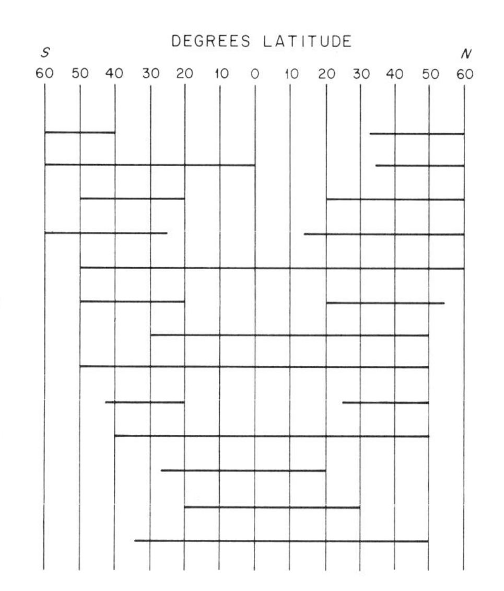

Fig. 84. Distribution of major foraminifera species in deep-sea sediments from the North Atlantic, South Pacific and Indian Oceans. After PARKER (1971)

(see below). In terms of both the area covered and carbonate deposited, planktonic foraminifera oozes undoubtedly are the most important modern carbonate deposit in the oceans.

The faunal composition of planktonic foraminifera assemblages changes markedly with latitude, apparently in response to latitudinal temperature gradients (Table 57 and Fig. 84). The planktonic zones shift seasonally in higher latitudes, with warmer water species moving poleward in summer and toward the equator in winter. Highest planktonic concentrations occur in areas of high organic productivity; lowest concentrations are found in central, oligotrophic areas. High productivity areas can contain 10000 individuals per 1000 cubic meters of surface water; low productivity areas may have only 100 individuals per 1000 cubic meters of water (BÉ and TOLDERLUND, 1971).

As discussed in Chapter 4, most planktonic foraminifera live within the upper 100 meters of the ocean. A few genera, such as *Globigerinoides* and *Globorotalia*, however, can live at depths greater than 500 meters.

In addition to species distribution, the coiling direction of many species also can reflect water temperature. *Globigerina pachyderma* coils to the left in higher latitudes and to the right in warmer climates. *Globorotalia truncatulinoides* also generally coils left in colder waters, but the distribution is not nearly as uniform as in *G. pachyderma* (PARKER, 1971). Species zonations and coiling directions of modern foraminifera have been helpful in determining Pleistocene oceanographic environments.

In addition to these ecologic factors, the distribution of planktonic foraminifera in oceanic sediments depends upon the resistance of individual tests to solution. The more delicate and smaller foraminifera will dissolve faster than the more massive types (BERGER, 1967, 1968; RUDDIMAN and HEEZEN, 1967). *Pulleniatina obliquiloculata*, *Globorotalia tumida* and *Sphaeroidinella dehiscens* are the most resistant to solution (PARKER, 1971), and therefore appear to be more common in bottom sediments than their presence in surface-water plankton would suggest.

Coccolith Ooze

Coccoliths are prominant contributors to the carbonate fraction finer than 6 microns in most deep-sea sediments. The distribution of coccoliths in oceanic sediments, however, is strongly dependent upon surface productivity, floral zonation of the coccolithophorids and dissolution. Recent studies of living coccolithophorids within the euphotic zone show latitudinal distributions similar to those found in planktonic foraminifera. MCINTYRE and BÉ (1967) recognized five floral zones within the Atlantic (Table 58) and OKADA and HONJO (1973) have found similar zones in the north Pacific.

Table 58. Major coccolithiphorid species in the various zones within the Atlantic Ocean (after MCINTYRE and BÉ, 1967)

Tropical	Transitional
Umbellosphaera irregularis	*Emiliania huxleyi*
Cyclolithella annulus	*Cyclococcolithus leptoporus*
Cyclococcolithus fragilis	*G. ericsonii*
U. tenuis	*R. stylifera*
Discosphaera tubifera	*G. oceanica*
Rhabdosphaera stylifera	*U. tenuis*
	Coccolithus pelagicus
Subtropical	Subarctic
U. tenuis	*Coccolithus pelagicus*
R. stylifera	*Emiliania huxleyi*
D. tubifera	*C. leptoporus*
C. annulus	Subantarctic
Gephyrocapsa oceanica	*E. huxleyi*
Umbilicosphaera mirabilis	*C. leptoporus*

Production of coccolithophorids is highest in subarctic and subantarctic waters, where upwelling causes high biologic productivity (Fig. 85). As a result of dilution by high diatom populations and the high dissolution rate of calcium carbonate, however, most polar sediments contain less than 1% coccoliths (MCINTYRE and MCINTYRE, 1971) (Fig. 85). Coccolithophorid productivity is lower, but the species diversity higher in the central portions of oceanic gyres, such as the central north Pacific.

The effect of solution by deep waters is particularly important in defining the distribution of coccoliths in bottom sediments. Because of their very small size

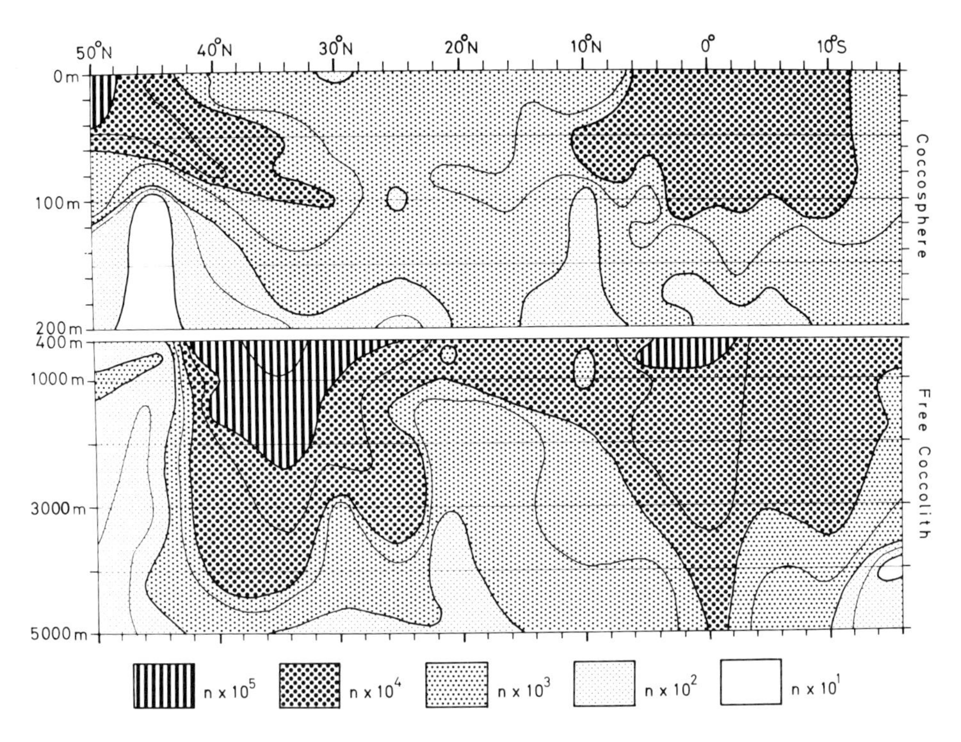

Fig. 85. Upper. Distribution of coccospheres in a N–S transect in the Pacific Ocean. Highest concentrations are in polar and equatorial regions. The number of coccospheres decreases markedly with depth, a result of organic disintegration. Lower. The decrease in the concentration of coccospheres coincides with a rapid increase in the number of free coccoliths within the water column. The paucity of coccoliths in polar waters and the general decrease with water depth in other regions reflect the dissolution during deposition. Figures courtesy of SUSUMU HONJO

(2 to 10 microns) individual coccoliths may require years to settle to the ocean bottom (LISITZIN, 1971)[1]. During this period, plus the time interval before final burial, dissolution can severely alter or destroy the coccoliths. The coccoliths most resistant to dissolution can survive (see p. 56ff.), but even these fragments can be altered sufficiently to prevent accurate species identification (MCINTYRE and MCINTYRE, 1971; GAARDER, 1971; SCHNEIDERMANN, 1971). One example of the effect of rapid dissolution is shown in the central north Pacific. Dissolution in this area is so rapid that although the coccolithophorid diversity is very high in the surface waters, coccolith diversity in the bottom sediments is very low, the less resistant species being removed by dissolution (OKADA and HONJO, 1973).

BRAMLETTE (1958) estimated that coccoliths comprise about 10 to 15% of the carbonate fraction in modern Pacific sediments. MCINTYRE and MCINTYRE

1 The effect of grazing by herbivorous animals, however, can have a marked effect on coccolith dissolution. The coccoliths often become incorporated into fecal pellets, which not only sink up to 10 times faster than individual coccoliths, but also provide protective organic coatings which can greatly deter dissolution (HONJO, 1972).

(1971) have made similar estimates for the Atlantic (14 to 30%), but found higher values in the Indian Ocean (10 to 50%). These estimates may be too low (S. HONJO, 1972, oral communication), but nevertheless coccoliths probably are not as important in modern oceanic sediments as are planktonic foraminifera. In some areas, however, such as the Mediterranean (BERNARD and LECAL, 1953) and the Black Sea (BUKRY and others, 1970; MÜLLER and BLASCHKE, 1969), coccoliths dominate the carbonate fraction (Plate XXVIII).

The present-day distribution of coccoliths in oceanic sediments does not fully indicate their sedimentological importance in older deposits. BRAMLETTE (1958) pointed out that coccolith deposition was greater during the early and mid Tertiary than was foraminiferal deposition. Coccoliths also appear to have been more abundant during interglacial periods than during glacial times (MCINTYRE and MCINTYRE, 1971; MCINTYRE and others, 1972), although BROECKER (1971) suggests that in the Caribbean the reverse was true.

Pteropod Ooze

Pteropods (planktonic gastropods, see p. 104 ff.) are common throughout the surface waters of the world oceans, generally living in the upper 100 m of the water column (Table 59). Due to their relatively soluble aragonitic shells, however,

Table 59. Latitudinal zonation of some common pteropod species (after SAITO and BÉ, 1969)

Tropical	Temperate
Creseis acicula	*Clio pyramidata*
Limacina trochiformis (plus 12 other species)	
	Subpolar
Subtropical	*Limacina retroversa*
Limacina inflata	
Limacina bulimoides	Polar
Styliola subula (plus 8 other species)	*Limacina helicina*

pteropod distribution in sediments is limited to tropical areas and to depths less than 3000 m in the Atlantic and considerably shallower in the Pacific (see above).

→

Plate XXVIII. Deep-Sea Carbonate Sediments. (a) Pteropod-planktonic foraminifera ooze, from the Blake Plateau. Water depth is 650 m. Scale is 5 mm. (b) Deep-water coral debris from the Blake Plateau. Water depth is 560 m; scale is 1 cm. (c) Coccolith ooze from the eastern Mediterranean Sea. Scale is 5 microns. (d) "Inorganically precipitated" magnesian calcite crystals in the deep-sea sediment from the eastern Mediterranean Sea. Compare the size and shape of these crystals with those from deep-sea limestone cements in Plate XXXVIII. Scale is 5 microns. (e) Photomicrograph of dolomite crystals from the deep-sea, recovered from JOIDES, Leg 11, North Atlantic Ocean. Scale is 25 microns. Photograph courtesy of J.C. HATHAWAY. (f) Dolomite crystals from the deep-sea sediments from the eastern Mediterranean Sea. Note the overgrowths in the upper middle dolomite crystal, suggesting authigenic growth. Scale is 20 microns

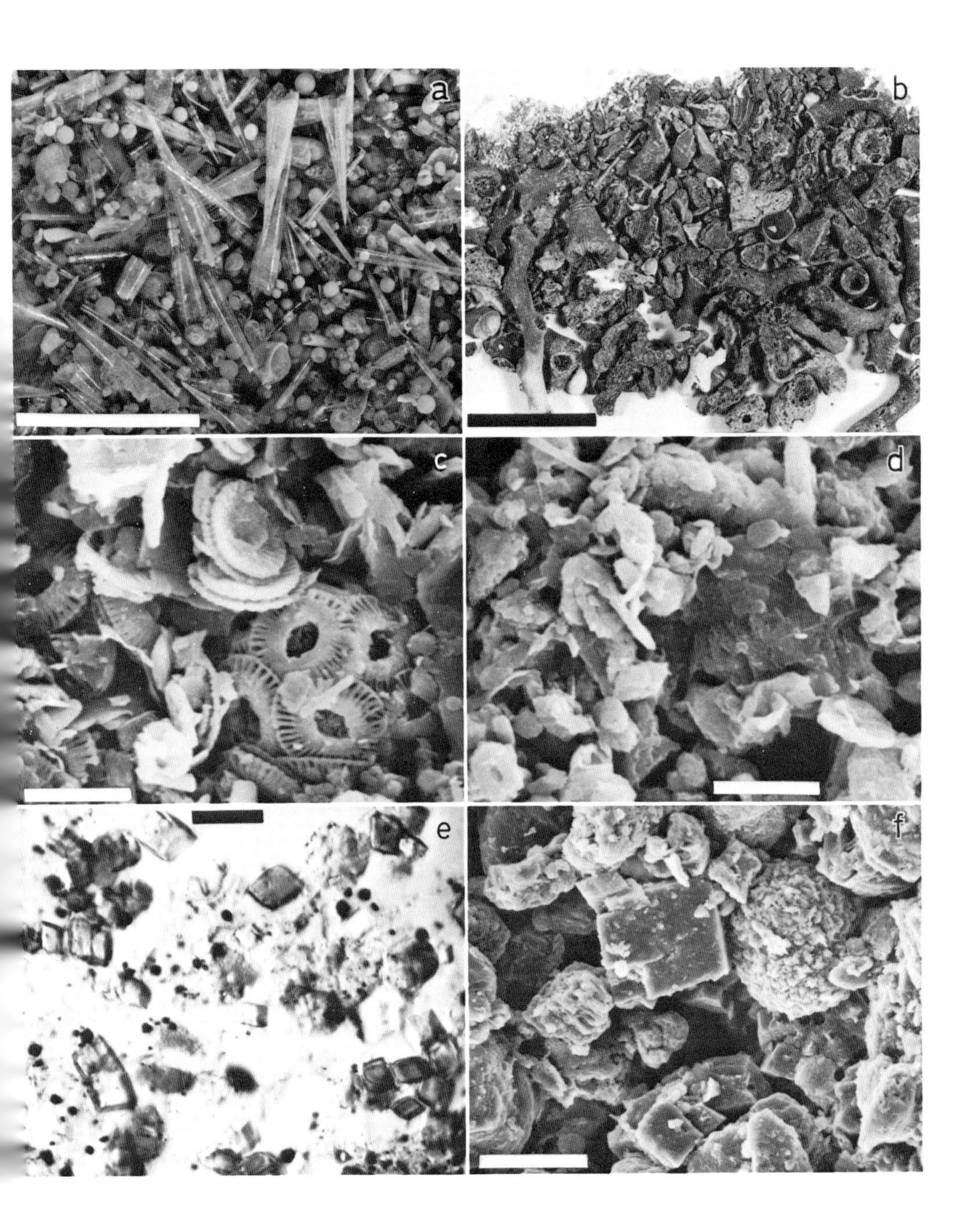

Plate XXVIII.

In Bermuda Pedestal sediments, maximum pteropod concentrations (700 specimens per mg of sediment) occur at water depths of about 2200 m (CHEN, 1964).

CHEN (1971) lists the following areas in which pteropod oozes occur:

North Atlantic: Mid-Atlantic Ridge, Blake Plateau, Azores Plateau, Bermuda Pedestal, Bahamas Outer Ridge, Gulf of Mexico, Caribbean Sea, Mediterranean Sea;

South Atlantic: Walvis Ridge, Rio Grande Rise, Mid-Atlantic Ridge north of 35° S;

Pacific: Tasman Sea, East Indian Archipelago, South China Sea;

Indian Ocean: Arabian Sea, Gulf of Aden, Chagos-Maldive Plateau, African slope, Australian Slope, Red Sea.

Pteropod oozes generally are very rich in carbonate (Plate XXVIII a). The pteropod oozes in the Bahamas and the Blake Plateau, for instance, contain more than 95% carbonate. Because of their mineralogical instability in the deep-sea, pteropod shells can decrease in abundance with increasing core depth (sediment age). Fluctuations in pteropod concentrations, however, also can have paleo-ecologic implications. CHEN (1968) reported that in the Gulf of Mexico and the Caribbean Sea, pteropods were more abundant during glacial intervals that during the interglacials. He explained this by a deeper aragonite compensation depth during glacial times. This point will be discussed further in a subsequent section.

Miscellaneous Deep-Sea Carbonate Components

Although planktonic foraminifera, coccoliths and pteropods contribute the bulk of deep-water carbonates, two other components can be of major importance in some areas. These are deep-sea corals and inorganically precipitated lutites; each will be discussed further in following paragraphs. In addition, there are a number of minor components which contribute to deep-sea sediments. Fish otoliths occur in most deep-sea sediments, but their abundance seldom exceeds 1% of the total carbonate fraction. Similarly, squid beaks are considered a very minor component, although BEKLEMISHEV (1971) reports that beak concentrations in some portions of the Pacific exceed 1000 per m^2 of bottom surface. Resting cysts of dinoflagellates are present in oceanic sediments, but only in trace amounts. Sponges are common, but most are siliceous (SAITO and BÉ, 1968). Planktonic ostracods lack calcareous layers and therefore do not contribute to deep-sea carbonates. In terms of biomass, deep-sea pelecypods are perhaps the major deep-sea benthic carbonate producers (SANDERS and HESSLER, 1969), but gastropods are represented by more species (CLARKE, 1962). Monoplacophorans are present only in the deep sea and scaphapods constitute relatively large populations compared to shallow waters (CLARKE, 1962).

Shallow-water carbonates can be introduced to the deep sea by turbidity currents, slumps and rafting. The influence of bioclastic turbidite sedimentation in the deep sea has been studied by many workers in the Bahamas (RUSNAK and NESTEROFF, 1964; PILKEY and RUCKER, 1966; RUCKER, 1968; BORNHOLD and

PILKEY, 1971). Accumulation rates can be several orders of magnitude greater than for normal deep-sea sediments. Generally, however, the horizontal extent of any one turbidite deposit will be restricted. BORNHOLD and PILKEY (1971), for instance, found that shallow-water components in the northern Columbus Basin (Bahamas) did not travel more than a few 10's of kilometers from their point of origin.

PILKEY and RUCKER (1966) and RUCKER (1968) explained increased calcite concentrations in the lower sequences of short cores from the Tongue of the Ocean and Exuma Sound as being due to Pleistocene sea-level changes. Shallow-water aragonite, eroded from the neighboring banks, is a prime sedimentary source during high stands of sea level. During low stands of sea level, the banks were subaerially exposed, and the aragonite altered to calcite, thus decreasing the aragonite and increasing the calcite concentrations in the deep-water sediments. However this theory would not explain why magnesian calcite, which is even less stable than aragonite to subaerial alteration, should be present in rather uniform amounts throughout the cores.

Deep-Water Coral Reefs

Ahermatypic corals are capable of constructing surprisingly large reefs in the deep sea. TEICHERT (1958) mentioned reefs off Norway that are as much as 60 m thick and cover areas greater than 4 km^2. According to TEICHERT (1958) and WELLS (1957b), maximum ahermatypic coral development occurs in temperature ranges between 4 and 10 °C and water depths between 180 and 360 m; but prominent reefs also are found at 850 meters on the Blake Plateau (STETSON and others, 1962, 1969) (Plate XXV). Dominant species in most deep-sea reefs are the scleractinians *Lophelia prolifera*, *Dendrophyllia profunda* and *Madrepora ramea*, as well as the hydrozoans *Stylaster gemmascens* and *Allopora norvegica* (TEICHERT, 1958; SQUIRES, 1959).

Deep-water coral reefs also provide the home for more than 100 species of hard shelled organisms that would not be present in normal deep-sea sediments (TEICHERT, 1958). The deep-reefs on the Blake Plateau are particularly rich in ophiuroids (more than 100 individuals per m^2 in places; MILLIMAN and others, 1967) with lesser amounts of echinoids, gastropods and decapods.

Deep coral reef sediments are petrographically unique. The gravel fraction, which is usually totally coral, can constitute more than 60 to 90% of the total sediment (Plate XXVIII b). The finer-grained material usually is composed of planktonic foraminifera, pteropods (in warmer climates) and skeletal debris from the various reef fauna. Because of the post-glacial rebound in northern Europe, many of the deep reefs off Norway have been elevated to depths where deep-water coral can no longer live. These dead reefs occasionally are shallow enough to allow coralline algae to live and encrust (TEICHERT, 1958). This coralline algae-deep coral assemblage, if preserved in the geologic record, could prove confusing, since one might assume that it represented a shallow-water coral reef. The superposition of the algae over the dead coral, plus the presence of many deep-water micro-fossils, however, should provide ample clues as to the reef's true origin.

Deep-Sea Precipitates

Inorganically precipitated calcium carbonate generally is considered to be restricted to aragonite-rich sediments in very shallow tropical waters, such as the Persian Gulf and the Bahama Banks (see Chapter 6). Deep-sea sediments are thought to be organically derived. CASPERS (1957) indicated that the calcite-rich lutites in the anoxic Black Sea might be inorganically precipitated. The reduction of nitrogen and sulfur compounds supposedly are responsible for this precipitation. However, subsequent studies have shown that the fine grained carbonate in the Black Sea is almost completely a coccolith ooze (BUKRY and others, 1970; MULLER and BLASCHKE, 1969).

This is not to say, however, that inorganic precipitation of calcium carbonate does not occur in the deep sea. In some semi-restricted basins, inorganically precipitated carbonates can dominate the carbonate fraction of the sediments. In the Red Sea magnesian calcite (12 mole % $MgCO_3$) is the dominant sedimentary component (MILLIMAN and others, 1969). Concentrations are greatest in the lutite fraction, although magnesian calcite also occurs as an internal cement as well as a limestone cement (GEVIRTZ and FRIEDMAN, 1966; MILLIMAN and others, 1969; see Chapter 10). Because of the similarly of this component with other deep-sea cements and also because of the lack of any known organic source, the magnesian calcite is considered to be inorganically precipitated, probably the result of the warm (20–21 °C) hypersaline (39–44‰) deep Red Sea waters (MILLIMAN and others, 1969). During Pleistocene periods, when the salinities presumably were even greater than at present (DEUSER and DEGENS, 1969; HERMAN, 1965), aragonite was a major sedimentary component and cement (MILLIMAN and others, 1969).

The deep waters in the eastern Mediterranean Sea have temperatures and salinities that are equidistant between normal deep-sea and Red Sea waters (13–14 °C and 38–39‰ at 2000 m). A considerable amount of the calcium carbonate in this area is inorganically precipitated (MÜLLER and FABRICIUS, 1970; MILLIMAN and MÜLLER, 1972), but the result is somewhere between normal ocean deep-sea sediments and those from the Red Sea (Fig. 86). Magnesian calcite (10–11 mole % $MgCO_3$) is present in the surficial deep-sea sediments, generally comprising between 15 and 40% of the total carbonate, but locally it can exceed 90% (Plate XXVIII d). In contrast to the Red Sea, however magnesian calcite was not been precipitated continuously throughout the Quaternary. Apparently during late glacial times (when the glaciers began their initial retreat) carbonate sedimentation was dominanted by the accumulation of coccoliths. These periods often coincided with the formation of sapropel layers, which are indicative of stagnant conditions in the bottom waters (OLAUSSON, 1965). No doubt these stagnant conditions were responsible for the lack of magnesian calcite precipitation (MILLIMAN and MÜLLER, 1972). Other areas of possible magnesian calcite deposition have been reported from deep-sea areas in the western Mediterranean (NORIN, 1956) and from the northwestern Indian Ocean (WISEMAN, 1965), but neither area has been studied sufficiently to warrant further discussion at this time.

The mode of precipitation of magnesian calcite in the deep sea is not known. EMELYANOV and SHIMKUS (1971) report "chemogenic" calcite crystals in the

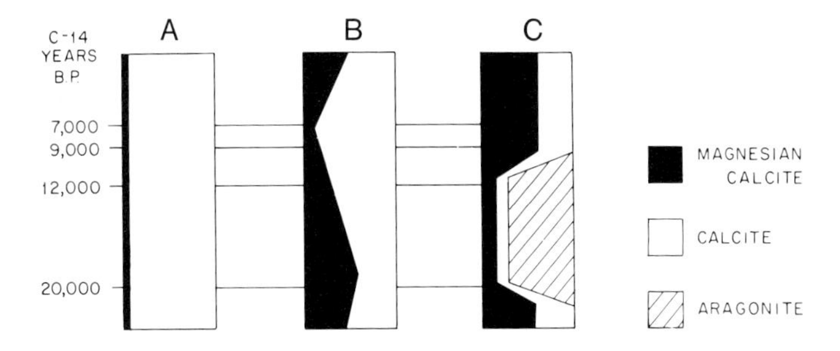

Fig. 86 A–C. Comparison of mineralogies in deep-sea sediments in the normal ocean (A), the eastern Mediterranean (B) and the Red Sea (C). From MILLIMAN and MÜLLER (1972)

suspended matter of the eastern Mediterranean, but quantiatively these crystals probably are unimportant. Most data suggest that precipitation occurs at the sediment-water interface rather than in the water column or after sediment burial (MILLIMAN and MÜLLER, 1972). FRIEDMAN (1965) concluded that the dominance of magnesian calcite over aragonite in the eastern Mediterranean and Red Sea indicates its greater stability. Would this suggest that either aragonite is not precipitated in the deep sea, or that if it is, it quickly alters to magnesian calcite? Clearly further work on the precipitation of deep-sea carbonates is needed.

World-Wide Distribution of Calcium Carbonate in Marine Sediments

The relative abundance of the various types of marine carbonate deposits in the surficial sediments throughout the world oceans can be calculated by knowing the average carbonate content in each of various types of carbonate deposits and the area covered by each type. For purposes of this calculation, marine carbonate sediments have been divided into shallow-water (reefs and shelves), slope (200 to 2000 m) and deep-sea (depths greater than 2000 m). Calcareous deep-sea sediments, composed primarily of planktonic foraminifera and coccoliths, occupy a total area of 128×10^6 km^2, with an average carbonate content of 65% (SVERDRUP and others, 1942). Siliceous oozes and red clays occupy somewhat smaller areas and contain far smaller amounts of calcium carbonate (Table 60; columns 6 and 7). Of the major shallow-water deposits, reef sediments occupy the smallest area, but contain the highest estimated carbonate content (80%). Shelves and slopes cover larger areas, but contain smaller carbonate values (Table 60).

When all these factors are calculated, the deep-sea oozes account for more than 89% of the total carbonate in the surface sediments of the world oceans. Shallow-water carbonates (reefs and shelves) account for only 4.8%, while slopes total 5.8% of the world carbonate. While these figures infer the dominance of deep-sea carbonate sedimentation, they do not take into account the relative rates of deposition. This factor will be discussed in a following section.

Table 60. Calculation of the distribution of $CaCO_3$ within the surface sediments in the world oceans. Many data are from SVERDRUP and others (1942)

	Average depth (m)	Percent area covered in each ocean			Total area covered ($\times 10^6$ km^2)	Estimated average weight percent $CaCO_3$	Relative total $CaCO_3$ content	Percent of total $CaCO_3$ in oceanic surface sediments	
		Indian Ocean	Pacific Ocean	Atlantic Ocean					
1. Pelagic sediments									
Calcareous ooze									
Foraminifera-coccolith	3600	48.5	31.6	50.1	126	65	83.2	78.2	89.5
Pteropod-coccolith	2000				2	(65)			
Siliceous ooze									
Diatom	3900	18.2	12.8	5.0	31	10	3.8	3.6	
Radiolarian	5300				7	10			
Red clay	5400	22.6	42.9	19.2	102	8	8.2	7.7	
2. Shallow-water sediments									
Reefs and banks	<20	–	–	–	1.4	80	1.1	1.0	4.8
Eulittoral-littoral (shelves)	20–200	4.2	5.7	13.3	26.9	15	4.0	3.8	
3. Slope sediments	200–2000	6.5	7.0	12.4	30.6	20	6.1	5.8 → 5.8	

Variations in Deep-Sea Carbonate Deposition with Geologic Time

The relative distribution of calcium carbonate sediments within the world oceans has not remained constant throughout geologic time. Changes in the patterns of deep-sea-sedimentation occurred during the Mesozoic, the mid-Tertiary and the Quaternary.

1. As mentioned elsewhere in this book, the site of calcium carbonate sedimentation shifted to the deep sea with the evolution of calcareous coccolithophorids and planktonic foraminifera in the early and mid Mesozoic. Prior to this time, most sedimentation probably occurred either in shallow water areas or in semi-enclosed basins.

2. The boundary marking the northern extent of carbonate-rich sediments in the north equatorial Pacific has migrated 1000 km towards the equator since the Miocene (ARRHENIUS, 1963; VAN ANDEL, 1971, oral communication). As a result, red clays of Quaternary age often overlie mid-Tertiary coccolith and foraminiferal oozes. HEATH (1969) reported that the compensation depth in the equatorial Pacific during the Oligocene was 5200 m, as compared with about 3000 to 4000 m at present. Similarly, GARTNER (1970) and SAITO (1971) report that the compensation depth in the Atlantic became shallower during the Quaternary. This shoaling of the compensation depth and its shift towards the equator may be related in part to the cooling of deep-sea bottom waters during the Oligocene and Miocene (EMILIANI, 1954), but it may also be partly due to the deepening of the oceanic basins during the mid and late Tertiary (MENARD, 1964).

3. Arrhenius (1952, 1963) found that carbonate accumulation in the equatorial Pacific during the last glaciation was some 2.3 times greater than at present. Similar trends have been noted in the equatorial Atlantic (WISEMAN, 1956). ARRHENIUS pointed out that such variations in carbonate content could be the result of changes in either carbonate production or carbonate dissolution; he assumed that carbonate productivity increased during glacial times in response to increased oceanic circulation and upwelling. MCINTYRE and others (1972) have suggested that carbonate productivity during the Pleistocene varied with latitude. According to this scheme, the zone of maximum productivity migrated towards the equator during glacial periods, thus resulting in lower carbonate concentration in high latitude waters and higher concentrations in low latitudes. During interglacial times, the zone of maximum productivity shifted poleward, along with carbonate maxima.

Other data also suggest that changes in dissolution rates may be an important factor in controlling deep-sea carbonate sedimentation. OBA (1969; cf. BROECKER, 1971) noted an increase in benthonic foraminifera fragments within interglacial sections of Indian Ocean cores, and CHEN (1968) found that aragonitic pteropod shells are present only in glacial facies of Caribbean cores. Both points indicate increased dissolution during interglacial times (BROECKER, 1971). Another possible mechanism for controlling glacial and interglacial fluctuations in deep-sea carbonate sedimentation is discussed in the following section.

Calcium Carbonate Budget for the World Oceans

The distribution of calcium carbonate in surficial sediments within various marine environments does not take into account relative rates of deposition. If this factor is considered, then a rough budget for calcium carbonate sedimentation can be computed. Until recently, the lack of sufficient data about the rates of sediment accumulation precluded such a calculation. However deep borings on coral reefs and the recent deep-drilling in ocean basins have overcome part of this lack of information. In order to obtain good long-term deposition rates, the time span of 6 million years (post-Miocene) has been used. In addition, Holocene (the past 15000 years) deposition rates have been calculated for comparison with the post-Miocene.

For purposes of this calculation the ocean has been divided into five depositional environments; reefs, shelves, slopes (200–2000 m), the deep sea and enclosed basins (such as the Mediterranean Sea). Each will be discussed in the following paragraphs.

1. Deep-Sea Sedimentation. Depositional rates for the Atlantic and north Pacific Oceans have been derived from preliminary JOIDES reports, volumes 1–7. These rates have been converted from cm/1000 years to gm $CaCO_3/cm^2/1000$ years by assuming a porosity of 70% (50% water by weight), averaging a $CaCO_3$ content from reported data, and assuming a dry sediment density of 2.6 g/cc.

Plots of accumulation rates of $CaCO_3$ versus water depth for the Atlantic and north Pacific JOIDES cores (post-Miocene) and assorted short cores (late Quaternary) are shown in Figs. 87 and 88. Although the late Quaternary rates show considerable variation, they seem to cluster around the broad post-Miocene trends. Clearly more data are needed before these curves can be quantified with confidence. Corresponding rates in the south Pacific and Indian Oceans are not known. It is assumed that rates in the south Pacific are about equal to those in the Atlantic (Fig. 87), and that the rates in the Indian Ocean are equidistant between those in the north Pacific and the Atlantic.

By knowing the area of ocean floor covered by various depth intervals (data are taken from Sverdrup and others, 1942 and Menard and Smith, 1966) and by knowing the approximate carbonate accumulation rates within these depth intervals, one can calculate the total accumulation of calcium carbonate within each depth interval and, in turn, within the entire deep sea (Table 61). The Pacific Ocean accounts for the greatest amount of carbonate deposition, followed closely by the Atlantic and Indian Oceans. The total carbonate deposited is computed to be 1.108×10^{15} grams $CaCO_3$ per year (Table 61).

2. Shelf Sedimentation. Many shelves have experienced little sediment accumulation during the post-Miocene. For example, Emery and Milliman (1970) estimated that the average thickness of post-Miocene sediment on the United States Atlantic continental shelf is about 40 meters. Assuming a 50% porosity, a sediment density of 2.6 g/cc, and an average calcium carbonate content of 15%, the carbonate accumulation rate is $0.04\ g/cm^2/10^3$ years. In view of the sketchy information available for other continental shelves, this value has been accepted as a world wide average, although in fact it may be too high (K.O. Emery, 1971, oral communication).

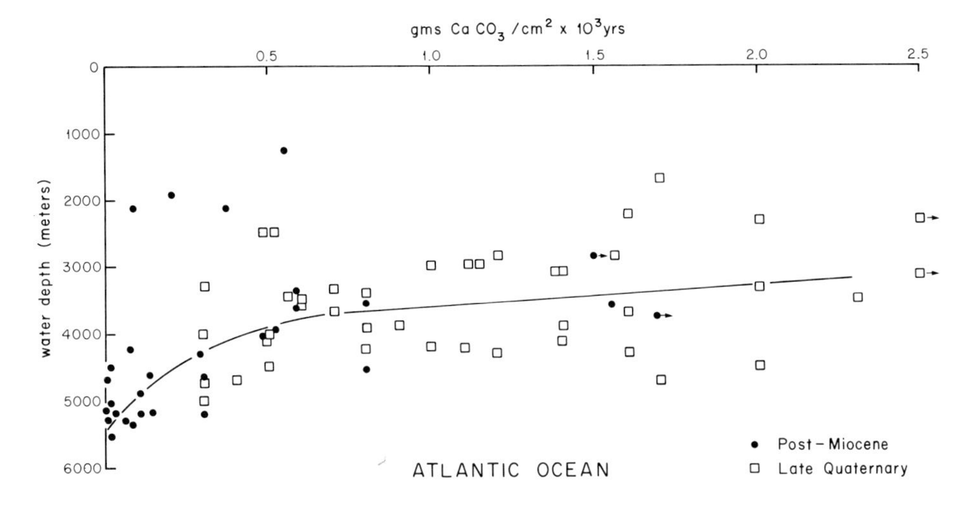

Fig. 87

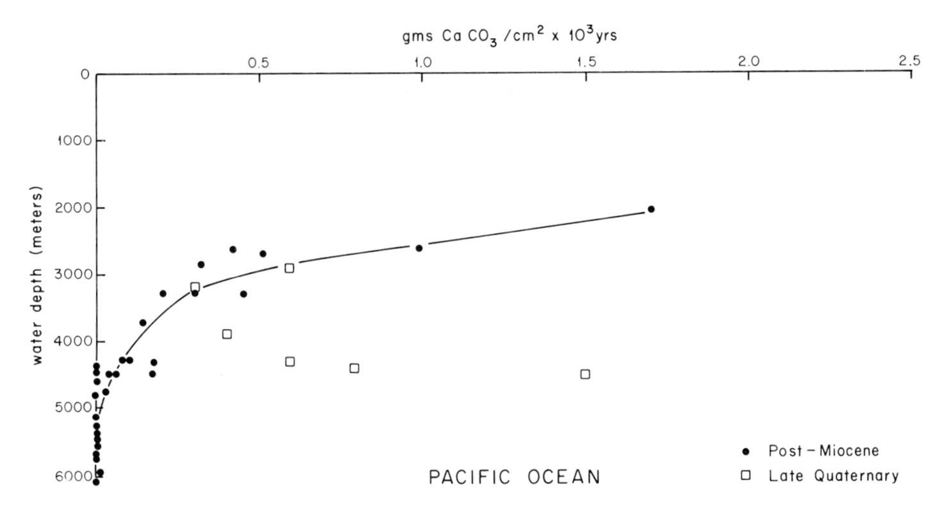

Fig. 88

Fig. 87 and 88. Variation in $CaCO_3$ accumulation (expressed as grams $CaCO_3$ per cm^2 per thousand years) with water depth in the Atlantic and Pacific Oceans. Post-Miocene values are derived from core descriptions and analyses made during JOIDES (Joint Oceanographic Institutions Deep-Earth Sampling) deep-sea drillings. The late Quaternary data come mostly from ERICSON and others (1961) and from BROECKER (1971). The wide scatter of data points is mostly a function of local environmental conditions as well as possible analytical errors

Table 61. Distribution of calcium carbonate accumulation in the deep sea

	Area ($\times 10^{16}$ cm^2)	Deposition rate (g/cm^2/10^3 yrs)	Total grams $CaCO_3$/yr ($\times 10^{15}$)
Atlantic			
2000–3000 m	7.0	2.0	0.140
3000–4000 m	16.7	1.2	0.202
4000–5000 m	27.3	0.2	0.055
		Subtotal	0.397
Pacific			
North Pacific			
2000–3000 m	5.7	1.1	0.063
3000–4000 m	18.1	0.22	0.040
4000–5000 m	29.0	0.05	0.014
South Pacific			
2000–3000 m	5.1	2.0	0.102
3000–4000 m	16.2	1.2	0.194
4000–5000 m	26.0	0.2	0.052
		Subtotal	0.465
Indian			
2000–3000 m	6.9	1.6	0.110
3000–4000 m	17.3	0.7	0.121
4000–5000 m	25.1	0.1	0.025
		Subtotal	0.246
		Total $CaCO_3$	1.108×10^{15} g/yr

One shelf area in which carbonate sedimentation has been significantly higher during the post-Miocene is the Sunda Shelf, in the western Pacific and eastern Indian Oceans. Sedimentation rates over this large (1.85×10^6 km^2) area have averaged about 8 cm per thousand years, with approximately 75% being carbonate (ZVI BEN AVRAHAM, 1971, oral communication). This one shelf, therefore, accounts for more than 10 times the carbonate accumulated on all the other world shelves (Table 62).

Studies of modern shelf sediments suggest that accumulation rates during the Holocene have been appreciably faster than during the entire post-Miocene period. Rates of 2.5 cm/10^3 years with average carbonate contents of 50% may not be unreasonable for the eastern United States. Assuming similar world-wide rates, the Holocene accumulation rates have been some 13 times higher than during the post-Miocene (Table 62).

3. *Reef Sedimentation.* The area covered by coral reefs is difficult to estimate without a lengthy calculation. For purposes of this compilation, the writer has assumed that the total area covered by post-Miocene reefs is twice the combined area of all the world's atolls plus the Great Barrier Reef, the Bahamas and Campeche Bank. The combined area is 800000 km^2. Campeche Bank had active accumulation during the post-Miocene, but at present it may be too deep to

Table 62. Calcium carbonate accumulation on shelves, reefs, enclosed ocean basins and slopes

	Area ($\times 10^{16}$ cm^2)	Deposition rate (g/cm^2 $\times 10^3$ yrs)	Total grams $CaCO_3$/yr($\times 10^{15}$)
Post Miocene			
Reefs	(0.890 × 2)	2.5	0.046
Shelves (except Sunda)	(24.46)	0.07	0.011
Sunda shelf	(1.85)	10.0	0.111
Enclosed basins			
Red Sea	0.45	4.9	0.022
Black Sea	0.5	2.9	0.128
Mediterranean	4.36	1.4	0.005
Slopes	(30.6)	1.5	0.458
Holocene			
Reefs	(0.72 × 2)	0.35	0.500
All shelves	(26.31)	0.9	0.143
Enclosed basins			
Red Sea	0.4	5.0	0.022
Black Sea	0.5	1.2	0.054
Mediterranean	4.36	1.9	0.009
Slopes	(30.6)	1.5	0.458

experience rapid deposition rates; therefore this bank was excluded from the Holocene calculations.

Post-Miocene reef sedimentation rates are approximated from the data available from seven reef borings. Although there is a considerable range between the various borings, a rate of 3.0 cm of sediment per thousand years appears reasonable (Table 63). Assuming a porosity of 60%, a primary density of 2.6 and and average carbonate content of 80% $CaCO_3$, the accumulation rate is 2.5 g $CaCO_3$/cm^2/10^3 years.

Holocene rates have been derived from various carbon-14 dates taken from reefs and carbonate banks, mostly from the Caribbean. Although the variation is great, the average rate of 35 g $CaCO_3$ per thousand years is more than an order of magnitude greater than the average post-Miocene rates (Tables 63 and 62).

4. *Semi-enclosed Deep-Sea Basins.* Three semi-enclosed basins that could influence the world-wide carbonate budget are the Red Sea, the Black Sea and the Mediterranean Sea. The Red Sea has an area of 0.453×10^6 km^2, a deposition rate of about 8 cm/10^3 years, and an average carbonate content of about 80% (MILLIMAN and others, 1969). Porosity of these sediments as well as those in other basins is assumed to be 70%. The Black Sea has a somewhat larger area (0.5×10^6 km^2) and accumulation rates of about 5 cm/10^3 years; carbonate contents have averaged 50% during the past 3000 years and 25% in former times (ROSS and others, 1970). In both the Red and Black Seas, late Quaternary sedimentation rates (derived from carbon-14 data) are assumed to be approximate post-Miocene

Table 63. Thickness and accumulation rates of post-Miocene sequences in various reef borings (a) and in various Holocene reef and bank sediments (b). (Data are from SCHLANGER, 1963; LADD and SCHLANGER, 1960; LADD and others, 1967; SPENCER, 1967; WANLESS, 1969; STOCKMAN and others, 1967; TAFT and HARBAUGH, 1964; GOODELL and GARMAN, 1969; GINSBURG and others, 1971; LAND and GOREAU, 1970; EBANKS, 1967; DAVIES, 1970a; and EASTON, 1969)

	Post-Miocene	
	Thickness (m)	Deposition rate (cm/10^3 yrs)
(a) *Post Miocene*		
Bikini	212	3.5
Eniwetok	186	3.1
Midway	140	2.3
Florida	136	2.3
Andros (Bahamas)	166	2.8
Kita-Daito-Jima	103	1.7
Funafuti	>360	>6.0
		Assume an average = 3.0 cm^3/10^3 yrs Assume 60% porosity, 3.6 g/cm^3 density and 80% $CaCO_3$ = 2.5 g/10^3 yrs

	Deposition rate (cm/10^3 yrs) (range in parentheses)
(b) *Holocene*	
Florida	
Biscayne Bay	265 (160–370)
Florida Bay	30 (8–73)
White Water Bay	19 (11–39)
Florida Reef Tract	44
Bahamas	70 (49–91)
Bermuda	120
Jamaica	120
British Honduras	26 (10–34)
Western Australia	12 (9–15)
Hawaii (Oahu)	500
	Assume an average of 40 cm^3/10^3 yrs × 0.6 porosity × 2.7 g/cm^3 = 35 g/10^3 yrs

rates. The Mediterranean Sea has an area of 4.357×10^6 km^2, an average post-Miocene depositional rate of 6 cm per cm^2 per thousand years, and a carbonate content of about 65% (unpublished JOIDES data). Late Quaternary carbonate content has been somewhat less, about 40% on the average, with a sedimentation rate about 4 cm per cm^2 per thousand years (MILLIMAN and MÜLLER, 1972) (Table 62).

5. *Continental Slope.* Sedimentation rates and carbonate values of slope sediments are not well documented. An average $CaCO_3$ deposition rate of 1.5 g per cm^2 per thousand years (the same rate as calculated for the 2000–3000 m depth interval) has been assumed in this calculation. This value, however, may be subject to considerable revision when further data become available.

Table 64. Computations of total accumulation of calcium carbonate within the various depositional environments in the oceans. Two Holocene calculations have been made; the first was made by totalling figures calculated in previous tables. This total, however, was considerably higher than the post-Miocene average, which suggests that if such rates were true, the oceans would be in disequilibrium. Assuming that such a state of disequilibrium would quickly correct itself by increased dissolution, the second Holocene calculation was made, assuming that total accumulation was equal to post-Miocene averages. The logical site of increased dissolution is the deep sea, that is, a shallower compensation depth

	$\times 10^{15}$ gm $CaCO_3$ year	Percent of total marine sedimentation
Post-Miocene		
Deep-Sea	1.108	59
Enclosed Basins	0.155	8
Slopes	0.458	24
Shelves (excl. Sunda Shelf)	0.011	1
Sunda Shelf	0.111	6
Reefs	0.046	2
	1.889 total	
Holocene (assuming disequilibrium with respect to post-Miocene averages)		
Deep-Sea	1.108 (assumed)	48
Enclosed Basins	0.085	4
Slopes	0.458 (assumed)	20
Shelves	0.143	6
Reefs	0.500	22
	2.314 total	
Holocene (assuming equilibrium with respect to post-Miocene averages)		
Deep-Sea	0.703 (assumed)	37
Enclosed Basins	0.085	4
Slopes	0.458 (assumed)	24
Shelves	0.143	8
Reefs	0.500	27
	1.889 total	

When the accumulation values of the various depositional areas are totalled (Table 64), the calcium carbonate deposited in the world oceans during the past 6 million years (post-Miocene) is calculated to have averaged about 1.9×10^{15} grams per year. One check on this calculation is to compare it with the average annual amount of $CaCO_3$ supplied to the oceans by modern rivers. Assuming that the oceans are in a steady state, amount of calcium carbonate deposited yearly should equal the amount brought into the oceans annually. Modern fluvial contributions of dissolved calcium are 48.8×10^{13} g per year (Table 5); this is equivalent to 1.2×10^{15} g of $CaCO_3$, which is reasonably close to 1.9×10^{15} grams calculated above. Of this total, 67% is calculated to be deposited in the deep-sea. Another 24% is deposited on continental slopes. Of the remaining 9%, only 2% has accumulated on reefs, the remainder being on shelves (mostly on the Sunda Shelf) (Table 64).

A markedly different pattern of carbonate deposition, however, has developed during the Holocene. The relatively rapid accumulation on reefs and shelves has

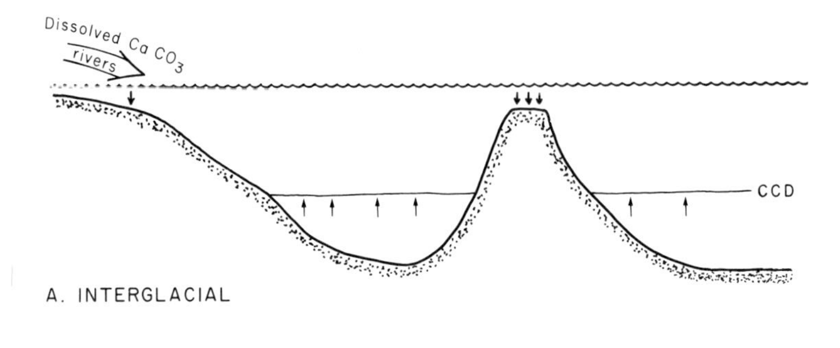

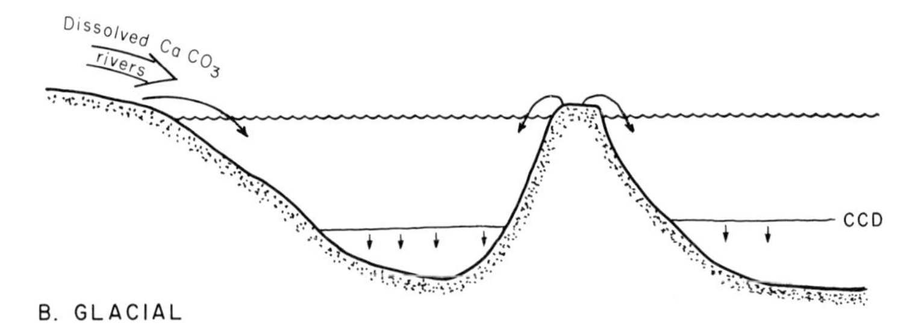

Fig. 89 A and B. Shift in the balance of carbonate deposition during interglacial and glacial times. During the interglacial periods (A) many banks and shelves are slightly submerged and thus reef and lagoonal accumulations are rapid. Eventually, as mentioned in the text, the rapid accumulation of shallow-water carbonates leads to a deficiency of $CaCO_3$ within the oceans waters; in order to compensate for this disequilibrium, the carbonate compensation depth (CCD) will shift upwards. During glacial times (B), the shallow-water areas are exposed to subaerial conditions. Not only do sediments no longer accumulate in these areas, but much of the material is leached away, thus providing an addition source of dissolved $CaCO_3$ in the oceans. These two factors will favor the lowering of the carbonate compensation depth during glacial periods

shifted the balance of calcium carbonate deposition towards shallower water. Presently, according to this calculation (Table 64), 28% of the carbonate deposition in the oceans occurs on modern reefs and on the continental shelf. Assuming that deposition in the other marine environments would remain more or less constant, the total carbonate deposition would increase from 1.9 to 2.3×10^{15} grams $CaCO_3$ per year (Table 64). This increased deposition exceeds the post-Miocene steady-state, but in terms of the calcium reservoir in oceanic waters, the deficit is small, less than 0.1% of the calcium within sea water per thousand years. If such a disequilibrium were allowed to continue, however, at some point (probably within the range of 10^4 to 10^5 years) the depletion of $CaCO_3$ within the oceans would result in undersaturation and thus increased dissolution. Presumably dissolution would increase to a point to offset the increased carbonate deposition in shallow waters, thus balancing the system again. The increased dissolution undoubtedly would be manifested by a shallowing of the carbonate compensation depth, thus decreasing accumulation rates in the deep sea (Table 64).

Assuming a return to equilibrium, reef and shelf sediments would account for about 35% of the total marine carbonate deposition (Table 64).

At some point, however, this rapid accumulation in shallow water will be halted, either when the rapidly accreting banks and reefs reach sea level, or when the sea level falls and present-day shallow-water areas are exposed to subaerial conditions. In the latter instance not only would shallow-water deposition decrease markedly (assuming an average drop in sea level of 120 m, shallow-water deposition would take place on what are presently continental slopes and which occupy a considerably smaller area than modern reefs and shelves), but ground waters probably would erode many of subaerially exposed deposits and contribute significant amounts of particulate and dissolved calcium carbonate back to the ocean (Fig. 89). These two factors would combine to shift the calcium budget back towards post-Miocene equilibrium and thus lower the compensation depth. Such a mechanism may help explain the observed fluctuations in compensation depths between interglacial and glacial times as described on p. 242.

Part IV. Carbonate Diagenesis

Diagenesis refers to the alteration and cementation of sediments during the interval between deposition and metamorphism. Because of their dependency upon the various chemical, physical and biological parameters affecting the depositional environment, carbonate sediments generally undergo more rapid diagenesis than most non-carbonate sediments. The type of diagenetic alteration depends upon the post-depositional environment, while the degree of alteration is related to the duration of exposure to that environment (PURDY, 1968).

Submarine alteration begins immediately after deposition (sometimes before), and continues until exposure to meteoric waters. Final alteration in subaerial conditions may remove most traces of submarine diagenesis, so that the absolute extent of submarine diagenesis in ancient carbonates may not be fully appreciated. In recent years, however, numerous studies have shown that such processes can be important factors in the transition of sediment into rock. Recognition of these processes within a limestone or its carbonate components can provide a valuable insight into the history of that rock.

Generally two separate but inter-related submarine diagenetic processes are recognized: destructive diagenesis and constructive diagenesis. The former includes biological, mechanical and chemical degradation of the substratum, while the latter involves the formation of new carbonate, as well as the preservation of the old (BATHURST, 1964a).

Chapter 9. Carbonate Degradation

Biological Erosion

Biological erosion of carbonate substrates serves two important purposes in carbonate diagenesis. Firstly it degrades and destroys carbonate components; secondly it creates secondary voids within the carbonate particles and substrate which can be important in subsequent recrystallization.

Carbonate substrates are attacked and altered by four major types of organisms: burrowers and borers, grazers, browsers and predators. CARRIKER and SMITH (1969) proposed that burrowers and borers be distinquished by their ability to penetrate the substrate for protection (burrowers) or food (borers). While this differentiation may be practical for biologists, most geologists continue to define borers as those organisms which penetrate a hard substrate and burrowers as excavating unconsolidated particles (ALEXANDERSSON, 1972a). Burrowing organisms have been discussed on p. 180. Grazers feed by scraping the bottom, usually in search of boring algae. Browsers (deposit-feeders) are non-selective sediment ingestors; they consume large quantities of sediment from which they remove the necessary nutrients, generally algae and detritus. Predators are primarily carnivores that actively hunt and consume their prey.

Borers

More than three plant and eight animal phyla, in addition to fungi and bacteria, are capable of penetrating carbonate substrates, for protection and food and possibly for other reasons that are not yet clear. The magnitude of research done on these various organisms is illustrated by the large annotated bibliography by CLAPP and KENK (1963). General reviews have been given by OTTER (1937), GINSBURG (1957) and YONGE (1963b). A recent symposium, organized by M.R. CARRIKER (American Zoologist, v. 9, p. 629–1020) brought together many of the current thoughts on burrowing and boring organisms and the mechanisms of penetration of hard mineralized substrata. ALEXANDERSSON (1972a) has presented an excellent discussion of the geologic effects of boring and the subsequent secondary precipitation of calcium carbonate.

Fungi and Bacteria. Fungi are known to be common agents in the penetration of carbonate substrates (BONAR, 1936; PORTER and ZEBROWSKI, 1937; KOHLMEYER, 1969) (Plate XXIX e). Small (1 to 4 microns in diameter) borings, perhaps fungal in origin, occur in shells from the North Carolina shelf and slope (to depths of 780 m) (HALSEY and PERKINS, 1970; PERKINS and HALSEY, 1971). The mode of penetration is unknown, but it is possible that the fungi are able to utilize the energy contained in organic matter within the carbonate. DISALVO (1969) has suggested a similar mode of destruction of coral heads by bacteria, and PURDY (1963) showed that decay of organic matrices can result in the comminution of coarser sediment into fine debris. Recent studies have shown that fungi may be important carbonate degraders. For example, ROONEY and PERKINS (1972) reported that fungi is the most wide-spread borer in the carbonate sediments at Arlington Reef (Australia) and that fungi generally are the first organisms to attact fresh carbonate debris. Apparently fungi preferentially attack mollusk fragments (ALEXANDERSSON, 1972a; ROONEY and PERKINS, 1972).

Algae. Green, blue-green and red algae have been mentioned as possible borers. For the most part, these algae are limited to relatively shallow depths, generally less than 25 to 50 m (HALSEY and PERKINS, 1970), although siphonaceous greens occur at depths as great as 195 m (PERKINS and HALSEY, 1971).

Filamentous green algae are abundant in many living corals. ODUM and ODUM (1955) estimated that some 70% of the biomass in corals is composed of filamentous greens living in the upper few mm of the carbonate skeleton. They concluded that these algae may be symbiotic rather than destructive, but their photosynthetic productivity seems to be minor (KANWISHER and WAINWRIGHT, 1967) suggesting that they may be more important as eroders.

Blue-green algae are one of the most important organic factors in the marine carbonate system. Not only are they important sediment binders (see Chapters 4 and 6) and nitrogen fixers (NEWHOUSE, 1954), but they also are perhaps the most important shallow-water agents of biologic erosion (DUERDEN, 1902; NADSON, 1927; FRÉMY, 1945; DOTY, 1954; BATHURST, 1966, 1971). Borings generally range from 5 to 15 microns in diameter (Plate XXXIII) but absolute dimensions depend upon the species of algae involved (GOLUBIC, 1969). Although the exact mode of erosion is unknown, a chemical process is inferred, perhaps one in which acid radicals are excreted by the algae (DUNCAN, 1876; KOSTER, 1939). Erosion may occur at night, when oxygen is consumed and pH decreases (NEWHOUSE, 1954). Growth expansions may push off excavated chips (YONGE, 1963b).

Blue-green algae are especially important in the littoral and sublittoral zones (GINSBURG, 1953a). The common black zone in the supratidal zone of rocky coasts (STEPHENSON and STEPHENSON, 1950) owes its color to the presence of the blue-green alga *Entophysalia* (PURDY and KORNICKER, 1958; RANSON, 1955). Although penetration is limited to only a few millimeters (BERTRAM, 1936), the burrowed carbonate can become so weakened that chips are flaked off by wind or waves (PURDY and KORNICKER, 1958). More important, however, is the fact that endolithic blue-greens serve as an important food source for many browsing animals, who in the process of extracting the algae also destroy the infested substrate. The effect of blue-green algal activities on sublittoral carbonates is even greater. Not only do endolithic algae bore extensively, but their post-mortum decomposition helps create a favorable environment for the subsequent precipitation of intragranular cement (see p. 272ff.).

The erosional effects by epilithic benthonic algae are not well documented. A recent study by BARNES and TOPINKA (1969), however, has shown that an "acidic material" excreted by the brown alga, *Fucus*, will partially dissolve barnacle shells upon which the *Fucus* grows. RANSON (1955) has proposed a somewhat similar process for the coralline alga, *Porolithon onkodes*, and it is possible that other benthonic algae also exhibit such erosive tendencies.

Sponges. Because of their ability to cause extensive damage to commercially valuable oysters boring sponges (especially *Cliona*) have received considerable attention since the early nineteenth century (GRANT, 1826) (Plate XXIX). Present evidence suggests that excavation is primarily chemical, since infestation is almost entirely limited to carbonate substrate. The chemical secretion is released by amoebocyte-like cells in the cytoplasm (COBB, 1969), but this secretion may not be acidic (WARBURTON, 1958). Whether the secretion contains carbonic anhydrase similar to that secreted by boring mollusks is not known. Chemical erosion occurs along the thin edge of each penetrating cell, the cell edge inserting itself into the shell until a chip is completely etched away from the substrate. The excavated chips are then transported to the excurrent canals and removed by currents set up by the sponge (COBB, 1969). The result is a distinctive faceted-burrow (Plate XXXIX b, c). Siliceous spicules do not play a role in either the excavation or the transport of the chips. Although earlier workers had suggested that *Clinoa* prefers substrate with certain mineralogies (HARTMAN, 1957 and references therein), NEUMANN (1966) found no evidence to substantiate this opinion.

Sponges seem capable of actively degrading carbonate substrates in water depths greater than 70 m (HARTMAN, 1957; GOREAU and HARTMAN, 1963). Most sponge borings are restricted to the outer 2 to 3 cm of the substrate's surface. On the basis of field experiments, NEUMANN (1966) calculated that *Cliona* can excavate 0.1 to 1.4 cm per year in the sublittoral zone. However, unless the eroded surficial layers can be removed relatively quickly, long-term erosion rates are probably considerably lower.

Mollusks. Boring mollusks are recognized as one of the major carbonate eroders in the seas. The rock boring bivalve, *Pholas*, for instance, can erode at rates of between 1.5 and 3 cm/yr (JEHU, 1918). Although most boring mollusks live in shallow waters, some have been reported from depths as great as 2300 m (CARRIKER, 1961).

Among the boring gastropods, primary penetration is accomplished by a chemical secreted by the accessory boring organ (ABO); a small portion of the weakened substratum is then scraped away by the radula and swallowed (CARRIKER, 1961, 1969, and references therein). Octopi also bore shells, possibly in similar ways (WODINSKY, 1969; ARNOLD and ARNOLD, 1969). The ABO secretion is acidic (DAY, 1969, has measured proboscus secretions with pH's as low as 2; and CARRIKER and others, 1967, measured pH's as low as 3.8 m in the ABO secretion) and contains enzymes such as carbonic anhydrase (CHETAIL and FOURNIE, 1969, and references therein). The secretion directly attacks both the carbonate and organic substrate, but differentially (CARRIKER, 1969). An alternative explanation of gastropod boring is that the secretant attacks the inorganic sheaths that encase the carbonate crystals; this results in the loosening of the crystals which then can be scraped off by radulae (TRAVIS and GONSALVES, 1969).

This latter explanation would not seem to apply to those mollusks which penetrate limestones free of most organic matrices.

Seven bivalve families are known to bore into substrate, but only one family, Mytilidae, is restricted to carbonate rocks (ANSELL and NAIR, 1969). Although this suggests that mytilids bore chemically, the fact that the absolute amount of carbonate within the substrate can vary from 5 to 100% (HODGKIN, 1962; WARME and MARSHALL, 1969) indicates that mechanical rasping also is utilized. Many burrowing mytilids also possess the ability to precipitate carbonate linings on the sides of the excavated cavities (BERTRAM, 1936; GOREAU and others, 1968).

Most other bivalves apparently burrow by mechanical means. ANSELL and NAIR (1969) state that a mollusk shell can abrade the substrate by movement of both the retractor and adductor muscles. Both the shell morphology of the borer and the rate at which the animal is able to bore are heavily dependent upon the hardness of the substrate (EVANS, 1968). On inactive substrates, boring is inwards, while on living substrate, bivalves must also bore outwards at a rate at least equal to the carbonate accretion (SOLIMAN, 1969).

Worms. Four genera of sipunculids are capable of carbonate degradation (CUTLER, 1968). NEWELL (1956) estimated that the activities of boring sipunculids (together with the barnacle *Lithotyra*) cause up to 20% porosity in the intertidal rocks at Raroia. Attachment of these worms must occur during the larval stages, as the adult is relatively immobile (CUTLER, 1968). Although OTTER (1937) and EBBS (1966) presented evidence suggesting that sipunculids are mechanical borers, RICE (1968) has noted that boring is limited to carbonate substrate, but that hardness does not seem critical. RICE has suggested that chemical secretions by the epidermal glands weaken the substrate; subsequent removal is facilitated by the papillae and outer setae.

Polychaetes are well-known borers. GARDINER (1903) and MCINTOSH (1902) estimated that polychaetes are one of the prime agents in breaking down tropical carbonates. Apparently, as in the sipunculids, enzymes or acids weaken the substrate and the setae remove some of it (HAIGLER, 1969). The excavated tubes can be either crooked or u-shaped, and are generally 0.2 to 0.5 mm in diameter (Plate XXIX d).

Echinoids. Rock-boring echinoids are restricted mainly to the littoral and sublittoral zones: the excavations offer protection from crashing surf and currents (OTTER, 1932). As such these organisms have been singled out as one of the prominent shoreline eroders in tropical areas (GARDINER, 1931; UMBGROVE, 1947; KAYE, 1959). Borings range from shallow excavations (in which the echinoid is partly encased) to deep holes (in which the opening is usually smaller than the echinoid, resulting in the animal being almost completely encased) (OTTER, 1932). OTTER thought that erosion results from the rotary motion by the spines and teeth; spines widening and teeth deepening the boring. The effect of enzymes in echinoid boring is not known.

Crustacea. Decapods (crabs), isopods and cirripeds (barnacles) are listed as borers. Crabs often form blisters on corals, but apparently do little actual destruction to the carbonate itself (UTINOMI, 1953). BARROWS (1919) found that intertidal

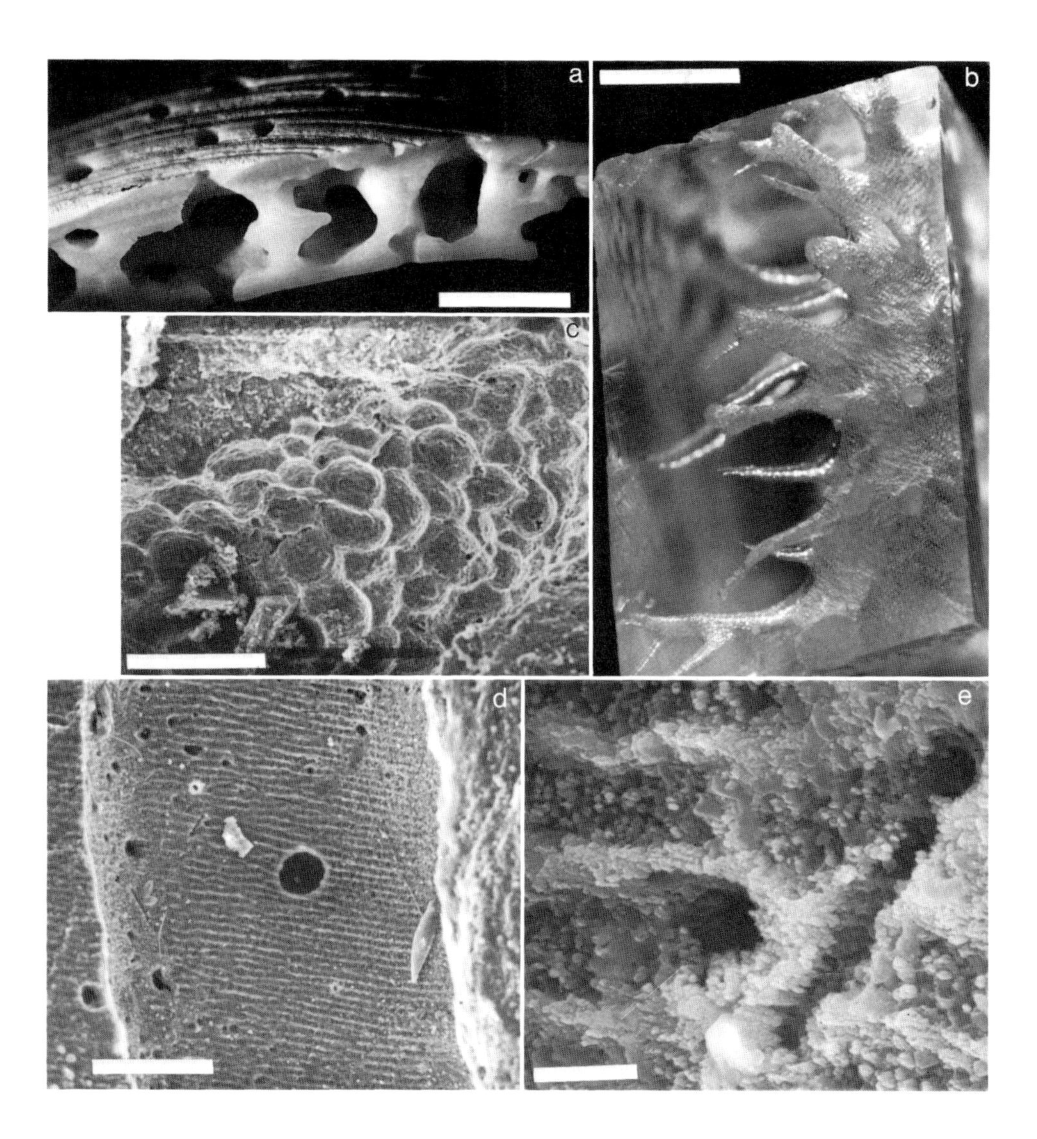

Plate XXIX. Borers in Carbonate Substrata. (a) Clam shell (*Mercenaria mercenaria*) bored by the sponge *Cliona celata*. Scale is 5 mm. (b) Impregnated block of calcite that has been bored by *C. celata*. Note the branching tunnels and serrated microtexture. Scale is 2 mm. Photographs (a) and (b) are courtesy of W. Cobb. (c) A closer view of the serrated micro-texture on a carbonate substrate made by a boring sponge. SEM; scale is 100 microns. (d) Worm boring in a mollusk shell. Note the daily growth bands and the smaller borings. SEM; scale is 100 microns. (e) A close-up of (d), showing a fungal boring across several laminae within the shell. SEM, scale is 5 microns

isopods bore into substrate but the erosional significance of these organisms seems slight. The most important crustaceans in biological erosion are boring barnacles, which are known to penetrate mollusks, corals, bryozoans, echinoids, barnacles and limestone (CANNON, 1935). The barnacle *Lithotrya* has been noted as a prominent agent in the destruction of the intertidal yellow zone at Raroia (NEWELL, 1956) and Jamaica (COLMAN, 1940). A recent article by AHR and STANTON (1973) discusses the geological significance of these boring barnacles. Cavity openings are generally less than 0.5 mm across and the interior cavities usually less than 5 mm in diameter (TOMLINSON, 1969). Barnacle larvae apparently bore by chemical means; the adult enlarges and deepens the hole by its chitinous teeth, aided perhaps by chemical softening of the substrate (TOMLINSON, 1969).

Miscellaneous. Several other phyla are capable of penetrating carbonate substrates, but they apparently are of only minor importance. Some brachiopods and species within the two or three ectoproct families are known to bore. (RUDWICK, 1965; SOULE and SOULE, 1969). The mode of boring is not known in the case of the brachiopods, but SILEN (1946) has reported traces of phosphoric acid in boring bryozoans, suggesting a chemical erosion.

Predators

Predators provide two important steps in carbonate destruction:

1. By eating shelled invertebrates, predators break the shells into smaller fragments (GINSBURG, 1957). Fish, including rays, break many shells while feeding. Most predatory fish appear to be nocturnal, while primary omnivores and herbivores are diurnal (HOBSON, 1965). Crabs, lobsters and octopi also can prey on shelled animals, but their importance is thought to be minimal.

2. Predators can also attack and kill living substrate, and thus hasten the post-mortum invasion of boring organisms. For instance, coral-eating polychaetes are well known (MARSDEN, 1962; GLYNN, 1962). ROBERTSON (1970) lists 5 phyla which can attack corals (fish, crustaceans, polychaetes, gastropods and asteroids). Attacks on Pacific corals by the starfish *Acanthaster planci* can wipe out large reef areas of living reef (CHESHER, 1969).

→

Plate XXX. Bioerosion of Carbonates. (a) The parrot fish (*Scarus*) is a prime eroder of tropical reef carbonate. This X-ray photograph of a scarid fish shows the sharp hard beak, the pharangyl teeth, and the diminuted carbonate sediment within the gut. This fish is approximately 25 cm long. (b) Many fish other than parrot fish are active in reef degradation. This photograph, taken in the southwestern Caribbean, shows a school of surgeon fish actively grazing the bottom. (c) Chitons (center right) are capable of a significant amount of intertidal erosion. Presumably these small depressions in the limestone were made by chitons. This chiton is about 1 cm long. (d) One result of the biogenic degradation of shallow-water carbonates is the formation of intertidal nips, such as this one on the Isle of Rhodes, eastern Mediterranean. Incutting is defined further by the horizontal accretion of coralline algae below the intertidal zone (see Plate XXV a). (e) An even more impressive example of intertidal erosion is shown in this photograph taken near Jidda, Saudi Arabia (Red Sea). The horizontal indentation in the intertidal nip exceeds 3 meters

Plate XXX.

Grazers

The most commonly cited grazers are gastropods, echinoids and fishes. All three attack calcareous substrate in search of epilithic and endolithic blue-green algae. Grazing gastropods are prominent members of the intertidal community; limpets, some of which possess goethite- and silica-reinforced radulae (LOWENSTAM, 1962a, RUNHAM, 1961), can erode the substrate at rates of 0.5 to 1.5 mm/yr (HAWKSHAW, 1878; SOUTHWARD, 1964). MCLEAN (1967) found that the erosion of beachrock at Barbados varied with the species; *Cittarium* is the most efficient rasper followed by *Nerita* and *Littorina*. He also found that erosion is strongly dependent upon the depth and degree of algal penetration into the substrate. EMERY (1946) estimated that small littorinids can excavate 550 kg of sandstone per year on a 30 by 100 m stretch of intertidal sandstone; this amounts to 0.1 mm of erosion per year, which is considerably lower than the 0.6 mm of littorinid erosion estimated by NORTH (1954).

The importance of chitons in intertidal erosion is not known but considering their abundant populations and hard magnetite denticle cappings (LOWENSTAM, 1962b), they probably are capable of considerable erosion (Plate XXX c). NORTH (1954) has found considerable amounts of sand grains in the guts of intertidal chitons, and STEPHENSON and coworkers (1958) concluded that the chiton *Acanthozostera* is the major substrate eroder in intertidal areas of southeastern Australia. This latter statement, however, was greatly tempered in a subsequent paper (STEPHENSON and SEARLES, 1960) in which the writers concluded that fish were the major eroders.

Echinoids (sea urchins) graze on rock, using their teeth to rasp away the carbonate. Although generally not regarded as major eroders, VERRILL (1900) and GARDINER (1931) both attributed extensive erosion to sea urchin activity.

Grazing fish are one of the most important erosional agents on tropical reefs. The best examples are the scarids (parrot fish) and the acantharids (surgeon fishes) which feed on microalgae that live within the carbonate substrate. In order to obtain endolithic alage, the fish must scrape the carbonate substrate (the scarids, with their parrot-like beaks, are especially well-adapted for this; Plate XXX a), or ingest organisms (such as coral polyps) or detritus that contain algae. Ingested grains are ground to finer sizes by pharyngal teeth within the fish's throat (Plate XXX a). Apparently the high carbonate content within the fish's digestive system negates the possibility of further reduction by stomach acids (GOHAR and LATIF, 1959).

Numerous other fish, such as chaetodontids, monocanthrids, and pomacentrids (CLOUD, 1952), are capable of carbonate degradation (Plate XXX b), as is evidenced by the large number of species of fish that contain carbonate grains within their guts (HIATT and STRASBURG, 1960; RANDALL, 1967). TALBOT (1965) estimated that "coral-eaters" constitute some 20% (by weight) of the Tanganyika reef fishes, and another 20%, classified as "herbivores", may also derive some of their food by rasping the substratum for endolithic algae.

CLOUD (1959) estimated that fish consume 0.4 to 0.6×10^3 m^3 of sand and fine gravel per km^2 of reef flat per year. BARDACH (1961), on the basis of the standing crop of fish, calculated that Bermuda reef fish transport about 2300 kg of sediment per hectare, or about 10^3 m^3 per km^2, a figure surprisingly close to Cloud's estimate.

In some instances the inter-relationships between the fish and their environment can be striking. For instance, the driving of grazing fish shoreward by reef predators can effect a large amount of erosion in the intertidal zone during high tide (STEPHENSON and others, 1958). Grazing fish may be responsible for the relative lack of barnacles in tropical intertidal zones (NEWMAN, 1960) and may even affect invertebrate evolution on the reef (BAKUS, 1966).

Browsers

Holothurians commonly are regarded as major sediment reworkers. The volume of sediment ingested by reef holothurians is enormous. Estimates range from about 16 kg (CROZIER, 1918) to 40 kg (BONHAM and HELD, 1963) per individual per year. BONHAM and HELD estimated that on Rongalap Atoll the *Holothuria atra* population collectively ingests and excretes 2×10^8 kg of sand annually. Workers have found that holothurians also have acidic stomaches, with pH's in the range of 5 to 7 (CROZIER, 1918; MAYOR, 1924; EMERY and others, 1954). EMERY (EMERY and others, 1954) found that the pH increased along the digestive tract inferring dissolution as the carbonate passes through the animal. All these data suggest that holothurians can be important dissolvers of shallow-water carbonates (VERRILL, 1900; GARDINER, 1931; DAPPLES, 1942); MAYOR (1924) calculated that an individual holothurian can dissolve between 234 and 414 grams of sediment annually.

Many workers, however, have been struck with the similarity of size of the material entering and exiting from holothurians and also with the general lack of corrosion of constituent particles after they have been excreted from the animals (FINCHK, 1904; CROZIER, 1918; BERTRAM, 1936; YAMANOUTI, 1939; EMERY and others, 1954; BONHAM and HELD, 1963; MILLIMAN and MULTER, unpublished data). For instance, TREFZ (1958) noted that even the delicate anchors and plates derived from other holothurians can pass through the gut without visable signs of dissolution. Thus, in light of available evidence, it is highly questionable whether holothurians are important carbonate eroders.

Browsing herbivorous gastropods, such as the conch *Strombus*, can ingest large quantities of sand grains, from which the epibenthic algae are removed (ROBERTSON, 1961); the possible abrasion or corrosion undergone by these particles has not been studied. Burrowing worms and crustaceans can ingest and defecate enormous quantities of sediment. DAVISON (1891) estimated the marine worm *Arenicola* reworks 1900 to 3100 tons of sediment per acre per year. Again, the erosive and corrosivie tendencies of these burrowing animals have not been demonstrated. More probably they are more important in altering sedimentary structures, in preferential sorting, in aggregating fine sediment into fecal pellets, and in forming potential fossil burrows (RHOADS, 1967; SHINN, 1968c). The chemical effects of the metabolic activity by bottom burrowing organisms however, may decrease the level of carbonate saturation within the sediment (see below).

Mechanical Erosion

Mechanical erosion of carbonates occurs on two scales. The first involves the removal and transport of large pieces of substrate, such as reef blocks; the second occurs on the particle-to-particle scale, and involves abrasion and fracturing.

Mechanical erosion is generally most active in depths shallower than 20 meters; however, many communities within this zone are structured to withstand normal wave action. The massive corals and encrusting coralline algae on the outer reef front of coral reefs are an obvious example. In the calmer areas, such as back reefs, lagoons and deeper sublittoral zones, the erosive forces of day-to-day waves and currents are probably insufficient for much destruction. Normal wave activity may account for the ridge-furrow systems found on beachrocks (McLean, 1967) and also spur and groove systems (see above). Probably the great bulk of large-scale mechanical erosion, however, occurs during cataclysmic events, such as storms, typhoons and hurricanes, or after biologic erosion has weakened the substratum to a point where normal waves and currents can remove it. Most storms probably are more important as sediment transporters than as substrate destroyers (see Chapter 6).

In particle to particle erosion, most erosion seems to be abrasional; some fracturing may also occur. Several generalizations can be made:

1. During fracturing, large fragments break into several smaller ones (Klähn, 1932; Swinchatt, 1965). In contrast, abrasion transforms coarse particles into very fine grains rather than intermediate sizes (Chave, 1960). *Halimeda* plates, for example, abrade into micron-size aragonitic needles (Folk, 1967). Force

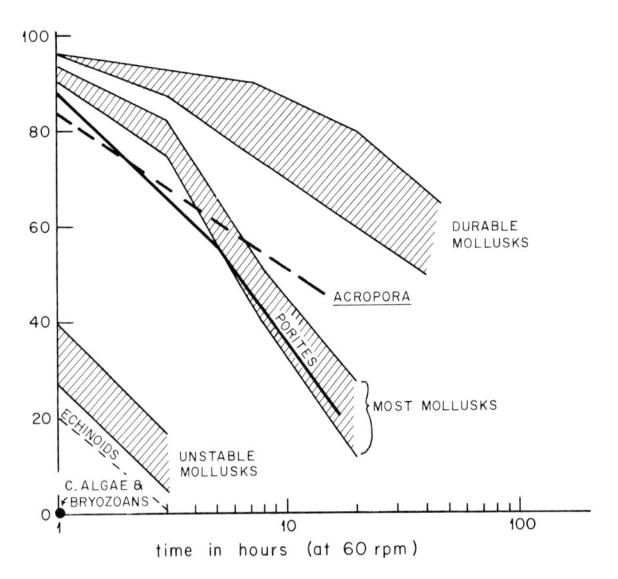

Fig. 90. Degradation of various carbonate components in a tumbling barrel at 30 rpm; data readjusted for 60 rpm. The abrasive was quartz sand. After Chave (1964)

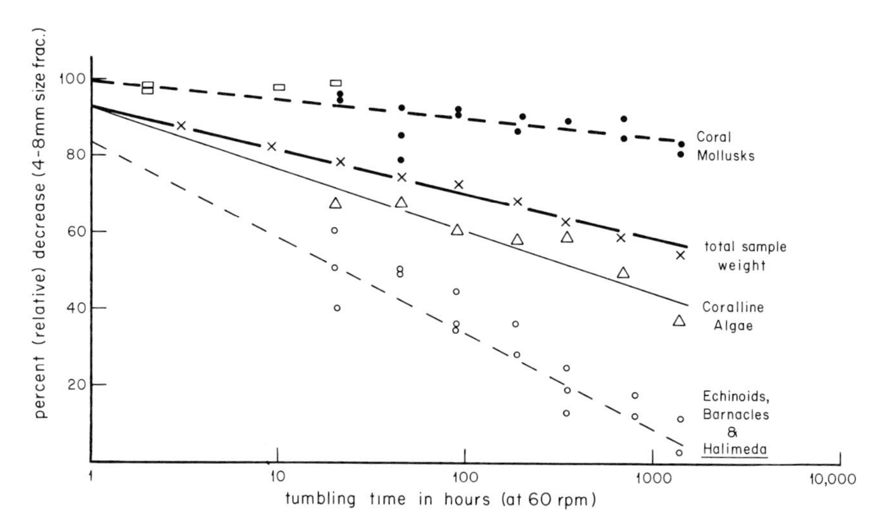

Fig. 91. Degradation of various carbonate components in a tumbling barrel run at 60 rpm. The abrasive was carbonate sand. The notations (□) refer to the rates of pelecypod abrasion measured on beaches by DRISCOL (1967)

(1969) found that mollusk shells abrade into three general size categories, 500 to 125 microns, 32 to 4 microns and 0.5 to 0.125 microns. Each size class is apparently controlled by organic or crystallographic constraints. In aragonitic shells, much of the abraded fine debris consists of aragonitic laths (CHAVE, 1960; FORCE, 1969).

2. Coarser grains erode faster than finer grains (CHAVE, 1960; DRISCOLL, 1967). In one experiment, for instance the present writer found that 4 to 8 mm carbonate grains abrade 3 to 4 times faster than 2 to 4 mm grains.

3. The rate of abrasion depends upon the skeletal microarchitecture of the individual particles (CHAVE, 1960, 1964) as well as the method by which the grains are abraded; carbonate mineralogy is of little or no importance. In a series of tumbling barrel experiments using chert as the abrasive, CHAVE (1960, 1964) found that porous and organic-rich carbonates (such as echinoids, bryozoans and articulated algae) abrade much faster than the more compact and laminated mollusk shells (Fig. 90). In similar experiments using pure carbonate sediments, the writer found that echinoids, barnacles and *Halimeda* degrade rapidly, while coral and mollusks tend to maintain themselves for considerable time; only 16% of the 4–8 mm coral and mollusk fragments were abraded after nearly 1400 hours of tumbling at 60 rpm, an equivalent transport of more than 3500 km. Coralline algae degrade at an intermediate rate. In general the rates of abrasion in total carbonate are about 2 orders of magnitude slower than those in which a quartz grit is used, but the slow rates seem more consistent with field measurements (DRISCOLL, 1967) (Fig. 91).

These data suggest than in very coarse grained sediments and areas of continual high energy, such as beaches, mechanical abrasion may be an important but relatively slow process. In other environments, however, it probably is not as important as other diagenetic processes.

Chemical Alteration

Calcium carbonate can dissolve in environments which are undersaturated with respect to one or more of the carbonate polymorphs. Environmental parameters conducive to undersaturation include low temperature, high partial pressures of CO_2 (low pH), and increased hydrostatic pressures (see Chapter 8). In intertidal and very shallow marine environments, dissolution is affected by reversible diurnal and seasonal changes in temperature and in the O_2—CO_2 system. Changes in deeper marine environments involve noncyclic processes, such as increased hydrostatic pressures and the oxidation of organic material.

Intertidal Dissolution

The possible importance of chemical dissolution of marine carbonates was suggested by Sir JOHN MURRAY, who theorized that reef waters become enriched with carbonic acid produced by biological and decompositional activities on the reef flats. As this "acid" water passes into the lagoon, argued MURRAY, it erodes the substrate, thus deepening the lagoon (MURRAY and IRVINE, 1891). GARDINER (1903), while recognizing the importance of biological erosion, considered chemical dissolution to be the major cause for the deepening of the Maldive lagoons and also for the production of carbonate muds in lagoonal sediments. Subsequent work in the early 20th century (LIPMAN, 1924; MAYOR, 1924), however, showed that most tropical waters are excessively supersaturated with respect to calcium carbonate. Thus any shallow-water chemical erosion must occur in local areas during diurnal changes in solubility or beneath the water-sediment interface.

The possible effects of intertidal dissolution have been discussed by many workers. MACFAYDEN (1930) showed that the intertidal nip in the Red Sea (Plate XXX e) could not be caused by mechanical erosion, since waves are small and too infrequent to erode effectively; instead he suggested chemical dissolution. KEUNEN (1933) also favored a "chemical sawing into the limestone" to explain the nipped shorelines found throughout the tropics. MAYER (1916) and WENTWORTH (1938) suggested that rain water accounts for much of the intertidal and supratidal erosion. The growth of salt crystals during the drying of sea spray apparently can erode some intertidal and supratidal substrates (GUILCHER and PONT, 1957; WELLMAN and WILSON, 1965). Most other workers, however, have felt that chemical dissolution in the intertidal zone results primarily from changes taking place within the salt-water system, not fresh water nor salt spray.

The most commonly suggested mechanism for shallow marine chemical dissolution involves diurnal changes in pH. Both plants and animals consume oxygen during evening hours, thus increasing the carbon dioxide content in the surrounding waters and decreasing pH. Although day and night pH values generally do not vary by more than 0.15 units (SCHMALZ and SWANSON, 1969), within restricted basins, they can vary by as much as 2 units (Table 65). Absolute values, however, rarely fall below 7 or 8. Diurnal variations in the degree of carbonate saturation can be considerable (SCHMALZ and SWANSON, 1969), but intertidal waters may remain supersaturated with respect to calcium carbonate

Table 65. Diurnal range of various physical and chemical parameters in intertidal pools and very shallow waters. Data from EMERY (1946), EMERY and others (1954), PARK and others (1958), EMERY (1962)

Area		Water depth (m)	Temp. range (°C)	Cl ‰ range	pH range	Specific alkalinity
La Jolla	solution basins	<0.1	13.0–25.0	18.60–22.07	7.62–9.00	0.086–0.127
Bikini	basins, reef flats	<0.1	26.4–34.8	18.58–19.34	8.00–8.50	0.122–0.127
	basin	<0.1	25.1–37.0	0.02–0.10	8.55–10.10	–
Guam	rimmed terraces	0.1–0.2	28.0–36.6	18.28–18.78	8.10– 9.39	0.100–0.145
	solution basins	<0.1	26.6–37.2	18.08–39.10	8.10– 9.39	0.090–0.170
	solution basins	<0.1	26.7–35.0	0.32–14.92	8.20– 9.50	0.172–0.514
	reef flat	0–0.3	29.0–33.0	18.2 –19.5	8.0 – 8.4	
Baffin Bay		1.5	19.2–32.0	31.8 –36.1	7.94– 8.62	0.108–0.128
Redfish Bay		0.3	18.0–35.0	15.7 –16.6	7.98– 8.91	0.147–0.188
Laguna Madre		0.6	14.4–32.0	17.6 –31.6	8.15– 8.49	0.090–0.162

throughout the night (REVELLE and EMERY, 1957). The only evidence suggesting diurnal dissolution is the increase in specific alkalinity during periods when the pH is lowest (REVELLE and EMERY, 1957) (Table 65). High alkalinity values infer carbonate dissolution and in turn may explain the lack of significant pH and carbonate saturation depressions. According to REVELLE and EMERY (1957) this alkaline water is flushed out of the tidal pools before reprecipitation can occur.

Available field data suggest that intertidal chemical dissolution is not an important erosional process. As NEUMANN (1966) has pointed out, if physio-chemical dissolution were important, then one would not expect the proliferation of carbonate-secreting organisms that occur in solution pools and intertidal nips. Chemical dissolution, however, may supplement biological erosion. For instance, CRAIG and others (1969) found that the erosion in supratidal pools by grazing gastropods is facilitated by the low pH of the ambient waters, which aids in the dissolution and loosening of the carbonate substate. The pH's, in turn, are dependent upon the physiological activities of the gastropods and other organisms within the pool.

Marine Dissolution

As discussed previously, only the surficial layers of the oceans are supersaturated with respect to calcium carbonate. At greater depths both aragonite and the various calcite phases tend to dissolve. Deep-sea dissolution has been discussed in Chapter 8. Several other aspects of marine dissolution, however, should be discussed; namely dissolution within restricted environments and the relative stability of the various carbonate minerals.

Dissolution in Restricted Environments—In restricted environments decaying organic material can lower pH values to levels where dissolution can begin. Algal mats, for instance, can form restricted micro-environments in which pH's can be depressed considerably (REVELLE and EMERY, 1957; BATHURST, 1967c). Alteration of organic material after burial can also produce severe environmental changes. Observations within many marine environments indicate that dissolution is not a significant factor in short-term periods; pH values in the sediments penetrated by most gravity and piston cores seldom fall more than 1 pH unit below normal values at the sediment-water interface. Studies of the interstitial water in short cores generally do not show marked variations in the Ca/Cl and Mg/Cl ratios with depth (SIEVER and others, 1965; BROOKS and others, 1968; PRESLEY and KAPLAN, 1968; FRIEDMAN and others, 1968).

During longer time intervals, such as those sampled by deep-drilling, dissolution of carbonate sediments can be significant. In the Mohole test-site core off Guadalupe, Mexico, the Ca/Cl ratio increased by about 50% along the 180 m core, and pH decreased to 7.3 at 140 m (RITTENBERG and others, 1963). Subsequent analyses of interstitial water from other deep-sea drillings have shown relatively wide fluctuations in calcium, magnesium and total CO_2 (JOIDES, 1969, 1970a, 1970b, 1970c). In some instances these changes can be related directly to solution or precipitation of calcium carbonate (for instance, F.T. MANHEIM, oral communication, reports an 8-fold increase in calcium content of the interstitial water associated with a lithified coccolith ooze from the northeastern Pacific), but in other instances, the adsorption by clays appears to be the dominant mechanism for decreasing the levels of Mg and Ca (MANHEIM, CHAN and SAYLES, in JOIDES, 1970c; F.T. MANHEIM, oral communication).

The Relative Stability of Carbonate Minerals—The instability of magnesian calcite and aragonite relative to calcite in the deep sea was discussed in the previous chapter. However, the stability of these two minerals relative to one another and the general stability of carbonate minerals in shallow water has been a subject of considerable debate in recent years. CHAVE (1962) and NEUMANN (1965) reported increasing calcite content with decreasing grain size in the shallow-water carbonate sediments from Bermuda and Campeche Bank. They interpreted these occurrences as evidence for the preferential dissolution of the less stable aragonite and magnesian calcite, even though the ambient waters appear saturated with respect to both minerals (SCHMALZ and CHAVE, 1963). Subsequent studies, however, found no direct evidence for shallow-water dissolution of either aragonite or magnesian calcite. The mineralogy of shallow-water carbonate sediments often remains constant with decreasing size, although slight *increases* of aragonite and magnesian calcite with decreasing grain size are not uncommon (BLACKMON, in CLOUD, 1962a; TAFT and HARBAUGH, 1964; PILKEY, 1964; MATTHEWS, 1966; DAVIES, 1970a; FALLS and TEXTORIS, 1970). The increased aragonite content is probably the result of the introduction of fine-grained aragonite needles (see p. 185ff.). Moreover, no evidence of dissolution has been noted in shallow-water cores, even though pH may drop considerably (TAFT and HARBAUGH, 1964; PILKEY, 1964; BERNER, 1966a). BERNER (1966a) found pH values in Florida cores as low as 7.3 and concluded that the interstitial water was in equilibrium with calcite but undersaturated with respect to magnesian calcite and aragonite. The fact that no mineralogic changes had occurred was explained by possible "surface protective or nucleation-inhibiting layers of adsorbed Mg" (BERNER, 1966a, b). Recent studies by BISCHOFF (1968a) together with earlier studies (Lippmann, 1960; Kitano and Hood, 1962) suggest that Mg does inhibit the inversion of aragonite to calcite (see below). The effect of protective organic coatings may also prevent dissolution (see above).

The relative stability of aragonite versus magnesian calcite in the oceans has been debated in recent years. Many workers (Stehli and Hower, 1961; Chave and others, 1962; Jansen and Kitano, 1963) have suggested that the sequence of stability is:

magnesian calcite < aragonite < calcite.

Friedman (1965), however, observed the persistence of magnesian calcite and dissolution of aragonitic shells in the deep-sea sediments of the Red Sea and eastern Mediterranean. He therefore suggested that:

aragonite < magnesian calcite < calcite.

More recent studies suggest an intermediate system. Weyl (1967) showed that a magnesian calcite with 12 to 13 mole % $MgCO_3$ has a solubility about equal to that of aragonite. Higher $MgCO_3$ contents result in greater solubilities and lower $MgCO_3$ contents have lower solubilities (Fig. 6). A similar scheme has been observed in deep-sea sediments. Most deep-sea magnesian calcite cements have $MgCO_3$ contents between 11 and 13 mole % (see Table 74). Magnesian calcites with higher $MgCO_3$ contents and aragonites tend to be infrequent. Thus a more plausible sequence of stability may be:

very high magnesian calcite ($\geqq$ 12 mole %) < aragonite
< magnesian calcite ($\leqq$ 12 mole %) < calcite.

Clearly the rate of dissolution or inversion depends upon the ambient environment. In shallow water, aragonite and magnesian calcite remain comparatively stable relative to calcite, but in the deep sea dissolution and inversion can occur at more rapid rates. Several workers (Murray, 1966; Kendall and Skipwith, 1969; Purdy, 1968; J. Müller, 1969) report possible alteration of very high magnesian calcite in foraminifera and coralline algae into aragonite (where intragranular cementation stops and actual inversion begins, however, may be difficult to define). Alexandersson (1972b) has presented evidence which suggests an inversion of 15 to 17 mole % magnesian calcite to aragonite in void-filling intragranular cements.

Aragonite does not "invert" to calcite, but rather dissolves. Magnesian calcite-calcite transformation occurs on scales small enough so that the grains slowly alter to calcite while not disturbing the original character; examples of this inversion can be found in the deep sea (Milliman, 1966; Gomberg and Bonatti, 1970).

Summary

The environment of deposition dictates the type and degree of carbonate alteration. For instance, carbonates in an area experiencing rapid sedimentation may undergo less degradation than carbonates in an area undergoing slow sedimen-

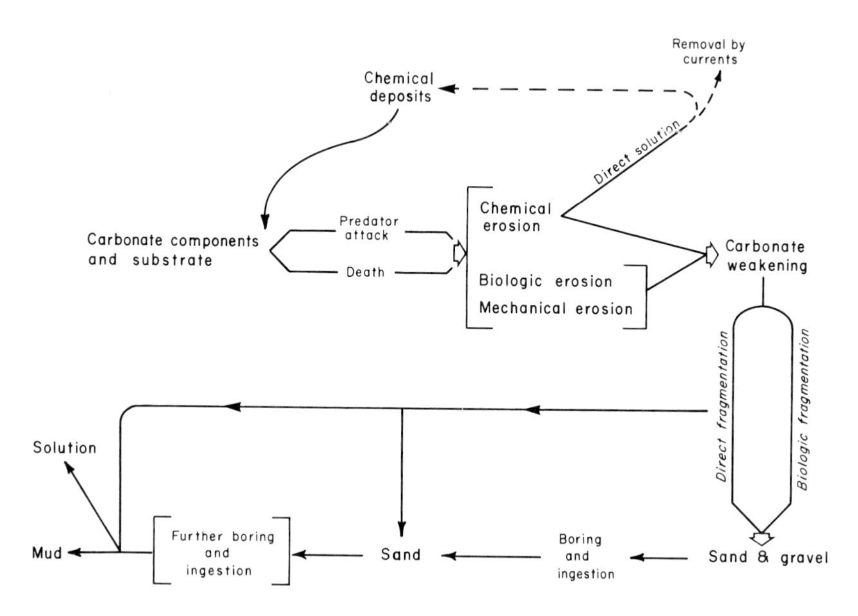

Fig. 92. The cycle of shallow-water carbonate degradation. Modified from OTTER (1937)

tation (TWENHOFEL, 1942). On the other hand, micro-environments conducive to chemical dissolution may be more likely to occur in areas with rapid sedimentation.

Generally, biological agents are dominant in shallow-water carbonate degradation (Plate XXX). Not only do biological eroders actively degrade the carbonate, but they also expose (and increase) surface areas to settlement by secondary borers (EVANS, 1968; WARME and MARSHALL, 1968) and to potential chemical dissolution. The most important biologic agents in shallow-water erosion are probably boring blue-green algae and fungi. Not only do these organisms degrade carbonates, but they also serve as food for grazing and browzing animals, who in the process of extracting the algae, destroy the weakened substrate. Because of their small diameter, algae and fungi can effectively erode finer size carbonates than can many of the other common biologic agents. Other important eroders include mollusks, sponges, echinoids, worms and fish; the former three groups of organisms are particularly active in the reduction of particle size (OTTER, 1937; SWINCHATT, 1965) (Fig. 92). Published observations indicate that the rate of biological erosion in shallow-water environments ranges from 0.3 and 30 mm per year (Table 66).

Prominent mechanical abrasion of carbonates occurs only in shallow-water, especially on beaches. Chemical dissolution probably is not an important factor in the erosion of either intertidal or shallow-water areas, but it can aid in the weakening of substrata which are degraded by organisms. Chemical dissolution becomes increasingly important in the deep sea; below depths of about 4000 m in the Pacific and 5000 m in the Atlantic carbonate sedimentation is rare.

Table 66. Intertidal and shallow-water erosion rates

Area	Rate of erosion (mm/yr)	Mechanism	Worker
Great Barrier Reef	0.5	fish	STEPHENSON (1961)
Norfolk Is. Australia	0.6 to 1.0	limpets (?)	HODGKIN
Point Peron, Australia	1	limpets (?)	HODGKIN
La Jolla, California	0.6	littorinids	NORTH
Puerto Rico	1	(total)	KAYE
La Jolla, California	0.3	littorinids and chem. and mech. erosion	EMERY (1946)
Harrington Sound, Bermuda	1 to 14	*Cliona*	NEUMANN (1966)
Scotland	15 to 30	pholads	JEHU
Barbados	7	*Nerita*	MCLEAN
Red Sea	2.5	*Lithophaga*	VITA-FINZI and CORNELIUS

Geochemical and petrographic data suggest that high magnesian calcite is more stable than aragonite in the sea. Both of these minerals, in turn, are less stable than calcite. In shallow-waters, there is little evidence that either mineral dissolves or alters to calcite. In the deep-sea however, we find that pteropods will dissolve and magnesian calcite, with time, will alter to calcite.

Chapter 10. Carbonate Cementation

Limestones have been classified in a variety of ways. Some classifications are generic, but most are descriptive. Among the parameters used in descriptive classifications are mineralogy, chemistry, pterography and texture. This latter parameter forms the basis for most classifications (HAM and PRAY, 1962). Limestones consist of three general textural components: 1. the framework, grains generally coarser than 62 microns (although LEIGHTON and PENDEXTER, 1962, assign a lower limit of 30 microns). 2. a matrix of finer material, usually less than 4 microns (FOLK, 1959, 1962); and 3. the cement. Although the distribution of these various components is a basic criterion for limestone classification, the emphasis varies with different classifications (see Tables 67–69). Such textural classifications, however, depend upon the ability of the investigator to distinquish between primary (depositional) and secondary (diagenetic) features, as well as to recognize organic- and energy-related features (HAM and PRAY, 1962). Clearly the complex subject of limestones cannot be treated completely within

Table 67. DUNHAM's (1962) classification of limestone types based on depositional texture

Increasing mud ↓

Boundstone	Components bound together during deposition
Grainstone	Grain supported components; not bound together during deposition
Packstone	Grain supported; some mud present
Wackestone	Mud supported; more than 10% framework grains
Mudstone	Mud supported; less than 10% framework grains

Table 68. LEIGHTON and PENDEXTER's (1962) classification of limestones according to texture

Percent grains	Grain type				Coated grains	Organic frame-builders
	Detrital	Skeletal	Pellets	Lumps		
~90	detrital ls.	skeletal ls.	pelletal ls.	lump ls.	oolitic ls. pisolitic ls.	coralline ls. (etc.)
~50	detrital-micritic ls.	skeletal-micritic ls.	pelletal-micritic ls.	lump-micritic ls.	oolitic-micritic ls. (etc.)	coralline-micritic ls.
~10	micritic-detrital is.	micritic-skeletal ls.	micritic-pelletal ls.	micritic-lump ls.	micritic-oolitic ls.	micritic-coralline ls.

Table 69. FOLK's classification of limestones is based on the relative abundances of three components, grains (allochems), cement, and matrix. The prefix comes from the dominant allochem. The suffix is determined by the dominance of cement or matrix. This table shows only the basic subdivisions of allochem rocks. More detailed descriptions can be found in FOLK (1959, 1962)

	Sparry calcite cement	Microcrystalline calcite matrix
Allochem composition		
Intraclasts	intrasparite	intramicrite
Ooids	oosparite	oomicrite
Fossils	biosparite	biomicrite
Pellets	pelsparite	pelmicrite

the short length of this chapter. Excellent discussions can be found in HAM (1962), PRAY and MURRAY (1965) and BATHURST (1971).

In discussing carbonate cementation, one must define the petrographic terms used in describing the various cements found in limestones. A myriad of petrographic classifications (Tables 67–69) and definitions (and jargon) (Table 70) have been used; many were collected into a published symposium (HAM, 1962), but many

Table 70. Terminologies for various carbonate cements and matrices

Acicular — Long, thin crystals, usually oriented normal to grain surface; usually aragonite

Blocky — Massive, equant grains

Cryptocrystalline — Light brown to opaque tiny crystals, less than 4 microns in diameter (FURDY, 1963). Similar to microcrystalline

Dentate — Drusy cement with jagged, tooth-like edges

Drusy — A type of sparry cement; crystalline to subhedral crystals, coarser than about 10 microns, generally forming in voids (drusy mosaic) or as thin coatings (BATHURST, 1958). In drusy mosaic, cement crystals increase in size away from the component grains

Fibrous — Similar to acicular

Granular — General term for cement crystals coarser than about 10 microns, but smaller than about 60 microns. Overlaps with drusy

Micrite — Microcrystalline calcite, less than 4 microns in diameter (FOLK, 1959). Less than 30 microns (LEIGHTON and PENDEXTER, 1962)

Microcrystalline — Subtranslucent crystals less than 4 microns in diameter (FOLK, 1959); between 4 and 30 microns (MACINTYRE, 1967a)

Microspar — Recrystallized matrix (FOLK, 1965)

Pelletal — Cement and matrix composed of numerous small pellets, 20 to 60 microns in diameter; the pellets are composed of cryptocrystalline grains

Submicrocrystalline — Less than 4 microns in diameter (MACINTYRE, 1967a)

Sucrosic — Surgary cement, common in dolomites

Sparry — Clear, coarse crystals, generally coarser than 10 microns (FOLK, 1959)

of the problems, unfortunately, remain. One problem of particular importance for this discussion may illustrate some of the confusion in commonly-used terminology: The term "micrite" was coined by FOLK (1959) as an abbreviation for the term "microcrystalline calcite", which referred to calcite crystals smaller than 4 microns; in the same paper he also defined it as a microcrystalline ooze. LEIGHTON and PENDEXTER (1962) re-defined micrite as being finer than 30 microns, MACINTYRE (1967a) subdivided fine grains into microcrystalline and submicrocrystalline, and BATHURST (1964b) referred to *aragonitic* micrite. Because of the many dangers and connotations presently inherent in the term "micrite", the writer prefers the terms cryptocrystalline (PURDY, 1963) or submicrocrystalline (MACINTYRE, 1967a). Other cement types are defined in Table 70.

Intragranular Cementation (Cryptocrystallization)

Cementation refers to the "process of precipitation of a binding material *around* grains or materials in rocks" (American Geological Institute, 1962, p. 77; Italics are the writer's). In addition to this intergranular cementation, a second type of carbonate cementation occurs—intragranular cementation in which precipitation occurs inside grain voids. In many cases the morphology, texture, mineralogy and chemistry of intragranular cements resemble those in intergranular cements, suggesting similar origins (see the summary at the end of this chapter).

The crystalline and (sometimes) compositional characteristics of individual carbonate grains can be altered in two separate manners: by the recrystallization of component crystals and by intragranular cementation. The former process, which has been discussed on p. 266ff. involves mineral "inversion" or dissolution and reprecipitation of carbonate on a micro-scale. The latter process involves the in-filling of voids, either primary or secondary, with authigenic cement. An excellent review on this subject has been given by ALEXANDERSSON (1972a).

BATHURST (1966) found that the formation of cryptocrystalline rinds and envelopes around mollusk shell fragments involves the filling of altered host material, rather than the precipitation of a new rind. BATHURST offered a simple scheme to explain this process: algae bore into the substrate, die, and the post-mortum alteration of the organic algal material presents micro-environments in which calcium carbonate can precipitate, thus filling the voids. Multiple gener-

→

Plate XXXI. Cryptocrystallization of *Halimeda* Plates. (a) and (b) Soon after death of the plant, *Halimeda* plates begin to loose their porous appearance and become increasingly cryptocrystalline with increasing diagenetic alteration. Polarized light; scales are 100 and 250 microns, respectively. (c) SEM photo of the utricle of a fresh, unaltered *Halimeda* plate. Scale is microns. (d) When viewed under an SEM, the nature of the cryptocrystalline cement becomes apparent. This utricle, which is beginning to fill intragranular cement is in a plate that came from a living plant, suggesting that internal cementation can begin before the death of the plant. Scale is 40 microns. (e) Close-up of the disk-like aragonite cement within the utricle shown in (d). Note the similarity between these disks and those seen in grapestones and shallow-water cements. SEM; scale is 10 microns. (f) Filled utricle; internal cement includes disks as well as more ill-defined carbonate cements. At this stage the *Halimeda* grain probably would be cryptocrystalline under refracted light. SEM; scale is 20 microns

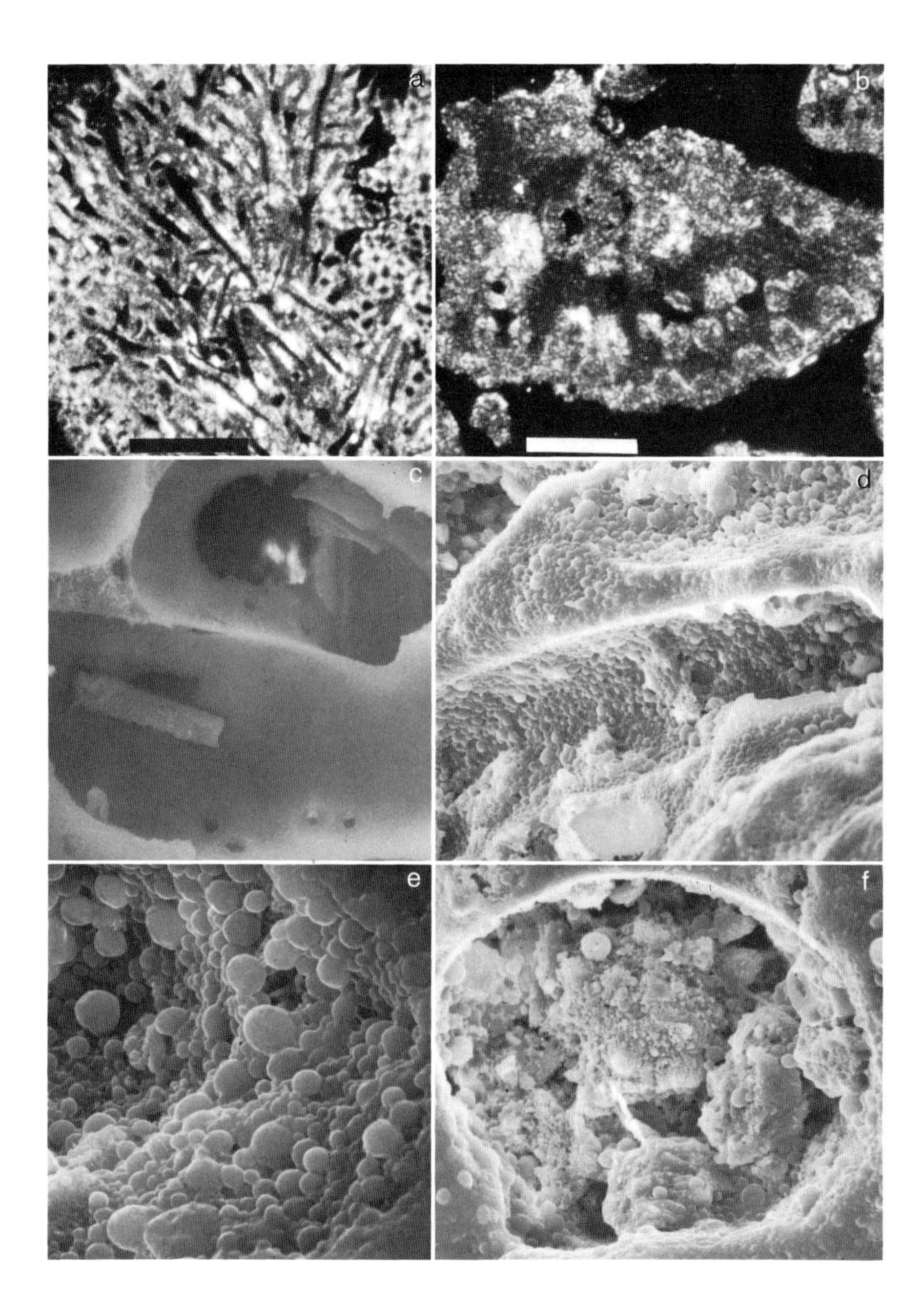

Plate XXXI

ations of algal borings can form complex cryptocrystalline rinds (Plate XXXIII). Similar rinds can form within other secondary voids, such as those formed by fungi and sponges. Perhaps much of the "recrystallization" of aragonite or magnesian calcite reported in various shallow-water carbonates (MURRAY, 1966, and other references quoted in the previous chapter) actually represents primary precipitation within micro-borings (KENDALL and SKIPWITH, 1969; MARGOLIS and REX, 1971).

Similar intragranular cements are found within the primary voids of various organisms: 1. In *Halimeda* plates, secondary precipitation begins shortly after the "death" of the individual plates. The utricles gradually fill with small (2 to 10 microns) round carbonate disks (Plate XXXI). The mineralogy of these disks is not fully known, but preliminary micro-scan analyses indicate that at least some are high-Sr aragonite. As the utricles fill, and as the cellular structure breaks down, the plate becomes grayish-tan and translucent under transmitted light, or, in the words of PURDY (1963), cryptocrystalline (Plate XXXI).

2. In the cryptocrystallization of coralline algae, carbonate rhombs are precipitate within individual cells (Plate XXXIII d, e). These rhombs are presumably magnesian calcite, although dolomite and aragonite also possibly can form within some corallines (SCHLANGER, 1957; MULLER, 1969) (Plate XXXIII f). SIBLEY and MURRAY (1972) state that with increased micritization of coralline algae in beach-rock and submarine limestone at Bonaire (Netherlands Antilles) the mineralogy changes from magnesian calcite to calcite.

3. Foraminiferal chambers also are common sites for secondary precipitation (Plate XXXII). Under transmitted light, the fillings within tropical foraminifera look much like those found in *Halimeda*, cryptocrystalline and brown in color. Scanning electron photomicrographs reveal two types of intragranular cements: scalenohedra, (probably magnesian calcite) and round disks (possibly aragonite) similar to those in *Halimeda* (Plate XXXII d).

4. The internal molds of planktonic organisms commonly occur in the sediments of enclosed basins such as the Red Sea and the Mediterranean (FRIEDMAN, 1965; EMELYANOV, 1965; MILLIMAN and others, 1969) as well as in submarine limestones (Plate XXXVII a, b). These molds are generally magnesian calcite. Pteropod molds (Plate XXXVIII d) are comprised of small, poorly formed rhombs (Plate XXXVIII e, f); some of this material may be detrital, although most is probably authigenic. Planktonic foraminifera molds, on the other hand, are composed of small fused crystals which have grown inside tests whose pores are sufficiently small to exclude most detritus, thereby suggesting that the carbonate was precipitated within the test. Many of the magnesian calcite foraminifera and pteropod molds in Mediterranean deep-sea sediments are not encased by the original test; presumably the tests have been dissolved. Preferential solution of the aragonite pteropod test can be understood (see Chapter 8), but the possibility of low magnesium calcite foraminiferal tests being dissolved and the seemingly more soluble magnesian calcite molds remaining, is difficult to explain. Similar dissolution of planktonic foraminifera and coccoliths from magnesian calcitic crusts is seen in Mediterranean deep-sea limestones (MÜLLER and FABRICIUS, 1970; MILLIMAN and MÜLLER, 1973).

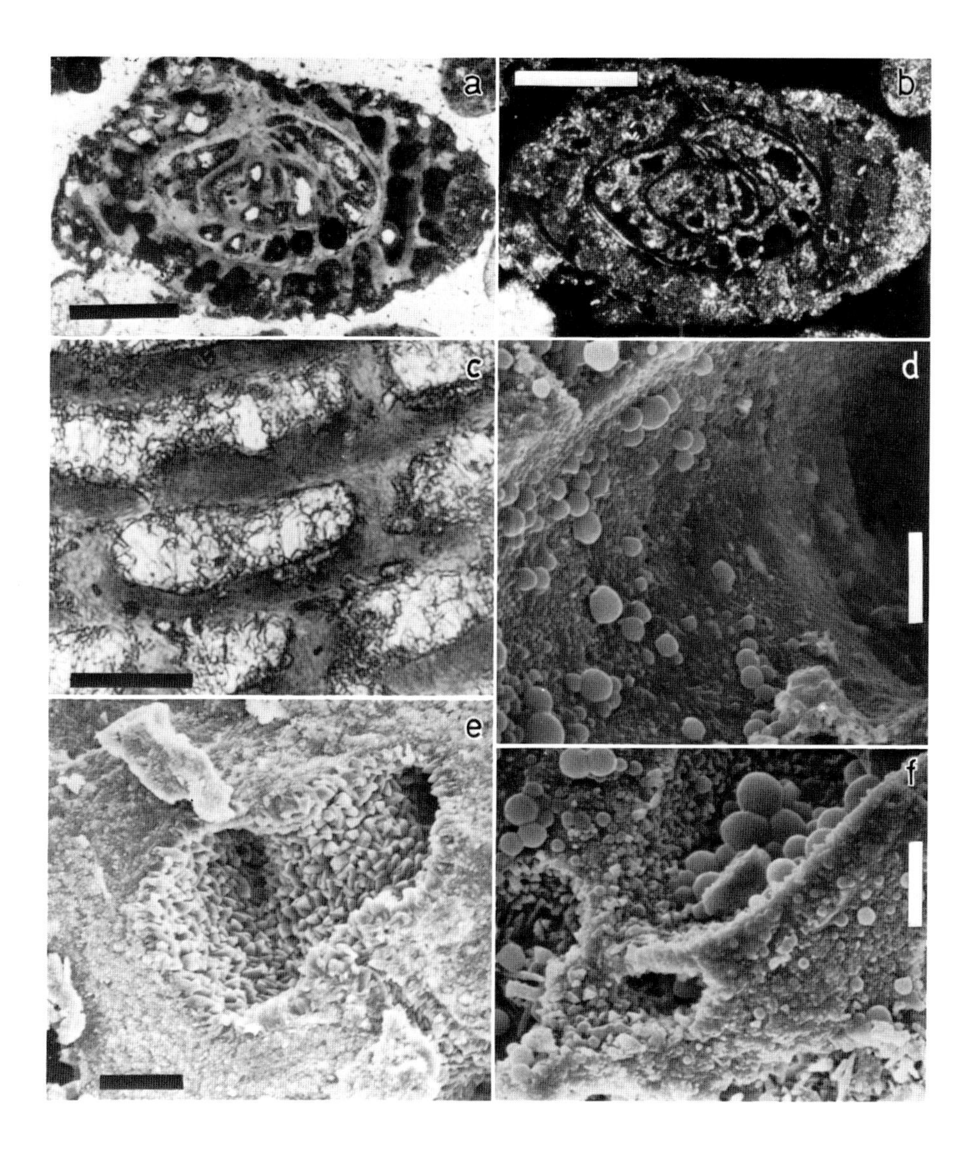

Plate XXXII. Cryptocrystalline Foraminifera. (a) and (b) Benthonic foraminifera (peneroplids) under refracted light, showing the degree of internal cementation within the chambers. Scales are 250 microns. (c) Close-up of the internal cement. Refracted light; scale is 50 microns. (d) SEM photomicrograph showing the initial stage of intragranular cementation within the chamber of a peneroplid. The disks probably are aragonite. Scale is 10 microns. (e) Intragranular cementation within a *Calcarina* test; cement is scalenohedral magnesian calcite. SEM; scale is 20 microns. (f) Intragranular cements in this well-altered foraminifera test apparently consist of both scalenohedral magnesian calcite and disk-like aragonite. SEM; scale is 10 microns

Table 71. Ranking of the susceptibility of various skeletal carbonates to cryptocrystallization (after PURDY, 1968)

Decreasing tendency towards cryptocrystallization
1. Coralline algae (magnesian calcite)
2. *Halimeda* (aragonite) and peneroplids (magnesian calcite)
3. Miliolids (magnesian calcite)
4. *Millepora* (aragonite), gastropods and pelecypods (aragonite and/or calcite)
5. Corals (aragonite)
6. Alcyonarian spicules (magnesian calcite) and echinoids (magnesian calcite)

PURDY (1968) proposed a susceptibility index for the cryptocrystallization (intragranular cementation) of shallow-water skeletal fragments (Table 71). This index, however, is open to revision: DAVIES (1970a) found that in Shark Bay (Australia) porcellaneous foraminifera and mollusks commonly are more altered than coralline algae. Rather than depending upon mineralogy, the susceptibility to cryptocrystallization probably is a function of the relative abundance of interconnected voids (permeability) within the various skeletons. This, in turn, depends upon the availability of primary voids as well as the susceptibility of the particle to boring and reworking. Thus coral and alcyonarian spicules, which contain few voids, exhibit a slower recrystallization than do porous coralline algae and *Halimeda*.

Although intragranular cementation is assumed to be relatively rapid, little is known about the actual time required. ALEXANDERSSON (1972b) reports an age of 890 ± 115 years B. P. for a chamber filling in the Mediterranean, but *Halimeda* can begin to fill before the entire plant is completely dead (Plate XXXI d).

Intragranular cements also occur in hardened non-skeletal fragments. Voids within fecal pellets often contain fillings of radiating aragonite needles (Plate I i), while in grapestone, round disks (probably aragonite) are common (Plate III). PURDY (1963, 1968) showed that ooids can alter into structureless pellets (Plate II f). The process begins on the outside of the ooid and proceeds inwards, with no apparent change in the aragonitic mineralogy (PURDY, 1963). Although it is possible that the recrystallization may involve secondary precipitation within microborings (see above), evidence at present is not conclusive.

BATHURST (1964b, 1966) found that cryptocrystalline fillings of algal filaments within mollusk shells are aragonite, but WINLAND (1968, 1971) showed that cryptocrystalline magnesian calcites (15 to 17 mole %, according to CHAVE'S 1952 curve) are more common. WINLAND (1969) suggested that these magnesian calcite fillings either could alter to aragonite (a distinct possibility in light of the greater stability of aragonite; see p. 267)[1], or could undergo incongruent solution to calcite, while aragonitic shells undergo solution and reprecipitation. This differential solution of magnesian calcite to calcite could explain why cryptocrystalline rinds are preserved in rocks while shell structures often are destroyed (WINLAND, 1968).

1 Recently confirmed by ALEXANDERSSON (1972b).

GLOVER and PRAY (1971) and ALEXANDERSSON (1972 b) have suggested that the mineralogy of the filling is dependent upon the mineralogy of the host. For instance, PUSEY (1964) and PURDY (1968) report that most benthonic foraminifera in the British Honduras reefs are filled with magnesian calcite (18 mole % $MgCO_3$). Similarly, most coralline algae have magnesian calcite fillings and *Halimeda* grains have aragonite. However, exceptions to this rule are not uncommon. PURDY (1968), MÜLLER (1969), DAVIES (1970 a) and WINLAND (1971) report aragonitic fillings in coralline algae and benthonic foraminifera (Plates XXXII d and XXXIII f), and KINSMAN and HOLLAND (1969) found magnesian calcite (12–13 mole % $MgCO_3$) within an aragonitic gastropod. Recent observations by ALEXANDERSSON (1972 a, b) indicate that aragonite is a successor to magnesian calcite and can nucleate directly on magnesian calcite crystals.

Several non-carbonate internal fillings also occur in altered carbonate fragments. Iron and manganese sulfides can give shells a speckled appearance; these are especially common in relict foraminiferal and pelletal sands (HOUBOLT, 1957; MAIKLEM, 1967; DAVIES, 1970 a). Similar fillings in large shell fragments result in gray or black colors (DOYLE, 1967). Oolitic shelf sands off the southeastern United States contain much more iron (up to 1% Fe) than fresh ooids (TERLECKY, 1967) and black pellets in the Persian Gulf have 1–7% Fe and 0.8% Mn (HOUBOLT, 1957). Presumably the precipitation of such ferrous and manganous sulfides occurs in post-depositional anoxic environments.

In summary, one can define four types of internal carbonate cements in shallow-water carbonates: aragonitic needles and laths, magnesian calcite scalenohedra, weakly crystalline magnesian calcite (deep sea), and round disks which probably are aragonite. These cements are similar to those in non-skeletal carbonates (pelletoids, ooids and grapestones) as well as to intergranular cements in some shallow-water and deep-water limestones (see below). Intragranular cements are far more common in carbonate sands than in carbonate muds (PURDY, 1963; PUSEY, 1964; MILLIMAN, 1969 b; BOYER, 1972). This may be related to the restricted flux of dissolved carbonate within carbonate muds or the incomplete decomposition of organic films on grains surrounded by muds (BOYER, 1972). Ultimate infilling, however, is dependent upon the degree of supersaturation of $CaCO_3$ in ambient sea water. In colder shallow waters, infilling generally is sparse or completely lacking (ALEXANDERSSON, 1972 a; ADEY and MACINTYRE, 1973).

LLOYD (1971) has found that the δC^{13} values of the micritic rinds on mollusk shells average about $+4.3^0/_{00}$, which is remarkably close to the C^{13} values in ooids and aggregates (Fig. 19), and substantially higher than values in most other carbonate components. LLOYD also measured similar C^{13} enrichments in Persian Gulf submarine limestones (Table 73). These data suggest a similarity in the precipitation of shallow-water intragranular cements and the intergranular cements of shallow-water limestones. This point will be discussed further at the end of this chapter.

Intergranular Cementation and Lithification

Marine cementation occurs within two general environments, intertidal and subtidal. Each will be discussed in the following paragraphs.

Intertidal Cementation

Numerous types of limestones have been reported from intertidal and supratidal areas. Beachrock, promenade rock and mangrove reef rock occur in the intertidal zone. Intertidal sabellarian reefs can form limestone masses remarkably similar to beachrock (MULTER and MILLIMAN, 1967). Supratidal rocks include cay sandstones, shingle conglomerates and dolomitic crusts. Elevated reef rock and older limestones, while present on many cays, represent pre-recent formations and therefore will not be discussed further.

Beachrock

Beachrock is the most common intertidal limestone on coral cays (Plate XXIII). It also occurs on many other tropical, subtropical and even temperate beaches. Beachrock has been reported from the Mediterranean, along much of the African and Brazilian coasts, in south Florida, the Caribbean and Pacific islands, in Japan, New Zealand, Australia, in California as far north as 33° N (RUSSELL, 1962b), and even in the Baltic Sea (MÜLLER and RUDOWSKI, 1967). However, in some carbonate-rich areas, such as Guam (EMERY, 1962) and Alacran Reef (FOLK, 1967), beachrock is notably absent.

Beachrock can be defined broadly as a layered calcarenite (not always layered, and sometimes calcirudite) cemented with calcium carbonate and which occurs in one or more bands along the intertidal zone (Plate XXXIII). Cementation is assumed to have occurred *in situ* or beneath a thin sediment cover. Strata generally dip at angles coincident with the slope of the beach. In places where outcrops are absent, subsurface probing often detects the beachrock lying under beach sand.

Modern beachrock can be confused with outcrops of older formations that occur in the intertidal zone. For example, the "beachrocks" that RUSSELL (1962b) mentions occurring along the eastern seaboard of the United States are actually outcrops of the Pleistocene (Sangamon?) Anastasia and Pamlico formations (DUBAR and JOHNSON, 1964; MULTER and MILLIMAN, 1967). Even the origin and age of the famous Brazilian stone reefs (BRANNER, 1904) are open to question (see MABESOONE, 1964).

Beachrock surfaces commonly are marked by three distinct erosional features. 1. Solution basins result from intertidal solution (WENTWORTH, 1944; EMERY, 1946; EMERY and COX, 1956; REVELLE and EMERY, 1957). 2. Cracks and channels are produced by mechanical erosion along elongate joints; the origin of the joints is not known, but may be related to thermal expansion and contractions (Plate XXIII e). 3. Potholes within beachrock are the result of mechanical abrasion by trapped cobbles within solution basins (EMERY and COX, 1956).

Sands within modern beachrocks reflect the composition of the ambient beach sands. Grains are usually calcareous, but detrital fragments can be common or even dominant. Caribbean beachrock tends to be rich in *Halimeda*, while Indo-Pacific rock usually contains many foraminifera and coralline algae (see Chapter 6). Rock with large quantities of coral cobble sometimes is termed reef (or beach) conglomerate.

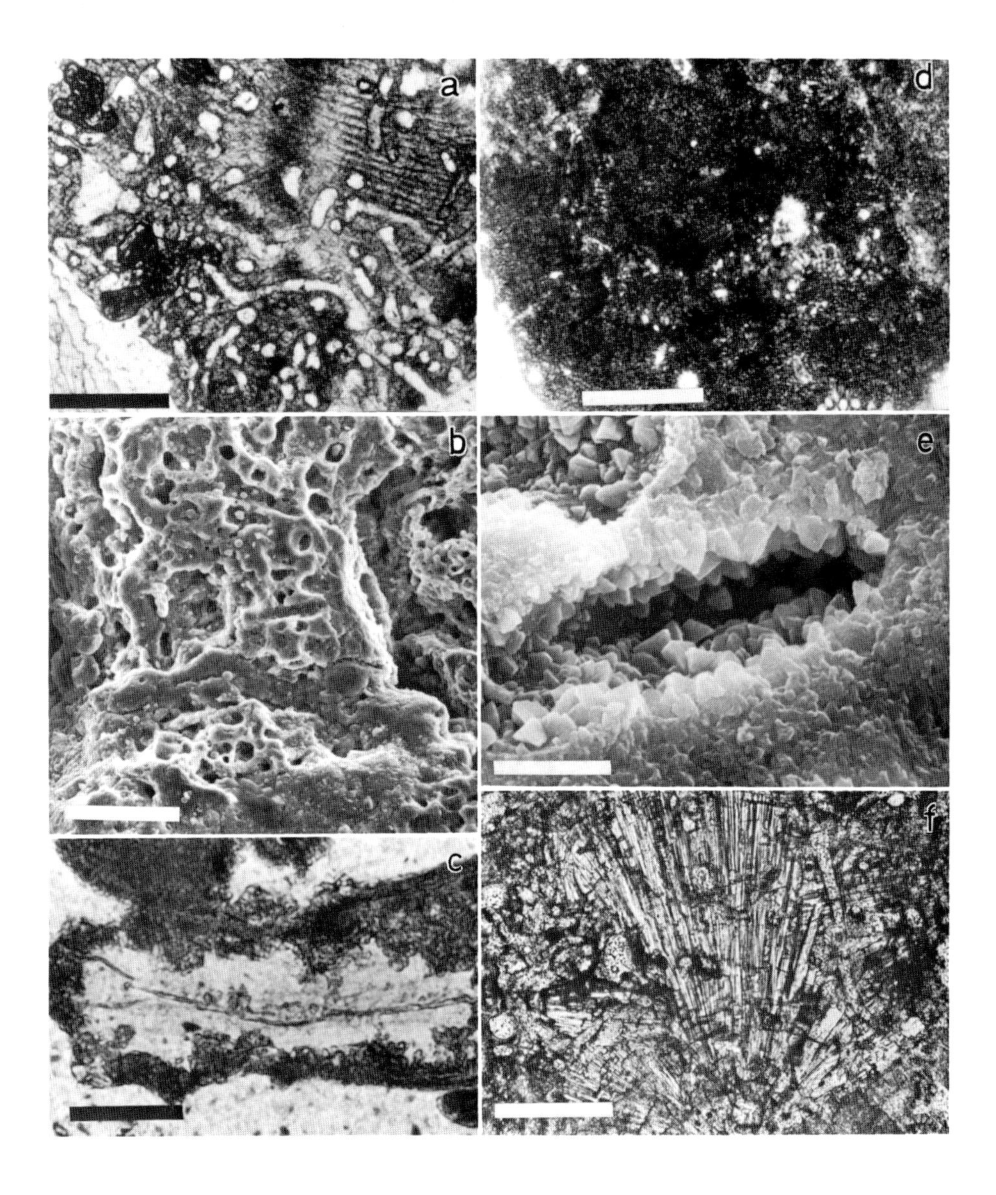

Plate XXXIII. Cryptocrystallization of Mollusks and Coralline Algae. (a) Extensively burrowed mollusk shell; the probable agents were blue-green algae. Refracted light; scale is 100 microns. (b) SEM photomicrograph of a thoroughly bored and burrowed mollusk shell. Scale is 70 microns. (c) Precipitation of internal cement within the algal burrows results in the formation of a cryptocrystalline rind surrounding the mollusk shell. Refracted light; scale is 50 microns. (d) Cryptocrystalline coralline algae in which the cellular structure is barely detectable. Compare this with the fresh coralline algae shown in Plate VI. Refracted light; scale is 100 microns. (e) SEM photomicrograph showing the growth of scalenohedral magnesian calcite in the individual algal cells. The precipitation of this carbonate results in the cryptocrystallization seen in (d). Scale is 5 microns. (f) "Recrystallization" of magnesian calcite coralline algae with aragonite. Refracted light; scale is 50 microns. Photograph courtesy of J. MÜLLER

Table 72. Properties of various beachrock cements

Worker (s)	Area	Cement	Mineralogy	Mole % $MgCO_3$ in calcite	Proposed origin
MOORE and BILLINGS; MOORE	Cayman Islands	cryptocrystalline acicular	arag. and Mg-calcite arag.,		Evaporation
TAYLOR and ILLING	Persian Gulf	drusy cryptocrystalline mud-derived matrix	arag. or Mg-calcite arag., Mg-calcite	15	Evaporation arag. alters to Mg-calcite
MULTER	Dry Tortugas, Fla.	needles normal to grain surface	arag.		
FRIEDMAN and GAVISH	Mediterranean and Gulf of Aqaba	fibrous semi-opaque, cryptocrystalline	arag. Mg-calcite		
ALEXANDERSSON	Mediterranean	cryptocrystalline envelope, 20–100 μ wide	Mg-calcite	13–15	
TAFT and others	Bahamas	acicular	arag.		Evaporation
MACINTYRE and MILLIMAN	Outer shelf, southeastern U.S.A.—submerged Pleistocene beachrock	pelletal and cryptocrystalline also some blocky dentate and dense acicular	Mg-calcite	12	

Some beachrocks are loosely cemented and can be disaggregated by rubbing between the fingers. Others are so well indurated that they ring when struck with a hammer. Often, as RUSSELL (1962b) and other workers have pointed out, the outer surface of the beachrock is case-hardened, while the interior is soft.

The crystal habit and mineralogy of the cements from various beachrocks are listed in Table 72. Either aragonite or magnesian calcite can be the cement (Plate XXXIV). The presence of this latter mineral has been noted only in recent years (TAFT and HARBAUGH, 1964; EBANKS, 1967). RUSSELL and his coworkers (RUSSELL, 1962b; DEBOO, 1962) had noted that most beachrock cements are

→

Plate XXXIV. Beachrock Cements. (a) Aragonitic cement from Grand Cayman beachrock. Refracted light; scale is 100 microns. (b) SEM photomicrograph of aragonitic needle clusters extending into a void from surrounding grains in a Grand Cayman beachrock. Scale is 40 microns. (c) Rhombohedral magnesian calcite cement from St. Croix (Virgin Islands) beachrock. Refracted light; scale is 100 microns. (d) Cryptocrystalline magnesian calcite from St. Croix beachrock. Refracted light; scale is 100 microns. (e) SEM photomicrograph of the magnesian calcite scalenohedra in St. Croix beachrock cements. Scale is 100 microns. (f) The encrusted tubules of blue-green (?) algae described by SCHROEDER and GINSBURG (1971) also are found in the magnesian calcite cements of the St. Croix beachrocks. SEM; scale is 100 microns. Photographs courtesy of C.H. MOORE, JR.

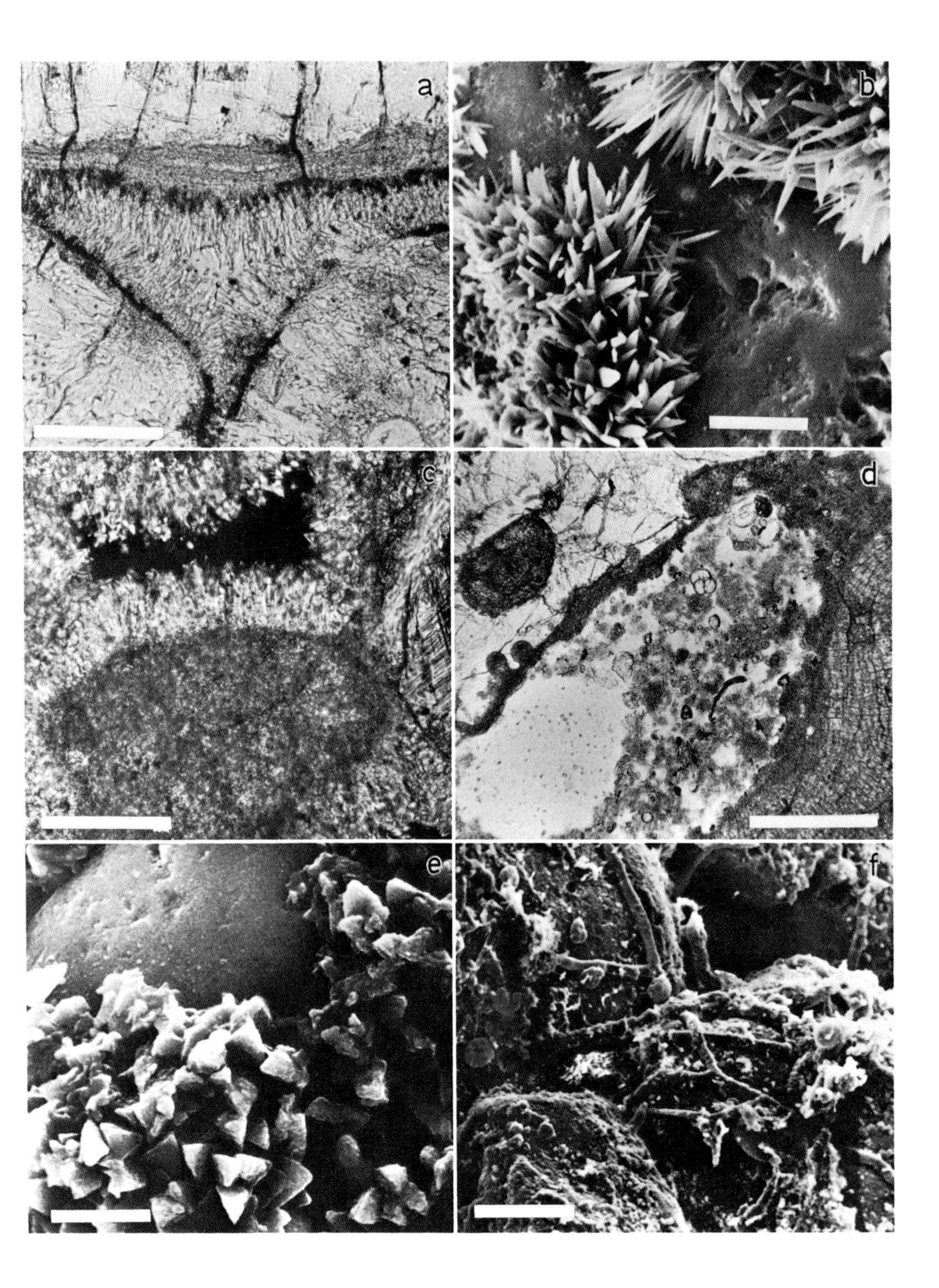

Plate XXXIV

calcite, but the fact that these cements occur in areas (such as the Caribbean) in which other workers have found only aragonite and magnesian calcite suggests that their findings were erroneous.

The aragonite cement is usually acicular or fibrous (Plate XXXIV a, b), although cryptocrystalline and drusy aragonite also have been noted. In most instances the magnesian calcite is present as semi-opaque cryptocrystalline cement (Plate XXXIV d), but rhombohedral cements also occur (Plate XXXIV c). Scanning-electron microscopy suggests that all magnesian calcite cements consist of small (4 to 8 microns) rhombohedra which nucleate upon existing crystalline surfaces during their growth (ALEXANDERSSON, 1972c), TAYLOR and ILLING (1971) have suggested that magnesian calcite might represent an alteration product of aragonite, but evidence for the opposite conclusion has been given in the preceding section.

Beachrock can lithify with startling rapidity, as evidenced by the incorporation of fishing boats, soft-drink bottles and World War II debris (including human skeletons) in tropical beachrocks. STODDART and CANN (1965) and DEBOO (1962) have suggested a two stage cementation process, in which the filling of voids by matrix occurs after initial cementation by rim cements. Void filling can decrease the porosity markedly and effectively case-harden the rock, but porosity is still higher than in most other limestones (DALY, 1924; GINSBURG, 1953b). Case-hardening can occur quickly, and probably happens as the beachrock is exposed to the air and salt spray. MORESBY (1835), perhaps the first European to describe beachrock and speculate upon its origin, noted that the Maldive Island natives mined beachrock (used in construction) by excavating the overlying beach sands. Once exposed to salt air, however, the soft beachrock had to be removed quickly, otherwise it would harden to a state where it could not be cut.

The origin of beackrock has been related to both organic and inorganic processes. FIELD (1920) and DALY (1920, 1924) suggested that as organic matter trapped in beach sands begins to decay, it releases ammonium carbonate (MURRAY and IRVINE, 1891) which precipitates calcium carbonate thus causing lithification. Daly mentioned several instances in which a quick burial of storm-derived organic debris resulted in sand lithification. NESTEROFF (1954), GUILCHER (1961) and PURI and COLLIER (1967) thought that microbial action could cause carbonate precipitation. KAYE (1959) leaned towards some type of biologic cementation of the Puerto Rican beachrocks, while CLOUD (1959) suggested precipitation by blue-green algae. Photomicrographs of encrusted tubules (Plate XXXIV f) are similar to the blue-green algal tubules described by SCHROEDER and GINSBURG (1971), suggesting that blue-green algae may be critically important.

There are numerous arguments against organic mechanisms. FIELD (1920) showed that decaying organic matter was not present in sufficient amounts in carbonate sands to cause cementation; for instance, EMERY and COX (1956) found that Hawaiian beach sands contain less than 0.2% organic carbon. The role of blue-green algae also must be examined critically; although blue-greens are common in the outer layers of beachrock, they are notably absent on the inner portions (EMERY and COX, 1956). KRAUS and GALLOWAY (1960) found no convincing chemical or biological evidence favoring an algal origin for beachrock. On the other hand, as NEWELL (1956) pointed out, blue-greens might serve an important role in beachrock formation by stabilizing the sand and thereby facilitating lithification.

Most workers who have investigated the beachrock problem have leaned towards a physiochemical origin, although opinion has been sharply divided as to whether the precipitation is caused by ground water or sea water. DAVID and SWEET (1904) theorized that humic acids in the ground water of Funafuti Island become saturated with calcium carbonate by the dissolution of surrounding sands. Seaward migration of the ground waters and subsequent contact with sea water could result in precipitation at the beach interface. MORESBY (1835) also favored a freshwater origin, but he did not specify the exact cause. FIELD (1920) thought that rain waters from tropical storms would quickly supersaturate with calcium carbonate as they permeated the carbonate beach sands, and precipitation would occur at depth.

Perhaps the most exhaustive study of beach rock was made by R.J. RUSSELL and his coworkers at the Louisiana Coastal Research Institute (RUSSELL, 1962a, 1962b; RUSSELL and MACINTYRE, 1965; JONES, 1961; DEBOO, 1962). On the basis of studies in many tropical Caribbean and Pacific areas Russell concluded that groundwater discharge is the basic cementing agent. Evidence includes the dominance of calcite cement together with the fact that beachrock occurs mainly on carbonate beaches or near limestone terraine. In higher latitudes, beachrock is usually associated with landward limestone formations from which the necessary calcium carbonate could be supplied (RUSSELL, 1962b; MÜLLER and RUDOWSKI, 1967). DEBOO (1962) suggested a two-stage cementation, in which the rim cement was added in the groundwater stage of lithification. Upon exposure to the sea, the internal matrix cement would be added.

The arguments against groundwater origin, however, outnumber those supporting it. First, as mentioned above, there is considerable question as to the validity of RUSSELL's cement identifications, and the presence of a calcite cement (as opposed to aragonite) is one of the main points of evidence used by RUSSELL and his coworkers to favor a groundwater origin. In addition RUSSELL labelled as beachrock many outcrops which are clearly pre-recent formations (see p. 278). Also, beachrock commonly occurs on cays that are too small to support a freshwater table (DALY, 1920; KUENEN, 1933; STODDART and CANN, 1965). Moreover, beachrocks in non-carbonate sands or on non-carbonate islands have no immediate landward source of calcium carbonate.

Most workers seem to favor the precipitation of cement by seawater; in a recent meeting on cementation held in Bermuda (1969) all but one of the workers (SCHMALZ, 1971) studying beachrocks supported a salt-water origin. The exact mode of lithification may involve perculation and lithification at depth or seawater evaporation and heating at the surface. DARWIN (1851) invoked the downward perculation of "calcareous matter". DANA (1875) and GARDINER (1903) favored the evaporation of splashing surf to increase the salinity to a point where precipitation would occur. GINSBURG (1953b) theorized that porous sands and high temperatures were necessary for the evaporation of seawater and the subsequent cementation of beachsands. Similar theories have been proposed by MOORE and BILLINGS (1971), TAFT and others (1968), and TAYLOR and ILLING (1971).

Mangrove Reefs

Limestones formed from lithified mangrove root casts have been found along the shores of several tropical islands, such as New Caledonia (AVIAS, 1950), Mauritius (RUSSELL and MACINTYRE, 1965, Fig. 13), and British Honduras (PUSEY, 1964). Perhaps the best documented occurrence, however, is the mangrove reef on Key Biscayne, Florida (HOFFMEISTER and MULTER, 1965). The reef is constructed of the fossilized root and pneumatophore casts of the black mangrove, *Avincennia nitida*, and extends for more than 1000 meters along the intertidal zone. Similar root casts at St. Croix, in the Virgin Islands, are composed of scalenohedral magnesian calcite crystals (MOORE and others, 1971).

HOFFMEISTER and MULTER (1965) have suggested that cementation was caused by the release of CO_2 by decaying mangrove roots, forming carbonic acid which dissolved ambient calcium carbonate grains. Perculation of the carbonate-rich waters resulted in lithification of the sands surrounding the roots, thereby forming the root casts. MORGAN and TREADWELL (1954) reached somewhat similar conclusions when they suggested that the acids from mangrove trees could form the cemented sandstone slabs found in the Chandeleur Islands of Louisiana.

Supratidal Crusts

Dolomitic crusts occur in the supratidal part of evaporitic-hypersaline environments (DEFFEYES and others, 1964, 1965; SHINN and others, 1965; EBANKS, 1967). Crusts tend to be thin, seldom more than 10 cm thick, and cream to brown in color. Many crusts are brecciated. Dolomite is commonly in a muddy matrix; preferential solution of some carbonate components is not uncommon (see EBANKS, 1967, Plate XXXV). The origin of these dolomitic crusts is discussed in Chapter 11.

Some sediments on algal flats are lithified with calcite or aragonite cements (RUSSELL, 1968; DAVIES, 1970b). These crusts are superficially similar to dolomitic crusts. The cements in the Shark Bay algal mats are cryptocrystalline (DAVIES, 1970b).

Submarine Lithification

Until recently, most geologists envisioned that the lithification of carbonates occurs exclusively in the intertidal and subaerial zones: "The only known example of thorough lithification of Recent carbonate sediments in the marine environment occurs in the intertidal zone of beaches in the coral seas" (GINSBURG, 1957, p. 95); "Sediments that remain in a marine environment almost always resist lithification, whereas those exposed subaerially to fresh water become lithified." (FRIEDMAN, 1964, p. 809). This notion of non-lithification in the oceans has been re-evaluated in recent years with the finding of extensive examples of submarine lithification. However, the concept of submarine lithification is not new; it was well documented before the turn of this century. For example, MURRAY and HJORT (1912) wrote:

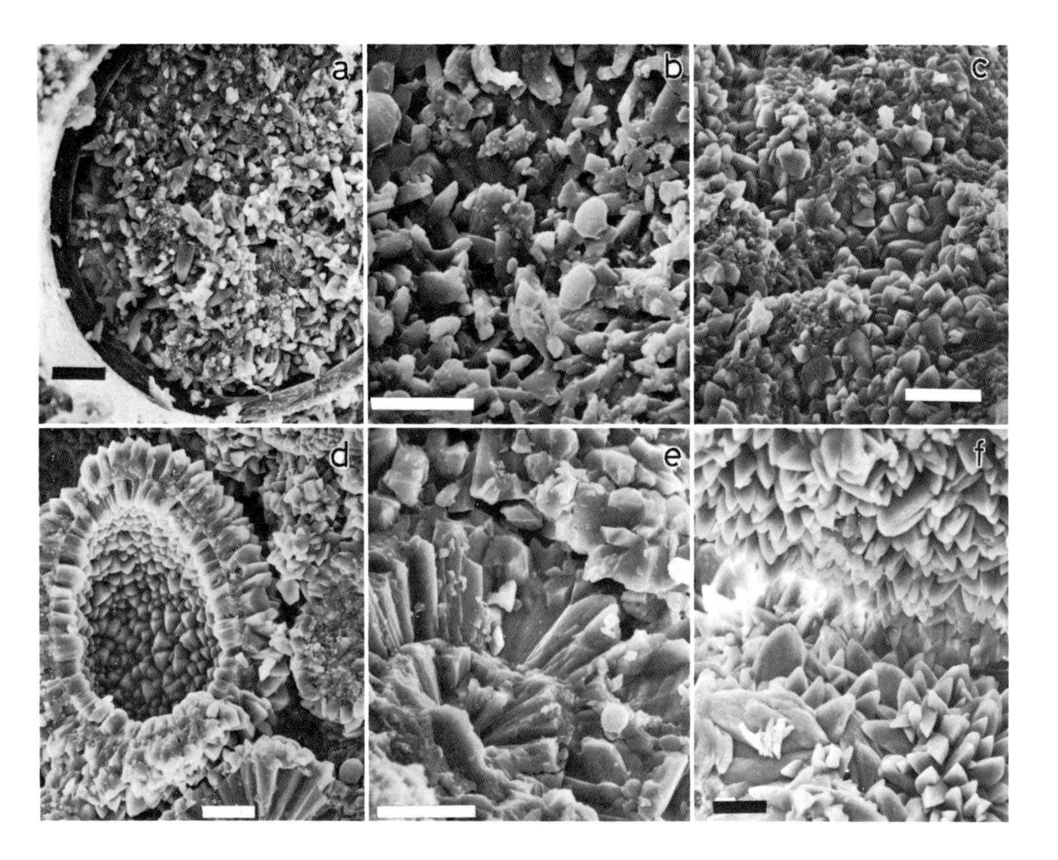

Plate XXXV. Shallow-water Limestones and Their Cements. (a) Internal sediment and matrix within a serpulid tube. The algal limestone from which this photomicrograph was taken is shown in Plate XXVII e. SEM; scale is 10 microns. (b) Infilling of a coralline algal limestone. Much of this material is authigenic magnesian calcite. SEM; scale is 10 microns. (c) Matrix of a reef rock from Serrana Bank. These scalenohedra are magnesian calcite. SEM; scale is 10 microns. (d) Intra- and intergranular magnesian calcite cements within a reef rock from Serrana Bank. Note the contacts between the intergranular cements of various particles. SEM; scale is 10 microns. (e) Close-up of the contact shown in the lower right corner of the previous photo. SEM; scale is 10 microns. (f) The pore space between two particles in an algal limestone, illustrating the growth of magnesian calcite cement towards the center of the void space. SEM; scale is 10 microns

"A limited amount of purely inorganic precipitation does, indeed, take place .... In the Mediterranean, for instance, stone-like crusts are plentiful, consisting of clay cemented by calcium carbonate, which is produced by ammonium carbonate arising from the decay of organic matter in the mud below bottom-level meeting with fresh sea-water from above. We have further the lime-concretions of the Pourtales, Argus and Seine Banks, the "Challenger" casts of shells from the Great Barrier Reef, and so on. But all these must be regarded as rarities." (p. 178).

Numerous other reports of submarine lithic crusts and limestones were reported from various early oceanographic cruises (MURRAY and IRVINE, 1891; NATTERER, 1894, 1898; BØGGILD, 1912), and the process of submarine lithification apparently was well accepted. Why this concept was rejected in subsequent years is not known.

Two distinct types of submarine limestones have been found: shallow-water and deep-water limestones. Generally the shallow-water limestones are restricted to depths of less than 100 m and are composed of shallow-water carbonate sediments. Deep-water limestones can occur at depths exceeding 3000 meters (although most form in much shallower water) and possess petrographic and compositional characteristics quite different from those in shallow-water limestones.

Shallow-Water Limestones

The occurrence of shallow-water limestones has been noted by numerous workers, some of whom are listed in Table 73. Two distinct types of shallow-water cementation have been found. One type is reef limestone in which primary cavities and biogenic burrows have been filled with internal sediment and subsequently lithified (Plates XXVII e, f; XXXV) (see p. 285, 291). Laminar layers of internal sediments can accumulate in large cavities; superficially these internal sediment layers can resemble vadose cement, which has been considered a diagnostic property of subaerial lithification (DUNHAM, 1969). In many instances, biologic boring continues after the initial cementation, so that the limestone contains a sequence of inter-connected and filled burrows. Examples of this type of limestone have been found in the Mediterranean (ALEXANDERSSON, 1969, 1971), Bermuda (GINSBURG and others, 1971 a, 1971 b) and Jamaica (LAND and GOREAU, 1970)[2].

In addition, algal limestones have been found on the upper slope and outer shelf off the southeastern United States (MACINTYRE and MILLIMAN, 1970). These algal limestones are mostly relict and were probably lithified during the last regression and subsequent transgression of sea level (MACINTYRE and MILLIMAN, 1970). However, the filling and cementing of internal cavities must still be active, since planktonic organisms comprise much of the lithified and unlithified internal sediment (Plates XXVII e, f; XXXV a, b).

The second type of shallow-water limestone formation involves the cementation of loose grains into a solid cohesive rock. Grains can be pellets, ooids or

2 See p. 160 and 168 for further discussions.

Table 73. Properties of various shallow-water limestones

Worker	Area	Water depth (m)	Rock type	Cement	Mineralogy	Mole % $MgCO_3$ in calcite	Stable isotopes	Age (yrs)
Alexandersson	Mediterranean	< 20	Algal limestones, internally lithified with much boring and internal sediment	cryptocrystalline fibrous clusters	Mg-calcite aragonite	15–17		890
Taft and others	Bahamas				aragonite			1425–5530
Garrison and others	Fraser River Delta, British Columbia	1–5	Irregular to platey nodules	fibrous rims	calcite			recent
Friedman and others	Red Sea	< 10	Reef rock	fibrous cryptocrystalline	aragonite Mg-calcite	30		
Ginsburg and others	Bermuda	at least 8	Reef rock, much internal sediment and boring; well lithified Fillings in shell voids	acicular to rhombohedral (xls, 5–30 long)	Mg-calcite	15		2980–3100
Shinn	Persian Gulf	2–25	Well-cemented limestones with much boring	acicular and packed fibrous cryptocrystalline and recrystallized muds	aragonite Mg-calcite		δO^{18}(+0.7 to 2.3) δC^{13}(+4.0 to 4.5)	
Land and Goreau	Jamaica	at least 70	Reef rock-well bored	internal pelletal sediment drusy, acicular	Mg-calcite Mg-calcite	18.5	δO^{18} +0.5 δC^{13} +3.0	130–8410
Pantin	New Zealand	125	Molluscan and quartzose limestone; heavily bored and with much matrix					< 19000
MacIntyre and others	Barbados	15	Skeletal limestone	pelletal cryptocrystalline matrix inner dark rind and outer dentate xls	Mg-calcite Mg-calcite	20–21		500–1500
MacIntyre and Milliman	Southeastern U.S.A.	to 120	Algal limestone, internally lithified and thoroughly bored; much internal sediment	cryptocrystalline, and pelletal	Mg-calcite	12–13	δO^{18}(+1.2 to 1.5) δC^{13}(+1.5 to 1.9)	(relict) 9000 to 13000 years

grapestone (ILLING, 1954; TAFT and others, 1963; SHINN, 1969), skeletal grains (GOULD and STEWART, 1953; MACINTYRE and others, 1968), or even terrigenous grains (JOHNSTON, 1921; GARRISON and others, 1969; NELSON, 1970). Both modern and relict limestones of this type also have been found on continental shelves (PANTIN, 1958; BLANC, 1968; MACINTYRE and MILLIMAN, 1970). These limestones generally contain both rim cements and matrix cements, and may be heavily burrowed.

Most of the cement in shallow-water limestones is magnesian calcite, although some aragonite also may be present (TAYLOR and ILLING, 1969; SHINN, 1969; TAFT and others, 1968; GINSBURG and others, 1971b; FRIEDMAN, 1972) (Table 73). The aragonite is present as acicular or packed fibrous needles, extending perpendicularly from the grain surfaces (Plate XXXVI a–c); crystal size may increase towards the center of the void. Some cements exhibit the same disk-like fabric found in grapestone and in *Halimeda* utricles; presumably these cements are aragonitic.

Magnesian calcites are generally cryptocrystalline, although some drusy magnesian calcite is found (Plate XXXV). Rim cements, sometimes associated with matrix cement, are diagnostic of many shallow-water limestones. Scanning-electron photomicrographs show that many of these cryptocrystalline matrix cements are actually composed of well-formed magnesian calcite scalenohedra, 2 to 8 microns in diameter (Plate XXXV). Another characteristic of shallow-water cryptocrystalline magnesian cements is the presence of a pelletal texture (Table 73). The pellets, generally smaller than 50 microns, are composed of small crystals up to 8 microns in diameter (MACINTYRE and MILLIMAN, 1970; ALEXANDERSSON, 1971). The pelletal texture appears to be the result of differential micro-growths rather than organic pellet-forming processes (ALEXANDERSSON, 1972b). The magnesian calcite contains from 12 to 21 mole % magnesium. In only one cited instance is the cement low magnesian calcite: GARRISON and others (1969) reported the occurrence of irregular to platy nodular masses in the shallow waters adjacent to the Fraser River Delta, British Columbia, in which the terrigenous grains are cemented by fibrous calcitic rims.

Available stable isotope data suggest that shallow-water limestones are not in equilibrium with ambient waters; δC^{13} values range from +3 to +4.5‰ (SHINN, 1969; LAND and GOREAU, 1970).

Deep-Water Limestones

Carbonates in the deep sea can be cemented by pyrite (FIELD and PILKEY, 1970), rhodochrosite and siderite (JOIDES, 1970a, c), but most are cemented by calcium carbonate. Many types of limestones have been dredged from the deep sea. FISCHER and GARRISON (1968) listed some 25 reported dredgings of deep-sea limestones, and numerous other examples have been cited (for example, BARTRUM, 1917; NIINO, 1931; STUBBINGS, 1939; PANTIN, 1958; KLENOVA and ZENKOVITCH, 1962; SAITO and others, 1966; CIFELLI and others, 1966; DUNCAN, 1970)[3]. All together, some 50 examples of deep-sea limestones have been reported in the

3 ZENKOVITCH (1967) has discussed crusts in the Black and Caspian Seas, as well as reviewing the Russian literature on this subject.

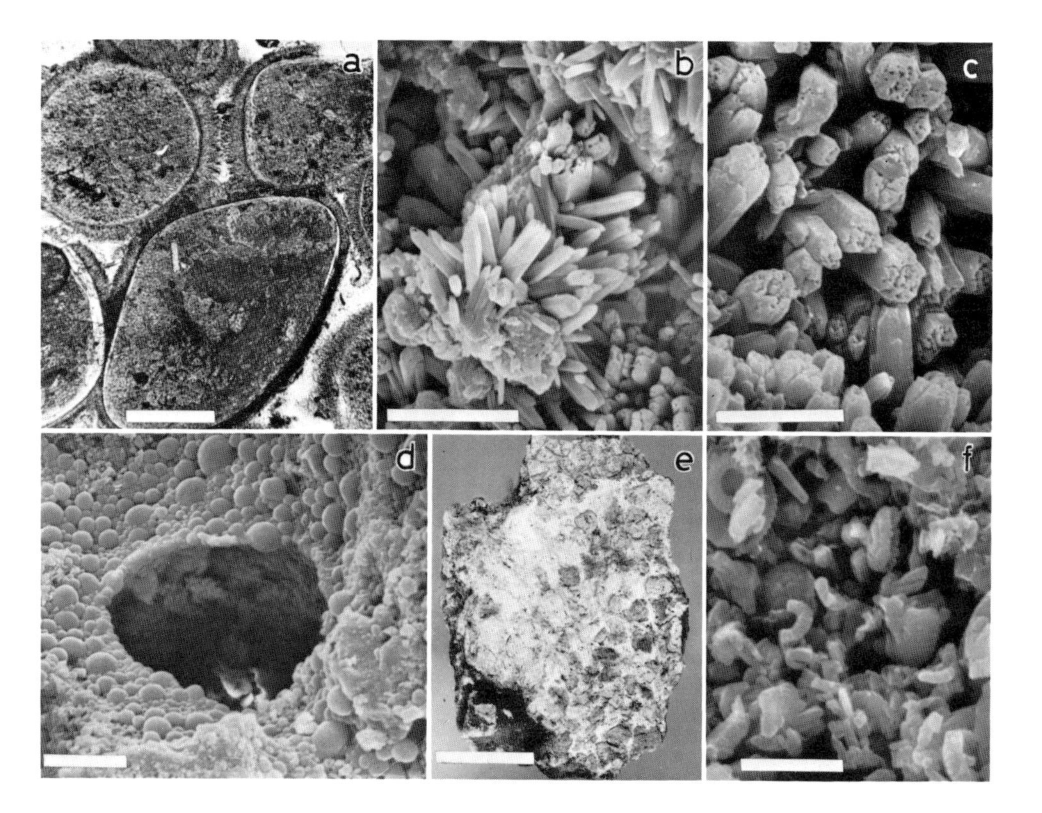

Plate XXXVI. Miscellaneous Limestones. (a) Oolitic limestone from the upper slope off central Florida. Carbon-14 age dates, together with the fabric of the cement, suggest that these limestones were deposited and cemented in shallow water, perhaps as a beachrock; compare the aragonite cement in this picture with that in Plate XXXIV. Refracted light; scale is 150 microns. (b) Aragonitic matrix and cement within an intertidal limestone from the Isle of Rhodes. SEM; scale is 10 microns. (c) A closer view of the aragonitic needles shows that they possibly are undergoing dissolution. SEM; scale is microns (d). Another type of cement in the Rhodes intertidal limestones is this disk-like carbonate that has been seen in other shallow-water carbonates. Published data by ALEXANDERSSON (1969) and unpublished X-ray data by the present writer suggest that much of the cement in the Rhodes limestones is magnesian calcite, but previous experience indicates that this particular type of cement is aragonite. SEM; scale is 10 microns. (e) Volcanic limestone from the Mid-Atlantic Ridge. Calcite is cemented around altered volcanic material. Scale is 3 cm. (f) SEM photomicrograph of the matrix of a volcanic limestone. This one is composed primarily of coccolith fragments, which have been joined, apparently, only at contract points. As a result, this limestone is very porous. Scale is 5 microns

literature, and judging from the lack of interest in carbonates shown by many geological oceanographers, there are probably many undocumented examples of submarine limestones stored in rock and core collections at various oceanographic institutions.

Five major types of submarine limestones have been recognized from the deep-sea: limestones occurring in non-depositional environments, volcanic environments, restricted oceanic basins, deep cemented layers, and limestones associated with methane-derived cements. Each will be discussed below.

1. Non-Depositional Environments. Many limestones have been dredged from the tops of seamounts, banks and plateaus, as well as from the sides of islands; some of these limestones are described in Table 74. Most were recovered from depths less than 1000 m but deeper than 300 m. These limestones often are heavily bored irregular nodules and slabs. Component grains are dominated by planktonic foraminifera and planktonic and benthic mollusks, but displaced shallow-water debris can be present. Many planktonic components are filled with intragranular calcite (Plate XXXVII). Similar infilling occurs in many of the carbonate grains in loose sediments surrounding the limestones, suggesting that intragranular cementation may represent the first step in lithification.

In late Tertiary or Quaternary limestones, the cement is mainly magnesian calcite, with an average of 12 mole % $MgCO_3$ (Table 74). Stable isotope analyses show that the limestones are in isotopic equilibrium with the ambient waters, therefore indicating *in situ* lithification (MILLIMAN, 1966, 1971). In older limestones, even those from areas that also contain younger magnesian calcite limestones, the cement is generally calcite. These older limestones contain more matrix cement, in contrast to the often porous younger limestones (Plate XXXVII d, e). Older limestones also may contain solution pits that have been filled with secondary cements (BARTLETT and GREGGS, 1970 b). The writer (MILLIMAN, 1966) has suggested that such limestones may precipitate as magnesian calcite, but later invert to calcite. The inversion of magnesian calcite to calcite has been discussed by LAND (1966) and examples have been documented in the deep sea by MILLIMAN (1966) and GOMBERG and BONATTI (1970). Subsequent diagenesis may involve phosphatization as well as the deposition of manganese coatings (MILLIMAN, 1966; MARLOWE, 1971 b).

MCFARLIN (1967) found that manganese nodules on the Blake Plateau contain veins filled with authigenic aragonite. The anomalous isotopic values of these fillings suggest that precipitation is related to local micro-environments. To the writer's knowledge, this is the only report of authigenic aragonite in the open ocean.

The mode of lithification of limestones found on seamounts, banks and island slopes is not known, but it is interesting to note that in all these environments, water movement (currents) is sufficiently strong to prevent sediment accumulation. NEUMANN and others (1972) have suggested that lithification of bottom ripples in the Straits of Florida have resulted in "lithoherms", 30 to 40 m high and 100's of m long. Most other examples of "non-depositional" limestones are thought to be represented by thin crusts; net accretion is probably minimal. The restriction of such limestones to areas of non-accumulation is somewhat reminiscent of the

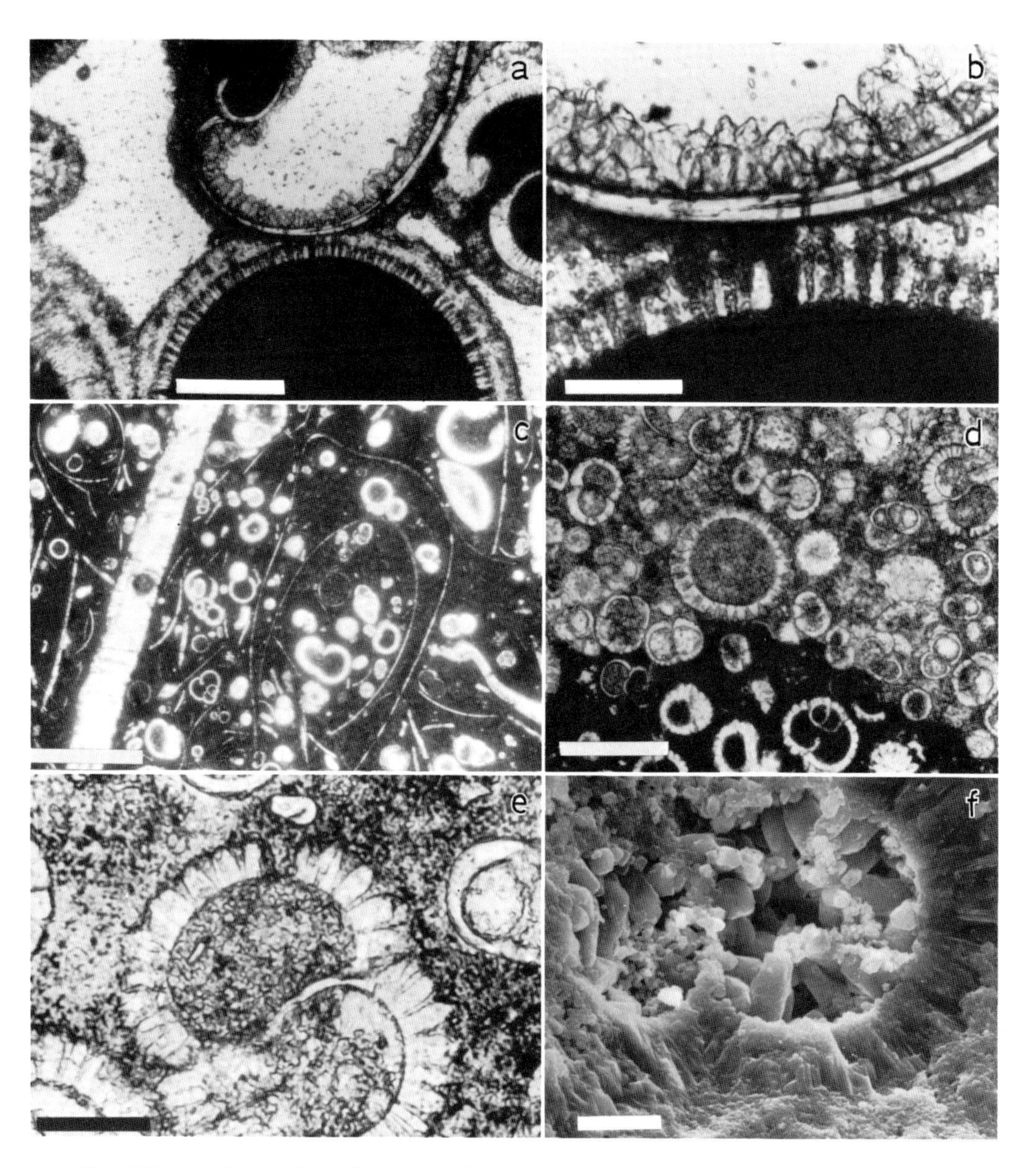

Plate XXXVII. Cementation of Deep-Sea Carbonates. (a) Porous limestone from Gerda Guyot, Southeastern Bahamas. Cement is magnesian calcite. Note the infilling of foraminifera and the contact rim cement. Refracted light; scale is 100 microns. (b) Close-up of a contact rim cement. Note the well formed scalenohedral magnesian calcite crystals. Refracted light; scale is 50 microns. (c) Magnesian calcite cemented limestone from Great Meteor Seamount. This limestone is not nearly as porous as that in the previous pictures, but is about the same age (Pleistocene). Refracted light; scale is 300 microns. (d) Miocene limestone from Gerda Guyot. The cement is calcite, with appreciable amounts of phosphorite (lower part of photograph). Refracted light; scale is 100 microns. (e) Close-up of the Miocene limestone, showing the complete filling of the foraminifera and the dense matrix. Refracted light; scale is 50 microns. (f) SEM photomicrograph showing the infilling of a void with magnesian calcite. The crystals are nearly as well-formed as those in shallow-water limestones. Scale is 20 microns

Table 74. Deep-sea limestones from non-depositional environments

Worker	Area	Water depth (m)	Petrography	Cement	Mineralogy (mole % $MgCO_3$)	Stable isotopes δO^{18}	Stable isotopes δC^{13}	Age
MILLIMAN	Gerda Guyot, Bahamas	700–800	Porous, planktonic foram. limestone	cryptocrystalline to drusy	Mg-calcite (12)	+1.87–+2.07	+1.26–+2.20	Pleistocene
			Dense, planktonic foram. limestone with phosphatic inclusions	cryptocrystalline to drusy matrix	Calcite	+1.47–+2.41	+0.52–+1.16	Miocene
FISCHER and GARRISON	Barbados	280–440	Dense, planktonic foram. limestone; also gastropod-rich limestone	rim cement together with matrix cement	Mg-calcite (13)			Late Cenozoic
MILLIMAN	Mid-Atlantic Ridge	350	Planktonic foram. limestone	rim cement and matrix	Mg-calcite (12)			
MILLIMAN	Courtown Cays SW Caribbean	450	Planktonic foram. limestone	rim cement and matrix	Mg-calcite (11)			
MILLIMAN	Great Meteor Seamount	600	Mollusk, planktonic foram. limestone	rim cement and matrix	Mg-calcite (12)	+2.64	+1.61	Pliocene-Holocene
		900	Mollusk, planktonic foram. limestone	rim cement and matrix	Calcite	+2.70–+3.08	+0.40–+1.05	Pliocene-Holocene
MARLOWE	Aves Swell	340–380	Porous pteropod, planktonic foram. limestone	Clear, fibrous rim cement	Mg-calcite (14)			Pliocene-Holocene
FRIEDMAN	Atlantis Seamount	270	Well-indurated gastropod-planktonic foram. limestone	cryptocrystalline	Mg-calcite			9000–12000 yrs
BARTLETT and GREGGS	Mid-Atlantic Ridge	1400–3000	Alternating layers of lithified/unlithified carbonates	sparite and "micrite"	Calcitic; some Mg-calcite			>38000 yrs
MILLIMAN	SW Caribbean island slopes	300–1000	Planktonic and shallow-water slump carbonate limestones	rim cement, some void fillings of cryptocrystalline	Mg-calcite (12)	−0.06–+2.97	+0.98–+2.11	
MCFARLIN	Blake Plateau	400–800	Vein fillings in manganese nodules	druse	Aragonite	+3.5–+4.0	−0.1–+1.9	
GEVIRTZ and FRIEDMAN	Blake Plateau	400–800	Planktonic foram.-pteropod limestone	cryptocrystalline	Mg-calcite (9.5–11.8)			
BRODIE	Capricorn Seamount SW Pacific	80	Porous limestone with mollusks, brachiopods (and large forams.)		Mg-calcite (12)			Miocene

Table 75. Deep-sea limestones from volcanic areas

Worker	Area	Water Depth (m)	Petrography	Cement	Mineralogy	Stable isotopes δO^{18}	Stable isotopes δC^{13}	Age
MILLIMAN	New England Seamounts	1800–2500	Calcite crystals and layers within altered volcanic tuffs	spar	calcite	+0.40– +0.70	−1.29– +2.44	?
MILLIMAN	Mid-Atlantic Ridge	4280	Cemented planktonic ooze with altered tuff fragments		calcite	+1.64	+2.21	
MECARINI and others	Mid-Atlantic Ridge		Friable, plantonic foram.—coccolith limestone	coccolith matrix	calcite			Quaternary
THOMPSON and others	Mid-Atlantic Ridge	1700–2700	Consolidated ooze; calcarous tuffs; foraminifera limestones	dense matrix	calcite	−4.2– +1.9	+4.1– +5.3	
MILLIMAN	Tuamotus	2600	Cemented ooze with altered tuff fragments		calcite			

Table 76. Chemical composition of various submarine limestones

Limestone type	Area	Water depth (m)	Percent carbonate	Carbonate mineralogy (A—MgC—C)	Percent Sr	Mg	Fe	Mn	P
Non-depositional	Gerda Guyot	700–800	95	tr–85–15	0.14(0.13–0.16)	3.4(3.7–2.8)		0.02(0.02–0.03)	0.06(0.06–0.07)
		700–800	<85	0–0–100	0.10(0.07–0.15)	0.6(0.5–0.8)		0.20(0.10–0.24)	>4.0
	Great Meteor Seamount	800	>95	0–70–30	0.11	3.3		0.02	0.06
		900	>95	0–10–90	0.12	0.65(0.2–1.1)		0.05(0.02–0.07)	0.04(0.03–0.05)
Semi-enclosed basins	Red Sea	800–2700	85	15–80–5	0.22(0.07–0.41)	2.69(2.16–3.41)	0.32(0.15–0.51)	0.20(tr–0.51)	
		800–2700	91	80–15–5	0.81(0.62–1.00)	0.56(0.27–0.73)	0.57(0.11–1.00)	0.20(0.05–0.33)	
Volcanic limestones	Mytilis Seamount	1800	99+	0–0–100	0.10	0.43	0.28	0.54	
	Balanus Seamount	2700	99+	0–0–100	0.10	0.20	0.18	<0.10	

environmental parameters necessary for the precipitation of shallow-water non-skeletal fragments (see Chapter 3). The exact relation between the non-accumulation of bottom sediments and carbonate precipitation however, is not understood. VAN STRAATEN (1967) stated that calcite cementation in the Mediterranean was caused by the interstitial dissolution of aragonite and the subsequent removal of overlying sediment (probably caused by turbidity currents). The subsurface pore waters, which were supersaturated with respect to calcium carbonate, thus were exposed to bottom waters, causing a precipitation of calcite. Perhaps a similar phenomenon occurs in areas of non-deposition: pore waters leach calcium carbonate from the surrounding sediments. When bottom currents remove the overlying sediment, the pore waters become supersaturated and carbonate is precipitated.

Lithification apparently can be rapid. FRIEDMAN (1964) reports well-indurated limestones from Atlantis Seamount which are 9000 to 12000 years old, and a calcarenite rock from Aves Swell is 13820 years old (MARLOWE, 1971 b). BARTLETT and GREGGS (1970a) found limestones on the Mid-Atlantic Ridge which were interlayed by unlithified sediments, suggesting that lithification can be periodic.

2. *Volcanic Limestones.* Limestones and calcite crystals embedded within volcanic tuffs have been recovered from the slopes of many submarine volcanoes (Plate XXXVI e, f). Some of the limestones appear to be baked sediments (SAITO and others, 1966), while others consist of dispersed crystals within an altered tuff. The few samples studied by the writer show no evidence of a precipitated matrix.

In all instances the cement in volcanic limestones is calcite[4]. The magnesium levels are generally below 0.5%, but manganese content may be relatively high (Tables 75, 76). Isotopic data show that the limestones are out of equilibrium with respect to the ambient ocean waters; THOMPSON and others (1968) report δO^{18} values from Mid-Atlantic Ridge limestones as low as $-4.2^0/_{00}$, or some $6^0/_{00}$ lower than expected at such depths in the deep sea.

Undoubtedly these limestones have been severely affected by submarine volcanism, but the exact method of lithification is not known. Submarine eruptions cause both increase in ambient water temperatures and locally can release

4 THOMPSON (1972) has reported aragonitic crusts and veins on and in weathered peridotite surfaces. The aragonite is high in Sr (1.0%) and low in Mg (0.09%); δO^{18} and δC^{13} values are $+5.4$ and $-1.3^0/_{00}$, respectively.

⟶

Plate XXXVIII. Submarine Limestones. (a) Well-crystallized matrix within a submarine limestone. Cement is magnesian calcite. SEM; scale is 10 microns. (b) This matrix is not nearly as well-crystallized as those shown in (a), and yet it comes from the same locality, Gerda Guyot. SEM; scale is 20 microns. (c) Magnesian calcite fragment from the Red Sea. These lithic fragments, which apparently form periodically in both the Red Sea and the eastern Mediterranean Sea, are characterized by a dense matrix and calcitic microfossils (foraminifera and coccoliths). Refracted light; scale is 500 microns. (d) Aragonitic shells often dissolve in the eastern Mediterranean, leaving magnesian calcitic casts (here a pteropod cast). SEM; scale is 200 microns. (e) SEM photomicrograph showing the infilling within a pteropod cast. Scale is 100 microns. (f) Close-up of the magnesian calcite matrix of a pteropod cast. Note the similarity of these crystals to those in other deep-sea limestones, and in some deep-sea sediments (Plate XXVIII). SEM; scale is 10 microns

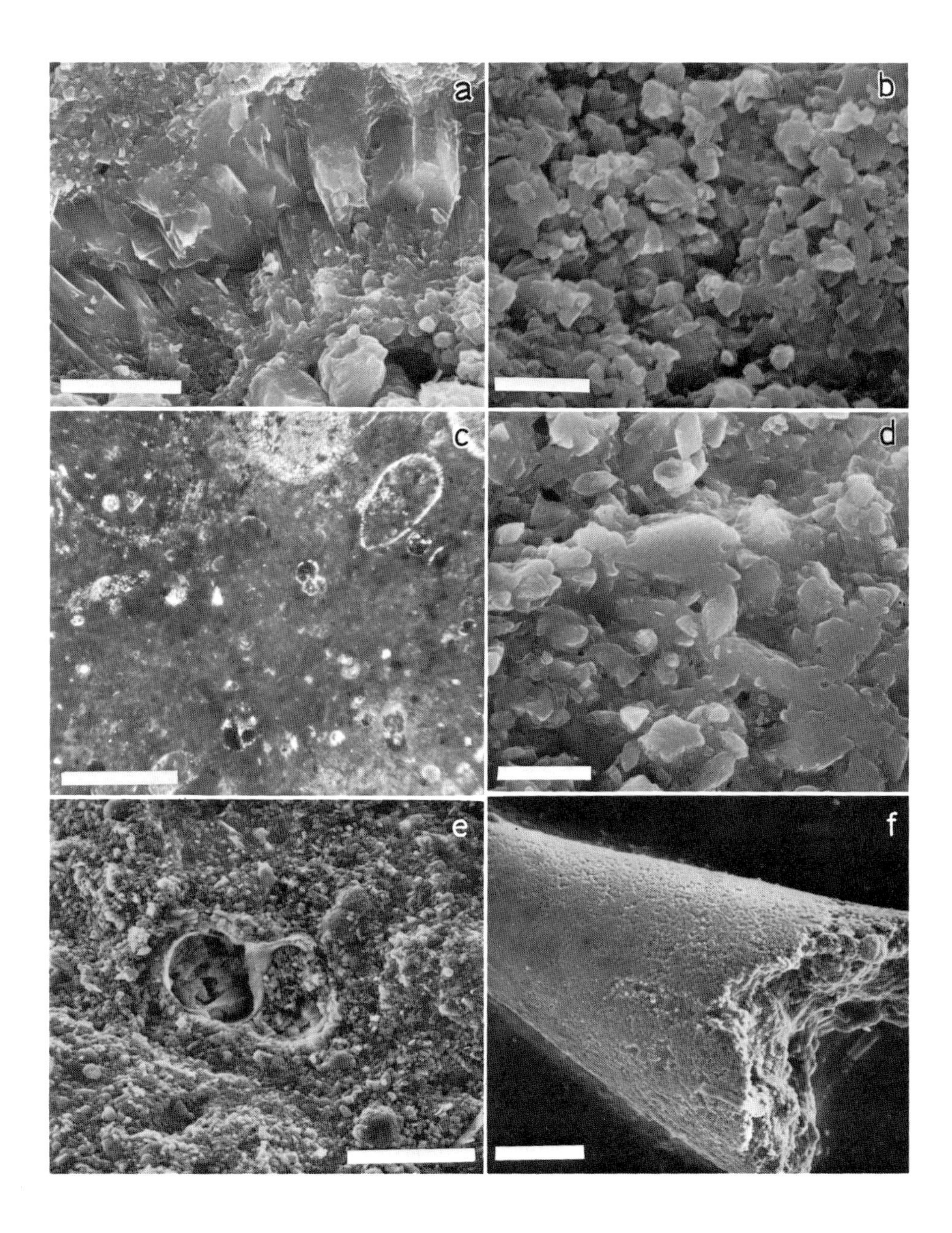

Plate XXXVIII

Table 77. Deep-sea limestones from semi-enclosed basins

Worker	Area	Water depth (m)	Petrography
FISCHER and GARRISON	Eastern Mediterranean	2055	heavily indurated planktonic limestone
MILLIMAN and MÜLLER	Eastern Mediterranean	1150	dense planktonic foram. limestone fragments
MILLIMAN and others	Red Sea	800–2700	dense planktonic foram. limestone
		800–2700	porous pteropod limestone; some matrix
MÜLLER and FABRICIUS	Ionian Sea	3115	dense planktonic foram.-pteropod crusts, 1 to 4 cm thick

highly acidic gases (such as CO_2) to the water. Whether an eruption would cause immediate carbonate precipitation (KANIA, 1929) or whether the acidic state of the surrounding waters would cause dissolution followed by reprecipitation, is not known. The fact that many "volcanic" limestones also contain abundant quantities of altered volcanic tuff suggests that precipitation may be related to alteration of the volcanic material (ZEN, 1959). Such a process is also inferred by the presence of calcite crystals found within the vesicles of altered volcanic tuffs dredged from several New England seamounts (Plate XXXVI e).

3. Semi-Enclosed Basins. Lithification of deep-sea sediments in open-ocean basins is mostly restricted to areas of non-deposition or areas exposed to volcanism; most Quaternary deep-sea sediments are unlithified. In at least two semi-enclosed basins, the Red Sea and the Mediterranean Sea, however, precipitation and lithification of carbonates have occurred.

Bottom waters of the Red Sea are supersaline (40–41‰) and warm (greater than 20 °C). These unusual conditions offer far different environments for carbonate sedimentation that those found in the open ocean. Several types of lithic crusts occur in the Red Sea (GERVITZ and FRIEDMAN, 1966; MILLIMAN and others, 1969). Pteropod layers, cemented by fibrous and cryptocrystalline aragonite, are found in most cores (Plate XXXIX), and are believed to represent a stratigraphic horizon, 11000 to 20000 years old (MILLIMAN and others, 1969) (Table 77). This layer probably precipitated during the last lower stand of sea level, during which the Red Sea became isolated from the Indian Ocean and the salinity rose substantially above its present 40‰ (HERMAN, 1965; BERGGREN and BOERSMA, 1969; DEUSER and DEGENS, 1969; MILLIMAN and others, 1969).

Other Red Sea lithic layers are rich in planktonic foraminifera and are cemented with a dense magnesian calcite matrix (12 mole % $MgCO_3$) (Plate XXXVIII c). These layers lie both above and below the aragonitic layer and are more or less in isotopic equilibrium with present-day ambient bottom waters (DEUSER and DEGENS, 1969). These data suggest that lithification of the magnesian calcite

Table 77. Continued

Cement	Mineralogy	Age
cryptocrystalline and sparry matrix	Mg-calcite (5.5–7.5)	Late Miocene
cryptocrystalline matrix and infillings	Mg-calcite (9–12)	Pleistocene
cryptocrystalline rim cement and matrix	Mg-calcite (12)	Late Pleistocene-Holocene
radiating and interlocking fibrous cement	Aragonite	11–20 thousand yrs
cryptocrystalline matrix cement; some druse	Mg-calcite (12)	Quaternary?

occurred in about present-day conditions. Similarly, inorganically precipitated magnesian calcite lutite is a dominant sedimentary component in the Red Sea deep-sea basin (MILLIMAN and others, 1969). Why magnesian calcite (and not aragonite) precipitates under such conditions is not known, but it is interesting to note that the aragonite cement in the 11–20 thousand year layer occurs in the presence of aragonitic pteropods, while the magnesian calcite occurs in those layers rich in calcitic planktonic foraminifera. This observation would agree with the suggestion by GLOVER and PRAY (1971) that mineralogy of an inorganically precipitated cement depends upon the mineralogy of the host (see p. 277).

Lithic crusts also have been reported from the Mediterranean Sea (NATTERER, 1894; BØGGILD, 1912; DE WINDTH and BERWERTH, 1904; FISCHER and GARRISON, 1967; BLANC, 1968; MILLIMAN and others, 1969; MÜLLER and FABRICIUS, 1970; MILLIMAN and MÜLLER, 1973). Most crusts are thin, usually less than 3 cm, and often break into small fragments when cored. Petrographically and chemically these crusts appear to be closely similar to the magnesian calcite crusts from the Red Sea (Table 77). Similarly, some of the lutite within the crust-rich layers is magnesian calcite, but in contrast to the Red Sea, much of the lutite in the non-lithified layers is composed of calcitic coccoliths (MILLIMAN and MÜLLER, 1973) (see Chapter 8). Thus, the lithification and precipitation of magnesian calcite in the Mediterranean appears to have been more sporadic than in the Red Sea. The mechanism causing such periodic lithification is not known, although it may be connected with fluctuating hydrographic conditions during the Quaternary (MILLIMAN and MÜLLER, 1973).

4. Deeply-Buried Deep-Sea Limestones. Although limestones in semi-enclosed basins may be locally important, many workers have down-graded the importance of lithification in the deep-sea. HAMILTON (1959) suggested that the second seismic layer in deep-sea ocean basins could be a lithified carbonate ooze, but this suggestion generally was not accepted. It has been only with the recovery of long cores by the Deep-Sea Drilling Project (JOIDES) that the lithification of deep layers has been documented. Although these lithified layers probably do not account for more than 5 to 10% of the total carbonate recovered in the three years of drilling, the obiquity of such occurrences is noteworthy.

Several types of limestones have been found and these probably reflect several types of lithification. For example, brecciated limestones recovered at Site 53 in the northwestern Pacific consist of nannofossils (coccoliths) within an anhedral to euhedral calcitic matrix (PIMM and others, 1971). Extensively recrystallized limestones contain poorly preserved nannofossils, resulting in a mosaic of secondary calcitic intergrowths. These brecciated limestones lie close to the basaltic basement, suggesting that they were lithified during thermal metamorphism (PIMM and others, 1971). Some deep basaltic layers contain cracks and crevices which are filled with primary calcite (W. BRYAN, 1971, oral communication); these may have undergone similar cementation. Stable isotope analysis of deep limestones from the central Caribbean also indicates lithification at elevated temperatures (ANDERSON and SCHNEIDERMANN, 1972).

Other limestone layers apparently have not been associated with igenous activity. For example, WISE and KELTS (1971) report an Oligocene limestone layer from the South Atlantic which is interbedded with unconsolidated calcareous oozes. The limestone is composed of coccoliths (mostly *Braarudosphaera rosa*) which have been cemented by overgrowths of euhedral calcite crystals, up to 10 microns in length. Stable isotope data infer that lithification occurred at ambient deep-sea temperatures, but no specific origin has been suggested. Perhaps overburden pressures and resulting solution-welding cementation may explain such occurrences (DAVIES and SUPKO, 1973). Other coccolith limestones are associated with dark layers and may have been lithified under reducing conditions (P. R. SUPKO, 1971, oral communication).

While most deeply-buried limestones have calcitic cements, PIMM and others (1971) report a fragment of Cenomanian cryptocrystalline limestone in which some of the cement may be aragonite. The crystals are in the form of rounded disks (PIMM and others, Plate XXVII, Fig. 3), similar to those noted elsewhere in this book.

5. *Methane-Derived Cements.* Quartzitic limestones with exceedingly low δC^{13} ratios have been dredged from the outer continental shelf and upper continental slope of the eastern United States (HATHAWAY and DEGENS, 1968, 1969; ALLEN

⟶

Plate XXXIX. Red Sea Aragonitic Limestones. (a) Multi-tiered layers of aragonite-cemented limestones occur throughout the Red Sea. They were deposited and cemented during the last lower stand of sea level, 12 to 20 thousand years ago, and as such, serve as an excellent stratigraphic horizon throughout the Red Sea. (b) Piston coring usually fragments these lithic layers, thereby recovering a mixture of fragments and mud. Scale is in centimeters. (c) Photomicrograph of the matrix within an aragonitic limestone from the Red Sea. Scale is 150 microns. (d) In contrast to the importance of calcite microfossils in magnesian calcite limestones, pteropods serve as the centers for aragonite cementation. Aragonite needles grow both inward and outward from the pteropod shells. SEM; scale is 200 microns. (e) SEM photomicrograph showing the intergrowth of aragonitic needles radiating from a pteropod shell (upper right) and the aragonitic matrix (lower left). Scale is 25 microns. (f) Close-up of the aragonite needles. SEM; scale is 10 microns

Plate XXXIX

and others, 1969). Cements are mostly aragonite, both rim and matrix cements, although magnesian calcitic cryptocrystalline cements are found in limestones dredged from Georges Bank (Table 78). The exceedingly light C^{13} composition of these limestones (δC^{13} values as low as $-80\,^0/_{00}$ have been reported) suggest a methane source for the carbon present in these limestones. This can occur either through the reduction of sulfate, such as has been suggested for calcites found in native sulfur and salt dome rocks (Thode, 1954; Feely and Kulp, 1957; Cheney and Jensen, 1967; Davis and Kirkland, 1970):

$$CaSO_4 + (C + 4\,H) \rightleftharpoons H_2S + CaCO_3 + H_2O;$$

or by the oxidation of organic matter (Spotts and Silverman, 1966). The absence of sulfate deposition in this area, together with the presence of relict shallow-water swamps (Allen and others, 1969), suggests that oxidation is more important than sulfate reduction. A further evidence for the relict source of the carbonate is the fact that the cement in the beachrock found by Allen and others (1969) is considerably older (15600 years B. P.) than the component organisms (4390 years). Similar origins can be offered for relict dolomites that occur in these same areas (Deuser, 1970; see below).

Summary

Submarine lithification occurs throughout the ocean, from the intertidal zone to the deep sea. The physical and chemical properties of the various types of lithified rocks, however, differ markedly (Table 79). Intertidal limestones tend to be relatively porous and cemented with either fibrous aragonite or cryptocrystalline magnesian calcite rim cements. Many workers have suggested that evaporation is the controlling mechanism involved in intertidal cementation.

Most shallow-water limestones have a fibrous rim or a pelletal matrix cement. Aragonitic cements are present (especially as a fibrous cement), but magnesian cements (usually cryptocrystalline) with ranges of 12 to 21 mole % $MgCO_3$, are more common. Although the cementing process is not understood, the fact that shallow-marine limestones tend to be out of isotopic equilibrium with ambient waters (Table 73) suggests that physiochemical processes may not control lithification. In this regard, it is interesting to note the similarity in δC^{13} values in shallow-water limestones, ooids and cryptocrystalline crusts (generally around $+4\,^0/_{00}$; Lloyd, 1971). Also, as mentioned above, the intergranular cements in shallow-water limestones seem similar to many intragranular cements in void fillings and in non-skeletal carbonates. Of special interest is the wide-spread occurrence of aragonitic disks (see Plates III, XXXI, XXXVI). Susumu Honjo (1970, oral communication) reports precipitating similar crystal clusters in sea water containing organic media. This observation, together with the petrographic, compositional and isotopic data presented throughout this book, suggest that the precipitation of non-skeletal carbonates, cryptocrystalline intragranular cements and at least some shallow-water cements may involve some form of organic complexing. Several workers have discussed the role of algal catalysis in cementation (Mitterer, 1968; Lloyd, 1971), and petrographic observations show that mucilagen-

Table 78. Methane-derived submarine cements

Worker	Area	Water depth (m)	Petrography	Cement	Mineralogy	Mole % $MgCO_3$ in calcite	Stable isotopes	Age (yrs)
HATHAWAY and DEGENS	Georges Bank	60	Massive terrigenous sandstones and nodules	cryptocrystalline	Mg-calcite	12	$\delta O^{18}(+1.2-+3.1)$ $\delta C^{13}(-53.6--60.6)$	
	northeastern U.S. slope	320–440	Massive terrigenous sandstones	thick layers of radiating crystals and cryptocrystalline cement	aragonite		$\delta O^{18}(+2.9-+4.9)$ $\delta C^{13}(-23.1--60.0)$	20000
ALLEN and others	Mid-Atlantic U.S. shelf	79	Shelly sandstone	rim cement and cryptocrystalline fillings	aragonite		$\delta O^{18}(+1.0)$ $\delta C^{13}(-44.8)$	10500

Table 79. Summary of various limestones formed in the marine environment

Rock type	Components	Cement	Mineralogy	Mole % $MgCO_3$ in calcite	Matrix	Isotopes	Possible mechanisms
Intertidal	shallow-water	fibrous micritic	— aragonite — Mg-calcite	12–15	some	disequilibrium	evaporation
Subtidal	shallow-water	fibrous	aragonite		frequent	disequilibrium	decomposition of organic matter
		cryptocrystalline, pelletal	Mg-calcite	12–21			
Deep-sea	planktonic and deep-water benthonic	drusy	Mg-calcite	11–13	infrequent to very much	equilibrium disequilibrium	1. non-deposition 2. elevated-salinities 3. volcanism
		cryptocrystalline to drusy	Mg-calcite, calcite	11–13			

ous algae are not only common in the matrices of recent limestones, but also are of major importance in ancient ones (SHEARMAN and SKIPWITH, 1965; PURDY, 1968; KENDALL and SKIPWITH, 1969; SCHROEDER and GINSBURG, 1971). Cryptocrystalline cements are suggestive of rapid precipitation (SCHMALZ, 1970), and LAND and GOREAU (1970) report that cementation of Jamaican sediments can be very rapid. DE GROOT (1969), however, finds that the interstitial waters in Persian Gulf sediments are undersaturated with respect to magnesian calcite, inferring that cementation is periodic, occurring only when the waters are sufficiently supersaturated.

Deep-water limestones tend to contain planktonic and deep-water benthonic components. Aragonitic cements are very rare in the deep sea; magnesian calcite and calcite predominate. The general absence of deep-sea aragonite may be due to the inhibiting effects of low temperatures and relatively low pH's on aragonite precipitation (WRAY and DANIELS, 1957; ZELLER and WRAY, 1956; GOTO, 1961; KINSMAN and HOLLAND, 1969). The nearly all magnesian calcites contain about 11 to 13 mole % $MgCO_3$, and many deep-sea limestones are in isotopic equilibrium with the ambient deep waters. Such limestones are found in areas of non-deposition or in semi-enclosed basins. Another type of deep-sea limestone, composed only of calcite, is found on and near submarine volcanoes. Of the few volcanic limestones analyzed, most are out of isotopic equilibrium with the deep ocean, suggesting that precipitation may have occurred as a direct result of the volcanism, or may have resulted from the subsequent subaqueous weathering of volcanic rocks.

One of the most important points to recognize in discussing submarine lithification is the great amount of sea water that must pass through the sediment. If all the calcium within a given volume of sea water were precipitated, it would take approximately 1000 volumes of sea water to precipitate one volume of $CaCO_3$. Since only a small portion of the available Ca^{+2} and $CO_3^=$ in sea water is utilized during precipitation, the amount of sea water washed through the voids must be considerably higher if the calcium and carbonate were derived only from sea water. Probably some or perhaps most of the $CaCO_3$ is derived from *in situ* dissolution and reprecipitation; but the fact that many deep-sea limestones are in isotopic equilibrium with respect to C^{13} and O^{18} suggest that much of the $CO_3^=$ is derived from ambient sea water. In beachrocks large volumes of water can seep through the porous grains, but in submarine limestones the passage of water is more difficult. Perhaps burrowing animals could provide such a mechanism for water passage. Many of these organisms are filter feeders and thus generate their own flushing currents. This would certainly explain the thoroughly burrowed and reworked nature of many deep-sea limestones.

Many workers have minimized the importance of submarine lithification (for example, DUNHAM, 1969; PURDY, 1968), and in the modern deep sea submarine lithification does appear to be quantitatively unimportant. JOIDES deep-sea drillings throughout the Pacific and Atlantic Oceans have encountered many limestone layers, but these probably do not constitute more than 5 to 10% of the total deep-sea carbonate oozes. In some holes, rhodochrosite, siderite and dolomite are more prominent cements than calcite. Thus, although submarine cementation is obiquitous, it probably never has been dominant in the deep sea.

The only areas in which modern submarine lithification is quantitatively important are the semi-enclosed Red Sea and eastern Mediterranean Sea basins. These basins, in which conditions are sufficiently restricted to facilitate inorganic precipitation and lithification of calcium carbonate, may be analogous to the epicontinental basins in which many of the Paleozoic and early Mesozoic carbonates were deposited. If one also considers that prior to the evolution of planktonic foraminifera and coccolithophorids the oceans were probably more supersaturated with respect to calcium carbonate (and temperatures were warmer; EMILIANI, 1954), then one can assume that submarine cementation would have been more likely to occur. Thus, the fact that submarine cementation is not common in many modern environments does not necessarily preclude it from having been an important diagenetic agent in ancient seas.

Chapter 11. Dolomitization

Few geological problems have received more attention and debate than the origin of dolomite. As a result of early studies, two "classic" schools of thought emerged concerning dolomite genesis: 1. primary precipitation of dolomite crystals, and 2. the secondary replacement of other minerals by dolomite. In recent years this controversy has been "... largely supplanted with the argument, penecontemporaneous or post-lithification replacement. For dolomitization is a diagenetic process" (MURRAY and PRAY, 1965). Such a strongly worded opinion probably has found almost unanimous support amongst most geologists, but as will be seen below, there is still reason to suspect that some dolomites are primary. In keeping with the theme of the book, this review is restricted to modern marine (together with some lacustrine) dolomites; for more complete discussions of the dolomite problem, the reader is referred to the excellent reviews of VAN TUYL (1918), FAIRBRIDGE (1957), FRIEDMAN and SANDERS (1967) and BATHURST (1971).

Until recently speculation into the origin of dolomite was limited to the study of ancient rocks since modern analogues were unknown (most workers seem to have overlooked the work of Mawson, who in 1929 reported dolomitic sediments in the lakes of South Australia). The failure to find modern dolomite can be partly explained by the very small size of most evaporitic dolomite crystals, thereby making petrographic recognition nearly impossible. As FRIEDMAN and SANDERS (1967) have pointed out, it was only with increased use of X-ray diffraction techniques in the 1950's and 1960's that widespread occurrences of modern dolomites were documented.

Three distinct types of dolomites are recognized: intertidal and supratidal, organic, and deep-sea. Each will be discussed in the following paragraphs.

Intertidal and Supratidal

Workers have known for many years that fossil dolomites often are associated with evaporitic deposits. KLEMENT (1895) realized that the association of dolomite with gypsum (and anhydrite) and aragonite must be critical in dolomite genesis. Similarly, CLOUD and BARNES (1948) hypothesized evaporitic environments for the dolomites found within the Ellenberger Formation in Texas. Thus it was not too surprising to find modern dolomites occurring in intertidal and supratidal zones within evaporitic areas.

Dolomites have been found in the beds of many dry lakes and on the banks of many highly saline lakes in such places as South Australia (MAWSON, 1929; ALDERMAN and SKINNER, 1957; ALDERMAN and VON DER BORCH, 1961), Russia

Table 80. Depositional environments and chemical properties of some modern marine dolomites. Data are from references quoted within the text

Location	Environment	$S^{0}/_{00}$	Maximum (molar) Mg/Ca in brines	Dolomite crystal size (μ)	Mole % $MgCO_3$ in dolomite	Age (yrs)
1. *Evaporitic*						
Pekelmeer, Bonaire	salt pan	>200		2	44–46	1480±140
Qatar, Persian Gulf	supratidal flat	>200		1–5	45–47	2450±130
Andros Island, Bahamas	supratidal flat	150–180	>40/1	1–2	44	0–160
Great Inagua Island	salt pan	>260	>600/1	<140 <5	40	<8(?) 2930–3420
Sugarloaf Key, Florida	supratidal flat	150–180	>40/1	2–3	30–44	250–600
Coorong, South Australia	supersaline lagoon and lakes	25–>200	4–16/1	<20	45–50	300±250
Abu Dhabi, Persian Gulf	supratidal flats	>250	>35/1	1–2		
Jarvis Atoll	brine pool		>35/1	?	calcium-rich	2650±200
2. *Biologic-Biochemical*						
Echinoid teeth					41	living animal
Coralline algae (?)	Canary Islands	?	?	?	40	recent
East Coast U.S.A.	continental shelf and slope	?	?	?	40	Quaternary
3. *Submarine*						
Mid Pacific Rise	deep-sea	?	?	50	?	?
Mid Atlantic Ridge	volcanic?	?	?	30–80	50	?

(TEODOROVICH, 1946; ZAHLMANSON, 1951), the western United States (EARDLEY, 1938; GRAF and others, 1961; JONES, 1961; PETERSON and others, 1963, and other references cited in FRIEDMAN and SANDERS, 1967) and Turkey (IRION and MÜLLER, 1968; MÜLLER and IRION, 1969). Marine evaporite dolomites have been found in the Persian Gulf (WELLS, 1962; ILLING and others, 1965; CURTIS and others, 1963), Bonaire Island (DEFFEYES and others, 1964, 1965; LUCIA, 1968), the Coorong area of South Australia (ALDERMAN and SKINNER, 1957, and references cited below), Andros Island (SHINN and others, 1965, 1969), South Florida (SHINN, 1968b;

ATWOOD and BUBB, 1970), British Honduras (EBANKS and TEBBUTT, 1966; EBANKS, 1967), Great Inagua Island (MILLER, 1961; BUBB and ATWOOD, 1968), the Canary Islands (MÜLLER and TIETZ, 1966), Baffin Bay, Texas (BEHRENS and LAND, 1970) and Jarvis Atoll in the southern Line Islands (SCHLANGER and TRACEY, 1970).

In each of these areas the sediment tends to be fine grained. Dolomite crystals are seldom greater than 20 microns in diameter, and usually less than 5 microns. Carbon-14 dates from various dolomites show that surficial dolomitic layers in many areas are less than 1000 years old (Table 80). Considering that at least some of the carbon in the $CO_3^=$ was derived from replaced carbonate minerals, these dolomites must be pene-contemporaneous. In the Persian Gulf, Bonaire and South Australia dolomite is associated with variable amounts of gypsum (and/or anhydrite), aragonite, magnesian calcite, calcite and halite. In Florida and the Bahamas the climate is not as arid, and gypsum and halite are absent. Other minerals commonly associated with evaporitic dolomite deposits include huntite ($CaMg_3(CO_3)_4$), hydromagnesite, magnesite and celestite (ALDERMAN and VON DER BORCH, 1960; KINSMAN, 1967; IRION and MÜLLER, 1968; ALDERMAN and VON DER BORCH, 1960, 1961; EVANS and SHEARMAN, 1964; EVANS and others, 1969; SKINNER, 1963); more rarely one may find polyhalite ($K_2MgCa(SO_4)_4 2H_2O$) (HOLSER, 1966; IRION, 1970).

The dolomitized mud is often pelletal and may be covered with algal mats. Leached shells and solution pits indicate that solution of calcium carbonate accompanies the dolomitization. Birdseye structures, the result of increased porosity by the weight-for-weight replacement of aragonite and calcite with denser dolomite, are characteristic of many evaporitic dolomites (SHINN, 1968a). Most evaporitic dolomites are poorly ordered ("protodolomites" of GRAF and GOLDSMITH, 1956, and GOLDSMITH and GRAF, 1958), with Mg/Ca ratios around 45/55. As a result of precipitation within saline waters, Na contents of evaporitic dolomites are much greater than in meteoric-water dolomites (LAND, 1972). Oxygen isotope values are considerably higher than those for skeletal carbonates; δC^{13} values are roughly similar to those in marine sediments (Fig. 93).

The depositional environments in which evaporitic dolomites occur are roughly similar. Water movement is restricted and evaporation exceeds precipitation. In supratidal zones and playas periodic flooring occurs during spring tides or storms (or in the case of playas, rains); salts can be marine or cyclic (VON DER BORCH, 1965b). Subsequent evaporation produces supersaline brines, either overlying the sediment or present as an interstitial water. Critical in dolomite formation is the high Mg/Ca ratio in ambient waters, the result of removal of calcium through gypsum and calcium carbonate (principally aragonite) precipitation. Mg/Ca ratios are usually greater than 10 and perhaps as great as 15 or 20 before dolomitization begins (FRIEDMAN and SANDERS, 1967; KINSMAN, 1969a; IRION, 1970) (Table 80). The abundance of sulfate within the pore waters may also aid dolomite formation (ZELLER and others, 1959). Although high CO_2 content and low pH's have been envisioned as conducive to dolomite precipitation (GRAF and GOLDSMITH, 1956; BISSELL and CHILINGAR, 1958) the pH of the interstitial water is generally greater than 8 and values can exceed 10. As pointed out by HSU (1967), there is no theoretical basis to feel that higher carbonate activities resulting from increased pH should favor dolomite precipitation, but low CO_2

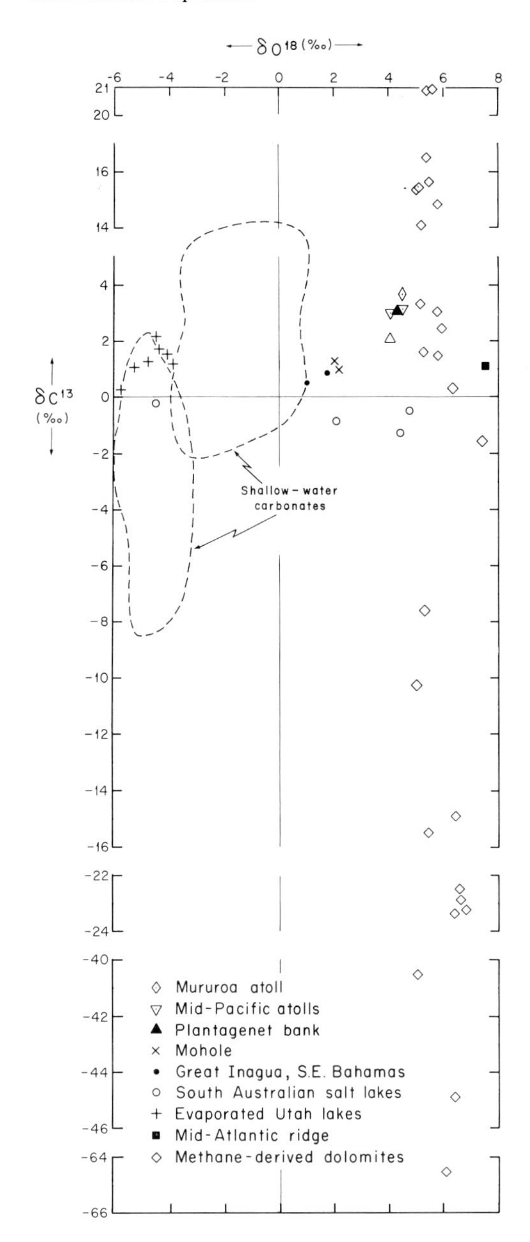

Fig. 93. Distribution of stable isotopes in various "modern" marine and evaporitic dolomites. Data are from DEGENS and EPSTEIN (1964), GROSS (1965), BERNER (1965), THOMPSON and others (1968), FONTES and others (1969) and DEUSER (1970). The fields defining the modern carbonate components come from Fig. 19

pressures may be important. LIEBERMANN (1967) has suggested the possible importance of diurnal processes, by which $CaCO_3$ is dissolved during night time (cooler temperatures and low pH) and dolomite precipitated during day time (warmer temperature and higher pH).

Although the conditions for dolomite formation appear to be similar throughout evaporitic intertidal and supratidal areas, three different modes of dolomitization have been envisioned:

1. Primary Precipitation. Except for one report of bacterially precipitated dolomite (OPPENHEIMER and MASTER, 1964) geologists have been unable to synthesize dolomite at normal earth surface temperatures and pressures. This lack of success, however, may be partly a problem of kinetics combined with the complex nature of the crystal structure of dolomite (GRAF and GOLDSMITH, 1956; HSU, 1967).

One of the main problems in delineating primary from secondary precipitates lies in the use of stable isotope compositions. Experimental data of CLAYTON and EPSTEIN (1958), EPSTEIN and others (1964) and O'NEIL and EPSTEIN (1966) suggest that precipitated dolomite should have δO^{18} values that are 4 to $7^0/_{00}$ higher than calcite co-precipitated at 25 °C. In mid-Tertiary dolomites from the mid-Pacific atolls (SCHLANGER, 1963, and references therein), BERNER (1965) found that dolomites are 6 to $8^0/_{00}$ richer in O^{18} than the co-existing calcite, and concluded that these dolomites had been precipitated from brine solutions. Until recently the fact that most dolomites seem to have δO^{18} values only slightly higher than co-existing dolomites (the richest concentrations have been reported from the salt pans of Great Inagua Island (Bahamas), where BUBB and ATWOOD, 1968, found dolomites which are 3.8 to $4.6^0/_{00}$ higher than co-existing aragonites and calcites) has been assumed to infer that either dolomite can precipitate without establishing isotopic equilibrium or that magnesium can substitute within the existing carbonate lattice, that is as a replacement rather than as a primary precipitate (EPSTEIN and others, 1964; DEGENS and EPSTEIN, 1964). However, FRITZ and SMITH (1970) have suggested that perhaps δO^{18} enrichments in dolomites are closer to $3^0/_{00}$ rather than the generally accepted 4 to $7^0/_{00}$. Extrapolation of the experimental data from TARUTANI and others (1969) would support this contention. If FRITZ and SMITH are correct, then the rejection of primary precipitation on the basis of stable isotope data must be reconsidered[1].

ALDERMAN and SKINNER (1957) and JONES (1961) reported finding primary dolomite in the salt lakes of southern Australia and the western United States. ALDERMAN and SKINNER suggested that dolomitization occurs during whitings, which are composed mostly of magnesian calcite and dolomite and which "instanteously" form near photosynthesizing plants (that presumably raise the pH and thus cause precipitation; SKINNER, 1963). Carbon-14 dates of both the south Australian and U. S. dolomites in question, however suggest that nucleation rates are much slower than what one would expect if dolomite were being nucleated as rapidly as suggested by ALDERMAN and SKINNER (PETERSON and others, 1963; SKINNER and others, 1963; VON DER BORCH and others, 1965).

In offering a compromise between primary precipitation and replacement, several workers have suggested an intermediate step by which dolomite forms through a process of slow nucleation and crystal growth. VON DER BORCH (1965a) pointed out that the maturity of the dolomites in southern Australia depends upon the degree of physical isolation of the lakes in which they occur; magnesian calcites predominate in those areas near marine influences, while the most isolated

1 Recently BEHRENS and LAND (1972) reported dolomites from Baffin Bay that contain δO^{18} concentrations approximately $3^0/_{00}$ richer than co-existing aragonite and calcite. These data, together with petrographic evidence, suggest primary precipitation.

inland lakes contain well-ordered dolomites. Perhaps the transition from magnesian calcite to dolomite is related to high pH's (as high as 10.2) found in some of these more isolated lakes (SKINNER, 1963; ALDERMAN, 1965; ALDERMAN and VON DER BORCH, 1963; VON DER BORCH, 1965a). GLOVER and SIPPEL (1967) precipitated poorly ordered magnesian calcite, which they suggested may alter into dolomite with sufficient time and wetting and drying. Such environmental parameters are common in the tidal flat and salt lakes of southern Australia.

PETERSON and others (1966) also invoked slow dolomitization by which each crystal in Deep Spring Lake (California) undergoes a progressive dolomitization, the innermost layers being dolomite, while the outermost layers are calcitic. More recently, however, CLAYTON and others (1968) reported no evidence of the outer calcite layer, nor did isotopic data suggest replacement of calcite. These workers therefore concluded that the dolomite in Deep Spring Lake may be primary.

If any one thing concerning the primary precipitation versus syngenetic replacement of dolomite is clear, it is the unresolved nature of this controversy. Isotopic data may well provide the answer, but in light of the findings of FRITZ and SMITH (1970) (see above) all isotopic data must await further re-evaluation.

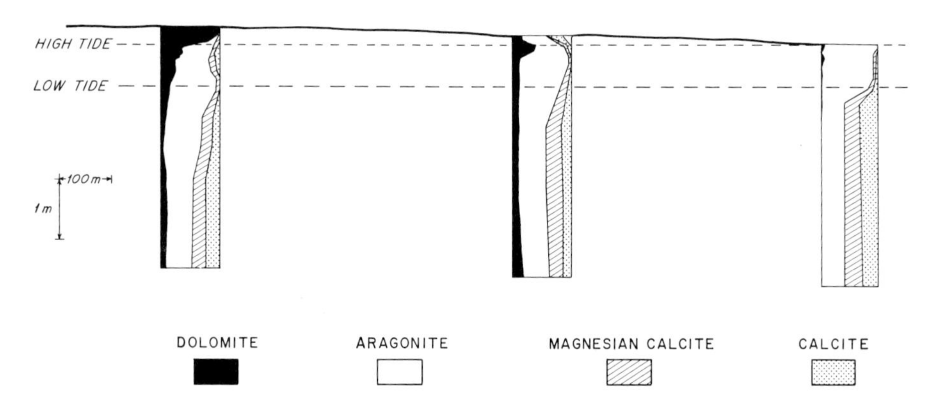

Fig. 94. Distribution of dolomite in cores across a Persian Gulf tidal flat. Note that the dolomite is almost completely restricted to the supratidal portions of the sediments. After Fig. 7 and accompanying data in ILLING and others (1965)

2. *Replacement by Interstitial Waters.* Most modern evaporitic dolomites occur on supratidal flats, such as those in the Persian Gulf, Florida and the Bahamas. In each case the dolomite is restricted to sediments above the high tide level (Fig. 94); in the Trucial Coast of the Persian Gulf this supratidal belt can be as wide as 15 km (BUTLER, 1969). In intertidal and subtidal depths the sediment is largely calcium carbonate. In Florida and the Bahamas algal mats cover most of the dolomite deposits (SHINN and others, 1965, 1969), but in the Persian Gulf sabkhas algal mats are not always present where dolomite is found (KENDALL and SKIPWITH, 1968). Another difference can be seen in the well ordered Qatar

Peninsula (Persian Gulf) dolomite, with Mg contents between 45 and 47 mole % (ILLING and others, 1965), as opposed to the dolomites found in Florida and the Bahamas, which are poorly ordered and have Mg contents generally less than 44 mole %. In most supratidal arcas dolomite is restricted to the surface sediment or crust, but CURTIS and others (1963) report that dolomite content at Abu Dhabi increases with depth, as aragonite decreases.

In spite of these apparent differences, the origins proposed by various workers for these dolomites are roughly similar. Periodic floodings supply water to the supratidal flats. The waters sink into the sediment, but subsequent capillary action, brought about by high evaporation rates at the surface, causes an upward movement of pore water. Subsequent evaporation results in brine formation with salinities commonly in excess of $200^0/_{00}$ (Table 80). Aragonite (or magnesian calcite) and gypsum are precipitated; when the Mg/Ca ratio reaches sufficiently high levels, dolomite diagenetically replaces the carbonate. As aragonite is replaced by dolomite, Mg content of the pore waters falls and Ca rises until the Mg/Ca ratio reaches levels less than that of sea water (KINSMAN, 1966, 1969a; BUTLER, 1969). Celestite, which is associated with many of these deposits, is probably precipitated as a result of the surplus Sr made available by the dolomitic replacement of Sr-rich aragonites (EVANS and SHEARMAN, 1964). Dolomitization generally is more rapid and complete in fine grained sediments. In coarser sediments, dolomitization may be inhibited and huntite precipitated (KINSMAN, 1966, 1967).

FRIEDMAN and SANDERS (1967) have suggested that "capillary concentration by evaporation of upward migrating interstitial waters is an important mechanism in dolomitization", and subsequent experiments by HSU and SIEGENTHALER (1969, 1971) have substantiated such an "evaporitic pumping". This process is similar to the pore water migration in caliche formation.

KINSMAN (1969a; KINSMAN and PARK, 1969) has made the observation that dolomitization is limited to those evaporitic areas in which carbonate sedimentation predominates. In areas where rapid terrigenous sedimentation is dominant, such as Peru (MORRIS and DICKEY, 1956) and the coastal lagoons of Baja California (PHLEGER and EWING, 1962; KINSMAN, 1969a), magnesite and magnesium calcite (or aragonite) may be the only carbonate minerals. Temperature also seems to be critical; Lake Barney in Antarctica has bottom salinities exceeding $300^0/_{00}$ and Mg/Ca ratios up to 20/1, but only gypsum and aragonite have been reported from the bottom sediments (ANGINO and others, 1964). Perhaps the very low temperatures (less than −1 °C near the bottom) have prevented dolomite precipitation[2]. The absence of dolomites in the tidal and supratidal flats in Hutchinson Embayment, western Australia, where pore water salinities reach nearly $300^0/_{00}$ and Mg/Ca ratios can be as high as 24, is more perplexing (HAGEN and LOGAN, 1971).

3. Seepage Refluxion. DEFFEYES and others (1964, 1965) have proposed that the seepage of dense lagoonal brines into the sea through porous limestones have dolomitized the pre-existing limestones on Bonaire. First the mud matrix,

2 Theoretical considerations together with available field data have led LOVERING (1969) to predict that at 25 °C, Mg/Ca ratios must exceed 5 for dolomite formation, while at 0 °C, the Mg/Ca ratio must exceed 16.

then the fecal pellets and finally the mollusks within the sediment are dolomitized (LUCIA, 1968). The replacement process is probably similar to that mentioned above, but the driving mechanism is different. This process is directly analogous to the seepage refluxion theory proposed by ADAMS and RHODES (1960). Apparently seepage refluxion is not applicable to all areas (or at least is not a rapid process); BUBB and ATWOOD (1968) found no evidence of the dolomitization of Pleistocene strata underlying commercial salt pans in Great Inagua Island.

Organic Dolomites

Many marine organisms precipitate magnesian calcite. During diagenesis the magnesium may be liberated and thus form magnesium-rich interstitial waters which subsequently may be precipitated as dolomite. At present, only one organism is known to precipitate dolomite: the dense axial zone of echinoid teeth contains protodolomite, with an average molar concentration of 41% $MgCO_3$ (SCHROEDER and others, 1969). Although WASKOWIAK (1962) reported Mg concentrations in serpulids exceeding 10% (in the range of protodolomite) subsequent analyses (BORNHOLD and MILLIMAN, 1973) indicate that these data are erroneous. The association of dolomites with coralline algae is thought to have diagenetic implications (SCHLANGER, 1957; GROSS, 1965; MARLOWE, 1971b). The finding of large quantities of poorly-ordered protodolomite (about 40 mole % Mg) in recent algal nodules in the Canary Islands (J. MÜLLER, oral communication) indicates that replacement may be penecontemporaneous.

Organic processes may also affect dolomite formation in other ways. Supratidal dolomites at Andros Island are usually associated with stromatolitic algal mats (SHINN and others, 1969). The blue-green algae within the mats are capable of concentrating Mg, bringing the Mg/Ca ratio to 3 or 4 times that of sea water, thus offering an additional source of Mg for dolomitization (GEBELEIN and HOFFMAN, 1971). This would also explain why some ancient stromatolites are only dolomitized in their darker layers. FRIEDMAN and others (1972) report algal mats in hypersaline marginal pools in the Gulf of Aqaba (Red Sea) which contain very high magnesian calcite (as high as 40 mole % $MgCO_3$). If the organically-complexed Mg is included, these laminites conceivably could reach 60 mole % $MgCO_3$. Dolomitization of such sediments is not difficult to imagine.

Several workers have suggested bacterial action in dolomitization (NEHER and ROHRER, 1958; INGERSON, 1962), but the mechanism of precipitation is not clear. At present the only reported laboratory precipitation of dolomite at room temperatures is that of OPPENHEIMER and MASTER (1964) who found that "algal mat" bacteria could cause such precipitation; particulars about this experiment, however, are lacking. LA LOU (1957a, 1957b) found that bacteria can produce magnesium-rich carbonates, but did not state whether this was exclusively magnesian calcite or a mixture of calcite and dolomite.

Dolomitic layers have been penetrated by piston cores on the upper continental slope off Oregon (RUSSELL and others, 1967). These dolomites are poor in magnesium (41 mole %) and have anomalous isotopic compositions (δO^{18}, -5.8‰; ∂C^{13}, -35.1‰). Similar calcium-rich dolomites have been dredged from the upper continental slope off the northeastern United States. These dolomites exhibit δC^{13} compositions ranging from -60 to $+21^0/_{00}$; δO^{18} values are mostly between $+5$ and $+7^0/_{00}$ (DEUSER, 1970) (Fig. 93; Table 80). DEUSER (1970) has suggested that methane produced by the oxidation of nearby organic matter (perhaps Pleistocene marsh peat) is the source of the carbon within the dolomite (see Spotts and SILVERMAN, 1966). The fact that these dolomites have ∂O^{18} values only slightly higher than co-existing methane-derived aragonites (HATHAWAY and DEGENS, 1968, 1969) may imply that the dolomites are an alteration product rather than a primary precipitate. The extreme range in ∂C^{13} values along the northeastern United States upper slope may represent a progressive utilization of heavier methane during dolomitization.

Deep-Sea Dolomites

Numerous occurrences of dolomite grains in deep-sea sediments have been reported (BØGGILD, 1912, CORRENS, 1939, FAIRBRIDGE, 1957; BRAMLETTE, 1961; FRIEDMAN, 1964; HATHAWAY and SACHS, 1965; MILLIMAN and MÜLLER, 1973), but the fact that these grains are commonly perfect rhombs (Plate XXVIII f) does not necessarily mean that they were formed *in situ*. For instance, the rhombic grains reported by TAFT (1961) from Florida Bay are deficient in Carbon-14, suggesting that the grains are detrital (DEFFEYES and MARTIN, 1962) Similarly many deep-sea dolomite crystals may be detritial, having been transported to open ocean areas by winds or currents.

In contrast, the dolomite crystals found by BONATTI (1966) are associated with a carbonate coze that is assumed to have been inorganically precipitated in association with submarine volcanism. These rather large rhombs (greater than 40 microns in diameter) probably formed *in situ*. Dolomitic-rich layers in the eastern Mediterrarean may be authigenic (MILLIMAN and MÜLLER, 1973). These grains, which range from 10 to 30 μ in size (Plate XXVIII f), probably are late Pleistocene in age.

Dolomite crystals found near the bottom of the exploratory Mohole test drill hole (RIEDEL and others, 1961; MURATA and ERD, 1964) lie directly above a basalt layer that may be intrusive, thus suggesting hydrothermal metamorphism. Recent Deep-Sea Drilling Project data show that dolomite-rich layers are ubiquitous but not common in either the Atlantic or Pacific Oceans (Plate XXVIII e) (FISCHER and GEALY, 1969). The world-wide distribution of such layers, however, suggests that some common diagenetic process may be critical in the formation of deep-sea dolomite.

PIERCE and MELSON (1967) have reported a dolomitic rock with large crystals (up to 50 microns) dredged from 1900 m depth off Southern California. Although it is possible that this rock may represent the seaward facies of a sedimentary formation, it seems improbable that lithification or dolomitization occurred subaerially. The best evidence for submarine dolomitization comes from THOMPSON and others (1968) who found dolomitic rocks on the Mid-Atlantic Ridge (depth somewhere between 1700 and 2500 m). The extremely high δO^{18} content ($+7.6^0/_{00}$) of the dolomite suggests formation within an extreme environment, perhaps submarine volcanism [3]. This dolomite is composed of large (30 to 80 microns), well-ordered crystals, with nearly equal molar concentrations of Mg and Ca.

In addition to deep-sea volcanism, DAVIES and SUPKO (1973) mention two other types of deep-sea dolomites. One type, occurring in the Mediterranean and in site 138 off western Africa, apparently is related to evaporitic conditions or to the transport of evaporitic fluids. The other dolomite type is associated with pyrite and probably formed in reducing conditions.

Summary

The three types of marine dolomites discussed in this chapter can be differentiated from one another on the basis of their crystal, chemical and isotopic properties. Evaporitic dolomites are composed of small crystals (commonly less than 5 microns), are generally poorly-ordered protodolomites and have widely variable but usually high δO^{18} values. In terms of both modern and ancient deposits, evaporitic dolomites appear to be the dominant type. "Biogenic" and bacterial dolomites are not well documented, but those dolomites which have been related to either biogenic deposition or bacterial catalysis are usually protodolomites with very low magnesium contents (40 to 43 %). Methane-derived dolomites have the widest spread in δC^{13} values of any depositional carbonate yet studied (Fig. 93). Some dolomites found in the deep sea appear to be formed by volcanic activity, although other processes may be important. Deep-sea dolomites can occur as individual sedimentary crystals or as dense rocks; in both instances the crystals tend to be large (20 to 80 microns) and well-ordered, with nearly equal amounts of magnesium and calcium. The large crystal size implies a slow rate of crystallization as opposed to the rapid rate implied for the smaller "evaporitic" dolomites (FÜCHTBAUER and MÜLLER, 1970).

Although the line delineating primary from penecontemporaneous is a thin one, most workers feel that intertidal and supratidal dolomites are replacements of existing carbonates rather than primary precipitates. Some data, however, support primary precipitation. Deep-sea dolomites also may be products of primary precipitation.

3 Subsequent analyses reported by THOMPSON (1972) show one deep-sea dolomite with a δC^{13} of $-16.1^{0}/_{00}$, suggesting an organic origin similar to that discussed in the previous section.

Appendix I

Key to the Identification of Carbonate Components in Thin Section

I. Opaque to semi-opaque under polarized light
- A. Swiss-cheese appearance (Plate IX) — *Halimeda*
- B. Cellular appearance; exact shape and configuration depends upon orientation of section. Individual cell size about 5 microns (Plate VI) — **Coralline algae**
- C. Cellular appearance with relatively large (5 to 25 microns) chambers. Shape may coil or be otherwise diagnostic (Plate X) — **Porcellanous foraminifera**

II. Birefringent under polarized light
- A. Characteristic petrographic properties under polarized light
  1. Laminations present
     - a) Well-defined, often varying during stage rotation (Plate XIV) — **Mollusks**
     - b) Not as well laminated (Plate XII; Fig. e–g); shape may be diagnostic — **Serpulids**
     - c) Possibly laminated, but poorly so; granular texture, gray under polarized light; longitudinal canals usually are prominent (Plate XV) — **Barnacles**
  2. Laminations absent
     - a) Unit extinction upon stage rotation; hollow or spongy in thin section (Plate XVI) — **Echinoderms**
     - b) Bundle extinction under polarized light; feathery appearance under plane light (Plate XII) — **Corals**
     - c) Extinction lines normal to the grain edge (Plates X, XI) — **Hyaline foraminifera**
     - d) Granular texture; exact identification depends upon other properties — **Mollusks, Barnacles or Bryozoans**
- B. No notable characteristics under polarized light
  1. Granular shell (usually) traversed by longitudinal canals (shape depends upon orientation of the section (Plate XV) — **Barnacles**
  2. Granular shell often with a "fractured" appearance; lower birefringence than B 1; cells (zooecia) characteristic (Plate XIII) — **Bryozoans**
  3. Tubes, weakly to strongly laminated (Plate XIII e–g) — **Serpulids**

III. Non-skeletal fragments
- A. Single grain, usually round and/or spherical; no matrix surrounding it
  1. Laminations surrounding a nucleus (Plate II) — **Ooids**
  2. No visable structure; often opaque (Plate I) — **Pelletoids**
- B. Two or more grains joined by matrix
  1. Grains dominant (Plate III) — **Aggregates (grapestone)**
  2. Matrix dominant (Plate III) — **Lumps**

Appendix II

Key to the Identification of Carbonate Components. Under Reflected Light

I. Skeletal features identifiable
- A. Cells and pores on outer surface
  - 1. Carbonate is relatively porous
    - a) Small pores (30–40 microns) on a flat (often hollow) plate; commonly shaped like a club, and chalk white (Plate VII) — *Halimeda*
    - b) Tiny pores (20–60 microns) on a spongy grain, with dull luster; sometimes spine sockets (about 200 microns in diameter) are visible (Plate XVI) — **Echinoids**
    - c) Two distinct pore sizes (50–70 microns and 100–160 microns); the larger pores are often star-shaped (Plate XII a) — *Millepora*
  - 2. Carbonate is more solid in appearance, but contains numerous pores or cells
    - a) Red to pink color; often encrusting other carbonates; pores about 50 microns in diameter (Plate XI b) — *Homotrema*
    - b) Individual corallites with septa often visible, each about 500 microns in diameter; smaller structural pores (100–150 microns) also seen (Plate XII b and c) — **Scleratinian corals**
    - c) Zooecia (80–150 microns in diameter) lacking the internal structure of scleractinians are diagnostic; growth forms can be encrusting or branching (Plate XIII a and b) — **Bryozoans**
- B. Curved shells
  - 1. Shell usually smooth on interior, possibly rough exterior — **Mollusks (Pelecypods and Gastropods)**
  - 2. Interior of shell usually has conspicuous longitudinal canals (Plate XV a and b) — **Barnacle**
  - 3. Valves are of unequal size, but equilateral; pedicle is diagnostic — **Brachiopod**
  - 4. Disk-like grains (0.5–3 mm diameter) usually more ornamented on concave side (Fig. 41) — **Otoliths**
- C. Tubular shape
  - 1. Straight or slightly curved
    - a) Hollow, open at both ends
      - 1. Tapered at one end; exterior usually smooth and polished — **Scaphopod**
      - 2. Equal diameter along tube length; exterior usually chalky and dull — **Serpulid**
    - b) Solid cylinder
      - 1. Dull luster; sometimes a freshly broken surface will show a cellular structure — **Coralline algae**
      - 2. Often ribbed; varies from 0.5–4 mm in length (Plate XVI) — **Echinoid spines**

2. Coiled shell (generally less than 2 mm)
   a) Internal chambers; often planispiral and can contain external pores — **Foraminifera**
   b) No internal chambers; outer shell smooth or ribbed; often spired — **Micro-gastropods**

D. Spicules
1. 100–150 microns in length; 3–30 microns in diameter; triradial or quadriradial (Fig. 24) — **Sponges**
2. 100–400 microns in length; 50–100 microns in width; monaxial with worts and spines (Fig. 25) — **Octocorals**
3. 40–125 microns in diameter; many various shapes (Fig. 36) — **Holothurians**
4. 25–75 microns in diameter; spherical, star-shaped or spiney (Fig. 40) — **Tunicates**

II. Distinct skeletal shape absent
A. Chalky white encrustation (stains as magnesian calcite) — **Coralline algae**
B. Non-skeletal grains
1. Aggregates of numerous constituent grains (Plate IV) — **Aggregates**
2. Round polished grains, often spheroids, usually 200–600 microns in diameter and milky white in color (Plate II) — **Ooids (or polished pelletoids)**
3. Pitted, dull ovoids, spheroids or cylinders; generally 300–800 microns in length (Plate I) — **Pelletoids**
4. Non-skeletal grains with no definite shape and lacking characteristics associated with other non-skeletal grains — **Cryptocrystalline grains**

Bibliography

ABBOTT, R.T. (1958): American Sea Shells. Fourth ed., 541 pp. New York: Van Nostrand.
ABE, N. (1944): Ecological observations on *Spirorbis*. Sci. Rept. Tohoku Univ. Ser. **17**, 327–351.
ABOLINS-KROGIS, A. (1968): Shell generation in *Helix pomatia* with special reference to the elementary calcifying particles. In: V. FRETTER (ed.), Studies in the structure, physiology and ecology of molluscs. Symp. Zool. Soc. London **22**, 75–92.
ADAMS, J.E., RHODES, M.L. (1960): Dolomitization by seepage refluxion. Am. Assoc. Petrol. Geologists Bull. **44**, 1912–1920.
ADEY, W.H. (1965): The genus *Clathromorphum* in the Gulf of Maine. Hydrobiol. **26**, 539–573.
ADEY, W.H. (1970): The effects of light and temperature on growth rates in boreal-subarctic crustose corallines. J. Phycol. **6**, 269–276.
ADEY, W.H., JOHANSEN, H.W. (1972): Morphology and taxonomy of corallinaceae with special reference to *Clathromorphum, Mesophyllum* and *Neopolyporolithon*. N. gen. Phycologia **11**, 159–180.
ADEY, W.H., MACINTYRE, I.G. (1973): Crustose coralline algae: a re-evaluation in the geological sciences. Bull. Geol. Soc. Am. **84**, 883–904.
AGASSIZ, A. (1894): A reconnaissance of the Bahamas and of the elevated reefs of Cuba. Bull. Museum of Comp. Zool., Harvard **26**, 203 pp.
AHR, W.M., STANTON, R.J., JR. (1973): The sedimentologic and paleoecologic significance of *Lithotrya*, a rock-boring barnacle. J. Sediment. Petrol. **43**, 20–23.
ALDERMAN, A.R. (1965): Dolomite sediments and their environment in the south-east of South Australia. Geochim. Cosmochim. Acta **29**, 1355–1365.
ALDERMAN, A.R., BORCH, C.C. VON DER (1960): Occurrence of hydromagnesite in sediments in South Australia. Nature **188**, 931.
ALDERMAN, A.R., BORCH, C.C. VON DER (1961): Occurrence of magnesite-dolomite sediments in South Australia. Nature **192**, 861.
ALDERMAN, A.R., BORCH, C.C. VON DER (1963): A dolomite reaction series. Nature **198**, 465–466.
ALDERMAN, A.R., SKINNER, H.C.W. (1957): Dolomite sedimentation in the south-east of South Australia. Am. J. Sci. **255**, 561–567.
ALEXANDERSSON, T. (1969): Recent littoral and sublittoral high-Mg calcite lithification in the Mediterranean. Sedimentology **12**, 47–61.
ALEXANDERSSON, T. (1971): Intragranular precipitation of aragonite and Mg-calcite in modern marine shallow-water sediments (abst.). Program, Internat. Sediment. Congr., 1971, Addendum, p. 1.
ALEXANDERSSON, T. (1972a): Micritization of carbonate particles: processes of precipitation and dissolution in modern shallow-marine sediments. Buil. Geol. Inst. Univ. Uppsala, N. S., **3**, 201–236.
ALEXANDERSSON, T. (1972b): Intragranular growth of marine aragonite and Mg-calcite: evidence of precipitation from supersaturated sea water. J. Sediment. Petrol. **42**, 441–460.
ALEXANDERSSON, T. (1972c): Shallow-marine carbonate diagenesis as related to the carbonate saturation level in sea water. Publ. Paleont. Inst. Univ. Uppsala, No. 126, 10 p.
ALLEN, J.A. (1963): Ecology and functional morphology of molluscs. In: H. BARNES (ed.), Oceanog. Marine Biol. Ann. Rev., vol. 1, p. 253–288.
ALLEN, J.R.L., WELLS, J.W. (1962): Holocene coral bank and subsidence in the Niger delta. J. Geol. **70**, 381–397.
ALLEN, R.C., GAVISH, E., FRIEDMAN, G.M., SANDERS, J.E. (1969): Aragonite-cemented sandstone from outer continental shelf off Delaware Bay: Submarine lithification mechanism yields product resembling beachrock. J. Sediment. Petrol. **39**, 136–149.
American Geological Institute (1962): Dictionary of geological terms, 545 pp. Garden City, New York: Doubleday & Co., Inc.

American Society for Testing Materials (1971): Methods of emission spectro-chemical analysis.

Andel, T. H. van, Veevers, J. J. (1967): Morphology and sediments of the Timor Sea. Dept. Nat. Development, Bureau Mineral Res., Geol. and Geophys. Bull., No. 83, 177 pp.

Anderson, T. F., Schneidermann, N. (1972): Isotope relationships in geologic limestones from central Caribbean, leg 15. Deep sea drilling project (Abst.) Trans. Am. Geophys. Union **53**, 555.

Andrews, P. B. (1964): Serpulid reefs, Baffin Bay, southeast Texas. In: A. J. Scott (ed.), Depositional environments south-central Texas Coast, p. 101–120. Gulf Coast Assoc. Geol. Sci. Field Trip Guidebook.

Angell, R. W. (1967): The test structure and composition of the foraminifera *Rosalina floridana*. J. Protozool. **14**, 299–307.

Angino, E. E., Armitage, K. B., Tash, J. C. (1964): Physico-chemical limnology of Lake Bonney, Antarctica. Limnol. Oceanog. **9**, 207–217.

Ansell, A. D., Nair, N. B. (1969): A comparative study of bivalves which bore mainly by mechanical means. Am. Zoologist **9**, 851–868.

Arnold, J. M., Arnold, K. O. (1969): Some aspects of hole-boring predation by *Octopus vulgaris*. Am. Zoologist **9**, 991–996.

Arrhenius, G. (1952): Sediment cores from the East Pacific. Rept. Swed. Deep-Sea Exped. **5**, 1–227.

Arrhenius, G. (1963): Pelagic Sediments. In: M. N. Hill (ed.), The sea, vol. 3, p. 655–727. New York: Interscience Press.

Arx, W. S. von (1954): Circulation systems of Bikini and Rongelap lagoons. U. S. Geol. Surv. Profess. Papers **260-B**, 265–273.

Atwood, D. K., Bubb, J. N. (1970): Distribution of dolomite in a tidal flat environment, Sugarloaf Key, Florida. J. Geol. **78**, 499–505.

Avias, J. (1950): Note preliminaire sur quelques phénomenes, actuels ou subactuels de pétrogenése et autres, dans les marais côtiers de Moindou et de Canala (Nouvelle Caledonie). C. R. Soc. Géol. France, No. 13, p. 277–280.

Baars, D. L. (1963): Petrology of carbonate rocks. Shelf Carbonates of the Paradox Basin, a Symposium, Fourth Field Conference. Four Corners Geol. Soc., p. 101–129.

Baars, D. L. (1968): Nature of calcification in codiacean algae (abst.). Am. Assoc. Petrol. Geologists Bull. **52**, 518.

Baas-Becking, L. G. M., Galliher, E. W. (1931): Wall structure and mineralization in coralline algae. J. Phys. Chem. **35**, 467–479.

Bakus, G. J. (1966): Some relationships of fishes to benthic organisms on coral reefs. Nature **210**, 280–284.

Bakus, G. J., Shinn, E. A., Stockman, K. W. (1967): The geologic effects of Hurricane Donna in South Florida. J. Geol. **75**, 583–597.

Ball, M. M. (1967): Carbonate sand bodies of Florida and the Bahamas. J. Sediment. Petrol. **37**, 556–591.

Bandy, O. L. (1960): General correlation of foraminiferal structure with environment. Rep. 21st International Geol. Congr., Copenhagen, Part 22, p. 7–19.

Bandy, O. L. (1964): General correlation of foraminiferal structure with environment. In: J. Imbrie and N. D. Newell (eds.), Approaches to paleoecology, p. 75–90. New York: John Wiley & Sons, Inc.

Bardach, J. E. (1961): Transport of calcareous fragments by reef fishes. Science **133**, No. 3446, p. 98–99.

Barnes, D. J. (1970): Coral skeletons: An explanation of their growth and structure. Science **170**, 1305–1308.

Barnes, H., Topinka, J. A. (1969): Effect of the nature of the substratum on the force required to detach a common littoral alga. Am. Zoologist **9**, 753–758.

Baron, G., Pesneau, M. (1956): Sur l'existence et un mode de préparation du monohydrate de carbonate de calcium. Compt. Rend. **243**, 1217–1219.

Barrows, A. L. (1919): The occurrence of a rock-boring isopod along the shore of San Francisco Bay, California. Univ. Calif. Publ. Zool. **10**, 299–316.

Bartlett, G. A., Greggs, R. G. (1970a): The Mid-Atlantic Ridge near 45° 00′ north. VIII. Carbonate lithification on oceanic ridges and seamounts. Can. J. Earth Sci. **7**, 257–267.

Bartlett, G. A., Greggs, R. G. (1970b): A reinterpretation of stylolitic solution surfaces deduced from carbonate cores from San Pablo Seamount and the Mid-Atlantic Ridge. Can. J. Earth Sci. **7**, 274–279.

BARTRUM, J. A. (1917): Concretions in the recent sediments of the Auckland Harbour, New Zealand. Trans. Roy. Soc. New Zealand **49**, 425–428.

BASAM, P. B. (1973): Aspects of sedimentation and development of a carbonate bank in the Baracuda Keys, South Florida. J. Sediment. Petrol. **43**, 42–53.

BASSETT-SMITH, P. W. (1899): On the formation of the coral-reefs on the N. W. coast of Australia. Proc. Zool. Soc., London, 157–159.

BATHURST, R. G. C. (1958): Diagenetic fabrics in some British Dinantian limestones. Liverpool Manchester Geol. J. **2**, 11–36.

BATHURST, R. G. C. (1964a): Diagenesis and paleoecology: A survey. In: J. IMBRIE and N. D. NEWELL (eds.), Approaches to paleoecology, p. 319–344. New York: John Wiley & Sons.

BATHURST, R. G. C. (1964b): The replacement of aragonite by calcite in the molluscan shell wall. In: J. IMBRIE and N. D. NEWELL (eds.), Approaches to paleoecology, p. 357–376. New York: John Wiley & Sons.

BATHURST, R. G. C. (1966): Boring algae, micrite envelopes and lithification of molluscan biosparites. Geol. J. **5**, 15–32.

BATHURST, R. G. C. (1967a): Oolitic films on low energy carbonate sand grains, Bimini Lagoon, Bahamas. Marine Geol. **5**, 89–109.

BATHURST, R. G. C. (1967b): Depth indicators in sedimentary environments. Marine Geol. **5**, 447–471.

BATHURST, R. G. C. (1967c): Subtidal gelatinous mat, sand stabilizer and food, Great Bahama Bank. J. Geol. **75**, 736–738.

BATHURST, R. G. C. (1968): Precipitation of ooids and other aragonitic fabrics in warm seas. In: G. MÜLLER and G. M. FRIEDMAN (eds.), Recent developments in carbonate sedimentology in central Europe, p. 1–10. Berlin-Heidelberg-New York: Springer.

BATHURST, R. G. C. (1971): Carbonate sediments and their diagenesis. Developements in sedimentology, vol. 12, 620 pp. New York: Elsevier Publ.

BAVENDAMM, W. (1932): Die microbiologische Kalkfällung in der tropischen See. Arch. Mikrobiol. **3**, 205–276.

BAYER, F. M. (1956): Octocorallia. In: R. C. MOORE (ed.), Treatise on invertebrate paleontology, p. 166–231. Geological Society of America, part F, Coelenterata.

BÉ, A. W. H. (1965): The influence of depth on shell growth in *Globigerinoides sacculifer* (Brady). Micropaleo. **11**, 81–87.

BÉ, A. W. H. (1968): Shell porosity of recent foraminifera as a climatic index. Science **161**, 881–884.

BÉ, A. W. H., ERICSON, D. B. (1963): Aspects of calcification in planktonic foraminifera (Sarcodina). Ann. N. Y. Acad. Sci. **109**, 65–81.

BÉ, A. W. H., HAMLEBEN, C. (1970): Calcification in a living planktonic foraminifer, *Globigerinoides sacculifer* (Brady). Neues Jahrb. Geol. Paleontol., Abhandl. **134**, 221–234.

BÉ, A. W. H., LOTT, L. (1964): Shell growth and structure of planktonic foraminifera. Science **145**, No. 3634, 823–824.

BÉ, A. W. H., TOLDERLUND, D. S. (1971): Distribution and ecology of living planktonic foraminifera in surface waters of the Atlantic and Indian Oceans. In: B. M. FUNRELL and W. R. RIEDEL (eds.), The micropaleontology of oceans, p. 105–149. London: Cambridge Univ. Press.

BEAVER, H. H., et al. (1967): Echinodermata. In: R. C. MOORE (ed.), Treatise on invertebrate paleontology. Geol. Soc. Am. Part S, **1**, 296 pp.

BEHRENS, E. W. (1968): Cyclic and current structures in a serpulid reef. Contrib. Marine Sci., Univ. Texas Marine Sci. Inst. **13**, 21–27.

BEHRENS, E. W. (1972): Subtidal holocene dolomite, Baffin Bay, Texas. J. Sediment. Petrol. **42**, 155–161.

BEHRENS, E. W., LAND, L. S. (1970): Subtidal, Holocene dolomite in Baffin Bay, Texas (abst.). Ann. Meeting, Geol. Soc. Am., p. 491–492.

BEKLEMISHEV, C. W. (1971): Distribution of plankton as related to micropaleontology. In: B. M. FUNNELL and W. R. RIEDEL (eds.), The micropaleontology of oceans, p. 75–87. London: Cambridge Univ. Press.

BERGER, W. H. (1967): Foraminiferal ooze: solution at depths. Science **156**, 383–385.

BERGER, W. H. (1968): Planktonic foraminifera: selective solution and paleoclimatic interpretation. Deep-Sea Res. **15**, 31–43.

BERGER, W. H. (1969): Ecologic patterns of living planktonic foraminifera. Deep-Sea Res. **16**, 1–24.

BERGER, W. H. (1970): Planktonic foraminifera: Differential production and expatriation off Baja California. Limnol. Oceanog. **15**, 183–204.

Berger, W.H., Soutar, A. (1967): Planktonic foraminifera: field experiment on production rate. Science **156**, 1495–1497.

Berggren, W.A., Boersma, A. (1969): Late pleistocene and Holocene planktonic foraminifera from the Red Sea. In: E.T. Degens and D.A. Ross (eds.), Hot brines and recent heavy metal deposits in the Red Sea, p. 282–298. Berlin-Heidelberg-New York: Springer.

Bernard, F. (1964): Le nannoplancton en zone aphotique des mers chauds. Pelagos. **2**, No. 2, 32 pp.

Bernard, F., Lecal, J. (1953): Role des Flagellés calcaires dans la sedimentation actuelle en Mediterranée. Publ. 19th Geol. Congr. Algiers, 1952, 4 (4), p. 11–23.

Berner, R.A. (1965a): Activity coefficients of bicarbonate, carbonate and calcium ions in sea water. Geochim. Cosmochim. Acta **29**, 947–965.

Berner, R.A. (1965b): Dolomitization of the mid-Pacific atolls. Science **147**, 1297–1299.

Berner, R.A. (1966a): Chemical diagenesis of some modern carbonate sediments. Am. J. Sci. **264**, 1–36.

Berner, R.A. (1966b): Diagenesis of carbonate sediments: Interaction of Mg^{++} in sea water with mineral grains. Science **153**, 188–191.

Berthois, L., Guilcher, A., Doumenge, F., Michel, A. (1963): Le renouvellement des eaux du lagoon dans l'atoll de Maupihaa-Mopelia (Îles de la Societé). Compt. Rend. **257**, 3992–3995.

Bertram, G.C.L. (1936): Some aspects of the breakdown of coral at Ghardaqa, Red Sea. Proc. Zool. Soc. London **106**, 1011–1026.

Bezrukov, P.L. (1964): Sedimentation in northern and central parts of the Indian Ocean. International Geological Congress 22nd Session, p. 41–51.

Birk, L.S. (1959): X-ray spectrochemical analysis, 137 pp. New York: Interscience Publ.

Birkenes, E., Braarud, T. (1952): Phytoplankton in the Oslo Fjord during a "*Coccolithus huxleyi*-Summer". Avh. Norske. Vidensk. Akad. Oslo. No. 2, p. 1–23.

Biscay, P.E. (1965): Mineralogy and sedimentation of recent deep-sea clay in the Atlantic Ocean and adjacent seas and oceans. Bull. Geol. Soc. Am. **76**, 803–832.

Bischoff, J.L. (1968a): Catalysis, inhibition, and the calcite-aragonite problem. I. The aragonite-calcite transformation. Am. J. Sci. **266**, 65–79.

Bischoff, J.L. (1968b): Catalysis, inhibition, and the calcite-aragonite problem. II. The vaterite-aragonite transformation. Am. J. Sci. **266**, 80–90.

Bischoff, J.L. (1968c): Kinetics of calcite nucleation: Magnesium ion inhibition and ionic strength catalysis. J. Geophys. Res. **73**, 3315–3322.

Bisque, R.E. (1961): Analysis of carbonate rocks for calcium, magnesium, iron and aluminum with EDTA. J. Sediment. Petrol. **31**, 113–122.

Bissell, H.J. (1957): Combined preferential staining and cellulose peel technique. J. Sediment. Petrol. **27**, No. 4, p. 417–420.

Bissell, H.J., Chilingar, G.V. (1958): Notes on diagenetic dolomitization. J. Sediment. Petrol. **28**, 490–497.

Black, M. (1933a): The precipitation of calcium carbonate on the Bahama Bank. Geol. Mag. **70**, 455–466.

Black, M. (1933b): The algal sediments of Andros Island, Bahamas. Phil. Trans. Roy. Soc. London, Ser. B **222**, 165–192.

Black, M. (1963): The fine structure of the mineral parts of Coccolithophoridae. Proc. Limn. Soc. London **174**, 41–46.

Black, M. (1965): Coccoliths. Endeavour **24**, 131–137.

Black, M. (1971): The systematics of coccoliths in relation to the paleontological record. In: B.M. Funnell and W.R. Riedel (eds.), The micropaleontology of oceans, p. 611–624. London: Cambridge Univ. Press.

Blackmon, P.D., Todd, R. (1959): Mineralogy of some foraminifera as related to their classification and ecology. J. Paleontol. **33**, 1–15.

Blanc, J.J. (1968): Sedimentary geology of the Mediterranean Sea. In: H. Barnes (ed.), Oceanog. Marine Biol. Ann. Rev. **6**, 377–454.

Blumenstock, D.I. (1961): A report on typhoon effects upon Jaluit Atoll. Atoll Res. Bull., No. 75, 105 pp.

Blumenstock, D.I., Fosberg, F.R., Johnson, C.G. (1961): The resurvey of the typhoon effects on Jaluit Atoll in the Marshall Islands. Nature **189**, 618–620.

BOARDMAN, R.S., CHEETHAM, A.H. (1969): Skeletal growth, intracolony variation and evolution in Bryozoa: a review. J. Paleontol. **43**, 205–233.

BOCK, W.D. (1967): Monthly variation in the foraminiferal biofacies on Thalassia and sediment in the Big Pine Key area, Florida. Unpubl. PH.D. THESIS, Univ. Miami, 291 pp.

BOCK, W.D., MOORE, D.R. (1969): The foraminifera and micromollusks of Hogsty Reef and Serrana Bank and their paleoecological significance. Caribbean Geol. Congr., preprint, 11 pp.

BØGGILD, O.B. (1912): The deposits of the sea bottom. Danish. Oceanog. Exped., 1908–1910, Rept. **1**, 255–269.

BØGGILD, O.B. (1930): The shell structure of the mollusks. Kgl. Danske Vidensk. Selsk. Skrifter Natur. Math. **9**, 231–326.

BÖHM, E.L. (1971): Calcification in the calcareous alga *Halimeda opuntia* (L) (Chlorophyta, Udoteaceae). Symposium on investigations and resources of the caribbean sea and adjacent regions, UNESCO, p. 357–361.

BOILLOT, G. (1964): Étude géologique de la Manche occidentale, fonds vocheux, dépôts Quaternaires, sédiments actuels. Ann. Inst. Oceanog. (Paris) **42**, 1–219.

BOILLOT, G. (1965): Organic gradients in the study of neritic deposits of biological origin: the example of the western English Channel. Marine Geol. **3**, 359–367.

BOLTOVSKOY, E. (1963): The littoral foraminiferal biocoenoses of Puerto Desado (Patagonia, Argentina). Contrib. Cushman Found. Foram. Res. **14**, 58–70.

BONAR, L. (1936): An unusual Ascomycete in the shells of marine animals. Univ. Calif. Publ. Bot. **19**, 187–194.

BONATTI, E. (1966): Deep-sea authigenic calcite and dolomite. Science **153**, 534–537.

BONHAM, K. (1965): Growth rate of giant clam *Tridacna gigas* at Bikini Atoll as revealed by radioautography. Science **149**, No. 3681, 300–302.

BONHAM, K., HELD, E.E. (1963): Ecological observations on the sea cucumbers *Holothuria atra* and *Holothuria leucospilata* at Rongelap Atoll. Pacific Sci. **17**, No. 3, 305–314.

BOOLOOTIAN, R.A. (ed.) (1966): The physiology of echinodermata, 772 pp. New York: John Wiley & Sons, Inc.

BORCH, C.C. VON DER (1965a): The distribution and preliminary geochemistry of modern carbonate sediments of the Coorong area, South Australia. Geochim. Cosmochim. Acta **29**, 781–799.

BORCH, C.C. VON DER (1965b): Source of ions for Coorong dolomite formation. Am. J. Sci. **263**, 684–688.

BORCH, C.C. VON DER, RUBIN, M., SKINNER, B.J. (1965): Modern dolomite from South Australia. Am. J. Sci. **262**, 1116–1118.

BORNHOLD, B.D., MILLIMAN, J.D. (1973): Generic and environmental control of carbonate mineralogy in serpulid (Polychaete) tubes. J. Geol. **81**, 363–373.

BORNHOLD, B.D., PILKEY, O.H. (1971): Bioclastic turbidite sedimentation in Columbus Basin, Bahamas. Bull. Geol. Soc. Am. **82**, 1341–1354.

BOSCHMA, H. (1956): Milleporina and Stylasterina. In: R.C. MOORE (ed.), Treatise on invertebrate paleontology, Geol. Soc. Am., vol. F, Coelenterata, p. 90–106.

BOSCHMA, H. (1959): The species problem in corals. Proc. Intern. Zool. Congr., 1958, p. 246–248.

BOSELLINI, A., GINSBURG, R.N. (1971): Form and internal structure of recent algal nodules (Rhodolites) from Bermuda. J. Geol. **79**, 669–682.

BOURCART, J. (1957): Géologie sous-marine de la Baie de Villefranche. Ann. Inst. Oceanog. Monaco **33**, 137–200.

BOWEN, H.J.M. (1956): Strontium and barium in seawater and marine organisms. Marine Biol. Assoc. United Kingdom J. **35**, 451–460.

BOYER, B.W. (1972): Grain accretion related phenonena in unconsolidated surface sediments of the Florida Reef tract. J. Sediment. Petrol. **42**, 205–210.

BRADY, H.B. (1884): Report on the foraminifera collected by H.M.S. CHALLENGER during the years 1873–1876. Chall. Rept. **9**, 814 pp.

BRAMLETTE, M.N. (1958): Significance of coccolithophorids in calcium carbonate deposition. Bull. Geol. Soc. Am. **69**, 121–126.

BRAMLETTE, M.N. (1961): Pelagic sediments. In: M. SEARS (ed.), Oceanography, Publ. 67, Am. Assoc. Adv. Sci., p. 345–366.

BRANNER, J.C. (1904): Stone reefs on the northeast coast of Brazil. Bull. Museum Comp. Zool. **44**, 175–285.

BRODIE, J.W. (1965): Capricorn Seamount, south-west Pacific Ocean. Trans. Roy. Soc. New Zealand, Geol. **3**, No. 10, 151–156.

BROECKER, W.S. (1963): A preliminary evaluation of uranium series inequilibrium as a tool for absolute age measurement of marine carbonates. J. Geophys. Res. **68**, 2817–2834.

BROECKER, W.S. (1971): Calcite accumulation rates and glacial to interglacial changes in oceanic mixing. In: K.K. TUREKIAN (ed.), Cenozoic glacial ages, p. 239–265. New Haven: Yale Univ. Press.

BROECKER, W.S., GERARD, R., EWING, M., HEEZEN, B.C. (1960): Natural radiocarbon in the Atlantic Ocean. J. Geophys. Res. **65**, 2903–2931.

BROECKER, W.S., TAKAHASHI, T. (1966): Calcium precipitation on the Bahama Banks. J. Geophys. Res. **71**, 1575–1602.

BROOKS, R., CLARK, L.M., THURSTON, E.F. (1950): Calcium carbonate and its hydrates. Phil. Trans. Roy. Soc. London, Ser. A **243**, 145–167.

BROOKS, R.R., PRESLEY, B.J., KAPLAN, I.R. (1968): Trace elements in the interstitial waters of marine sediments. Geochim. Cosmochim. Acta **32**, 397–414.

BROOKS, R.R., RUMSBY, M.G. (1965): The biogeochemistry of trace element uptake by some New Zealand bivalves. Limnol. Oceanog. **10**, 521–527.

BROWN, T.C. (1914): Origin of oolites and the oolitic texture in rocks. Bull. Geol. Soc. Am. **25**, 745–778.

BRUYEVICH, S.U. (1963): Rates of mineralization of suspended organic matter in the low latitudes of the Pacific during the predepositional stage. Geochim. Cosmochim. Acta **3**, 349–352.

BRYAN, E.H., JR. (1953): Check list of atolls. Atoll Res. Bull., No. 19, 38 pp.

BUBB, J.N., ATWOOD, D.K. (1968): Recent dolomitization of Pleistocene limestone by hypersaline brines, Great Inagua Island, Bahamas (abst.). Am. Assoc. Petrol. Geologists Bull. **52**, 522.

BUCH, K. (1933): Der Borsäuregehalt des Meerwassers und seine Bedeutung bei der Berechnung des Kohlensäuresystems im Meerwasser. Rappt. Cons. Explor. Mer. **85**, 71–75.

BUCH, K. (1938): New determination of the second dissociation constant of carbonic acid in sea water. Acad. Aboensis Acta, Math. et Phys. **11**, 1–18.

BUCH, K., GRIPENBERG, S. (1932): Über den Einfluß des Wasserdruckes auf pH und das Kohlensäuregleichgewicht in großen Meerestiefen. J. Cons. Int. Explor. Mer. **7**, 233–245.

BUEHLER, E.J. (1948): The use of peels in carbonate petrology. J. Sediment. Petrol. **18**, 71–73.

BUKRY, D., KLING, S.A., HORN, M.K., MANHEIM, F.T. (1970): Geological significance of coccoliths in fine-grained carbonate bands of postglacial Black Sea sediments. Nature **226**, 156–158.

BURAKOV, V.S., IANKOVSKII, A.A. (1964): Practical handbook on spectral analysis, 190 pp. [translated from Russian]. New York: MacMillian.

BURNS, J.H., BREDIG, M.A. (1956): Transformation of calcite to aragonite by grinding. J. Chem. Phys. **25**, 1281.

BUTLER, G.P. (1969): Modern evaporite deposits and geochemistry of coexisting brines, the Sabkha, Trucial Coast, Arabian Gulf. J. Sediment. Petrol. **39**, 70–89.

CADOT, H.M., SCHMUS, W.R. VAN, KAESLER, R.L. (1972): Magnesium in calcite of marine Ostracoda. Bull. Geol. Soc. Am. **83**, 3519–3522.

CAMPBELL, R.B. (1929): Fish otoliths, their occurrence and value as stratigraphic markers. J. Paleontol. **3**, 254–279.

CANNON, H.G. (1935): On the rock-boring barnacle, *Lithotrya valentiana*. Great Barrier Reef Exped. Sci. Rept. **5**, 1–17.

CAROZZI, A.V. (1957): Contribution à l'étudé des propriétés géométriques des oolites — L'example du Grand Lac Salé, Utah, U.S.A. Bull. Inst. National Geneve **58**, 1–51.

CAROZZI, A.V. (1960): Microscopic sedimentary petrography, 485 pp. New York: John Wiley & Sons, Inc.

CARPENTER, R. (1969): Factors controlling the marine geochemistry of fluorine. Geochim. Cosmochim. Acta **33**, 1153–1167.

CARRIGY, M.A., FAIRBRIDGE, R.W. (1954): Recent sedimentation, physiography and structure of the continental shelves of Western Australia. J. Roy. Soc. W. Australia **38**, 65–95.

CARRIKER, M.R. (1961): Comparative functional morphology of boring mechanisms in gastropods. Am. Zoologist **1**, 263–266.

CARRIKER, M.R. (1969): Excavation of bore holes by the gastropod *Urosalpinx*: an analysis by light and scanning electron microscopy. Am. Zoologist **9**, 917–933.

CARRIKER, M.R., CHARLTON, G., VAN ZANDT, D. (1967): Gastropod *Urosalpinx*: pH of accessory boring organ while boring. Science **158**, 920–922.

CARRIKER, M.R., SMITH, E.H. (1969): Comparative calcibiocavitology: Summary and conclusions. Amer. Zoologist **9**, 1011–1020.

CARROLL, J.J., GREENFIELD, L.J., JOHNSON, R.F. (1965): The mechanism of calcium and magnesium uptake from sea water by marine bacterium. Jour. Cell. and Comp. Physiol. **66**, 109–118.

CARY, L.R. (1918): The Gorgonaceae as a factor in the formation of coral reefs. Carnegie Inst. Wash. Publ. 213, Papers Dept. Marine Biol. **9**, 341–362.

CASPERS, H. (1957): Black Sea and Sea of Azov. In J.W. HEDGPETH (ed.), Treatise on Marine Ecology, Geol. Soc. Amer. Mem. 67, **1**, 801–890.

CAULET, J.P. (1972): Recent biogenic calcareous sedimentation on the Algerian continental shelf. In D.J. STANLEY (ed.), The Mediterranean Sea: A natural sedimentation laboratory, p. 261–277. Stroudsburg: Pa. Dowden, Hutchinson and Ross.

CAYEUX, L. (1935): Les roches sedimentaires de France, Roches carbonatees (calcium et dolomies). Masson, Paris, 463 pp.

CERAME-VIVAS, M.A., GRAY, I.E. (1966): The distribution pattern of benthic invertebrates on the continental shelf off North Carolina. Ecology **47**, 260–270.

CHAPMAN, F., MAWSON, D. (1906): On the importance of *Halimeda* as a reef forming organism: with a description of Halimeda-limestones of the New Hebrides. Quart. Jour. Geol. Soc. London **62**, 702–711.

CHAVE, K.E. (1952): A solid solution between calcite and dolomite. Jour. Geol. **60**, 190–192.

CHAVE, K.E. (1954): Aspects of the biogeochemistry of magnesium. Jour. Geol. **62**, 266–283, 587–599.

CHAVE, K.E. (1960): Carbonate skeletons to limestones: problems. Trans. N.Y. Acad. Sci., Ser. II, **23**, 14–24.

CHAVE, K.E. (1962): Factors influencing the mineralogy of carbonate sediments. Limnol. Oceanogr. **7**, 218–223.

CHAVE, K.E. (1964): Skeletal durability and preservation. In J. IMBRIE and N.D. NEWELL (eds.), Approaches to Paleoecology, p. 377–387. New York: John Wiley & Sons, Inc.

CHAVE, K.E. (1965): Carbonates: association with organic matter in surface seawater. Science **148**, 1723–1724.

CHAVE, K.E., DEFFEYES, K.S., WEYL, P.K., GARRELS, R.M., THOMPSON, M.E. (1962): Observations on the solubility of skeletal carbonates in aqueous solution. Science **137**, 33–34.

CHAVE, K.E., SCHMALZ, R.F. (1966): Carbonate-seawater interactions. Geochim. Cosmochim. Acta **30**, 1037–1048.

CHAVE, K.E., SUESS, E. (1970): Calcium carbonate saturation in seawater: effects of dissolved organic matter. Limnol. Oceanog. **15**, 633–637.

CHAVE, K.E., SMITH, S.V., ROY, K.I. (1972): Carbonate production by coral reefs. Marine Geol. **12**, 123–140.

CHAVE, K.E., WHEELER, B.D. (1965): Mineralogic changes during growth in the red alga *Clathromorphum compactum*. Science **147**, 621.

CHEETHAM, A.H., RUCKER, J.B., CARVER, R.E. (1969): Wall structure and mineralogy of the cheilostome bryozoan *Metrarabdotus*. J. Paleontol. **43**, 129–135.

CHEN, C. (1964): Pteropod ooze from Bermuda Pedestal. Science **144**, No. 3614, 60–62.

CHEN, C. (1966): Calcareous zooplankton in the Scotia Sea and Drake Passage. Nature **212**, 678–681.

CHEN, C. (1968): Pleistocene pteropods in pelagic sediments. Nature **219**, 1145–1149.

CHEN, C. (1971): Occurrence of pteropods in pelagic sediments (abst.). In: B.M. FUNNEL and W.R. RIEDEL (eds.), The micropaleontology of oceans, p. 351. London: Cambridge Univ. Press.

CHENEY, E.S., JENSEN, M.L. (1967): Corrections to carbon isotopic data of Gulf Coast salt dome rock. Geochim. Cosmochim. Acta **31**, 1345–1346.

CHESHER, R.H. (1969): Destruction of Pacific corals by the sea star *Acanthaster planci*. Science **165**, 280–283.

CHÉTAIL, M., FOURNÍE, J. (1969): Shell-boring mechanism of the gastropod, *Purpura (Thais lapillus*: a physiological demonstration of the role of carbonic anlydiase in the dissolution of $CaCO_3$. Am. Zoologist **9**, 983–990.

Chief Office of Geodesy and Cartography, USSR (Glavnoye Upravleniiye Geodezii i Kartograffi): Carbonate distribution in the Atlantic Ocean sediments. Moscow, 1 chart.

CIFELLI, R., BOWEN, V.T., SIEVER, R. (1966): Cemented foraminiferal oozes from the Mid-Atlantic Ridge. J. Marine Res. **26**, 105–109.

CLAPP, W. F., KENK, R. (1963): Marine borers, an annotated bibliography. Office of Naval Research, 1136 pp.

CLARK, S. P. (1957): A note on calcite-aragonite equilibrium. Am. Mineralogist **42**, 564–566.

CLARKE, A. H., JR (1962): Annotated list and bibliography of the abyssal marine molluscs of the world. Nat. Mus. Canada Bull. No. 181, 142 pp.

CLARKE, F. W., WHEELER, W. C. (1917): The inorganic constitutents of marine invertebrates. U. S. Geol. Surv. Profess. Papers **102**, 56 pp.

CLARKE, F. W., WHEELER, W. C. (1922): The inorganic constitutents of marine invertebrates. U. S. Geol. Surv. Profess. Papers **124**, 62 pp.

CLAYTON, R. N., EPSTEIN, S. (1958): The relationship between O^{18}/O^{16} ratios in coexisting quartz, carbonate and iron oxides from various geological deposits. J. Geol. **66**, 352.

CLAYTON, R. N., JONES, B. F., BERNER, R. A. (1968): Isotope studies of dolomite formation under sedimentary conditions. Geochim. Cosmochim. Acta **32**, 415–432.

CLOUD, P. E., JR. (1952): Preliminary report on geology and marine environments of Onotoa Atoll, Gilbert Islands. Atoll Res. Bull., No. 12, 73 pp.

CLOUD, P. E., JR. (1959): Geology of Saipan, Mariana Islands; Part 4, Submarine topography and shoal water ecology. U. S. Geol. Surv. Profess. Papers **280-K**, 361–445.

CLOUD, P. E., JR. (1962a): Environment of calcium carbonate deposition west of Andros Island, Bahamas. U. S. Geol. Surv. Profess. Papers **350**, 138 pp.

CLOUD, P. E., JR. (1962b): Behaviour of calcium carbonate in sea water. Geochim. Cosmochim. Acta **26**, 867–884.

CLOUD, P. E., JR., BARNES, V. E. (1948): The Ellenburger Group of Central Texas, 473 pp. Texas: Univ. Texas, Austin.

CLOUD, P. E., JR., SCHMIDT, R. G., BURKE, H. W. (1956): Geology of Saipan, Mariana Islands. Part 1. General Geology. U. S. Geol. Surv. Profess. Papers **280-A**, 1–126.

COBB, W. R. (1969): Penetration of calcium carbonate substrates by the boring sponge, *Cliona. Am.* Zoologist **9**, 783–790.

COLES, S. L. (1969): Quantitative estimates of feeding and respiration for three scleractinian corals. Limnol. Oceanog. **14**, 949–953.

COLINVAUX, L. H., WILBUR, K. M., WATABE, N. (1965): Tropical marine algae: growth in laboratory culture. J. Phycol. **1**, 69–78.

COLLINS, A. C. (1958): Foraminifera. Great Barrier Reef Exped. Sci. Rept. **6**, No. 6, 335–437.

COLMAN, J. (1940): Zoology of the Jamaican shoreline. Geogr. J. **46**, 323–327.

CONRAD, E. H. (1968): The precipitation of metastable carbonate minerals at low temperature and pressure. Southeastern Geol. **9**, 1–7.

CORNWALL, I. E. (1962): The identification of barnacles with further figures and notes. Can. J. Zool. **40**, 621–629.

CORRENS, C. W. (1939): Pelagic sediments of the North Atlantic Ocean. In: P. D. TRASK (ed.), Recent Marine Sediments. Am. Assoc. Petrol. Geologists 373–395.

CORRENS, C. W. (1941): Beiträge zur Geochemie des Eisens und Mangans. Nachr. Akad. Wiss. Göttingen Math.-Phys. Kl., IIa, No. 5, p. 219–230.

COSTLOW, J. D., JR., BOOKHOUT, C. G. (1952): Molting and growth in *Balanus improvisus*. Biol. Bull. **105**, 420–433.

COSTLOW, J. D., JR., BOOKHOUT, C. G. (1956): Molting and shell growth in *Balanus amphitrite niveus*. Biol. Bull. **110**, 107–116.

COX, L. R. (1960): Gastropoda—general characteristics of gastropoda. In: J. B. KNIGHT et al. (eds.), Mollusco 1, Treatise on invertebrate paleontology, Geol. Soc. Am., part I, p. 84–168.

CRAIG, A. K., DOBKIN, S., GRIMM, R. B., DAVIDSON, J. B. (1969): The gastropod, *Siphonaria pectinata:* A factor in destruction of beach rock. Am. Zoologist **9**, 895–901.

CRAIG, H. (1953): The geochemistry of the stable carbon isotopes. Geochim. Cosmochim. Acta **3**, 53–92.

CRAIG, H. (1957a): Isotopic standards for carbon and oxygen and correction factors for mass spectrographic analysis of carbon dioxide. Geochim. Cosmochim. Acta **12**, 133–149.

CRAIG, H. (1957b): The natural distribution of radiocarbon and the exchange time of carbon dioxide between atmosphere and sea. Tellus **9**, No. 1, 1–17.

CRAIG, H. (1961): Standard for reporting concentrations of deuterium and oxygen-18 in natural waters. Science **133**, 1833–1834.

CRAWFORD, W.A., FYFE, W.S. (1964): Calcite-aragonite equilibrium at 100 °C. Science **144**, 1569–1570.

CRENSHAW, M.A., NEFF, J.M. (1969): Decalcification at the mantle-shell interface in molluscs. Am. Zoologist **9**, 881–885.

CRISP, D.J. (1964): An assessment of plankton grazing by barnacles. In: D.J. CRISP (ed.), Grazing in terrestrial and marine environments, p. 251–264. Oxford: Blackwell Scientific Publ.

CRISP, D.J. (1972): Salinity, species, and age effect of the trace chemistry of nine molluscan species (abst.). Program, Geol. Soc. Am. Ann. Meet. **4**, 716–717.

CRONEIS, C., MCCORMACK, J.T. (1932): Fossil Holothuria. J. Paleontol. **6**, 111–148.

CROZIER, W.J. (1918): The amount of bottom material ingested by Holothurians *(Stichopus)*. J. Exp. Zool. **26**, No. 2, 379–389.

CUENOT, L. (1948): Anatomie, éthologie et systématique des Échinodermes. Traité de Zoologie **11**, 3–363.

CULBERTSON, C., PYTKOWICZ, R.M. (1968): Effect of pressure on carbonic acid, boric acid, and the pH in seawater. Limnol. Oceanog. **13**, 403–417.

CULKIN, F. (1965): The major constituents of sea water. In: J.P. RILEY and G. SKIRROW (eds.), Chemical oceanography, vol. 1, p. 121–161. London: Academic Press.

CULLITY, B.D. (1956): Elements of X-ray diffraction, 514 pp. Reading, Mass.: Addison-Wesley Publ. Co., Inc.

CURTIS, R., EVANS, G., KINSMAN, D.J.J., SHEARMAN, D.J. (1963): Association of dolomite and anhydrite in recent sediments of the Persian Gulf. Nature **197**, 679–680.

CUSHMAN, J.A., TODD, R., POST, R.J. (1953): Recent foraminifera of the Marshall Islands. U. S. Geol. Surv. Profess. Papers **260-H**, 319–384.

CUTLER, E.B. (1968): A review of coral-inhabiting Sipuncula (abst.). Am. Zoologist **8**, 796.

DACHILLE, F., ROY, R. (1960): High-pressure phase transformations in laboratory mechanical mixers and mortars. Nature **186**, 3471.

DAETWYLER, C.C., KIDWELL, A.L. (1959): The Gulf of Batabano, a modern carbonate basin. Fifth World Petroleum Congress Section 1, p. 1–21.

DALRYMPLE, D.W. (1964): Recent sedimentary facies of Baffin Bay, Texas. Ph. D. Dissertation, Rice Univ.

DALRYMPLE, D.W. (1965): Calcium carbonate deposition associated with blue-green algal mats, Baffin Bay, Texas. Publ. Inst. Mar. Sci., Port Arasanas **10**, 187–200.

DALY, R.A. (1910): Pleistocene glaciation and the coral reef problem. Am. J. Sci. **30**, 297–308.

DALY, R.A. (1920): Origin of beachrock. Carnegie Inst. Wash. Yearbook (1919) **18**, 192.

DALY, R.A. (1924): The geology of American Samoa. Carnegie Inst. Wash. Papers Geophys. Lab. **340**, 93–143.

DANA, J.D. (1875): Corals and coral islands, 348 pp. London: Sampson Low, Marston Low & Searle.

DANA, T.F. (1971): On the reef corals of the world's most northern atoll (Kure: Hawaiian Archipelago). Pacific Sci. **25**, p. 80–87.

DANEGARD, L. (1935): Étude des calcaires par coloration et décalcification. Bull. Soc. Geol. France **6**, 237–245.

DAPPLES, E.C. (1942): Effect of macro-organisms on nearshore marine sediments. J. Sediment. Petrol. **12**, 118–126.

DARWIN, C. (1851): The structure and distribution of coral reefs, 214 pp. London: Smith, Elder & Co. (Reprinted, 1962, by Univ. Calif. Press, Berkeley.)

DAVID, T.W.E., SWEET, G. (1904): The geology of Funafuti. In the atoll of Funafuti. Roy. Soc. London. Rept. Coral Reef Comm., p. 61–124.

DAVIES, G.R. (1970a): Carbonate bank sedimentation, eastern Shark Bay, Western Australia. In: Carbonate Sedimentation and Environments, Shark Bay, Western Australia. Am. Assoc. Petrol. Geologists, Mem. **13**, 85–168.

DAVIES, G.R. (1970b): Algal-laminated sediments, Gladstone Embayment, Shark Bay, Western Australia. In: Carbonate Sedimentation and Environments, Shark Bay, Western Australia. Am. Assoc. Petrol. Geologists, Mem. **13**, 169–205.

DAVIES, P.J., TILL, R. (1968): Stained dry cellulose peels of ancient and recent impregnated carbonate sediments. J. Sediment. Petrol. **38**, 234–236.

DAVIES, T.A., SUPKO, P.R. (1973): Oceanic sediments and their diagenesis: Some examples from deep-sea drilling. J. Sediment. Petrol. **43**, 381–390.

DAVIES, T.T. (1965): Effect of environmentally induced growth rate changes in *Mytilus edulis* shell (abs.). Geol. Soc. Am. Spec. Papers **87**, 42.

DAVIES, T.T., CRENSHAW, M.E., HEATHFIELD, B.M. (1971): The effect of temperature on the chemistry and structure of echinoid spine regeneration (abst.). Program, S. E. Geol. Soc. Am. Ann. Meeting, p. 307.

DAVIES, T.T., HOOPER, P. R. (1963): The determination of the calcite: aragonite ratio in mollusc shells by x-ray diffraction. Mineral. Mag. **33**, 608.

DAVIES, T.T., SAYRE, J.G. (1970): The effect of environmental stress on pelecypod shell ultrastructure (abst.). Program, Geol. Soc. Am., S. E. Section, p. 204–205.

DAVIS, J.B., KIRKLAND, D.W. (1970): Native sulfur deposition in the Castile Formation, Culberson County, Texas. Econ. Geol. **65**, 107–121.

DAVIS, J.H. (1940): The ecology and geologic role of mangroves in Florida. Pap. Tortugas Lab., Carnegie Inst. Wash. **32**, 302–412.

DAVIS, W.M. (1928): The coral reef problem. Am. Geographical Soc., Special Publ. **9**, 596 pp.

DAVISON, C. (1891): On the amount of sand brought up by lobworms to the surface. Geol. Mag. **8**, 489–493.

DAWSON, E.Y. (1961): The rim of the reef. Calcareous algae occupy a major role in the growth of atolls. Nat. Hist. **70**, 6, 8–17.

DAY, J. (1968): The proboscis and proboscic gland of the cymatiid megogastropod *Argobuccinum argus* (GEMLIN, 1793) (abst.). Am. Zoologist **8**, 801.

DEAN, J.A. (1960): Flame photometry, 354 pp. New York: McGraw-Hill Book Co., Inc.

DEBOO, P.B. (1962): A preliminary petrographic study of beach rock. In: Proc. Natl. Coastal & Shallow Water Res. Conf., 1961, p. 456–458.

DEER, W.A., HOWIE, R.A., ZUSSMAN, J. (1962): Rock-forming minerals, vol. 5, Non-Silicates, 371 pp. New York: John Wiley & Sons, Inc.

DEFFEYES, K.S., LUCIA, F.J., WEYL, P.K. (1964): Dolomitization: observations on the island of Bonaire, Netherlands Antilles. Science **143**, No. 3607, 678–679.

DEFFEYES, K.S., LUCIA, F.J., WEYL, P.K. (1965): Dolomitization of Recent and Plio-Pleistocene sediments by marine evaporite waters on Bonaire, Netherlands Antilles. In: L.C. PRAY and R.C. MURRAY (eds.), Dolomitization and limestone diagenesis, Soc. Econ. Paleontologists and Mineralogists, Special Publ. No. 13, p. 71–88.

DEFFEYES, K.S., MARTIN, E.L. (1962): Absence of carbon-14 activity in dolomite from Florida Bay. Science **136**, 782.

DEFLANDRE, G. (1948): Les Calciodinellidés dinoflagelles fossiles a thèque calcaire. Botaniste **34**, 191–219.

DEGENS, E.T. (1965): Geochemistry of sediments: A brief survey, 342 pp. Englewood Cliffs, N. J.: Prentice Hall, Inc.

DEGENS, E.T. (1967): Evolutionary trends inferred from the organic tissue variation of mollusc shells. Medd. fra Dans Geol. Forening. Kobenhavn **17**, 112–124.

DEGENS, E.T., DEUSER, W.G., HAEDRICH, R.L. (1969): Molecular structure and composition of fish otoliths. Marine Biol. **2**, 102–113.

DEGENS, E.T., EPSTEIN, S. (1964): Oxygen and carbon isotope ratios in coexisting calcites and dolomites from recent and ancient sediments. Geochim. Cosmochim. Acta **28**, 23–44.

DEGENS, E.T., SPENCER, D.W., PARKER, R.H. (1967): Paleobiochemistry of molluscan shell proteins. Comp. Biochem. Physiol. **20**, 553–579.

DEGROOT, K. (1965): Inorganic precipitation of calcium carbonate from sea-water. Nature **207**, 404–405.

DEGROOT, K. (1969): The chemistry of submarine cement formation at Rabat Hussain in the Persian Gulf. Sedimentology **12**, 63–68.

DELAUBENFELS, M.W. (1955): Porifera. In: Treatise on invertebrate paleontology. Geol. Soc. Am. Part E, 21–112.

DEUSER, W.G. (1970): Extreme $^{13}C/^{12}C$ variations in Quaternary dolomites from the continental shelf. Earth Planet. Sci. Letters **8**, 118–124.

DEUSER, W.G., DEGENS, E.T. (1967): Carbon isotope fractionation in the system CO_2 (gas)—CO_2 (aqueous)—HCO_3^- (aqueous). Nature **215**, 1033–1035.

DEUSER, W.G., DEGENS, E.T. (1969): O^{18}/O^{16} and C^{14}/C^{12} ratios of fossils from the hot brine deep area of the central Red Sea. In: E.T. DEGENS and D.A. ROSS (eds.), Hot brines and recent heavy metal deposits in the Red Sea, p. 336–347. Berlin-Heidelberg-New York: Springer.

DeWindt, J., Berwerth, F. (1904): Untersuchung von Grundproben der I., II. und IV. Reise von S.M. "Pola" in den Jahren 1890, 1892 und 1893. Denkschr. Akad. Wiss. Wien, Math.-naturw. Kl. **74**, 285–294.

Dickson, J.A.D. (1966): Carbonate identification and genesis as revealed by staining. J. Sediment. Petrol. **36**, 491–505.

Digby, P.S.B. (1968): The mechanism of calcification in the molluscan shell. In: V. Fretter (ed.), Studies in the structure, physiology and ecology of Molluscs. Symp. Zool. Soc. London **22**, 93–107.

Dill, R.F. (1969): Submerged barrier reefs on the continental slope north of Darwin, Australia (abs.). Ann. Meeting Geol. Soc. Am., p. 264–266.

DiSalvo, L.H. (1969): Isolation of Bacteria from the corallum of *Porites lobata* (Vaughan) and its possible significance. Am. Zoologist **9**, 735–740.

Dittmar, H., Vogel, K. (1968): Die Spurenelemente Mangan und Vanadium in Brachiopodenschalen in Abhängigkeit vom Biotop. Chem. Geol. **3**, 95–110.

Dittmar, W. (1884): Report on researches into the composition of ocean water collected by H.M.S. Challenger. Challenger Reports, Phys. and Chem. **1**, 1–251.

Dodd, J.R. (1963): Paleoecological implications of shell mineralogy in two pelcypod species. J. Geol. **71**, 1–11.

Dodd, J.R. (1965): Environmental control of strontium and magnesium in *Mytilus*. Geochim. Cosmochim. Acta **29**, 383–398.

Dodd, J.R. (1966): Diagenetic stability of temperature-sensitive skeletal properties in *Mytilus* from the Pleistocene of California. Bull. Geol. Soc. Am. **77**, 1213–1224.

Dodd, J.R. (1967): Magnesium and strontium in calcareous skeletons: A review. J. Paleontol. **41**, 1313–1329.

Doderlein, L. (1898): Über die Lithonina, eine neue Gruppe von Kalkschwämmen. Zool. Jahrb. Abt. Syst. Geog. Biol. Tiere, **10**, 15–32.

Donahue, J. (1965): Laboratory growth of pisolite grains. J. Sediment. Petrol. **35**, 251–256.

Donahue, J. (1969): Genesis of oolite and pisolite grains: an energy index. J. Sediment. Petrol. **39**, 1399–1411.

Donn, W.L., Shaw, D.M. (1967): The generalized temperature curve for the past 425000 years: a discussion. J. Geol. **75**, 497–503.

Donnay, G., Pawson, D.L. (1969): X-ray diffraction studies of echinoderm plates. Science **166**, 1147–1150.

Doran, E. (1955): Land forms of the southeast Bahamas. Dept. Geogr. Univ. Texas, Austin, 38 pp.

Doty, M.S. (1954): Floristics and plant ecology of Raroia Atoll, Tuamotu. Part I. Floristic and ecological notes on Raroia. Atoll Res. Bull., No. 33, p. 1–41.

Doty, M.S., Newhouse, J., Miller, H.A., Wilson, K. (1954): Floristics and plant ecology of Raroia Atoll, Tuamotus. Atoll Res. Bull., No. 33, 58 pp.

Doyle, L.J. (1967): Black shells. Unpubl. thesis. Duke Univ., 69 pp.

Drew, G.H. (1911): The action of some denitrifying bacteria in tropical and temperate seas, and the bacterial precipitation of calcium carbonate in the sea. J. Marine Biol. Assoc. U. K. **9**, 142–155.

Driscol, E.G. (1967): Experimental field study of shell abrasion. J. Sediment. Petrol. **37**, 1117–1123.

Duane, D.B., Meisburger, E.P. (1969): Geomorphology and sediments of the nearshore continental shelf, Miami to Palm Beach, Florida. U. S. Army Corps Engineers Coastal Eng. Research Center Tech. Memo. 29, 47 pp.

DuBar, J.R., Johnson, H.S. (1964): Pleistocene "Coquina" at 20th Avenue South, Myrtle Beach, South Carolina, and other similar deposits. Southeastern Geol. **5**, 79–100.

Duerden, J.E. (1902): Boring algae as agents in the disintegration of corals. Bull. Am. Mus. Nat. Hist. **16**, 323–332.

Duncan, A.R. (1970): Petrology of rock samples from seamount near White Island, Bay of Plenty. New Zealand J. Geol. Geophys. **13**, 690–696.

Duncan, P.M. (1876): On some thallophytes parasitic within recent Madreporaria. Proc. Roy. Soc. London **25**, 238–257.

Dunham, R.J. (1962): Classification of carbonate rocks according to depositional texture. In: W.E. Ham (ed.), Classification of carbonate rocks. Am. Assoc. Petrol. Geologists Mem. **1**, 108–121.

Dunham, R.J. (1969): Early vadose silt in Townsend Mound (reef) New Mexico. In: G.M. Friedman (ed.), Depositional environments in carbonate rocks. Soc. Econ. Paleontologists and Mineralogists Special Publ. No. 14, p. 139–181.

DUPLESSY, J. C., LALOU, C., VINOT, A. C. (1970): Differential isotopic fractionation in benthic foraminifera and paleotemperatures reassessed. Science **168**, 250–251.

DURHAM, J. W., et al. (1966): Echinoids. In: R. C. MOORE (ed.), Echinodermata 3 (1), Treatise on invertebrate paleontology, Geol. Soc. Am., part U, p. 211–640.

EARDLEY, A. J. (1939): Sediments of the Great Salt Lake, Utah. Am. Assoc. Petrol. Geologists Bull. **22**, 1305–1411.

EASTON, W. H. (1969): Radiocarbon profile of Hanauma Reef, Oahu (abst.). Geol. Soc. Am. Special Papers **121**, 86.

EBANKS, W. J., JR. (1967): Recent carbonate sedimentation and diagenesis, Ambergris Cay, British Honduras. Ph. D. Thesis, Rice Univ., 189 pp.

EBANKS, W. J., JR., TEBBUTT, G. E. (1966): Diagenetic modification of recent sediments associated with a limestone island (abst.). Am. Assoc. Petrol. Geologists Bull. **50**, 611–612.

EBBS, N. K. (1966): The coral-inhabiting polycaetes of the northern Florida reef tract, part I: Aphroditidae, Polynoidae, Amphinomidae, Eunicidae and Lysaredtidae. Bull. Marine Sci. **16**, 485–555.

EBERT, T. A. (1968): Growth rates of the sea urchin *Strongylocentrotus purpuratus* related to food availability and spine abrasion. Ecology **49**, 1075–1091.

EDGINGTON, D. N., GORDON, S. A., THOMMES, M. M., ALMODOVAR, L. R. (1970): The concentration of radium, thorium, and uranium by tropical marine algae. Limnol. Oceanog. **15**, 945–955.

EDMONDSON, C. H. (1929): Growth of Hawaiian corals. Bull. Bernice P. Bishop Mus. **58**, 1–38.

EICHLER, R., RISTEDT, H. (1966): Isotopic evidence on the early life history of *Nautilus pompilius* (Linne). Science **153**, 734–736.

EISMA, D. (1966): The influence of salinity on mollusk shell mineralogy: a discussion. J. Geol. **74**, 89–94.

EKMAN, S. (1953): Zoogeography of the sea, 417 pp. London: Sidgwick & Jackson.

EMELYANOV, E. M. (1965): Karbonatnost sovremennych donnykh otlozhenii Sredizemnogo morya (Carbonate content of recent sediments of the Mediterranean Sea). In: Osnovonge cherty geologicheskogo stoeniya gidrologicheskogo rezhima ibiologii Sredizemnogo morya, p. 71–83. Moscow: "Nauka" Publ. House.

EMELYANOV, E. M., SHIMKUS, K. M. (1971): Suspended matter in the Mediterranean Sea (abst.). 8th Intern. Sediment. Congress, Heidelberg, p. 27.

EMERY, K. O. (1946): Marine solution basins. J. Geol. **54**, 209–228.

EMERY, K. O. (1956a): Marine geology of Johnston Island and its surrounding shallows, Central Pacific Ocean. Bull. Geol. Soc. Am. **67**, 1505–1520.

EMERY, K. O. (1956b): Sediments and water of Persian Gulf. Am. Assoc. Petrol. Geologists Bull. **40**, 2354–2383.

EMERY, K. O. (1962): Marine geology of Guam. U. S. Geol. Surv. Profess. Papers **403-B**, 76 pp.

EMERY, K. O. (1968): Relict sediments on continental shelves of the world. Am. Assoc. Petrol. Geologists Bull. **52**, 445–464.

EMERY, K. O., COX, D. C. (1956): Beachrock in the Hawaiian Islands. Pacific Sci. **10**, 382–402.

EMERY, K. O., MERRILL, A. S., TRUMBULL, J. V. A. (1965): Geology and biology of the sea floor as deduced from simultaneous photographs and samples. Limnol. Oceanog. **101**, 1–21.

EMERY, K. O., MILLIMAN, J. D. (1970): Quaternary sediments of the Atlantic continental shelf of the United States. Quaternaria **12**, 3–18.

EMERY, K. O., TRACEY, J. I., JR., LADD, H. S. (1954): Geology of Bikini and nearby atolls. U. S. Geol. Surv. Profess. Papers **260-A**, 265 pp.

EMILIANI, C. (1954): Temperatures of Pacific bottom waters and polar surficial waters during the Tertiary. Science **119**, 853–855.

EMILIANI, C. (1955a): Mineralogical and chemical composition of the tests of certain pelagic foraminifera. Micropaleontology **1**, 377–380.

EMILIANI, C. (1955b): Pleistocene temperatures. J. Geol. **63**, 538–578.

EMILIANI, C. (1967): The generalized temperature curve for the past 425000 years: a reply. J. Geol. **75**, 504–510.

EMILIANI, C. (1970): Pleistocene paleotemperatures. Science **168**, 822–825.

EMILIANI, C. (1971): Depth habitats of growth stages of pelagic foraminifera. Science **173**, 1122–1124.

EMILIANI, C., FLINT, R. F. (1963): The Pleistocene record. In: M. N. HILL (ed.), The sea, vol. 3, p. 888–927. New York: Interscience Publ.

EMRICH, K., EHHALT, D. H., VOGEL, J. C. (1970): Carbon isotope fractionation during the precipitation of calcium carbonate. Earth and Planet. Sci. Letters **8**, 363–371.

ENDEAN, R., JONES, O.A. (1972): Biology and geology of Coral Reefs. New York: Academic Press. In press.

EPSTEIN, S. (1959): The variations of the O^{18}/O^{16} ratio in nature and some geologic implications. In: ABELSON, P.H. (ed.), Researches in geochemistry, p. 217–240. New York: John Wiley & Sons, Inc.

EPSTEIN, S., BUCHSBAUM, R., LOWENSTAM, H., UREY, H.C. (1951): Carbonate-water isotope temperature scale. Bull. Geol. Soc. Am. **62**, 417–425.

EPSTEIN, S., BUCHSBAUM, R., LOWENSTAM, H., UREY, H.C. (1953): Revised carbonate-water isotopic temperature scale. Bull. Geol. Soc. Am. **64**, 1315–1325.

EPSTEIN, S., GRAF, D.L., DEGENS, E.T. (1964): Oxygen isotope studies on the origin of dolomites. In: H. CRAIG, S.L. MILLER, and G.J. WASSERBURG (eds.), Isotopic and cosmic chemistry, p. 169–180. Amsterdam: North-Holland Publ. Co.

EPSTEIN, S., LOWENSTAM, H.A. (1953): Temperature-shell-growth relations of Recent and interglacial Pleistocene shoal-water biota from Bermuda. J. Geol. **61**, 424–438.

EPSTEIN, S., MAYEDA, T. (1953): Variation of O^{18} content of waters from natural sources. Geochim. Cosmochim. Acta **4**, 213–224.

ERICSON, D.B., EWING, M., WOLLIN, G., HEEZEN, B.C. (1961): Atlantic deep-sea sediment cores. Bull. Geol. Soc. Am. **72**, 193–286.

ERWE, W. (1913): Holothurioidea. In: W. MICHAELSON and R. HARTMEYER (eds.), Die Fauna Südwest-Australiens **4**, No. 9, 351–402.

EVANS, G. (1966): The recent sedimentary facies of the Persian Gulf. In, A discussion concerning the floor of the northwest Indian Ocean. Phil. Trans. Roy. Soc. London Ser. A. **259**, 291–298.

EVANS, G., SCHMIDT, V., BUSH, P., NELSON, H. (1969): Stratigraphy and geologic history of the Sabkha, Abu Dhabi, Persian Gulf. Sedimentology **12**, 145–159.

EVANS, G., SHEARMAN, D.J. (1964): Recent celestite from the sediments of the Trucial Coast of the Persian Gulf. Nature **202**, 385–386.

EVANS, J.W. (1968): Growth rate of the rock-boring clam *Penitella penita* (CONRAD, 1937) in relation to hardness of rock and other factors. Ecology **49**, 619–628.

FABRICIUS, F.H., BERDAU, D., MUNNICH, K.O. (1970a): Early Holocene ooids in modern littoral sands reworked from a coastal terrace, southern Tunisia. Science **169**, 757–760.

FABRICIUS, F.H., RAD, U., VON, HESSE, R., OTT, W. (1970b): Die Oberflächensedimente der Straße von Otranto (Mittelmeer). Geol. Rundschau **60**, 164–192.

FAIRBRIDGE, R.W. (1950): Recent and Pleistocene coral reefs of Australia. J. Geol. **58**, 330–401.

FAIRBRIDGE, R.W. (1957): The dolomite question. In: R.J. LEBLANC and J.G. BREEDING (eds.), Regional aspects of carbonate deposition. Soc. Econ. Paleontologists and Mineralogists Special Publ. **5**, 125–178.

FALLS, D.L., TEXTORIS, D.A. (1970): Size, grain type, and mineralogical relationship in recent marine calcareous beach sands (abst.). Progr. Geol. Soc. Am. S. E. Section, p. 208.

FARROW, G.E. (1971): Back-reef and lagoonal environments of Aldabra Atoll distinguished by their crustacean borrows. In: D.R. STODDART and M. YONGE (eds.), Regional variation in Indian Ocean coral reefs, Symp. Zool. Soc. London, No. 28, p. 455–500.

FEELY, H.W., KULP, J.L. (1957): Origin of Gulf Coast salt dome sulfur deposits. Am. Assoc. Petrol. Geologists Bull. **41**, 1802–1853.

FIELD, M.F., PILKEY, O.H. (1970): Lithification of deep-sea sediments by pyrite. Nature **226**, 836–837.

FIELD, R.M. (1920): Investigations regarding the calcium carbonate oozes at Tortugas and the beach-rock at Loggerhead Key. Carnegie Inst. Wash. Yearbook **18**, 197–198.

FINCKH, A.E. (1904): The biology of the reef-forming organisms at Funafuti Atoll. In: The atoll of Funafuti, p. 125–150. Roy. Soc. London.

FISCHER, A.G., GARRISON, R.E. (1967): Carbonate lithification on the sea floor. J. Geol. **75**, 488–497.

FISCHER, A.G., GEALY, E.L. (1969): Summary and comparison of lithology and sedimentary sequence in northwest Atlantic and northwest Pacific (abst.). Ann. Meeting Geol. Soc. Am., p. 65.

FLEECE, J.H. (1962): The carbonate geochemistry and sedimentology of the Keys of Florida Bay, Florida. Contrib. No. 5, Sed. Res. Lab. Dept. Geol. Fla. State Univ., Tallahassee.

FOLK, R.L. (1959): Practical petrographic classification of limestones. Am. Assoc. Petrol. Geologists Bull. **43**, 1–38.

FOLK, R.L. (1962): Spectral subdivision of limestone types. In: W.E. HAM (ed.), Classification of carbonate rocks. Am. Assoc. Petrol. Geologists Mem. **1**, 62–84.

FOLK, R.L. (1965): Some aspects of recrystallization in ancient limestones. In: L.C. PRAY and

R.C. MURRAY (eds.), Dolomitization and limestone diagenesis. Soc. Econ. Paleontologists and Mineralogists Special Publ. **13**, 14–48.

FOLK, R.L. (1967): The sand cays of Alacran Reef, Yucatan, Mexico: morphology. J. Geol. **75**, 412–437.

FOLK, R.L., ROBLES, R. (1964): Carbonate sands of Isla Perez, Alacran Reef Complex, Yucatan. J. Geol. **72**, 255–292.

FONTES, J.C., KULBICKI, G., LETOLLE, R. (1969): Les sondages de l'atoll de Mururoa; apercu geochimique et isotopique de la serie carbonate. Cahiers. Pacif. No. 27, p. 69–74.

FORCE, L.M. (1969): Calcium carbonate size distribution on the west Florida shelf and experimental studies on the microarchitectural control of skeletal breakdown. J. Sediment. Petrol. **39**, 902–934.

FOSBERG, F.R. (1962): A brief study of the cays of Arrecife Alacran, a Mexican atoll. Atoll Res. Bull. No. 93, 25 pp.

FOSBERG, F.R., CARROLL, D. (1965): Terrestrial sediments and soils of the northern Marshall Islands. Atoll Res. Bull. No. 113, 156 pp.

FOSLIE, M. (1895): The Norwegian forms of *Lithothamnium*. Det. Kgl. Norske Vidensk. Selsk. Skrifter (1894), p. 29–208.

FOSLIE, M., PRINZ, H. (1929): Contributions to a monograph of the Lithothamnia. Aktietrykkeriet I Trondhjem, 60 pp.

FOSTER, G.L., BENSON, R.H. (1958): Constituents and structural arrangement in ostracode carapaces (abst.). Bull. Geol. Soc. Am. **69**, 1565.

FOWLER, M.L., DODD, J.R. (1969): Magnesium and strontium variation within echinoid skeletons (abst.). Program, Ann. Meeting Geol. Soc. Am. S. E. Section p. 24–25.

FRANKENBERG, D., COLES, S.L., JOHANNES, R.E. (1967): The potential trophic significance of *Callianassa major* fecal pellets. Limnol. Oceanog. **12**, 113–120.

FREEMAN, T. (1962): Quiet water oolites from Laguna Madre, Texas. J. Sediment. Petrol. **32**, 475–483.

FRÉMY, P. (1945): Contribution à la physiologie des Thallophytes marins perforant et cariant les roches calcaires et coquilles. Ann. Inst. Oceanog. (Paris). **22**, 107–143.

FRIEDMAN, G.M. (1959): Identification of carbonate minerals by staining methods. J. Sediment. Petrol. **29**, 87–97.

FRIEDMAN, G.M. (1964): Early diagenesis and lithification in carbonate sediments. J. Sediment. Petrol. **34**, 777–813.

FRIEDMAN, G.M. (1965): Occurrence and stability relationships of aragonite, high-magnesian calcite, and low-magnesian calcite under deep-sea conditions. Bull. Geol. Soc. Am. **76**, 1191–1196.

FRIEDMAN, G.M. (1969): Trace elements as possible environmental indicators in carbonate sediments. In: G.M. FRIEDMAN (ed.), Depositional environments in carbonate rocks. Soc. Econ. Paleontologists and Mineralogists, Special Publ. No. 14, p. 193–198.

FRIEDMAN, G.M. (1972): Coral reef rock from Red Sea: Sequence and time scale for progressive diagenesis and its effect on porosity and permeability (abst.). Am. Assoc. Petrol. Geologists Bull. **56**, 618.

FRIEDMAN, G.M., AMIEL, A.J., SCHNEIDERMANN, N. (1970): Submarine cements in modern Red Sea reef rock (abst.). Ann. Meeting Geol. Soc. Am. p. 554–555.

FRIEDMAN, G.M., AMIEL, A.S., BRAUN, M., MILLER, D.S. (1972): Algal mats, carbonate laminites, ooids, oncolites, and cements in hypersaline sea-marginal pool, Gulf of Aqaba, Red Sea (abst.). Am. Assoc. Petrol. Geologists Bull. **56**, 618.

FRIEDMAN, G.M., FABRICAND, B.P., IMBIMBO, E.S., BREY, M.E., SANDERS, J.E. (1968): Chemical changes in interstitial waters from continental shelf sediments. J. Sediment. Petrol. **38**, 1313–1319.

FRIEDMAN, G.M., GAVISH, E. (1971): Mediterranean and Red Sea (Gulf of Aqaba) beachrocks. In: O.P. BRICKER (ed.), Carbonate cements. The Johns Hopkins Univ. Studies in Geology, No. 19, p. 13–16.

FRIEDMAN, G.M., SANDERS, J.E. (1967): Origin and Occurrence of dolostones. In: G.V. CHILINGAR, H.J. BISSELL and R.W. FAIRBRIDGE (eds.), Carbonate rocks (a), p. 267–348. New York: Elsevier Publ. Co.

FRISHMAN, S.A., BEHRENS, E.W. (1969): Geochemistry of oolites, Baffin Bay, Texas (abst.). Program, Ann. Meeting Geol. Soc. Am. p. 71.

FRITSCH, F.E. (1945): Structure and reproduction of the algae, vol. 2, 939 pp. London: Cambridge Univ. Press.

FRITZ, P., SMITH, D.G.W. (1970): The isotopic concentration of secondary dolomites. Geochim. Cosmochim. Acta **34**, 1161–1173.

FRIZZELL, D.L., EXLINE, H. (1955): Micropaleontology of holothurian sclerites. Micropaleontology **1**, 335–342.

GAARDER, K.R. (1971): Comments on the distribution of coccolithophorids in the oceans. In: B.M. FUNNELL and W.R. RIEDEL (eds.), The micropaleontology of oceans, p. 97–103. London: Cambridge Univ. Press.

GALTSOFF, P.S. (1964): The American oyster *Crassostrea virginica* Gmelin. Fishery Bull. Fish & Wildlife Serv. **64**, 480 pp.

GAMULIN-BRIDA, H. (1967): The benthonic fauna of the Adriatic Sea. In: H. BARNES (ed.), Oceanog. Marine Biol. Ann. Rev., vol. 5, p. 535–568. London: George Allen & Unwin Ltd.

GARDINER, J.S. (1903): The Maldive and Laccadive Groups, with notes on other coral formations in the Indian Ocean. In: J.S. GARDINER (ed.), The fauna and geography of the Maldive and Laccadive Archipelagoes, vol. 1, p. 313–346. London: Cambridge Univ. Press.

GARDINER, J.S. (1931): Coral reefs and atolls, 181 pp. London: MacMillan & Co., Ltd.

GARRELS, R.M. (1965). Silica: role in the buffering of natural waters. Science **148**, 69.

GARRELS, R.M., THOMPSON, M.C. (1962): A chemical model for sea water at 20 °C and one atmosphere total pressure. Am. J. Sci. **260**, 57–66.

GARRETT, P., SMITH, D.L., WILSON, A.O., PATRIQUIN, D. (1971): Physiography, ecology, and sediments of two Bermuda patch reefs. J. Geol. **79**, 647–668.

GARRISON, R.E., LUTERNAUER, J.L., GRELL, E.V., MACDONALD, R.D., MURRAY, J.W. (1961): Early diagenetic cementation of recent sands, Fraser River Delta, British Columbia. Sedimentology **12**, 27–46.

GARTNER, S., JR. (1970): Sea-floor spreading, carbonate dissolution level, and the nature of Horizon A. Science **169**, 1077–1079.

GEBELEIN, C.D. (1969): Distribution, morphology, and accretion rate of recent subtidal algal stromatolites, Bermuda. J. Sediment. Petrol. **39**, 49–69.

GEBELEIN, C.D., HOFFMAN, P. (1968): Intertidal stromatolites and associated facies from Lake Ingraham, Cape Sable, Florida (abst.). Ann. Meeting Geol. Soc. Am. p. 109.

GEBELEIN, C.D., HOFFMAN, P. (1971): Algal origin of dolomite in interlaminated limestone-dolomite sedimentary rocks. In: O.P. BRICKER (ed.), Carbonate cementation, The Johns Hopkins Univ. Studies in Geology, No. 19, p. 319–326.

GEE, H. (1934): Lime deposition and the bacteria. I. Estimate of bacterial activity at the Florida Keys. Carnegie Inst. Wash. Publ. **435**, p. 67–82.

GEE, H., MOBERG, E.G., GREENBERG, D.M., REVELLE, R. (1932): Calcium equilibrium in sea water. Scripps Inst. Oceanogr. Univ. Calif. Bull. Tech. Ser. **3**, 145–200.

GEVIRTZ, J.L., FRIEDMAN, G.M. (1966): Deep-sea carbonate sediments of the Red Sea and their implications on marine lithification. J. Sediment Petrol. **36**, 143–151.

GHISELIN, M.T., DEGENS, E.T., SPENCER, D.W., PARKER, R.H. (1967): A phylogenetic survey of molluscan shell matrix proteins. Breviora No. 262, 35.

GILMARTIN, M. (1960): The ecological distribution of the deep water algae of Eniwetok Atoll. Ecology **41**, 209–221.

GINSBURG, R.N. (1953a): Intertidal erosion on the Florida Keys. Bull. Marine Sci. **3**, 55–69.

GINSBURG, R.N. (1953b): Beachrock in South Florida. J. Sediment. Petrol. **23**, 89–92.

GINSBURG, R.N. (1956): Environmental relationships of grain size and constituent particles in some South Florida carbonate sediments. Am. Assoc. Petrol. Geologists Bull. **40**, 2384–2427.

GINSBURG, R.N. (1957): Early diagenesis and lithification of shallow-water carbonate sediments in South Florida. In: R.J. LEBLANC and J.G. BREEDING (eds.), Regional aspects of carbonate sedimentation. Soc. Econ. Paleontologists and Mineralogists Special Publ. No. 5, p. 80–100.

GINSBURG, R.N. (1960): Ancient analogues of recent stromatolites. Rept. 21st International Geol. Congr. Copenhagen Part 22, p. 26–35.

GINSBURG, R.N. (1964): South Florida carbonate sediments. Ann. Meeting Geol. Soc. Am., Guidebook Field Trip No. 1, 72 pp.

GINSBURG, R.N., BERNARD, H.A., MOODY, R.A., DAIGLE, E.E. (1966): The Shell method of impregnating cores of unconsolidated sediments. J. Sediment. Petrol. **36**, 1118–1125.

GINSBURG, R.N., BRICKER, O.P., WANLESS, H.R., GARRETT, P. (1970): Exposure index and sedimentary structures of a Bahama tidal flat (abst.). Ann. Meeting Geol. Soc. Am., p. 744–745.

GINSBURG, R.N., ISHAM, L.B., BEIN, S.J., KUPERBERG, J. (1954): Laminated algal sediments of south Florida, and their recognition in the fossil record. Unpubl. Rept. No. 54.21, Marine Laboratory Univ. Miami, 33 pp.

GINSBURG, R. N., JAMES, N. P. (1973): British Honduras by submarine. Geotimes **18**, 23–24.

GINSBURG, R. N., LLOYD, R. M., STOCKMAN, K. W., MCCALLUM, J. S. (1963): Shallow-water carbonate sediments. In: M. N. HILL (ed.), The seas, vol. 3, p. 554–582. New York: Interscience Publ.

GINSBURG, R. N., LOWENSTAM, H. A. (1958): The influence of marine bottom communities on the depositional environment of sediments. J. Geol. **66**, 310–318.

GINSBURG, R. N., MARSZALEK, D. S., SCHNEIDERMANN, N. (1971a): Ultrastructure of carbonate cements in a Holocene algal reef of Bermuda. J. Sediment. Petrol. **41**, 472–482.

GINSBURG, R. N., SCHROEDER, J. H., SHINN, E. A. (1971b): Recent synsedimentary cementation in subtidal Bermuda reefs. In: O. P. BRICKER (ed.), Carbonate cements. The Johns Hopkins Univ. Studies in Geology, No. 19, p. 54–58.

GLOVER, E. D., PRAY, L. C. (1971): High magnesium calcite and aragonite cementation within modern subtidal carbonate sediment grains. In: O. P. BRICKER (ed.), Carbonate cements. The Johns Hopkins Univ. Studies in Geology, No. 19, p. 80–87.

GLOVER, E. D., SIPPEL, R. F. (1967): Synthesis of magnesium calcite. Geochim. Cosmochim. Acta **31**, 503–613.

GLYNN, P. W. (1962): *Hermodice carunculata* and *Mithraculus sculptus*, two hermatypic coral predators. Assoc. Island Marine Labs. Caribb. 4th Meeting, Curacao, p. 16–17.

GLYNN, P. W. (1963): Species composition of *Porites furcata* reefs in Puerto Rico with notes on habitat niches. Assoc. Island Marine Labs Caribb. 5th Meeting, p. 6–9.

GLYNN, P. W. (1973): Aspects of the ecology of coral reefs in the western Atlantic region. In: R. ENDEAN and O. A. JONES (eds.), Biology and geology of coral reefs. New York: Academic Press (in press).

GOHAR, H. A. F., LATIF, A. F. A. (1959): Morphological studies on the gut of some scarid and labrid fishes. Publ. Mar. Biol. Sta. Al-Ghardaqa (Red Sea), No. 10, p. 145–190.

GOLDBERG, E. D. (1965): Minor elements in sea water. In: J. P. RILEY and G. SKIRROW (eds.), Chemical oceanography, p. 163–196. London: Academic Press.

GOLDMAN, M. I. (1926): Proportions of detrital organic calcareous constituents and their chemical alteration in a reef sand from the Bahamas. Papers Tortugas Lab. Carnegie Inst. Wash. **23**, 37–66.

GOLDSMITH, J. R. (1959): Some aspects of the geochemistry of carbonates. In: P. H. ABELSON (ed.), Researches in geochemistry, p. 336–358. New York: John Wiley & Sons, Inc.

GOLDSMITH, J. R., GRAF, D. L. (1958a): Relation between lattice constants and composition of the Ca-Mg carbonates. Am. Mineralogist **43**, 84–101.

GOLDSMITH, J. R., GRAF, D. L. (1958b): Structural and compositional variations in some natural dolomites. J. Geol. **66**, 678–693.

GOLDSMITH, J. R., GRAF, D. L., HEARD, H. C. (1961): Lattice constants of the calcium-magnesium carbonates. Am. Mineralogist **46**, 453–457.

GOLDSMITH, J. R., GRAF, D. L., JOENSUU, O. (1955): The occurrence of magnesian calcite in nature. Geochim. Cosmochim. Acta **7**, 212–230.

GOLUBIC, S. (1969): Distribution, taxonomy and boring patterns of marine endolithic algae. Am. Zoologist **9**, 747–751.

GOMBERG, D. N., BONATTI, E. (1970): High-magnesian calcite: Leaching of magnesium in the deep sea. Science **168**, 1451–1453.

GOODELL, H. G., GARMAN, R. K. (1969): Carbonate geochemistry of Superior deep test well, Andros Island, Bahamas. Am. Assoc. Petrol. Geologists Bull. **53**, 513–536.

GOODELL, H. G., KUNZLER, R. H. (1965): Thermal inversion of aragonite to calcite (abst.). Geol. Soc. Am. Special Paper, No. 82, p. 300.

GORDON, C. M., CARR, R. A., LARSON, R. E. (1970): The influence of environmental factors on the sodium and manganese content of barnacle shells. Limnol. Oceanog. **15**, 461–466.

GOREAU, T. F. (1959a): The ecology of Jamaican coral reefs. I. Species composition and zonation. Ecology **40**, 67–89.

GOREAU, T. F. (1959b): The physiology of skeleton formation in corals. I. A method for measuring the rate of calcium deposition by corals under different conditions. Biol. Bull. **116**, 59–75.

GOREAU, T. F. (1961): On the relation of calcification to primary productivity in reef-building organisms. In: The biology of hydra, p. 269–285. Miami: Univ. Miami Press.

GOREAU, T. F. (1963): Calcium carbonate deposition by coralline algae and corals in relation to their roles as reef builders. Ann. N. Y. Acad. Sci. **109**, 127–167.

GOREAU, T. F. (1964): Mass expulsion of zooxanthellae from Jamaican reef communities after Hurricane Flora. Science **145**, No. 3630, 383–386.

GOREAU, T. F., GOREAU, N. I. (1960a): Distribution of labelled carbon in reef-building corals with and without zooxanthellae. Science **131**, 668–669.

GOREAU, T. F., GOREAU, N. I. (1960b): The physiology of skeleton formation in corals. IV. On isotopic equilibrium exchanges of calcium between corallum and environment in living and dead reef-building corals. Biol. Bull. **119**, 416–427.

GOREAU, T. F., GOREAU, N. I., NEUMANN, Y., YONGE, C. M. (1968): *Fungiacava eilatensis* n. gen., n. sp. (Bivalvia, Mytilidae), a boring bivalve commensal in reef corals (abst.). Am. Zoologist **8**, 799.

GOREAU, T. F., GRAHAM, E. A. (1967): A new species of *Halimeda* from Jamaica. Bull. Marine Sci. **17**, 432–441.

GOREAU, T. F., HARTMAN, W. D. (1963): Boring sponges as controlling factors in the formation and maintenance of coral reefs. In: R. F. SOGNNAES (eds.), Mechanisms of hard tissue destruction, Publ. 75, Am. Assoc. Adv. Sci., p. 23–54.

GOREAU, T. F., LANG, J. C., GRAHAM, E. A., GOREAU, P. D. (1972): Structure and ecology of Saipan reefs in relation to predation by *Acanthaster planci* (Linnaeus). Bull. Marine Sci. **25**, 113–152.

GOREAU, T. F., WELLS, J. W. (1967): The shallow-water Scleractinia of Jamaica: Revised list of species and their vertical distribution range. Bull. Marine Sci. **17**, 442–453.

GORSLINE, D. S. (1963): Environments of carbonate deposition in Florida Bay and the Florida Straits. Shelf carbonates of the Paradox Basin. A Symposium, Fourth Field Conference, 1963 Four Corners Geological Society, p. 130–143.

GOTO, M. (1961): Some mineral-chemical problems concerning calcite and aragonite with special reference to the genesis of aragonite. J. Faculty Science, Hokkaido Univ. Japan **10**, 571–640.

GOULD, H. R., STEWARD, R. H. (1955): Continental terrace sediments in the northeastern Gulf of Mexico. In: Finding ancient shorelines. Soc. Econ. Paleontologists and Mineralogists Special Publ. **5**, 2–19.

GRAF, D. L. (1960): Geochemistry of carbonate sediments and sedimentary carbonate rocks. Div. Ill. State Geol. Survey Circ. **297**, Parts 1–5.

GRAF, D. L., EARDLEY, A. J., SHIMP, N. F. (1961): A preliminary report on magnesium carbonate formation in Glacial Lake Bonneville. J. Geol. **69**, 219–223.

GRAF, D. L., GOLDSMITH, J. R. (1956): Some hydrothermal synthesis of dolomite and protodolomite. J. Geol. **64**, 137–186.

GRANT, R. E. (1826): Notice of a new zoophyte (*Cliona celata* Gr.) from the Firth of Forth. Edinburgh New Philos. J. p. 78–81.

GRAVE, B. H. (1933): Rate of growth, age at sexual maturity and duration of life of certain sessile organisms at Woods Hole, Massachusetts. Biol. Bull. **65**, 375–386.

GREENFIELD, L. J. (1963): Metabolism and concentration of calcium and magnesium and precipitation of calcium carbonate by a marine bacterium. Ann. N. Y. Acad. Sci. **109**, 23–45.

GREZE, I. I. (1967): On the amount of chitin and calcite in shells of Amphipoda (Gammaridea). Zool. Sh., **46**, 1655–1658.

GRIFFIN, J. J., WINDOM, H., GOLDBERG, E. D. (1968): The distribution of clay minerals in the world ocean. Deep-Sea Res. **15**, 433–459.

GROSS, M. G. (1964): Variations in the O^{18}/O^{16} and C^{13}/C^{12} ratios in diagenetically altered limestones in the Bermuda Islands. J. Geol. **72**, 170–194.

GROSS, M. G. (1965): Carbonate deposits on Plantagenet Bank near Bermuda. Bull. Geol. Soc. Am. **76**, 1283–1290.

GROSS, M. G., MILLIMAN, J. D., TRACEY, J. I., JR., LADD, H. S. (1969): Marine geology of Kure and Midway Atolls, Hawaii: a preliminary report. Pacific Sci. **23**, 17–25.

GROSS, M. G., TRACEY, J. I., JR. (1966): Oxygen and carbon isotopic composition of limestones and dolomites, Bikini and Eniwetok Atolls. Science **151**, 1082–1084.

GUILCHER, A. (1961): Le "beach-rock" ou grès de plage. Ann. Geog. **70** (378), 113–125.

GUILCHER, A. (1964): La sédeimentation sous-marine dans la partie orientale de la Rade de Brest, Bretagne. In: L. M. J. U. VAN STRAATEN (ed.), Developments in sedimentology, vol. 1, p. 148–156.

GUILCHER, A. (1965): Coral reefs and lagoons of Mayotte Island, Comoro Archipelago, Indian Ocean, and of New Caledonia, Pacific Ocean. In: Proc. 17th Symposium Colston Res. Soc. Butterworths Scientific Publ., London, p. 21–45.

GUILCHER, A., PONT, P. (1957): Étude experimentale de la corrosion littorale des calcaires. Bull. Assoc. Geogr. Franc., No. 265–266, p. 48–62.

HAAS, P., HILL, T. G., KARSTENS, W. K. H. (1935): The metabolism of calcareous algae. II. The seasonal variation in certain metabolic products of *Corallina squamata* Ellis. Ann. Botany **49**, 609–619.

HAGEN, G.M., LOGAN, B.W. (1973): Tidal flat history and sedimentation, Hutchinson Embayment, Western Australia. Am. Assoc. Petrol. Geologists Bull., in press.

HAIGLER, S.A. (1969): The boring mechanism of *Polydora websteri* inhabiting *Crassostrea virginica*. Am. Zoologist **9**, 821–828.

HALLAM, A., PRICE, N.B. (1966): Strontium contents of recent and fossil aragonitic cephalopod shells. Nature **212**, 25–27.

HALLAM, A., PRICE, N.B. (1968): Environmental and biochemical control of strontium in shells of *Cardium edule*. Geochim. Cosmochim. Acta **32**, 319–328.

HALLDAL, P., MARKALI, J. (1955): Electron microscope studies on coccolithophorids from the Norwegian Sea, the Gulf Stream and the Mediterranean. Skrifter Norske Videnskaps-Akad., Oslo **1**, p. 5–29.

HALSEY, S.D., PERKINS, R.D. (1970): Microborings in relict sediments: A possbile key to Carolina shelf history (abst.). Ann. Meeting, Geol. Soc. Am. 565.

HAM, W.E. (ed.) (1962): Classification of carbonate rocks. Am. Assoc. Petrol. Geologists, Mem. **1**, 279 pp.

HAM, W.E., PRAY, L.C. (1962): Modern concepts and classifications of carbonate rocks. In: W.E. HAM (ed.), Classification of carbonate rocks. Am. Assoc. Petrol. Geologists, Mem. **1**, 2–19.

HAMAI, I. (1935): On the growth of the shell *Meretrix meretrix*, especially with regard to periodicity of growth relative to seasonal variations in the environment. Sci. Repts., Tohuku Univ., 4th Ser. **9**, 339–371.

HAMILTON, E.L. (1959): Thickness and consolidation of deep-sea sediments. Bull. Geol. Soc. Am. **70**, 1399–1424.

HAMPTON, J.S. (1958): Chemical analysis of holothurian sclerites. Nature **181**, 1608–1609.

HANSON, J. (1948): Formation and breakdown of serpulid tubes. Nature **161**, 610.

HANTZSCHEL, W., EL-BAZ, F., AMSTUTZ, G.C. (1968): Coprolites. An annotated bibliography. Geol. Soc. Am. Mem. **108**, 132 pp.

HARE, P.E. (1963): Amino acids in the proteins from aragonite and calcite in the shells of *Mytilus californianus*. Science **139**, 216–217.

HARE, P.E., ABELSON, P.H. (1964): Comparative biochemistry of the amino acids in molluscan shell structures (abst.). Geol. Soc. Am. Special Paper **82**, 84.

HARE, P.E., ABELSON, P.H. (1965): Amino acid composition of some calcified proteins. Carnegie Inst. Wash. Yearbook **64**, 223–232.

HARE, P.E., MEENAKSHI, V.R. (1968): Organic composition of some molluscan shell structures, including periostracum (abst.). Am. Zoologist **8**, 792.

HARRISS, R.C. (1965): Trace element distribution in molluscan skeletal material. I. Magnesium, iron, manganese, and strontium. Bull. Marine Sci. **15**, 265–273.

HARRISS, R.C., ALMY, C.C. (1964): A preliminary investigation into the incorporation and distribution of minor elements in the skeletal material of scleractinian corals. Bull. Marine Sci. **14**, 418–423.

HARRISS, R.C., PILKEY, O.H. (1966): Temperature and salinity control of the concentration of skeletal Na, Mn, and Fe in *Dendraster excentricus*. Pacific Sci. **20**, 235–238.

HART, G.F., PIENAAR, R.N., CAVENEY, R. (1966): An aragonitic coccolith from South Africa. S. African J. Sci. **61**, 425–426.

HARTMAN, O. (1954): Marine annelids from the northern Marshall Islands. U.S. Geol. Surv. Profess. Paper **260-Q**, 619–644.

HARTMAN, W.D. (1957): Ecological niche differentiation in the boring sponges (Clionidae). Evolution **11**, 294–297.

HARTMAN, W.D., GOREAU, T.F. (1970a): Jamaican coralline sponges: their morphology, ecology and fossil relatives. Symposia, Zool. Soc. London **25**, p. 205–243.

HARTMAN, W.D., GOREAU, T.F. (1970b): A new Pacific sponge: Homeomorph or descendent of the tabulate "corals"? (abst.). Ann. Meeting, Geol. Soc. Am. p. 570.

HARVEY, H.W. (1960): The chemistry and fertility of sea water, 240 pp. London: Cambridge Univ. Press.

HASKIN, H.H. (1954): Age determinations in mollusks. Trans. N.Y. Acad. Sci. **16**, 300–304.

HATHAWAY, J.C., DEGENS, E.T. (1968): Methane-derived marine carbonates of Pleistocene age (abst.). Ann. Meeting Geol. Soc. Am., p. 129–130.

HATHAWAY, J.C., DEGENS, E.T. (1969): Methane-derived marine carbonates of Pleistocene age. Science **165**, 690–692.

Hathaway, J.C., Sachs, P.L. (1965): Sepiolite and clinoptilolite from the Mid-Atlantic Ridge. Am. Mineralogist **50**, 852–867.

Hawkshaw, C. (1878): On the action of limpets in sucking pits and abrading the surface of chalk at Dover. J. Limnol. Soc. (Zool.) **14**, 406–411.

Hawley, J., Pytkowicz, R.M. (1969): Solubility of calcium carbonate in seawater at high pressures and 2 °C. Geochim. Cosmochim. Acta **33**, 1557–1561.

Haynes, J. (1965): Symbiosis, wall structure and habitat in Foraminifera. Contrib. Cushman Found. Foram. Res. **16**, 40–44.

Heath, G.R. (1969): Carbonate sedimentation in the abyssal equatorial Pacific during the past 50 million years. Bull. Geol. Soc. Am. **80**, 689–694.

Hedley, R.H. (1956a): Studies on serpulid tube formation. I. The secretion of the calcareous and organic components of the tube by *Pomatoceros triqueter*. Quart. J. Microscop. Sci. **97**, 411–419.

Hedley, R.H. (1956b): Studies on serpulid tube formation. II. The calcium secreting glands in the peristomium of *Spirorbis*, *Hydroides* and *Serpula*. Quart. J. Microscop. Sci. **97**, 421–427.

Hedley, R.H. (1958): Tube formation by *Pomatoceros triqueter* (Polychaeta). J. Marine Biol. Assoc., U. K. **37**, 315–322.

Helfrich, P., Townsley, S.J. (1963): In: F.R. Fosberg (ed.), Man's place in the island ecosystem. p. 39–56. Honolulu, Bishop Museum Press.

Herdman, W.A. (1885): The presence of calcareous spicules in the Tunicata. Proc. Liverpool Geol. Soc. **5**, 46–51.

Herdman, W.A., Lomas, J. (1898): On the floor deposits of the Irish Sea. Proc. Liverpool Geol. Soc. **13**, 205–232.

Herman, Y.R. (1965): Études des sediments Quaternaires de la Mer Rouge. Ph. D. Thesis, Univ. Paris, Masson & Cie. Editeurs, Paris, p. 341–415.

Hiatt, R.W., Strasburg, D.W. (1960): Ecological relationships of the fish fauna on coral reefs of the Marshall Islands. Ecol. Monographs **30**, 65–127.

High, L.R., Jr. (1969): Storms and sedimentary processes along the northern British Honduras coast. J. Sediment. Petrol. **38**, 235–245.

Hillis, L.W. (1959): A revision of the genus *Halimeda* (order Siphonales). Inst. Marine Sci., Univ. Texas **6**, 321–403.

Hilmy, M.E. (1951): Beach sands of the Mediterranean coast of Egypt. J. Sediment. Petrol. **21**, 109–120.

Hobson, E.S. (1965): Diurnal-nocturnal activity of some inshore fishes in the Gulf of California. Copeia No. 3, p. 291–302.

Hodgkin, N.M. (1962): Limestone boring by the mytilid *Lithophaga*. Veliger **4**, 123–129.

Hoffman, P.F., Logan, B.W., Gebelein, C.D. (1968): Biological versus environmental factors governing the morphology and internal structures of recent algal stromatolites in Shark Bay, Western Australia (abt.). Program, N. E. Ann. Meeting Geol. Soc. Am., p. 28–29.

Hoffmeister, J.E., Ladd, H.S. (1935): The foundations of atolls: a discussion. J. Geol. **43**, 653–665.

Hoffmeister, J.E., Ladd, H.S. (1944): The antecedent-platform theory. J. Geol. **52**, 338–402.

Hoffmeister, J.E., Multer, H.G. (1965): Fossil mangrove reef of Key Biscayne, Florida. Bull. Geol. Soc. Am. **76**, 845–852.

Hoffmeister, J.E., Multer, H.G. (1968): Geology and origin of the Florida Keys. Bull. Geol. Soc. Am. **79**, 1487–1502.

Hoffmeister, J.E., Stockman, K.W., Multer, H.G. (1967): Miami limestone of Florida and its recent Bahamian counterpart. Bull. Geol. Soc. Am. **78**, 175–190.

Hofmann, H.J. (1969): Attributes of stromatolites. Geol. Surv. Canada Paper **69-39**, 58 pp.

Holser, W.T. (1966): Diagenetic polyhalite in recent salt from Baja California. Am. Mineralogist **51**, 99–109.

Hommeril, P., Rioult, M. (1965): Étude de la fixation des sédiments meubles par deux algues marines: *Rhodothamniella floridula* (Dillwyn) J. Feldm. et *Microcoleus chtonoplastes* Thur. Marine Geol. **3**, 131–155.

Honjo, S. (1963): New serial micropeel technique. Kansas Geol. Surv. Bull. **165**, Part 6.

Honjo, S. (1972): Suspension and dissolution of coccoliths in the Pacific water column. In: The paleontology of oceans. Amsterdam: Elsevier Publ. Co. In press.

Hood, D.W., Park, K., Smith, J.B. (1959): Calcium carbonate solubility equilibrium in sea water. Texas A & M, Dept. Oceanography and Meteorology, Ref. 59-13 F.

Horowitz, A.S., Potter, P.E. (1971): Introductory petrography of fossils, 302 pp. Berlin-Heidelberg-New York: Springer.

HOSKIN, C. M. (1963): Recent carbonate sedimentation on Alacran Reef, Yucatan, Mexico. Nat. Acad. Sci.-Nat. Res. Coun. Publ. No. 1089, 160 pp.

HOSKIN, C. M. (1966): Coral pinnacle sedimentation, Alacran Reef lagoon, Mexico. J. Sediment. Petrol. **36**, 1058–1074.

HOSKIN, C. M. (1968): Magnesium and strontium in mud fraction of Recent carbonate sediment, Alacran Reef, Mexico. Am. Assoc. Petrol. Geologists Bull. **52**, 2170–2177.

HOSKIN, C. M. (1971): Biogenic carbonate sediment from an intertidal encrusting community, Sitka Sound, Alaska (abst.). 1971 Ann. Meeting Geol. Soc. Am., p. 604–605.

HOSKIN, C. M., NELSON, R. V., JR. (1969): Modern marine carbonate sediment, Alexander Archipelago, Alaska. J. Sediment. Petrol. **39**, 581–590.

HOUBOLT, J. J. H. C. (1957): Surface sediments of the Persian Gulf near the Qatar Peninsula. Dissertation, Univ. Utrecht. Mouton & Co., The Hague, 113 pp.

HSU, K. J. (1967): Chemistry of dolomite formation. In: G. V. CHILINGAR, H. J. BISSELL, and R. W. FAIRBRIDGE (eds.), Carbonate rocks, Developments in sedimentology, vol. 9 (B), p. 169–191. Elsevier Publ. Co.

HSU, K. J., SIEGENTHALER, C. (1969): Preliminary experiments on hydrodynamic movement induced by evaporation and their bearing on the dolomite problem. Sedimentology **12**, 11–25.

HSU, K. J., SIEGENTHALER, C. (1971): Preliminary experiments on hydrology of supratidal dolomitization and cementation. In: O. P. BRICKER (ed.), Carbonate cements, The Johns Hopkins Univ. Studies in Geology, No. 19, p. 315–318.

HULSEMANN, J. (1966): On the routine analysis of carbonates in unconsolidated sediments. J. Sediment. Petrol. **36**, 622–625.

HUMM, H. J. (1964): Epiphytes of the seagrass *Thalassia testudinum*. Bull. Marine Sci. **14**, 306–341.

HUTTON, C. O. (1936): Mineralogical notes from the University of Otago. Trans. Roy. Soc. New Zealand **66**, 35.

HUVE, P. (1954): Étude experimentale de la reinstallation d'un "trottoir à *Tenarea*", en Méditerranée occidentale. Compt. Rend. **239**, 323–325.

HUVE, P. (1956): Contribution à l'étude des fonds à *Lithothamnium*(?) *solutum* Foslie (= *Lithophyllum solutum* (Foslie) Lemoine) de la région de Marseille. Rec. Travaux Sta. Mas. d'Endoume Bull. **11**, 105–134.

HYMAN, L. H. (1940): The invertebrates: Protozoa through Ctenophora, 726 pp. New York: McGraw-Hill Book Co.

HYMAN, L. H. (1955): The invertebrates: Echinodermata, 763 pp. New York: McGraw-Hill Book Co.

ILLING, L. V. (1954): Bahaman calcareous sands. Am. Assoc. Petrol. Geologists Bull. **38**, 1–95.

ILLING, L. V., WELLS, A. J., TAYLOR, J. C. M. (1965): Penecontemporary dolomite in the Persian Gulf. In: L. C. PRAY and R. C. MURRAY (eds.), Dolomitization and limestone diagenesis, Soc. Econ. Paleontologist Mineralogist Special Publ. No. 13, p. 89–111.

ILLING, M. A. (1950): The mechanical distribution of recent Foraminifera in Bahama Banks sediments. Ann. Mag. Nat. Hist. Ser. 12, **3**, 757–761.

INGERSON, E. (1962): Problems of the geochemistry of sedimentary carbonate rocks. Geochim. Cosmochim. Acta **26**, 815–847.

IRION, G. (1970): Mineralogisch-sedimentpetrographische und geochemische Untersuchungen am Tuz Gölü („Salzsee"), Türkei. Unpubl. Ph. D. Dissertation, Heidelberg Univ., 68 pp.

IRION, G., MULLER, G. (1968): Huntite, dolomite, magnesite and polyhalite of Recent age from Tuz Gölü, Turkey. Nature **220**, 1309–1310.

ISENBERG, H. D., DOUGLAS, S. D., LAVINE, L. S., WEISSFELLNER, H. (1967): Laboratory studies with coccolithophorid calcification. Studies in Trop. Oceanogr., Univ. Miami **5**, 155–177.

IVANENKOV, V. N. (1966): Karbonatnaya sistema (Carbonate system). In: S. V. BRUEVICH (ed.), Khimiya tikhogo okean, Chapt. 3, p. 57–81. Moscow: "Nauka".

JACKSON, J. B. C., GOREAU, T. F., HARTMAN, W. D. (1971): Recent brachiopod-coralline sponge communities and their paleoecological significance. Science **173**, 623–625.

JACKSON, T. A., BISCHOFF, J. L. (1971): The influence of amino acids on the kinetics of the recrystallization of aragonite to calcite. J. Geol. **79**, 493–497.

JACQUOTTE, R. (1962): Étude des fonds de maerl en Méditerranée. Rec. Travaux Sta. Mar. d'Endoume Bull. **26**, 141–230.

JAMIESON, J. C. (1953): Phase equilibrium in the system calcite-aragonite. J. Chem. Phys. **21**, 1385–1390.

JAMIESON, J. C., GOLDSMITH, J. R. (1960): Some reactions produced in carbonates by grinding. Am. Mineralogist **45**, 818–827.

JANSEN, J.F., KITANO, Y. (1963): The resistance of recent marine carbonate sediments to solution. J. Ocean Soc. Japan **18**, 208–219.

JEHU, T.J. (1918): Rock-boring organisms as agents in coast erosion. Scottish Geograph. Mag. **34**, 1–10.

JELL, J.S., MAXWELL, W.H.G., MCKELLAR, R.G. (1965): The significance of larger Foraminifera in the Heron Island reef sediments. J. Paleontol. **39**, 273–279.

JINDRICH, V. (1969): Recent sedimentation by tidal currents in lower Florida Keys. J. Sediment. Petrol. **39**, 531–553.

JOHANNES, R.E., COLES, S.L., KUENZEL, N.T. (1970): The role of zooplankton in the nutrition of some scleractinian corals. Limnol. Oceanog. **15**, 579–586.

JOHANSEN, H.W., AUSTIN, L.F. (1970): Growth rates in the articulated coralline, *Calliarthron* (Rhodophyta). Can. J. Botany **48**, 125–132.

JOHNSON, J.H. (1951): An introduction to the study of organic limestones. Quart. Colo. School Mines **46**, No. 2, 185 pp.

JOHNSON, J.H. (1957): Geology of Saipan, Mariana Islands, Part 2. Petrology and soils. Petrography of the limestones. U.S. Geol. Surv. Profess. Paper **280-C**, 177–187.

JOHNSON, J.H. (1961): Limestone-building algae and algal limestones. Quart. Colo. School Mines, 297 pp.

JOHNSON, R.F., CARROLL, J.J., GREENFIELD, L.J. (1964): Some sources of carbonate in molluscan shell formation. Limnol. Oceanog. **9**, 377–384.

JOHNSTON, J., MERWIN, H.E., WILLIAMSON, E.D. (1916): The several forms of calcium carbonate. Am. J. Sci. **41**, 473–512.

JOHNSTON, W.A. (1921): The occurrence of calcareous sandstones in the recent delta of the Fraser River, British Columbia, Canada. Am. J. Sci. **1**, 447–453.

Joint Committee on Powder Diffraction Standards, 1970. Index to the Powder Diffraction File 1970, Philadelphia, Pa.

(JOIDES) Joint Oceanographic Institutions for Deep Earth Sampling (1969): Initial Reports of the Deep-Sea Drilling Project, vol. I, Orange, Texas to Hoboken, N. J., 672 pp.

JOIDES (1970a): Initial Reports of the Deep-Sea Drilling Project, vol. II. Hoboken, N. J. to Dakar, Senegal, 501 pp.

JOIDES (1970b): Initial Reports of the Deep-Sea Drilling Project, vol. IV. Rio de Janeiro, Brazil to San Cristobal, Panama, 753 pp.

JOIDES (1970c): Initial Reports of the Deep-Sea Drilling Project, vol. V. San Diego, Calif. to Honolulu, Hawaii, 827 pp.

JONES, B.F. (1961): Zoning of saline minerals at Deep Spring Lake. U. S. Geol. Surv. Profess. Paper **424B**, 199–209.

JONES, C.L. (1961): Petrography of some Recent beachrock from the Caribbean. Unpubl. MS. Thesis, Louisiana State Univ.

JONES, W.C. (1967): Sheath and axial filament of calcareous sponge spicules. Nature **214**, 365–368.

JONES, W.C. (1970): The composition, development, form and orientation of calcareous sponge spicules. Symposia, Zool. Soc. London, No. 25, p. 91–123.

JONES, W.C., JAMES, D.W.F. (1969): An investigation of some calcareous sponge spicules by means of electron probe microanalysis. Micron **1**, 34–39.

JONES, W.C., JENKINS, D.A. (1970): Calcareous sponge spicules: a study of magnesian calcites. Calc. Tiss. Res. **4**, 314–329.

JOPE, H.M. (1965): Composition of brachiopod shells. In: R.C. MOORE (ed.), Treatise on invertebrate paleontology. Geol. Soc. Am. Part H, p. 57–155.

JORDAN, C.F., JR. (1971): Bioclastic sediment dispersion off Bermuda patch reefs (abst.). Am. Assoc. Petrol. Geologists Bull. **55**, 346.

JORDAN, G.F. (1952): Reef formation in the Gulf of Mexico off Apalachicola Bay, Florida. Bull. Geol. Soc. Am. **63**, 741–744.

JORDAN, G.F., STEWART, H.B., JR. (1959): Continental slope off southwest Florida. Am. Assoc. Petrol. Geologists Bull. **43**, 974–991.

JOUBIN, L. (1922): Distribution géographique de quelques coraux abyssaux dans les mers occidentales Europeenes. Compt. Rend. **175**, 930–933.

JUDD, J.W. (1904): Report on the materials from the borings at the Funafuti Atoll. Report Coral Reef Comm. Roy. Soc. London, p. 167–185.

KAHLE, C.F. (1965): Strontium in oolitic limestone. J. Sediment. Petrol. **35**, 846–856.

KALKOWSKI, E. (1908): Oolith und Stromatolith in norddeutschem Buntsandstein. Z. Deut. Geol. Ges. **60**, 68–125.

KANIA, J. E. A. (1929): Precipitation of limestone by submarine vents. Am. J. Sci. **37**, 347–359.

KANWISHER, J. W. (1960): pCO_2 in sea water and its effect on the movement of CO_2 in nature. Tellus **12**, 209–215.

KANWISHER, J. W., WAINWRIGHT, S. A. (1967): Oxygen balance in some reef corals. Biol. Bull. **133**, 378–390.

KATZ, A., FRIEDMAN, G. M. (1965): The preparation of stained acetate peels for the study of carbonate rocks. J. Sediment. Petrol. **35**, 248–249.

KAYE, C. A. (1959): Shoreline features and Quaternary shoreline changes, Puerto Rico. U. S. Geol. Surv. Profess. Paper **317-B**, 49–140.

KEARY, R. (1967): Biogenic carbonate in beach sediments of the west coast of Ireland. Sci. Proc. Roy. Dublin Soc., Ser. A **3**, 75–85.

KEEN, A. M. (1961): A proposed reclassification of the gastropod family Vermetidae. Bull. Brit. Mus. Zool. **7**, 183–213.

KEITH, M. L., ANDERSON, G. M., EICHLER, R. (1964): Carbon and oxygen isotopic composition of mollusk shells from marine and freshwater environments. Geochim. Cosmochim. Acta **28**, 1757–1786.

KEITH, M. L., PARKER, R. H. (1965): Local variation of ^{13}C and ^{18}O content of mollusk shells and the relatively minor temperature effect in marginal marine environments. Marine Geol. **3**, 115–129.

KEITH, M. L., WEBER, J. N. (1965): Systematic relationships between carbon and oxygen isotopes in carbonates deposited by modern corals and algae. Science **150**, 498–501.

KELLERMAN, K. F., SMITH, N. R. (1914): Bacterial precipitation of calcium carbonate. J. Wash. Acad. Sci. **4**, 400–402.

KEMPF, M. (1970): Notes on the benthic bionomy of the N-NE Brazilian shelf. Marine Biol. **5**, 213–224.

KEMPF, M., LABOREL, J. (1968): Formations de Vermets et d'algues calcaires sur les côtes du Brasil Rec. Trav. Sta. Marine End. Bull. **43**, 9–24.

KENDALL, C. G. ST. C., SKIPWITH, P. A. D'E. (1968): Recent algal mats of a Persian Gulf lagoon. J. Sediment. Petrol. **38**, 1040–1058.

KENDALL, C. G. ST. C., SKIPWITH, P. A. D'E. (1969): Holocene shallow-water carbonate and evaporite sediments of Khor al Baxam, Abu Dhabi, Southwest Persian Gulf. Am. Assoc. Petrol. Geologists Bull. **53**, 841–869.

KENNEDY, W. J., TAYLOR, J. D., HALL, A. (1969): Environmental and biological controls on bivalve shell mineralogy. Biol. Rev. **44**, 499–530.

KENNETT, J. P. (1966): Foraminiferal evidence of a shallow calcium carbonate solution boundary, Ross Sea, Antarctica. Science **153**, 191–193.

KERN, D. M. (1960): The hydration of carbon dioxide. J. Chem. Educ. **37**, 14–23.

KESSEL, E. (1936): Über Verfärbung mariner Molluskenschalen durch Einlagerung von Eisen. Zool. Anz. **115**, 129–139.

KILHAM, S. S. (1970): Deep sea bivalve molluscs: shell morphology, mineralogy and geochemistry. Unpubl. Ph. D. Dissertation, Duke Univ., 197 pp.

KINSMAN, D. J. J. (1964a): Recent carbonate sedimentation near Abu Dhabi, Trucial Coast, Persian Gulf. Unpubl. Ph. D. Thesis Univ. London, 302 pp.

KINSMAN, D. J. J. (1964b): Reef coral tolerance of high temperatures and salinities. Nature **202**, 1280–1282.

KINSMAN, D. J. J. (1966): Gypsum and anhydrite of recent age, Trucial Coast, Persian Gulf. Symp. Salt Northern Ohio Geol. Soc., 2nd 1966 Cleveland Ohio, vol. 1, p. 302–326.

KINSMAN, D. J. J. (1967): Huntite from a carbonate-evaporite environment. Am. Mineralogist **52**, 1332–1340.

KINSMAN, D. J. J. (1969a): Modes of formation, sedimentary associations and diagnostic features of shallow-water and supratidal evaporites. Am. Assoc. Petrol. Geologists Bull. **53**, 830–840.

KINSMAN, D. J. J. (1969b): Interpretation of Sr^{+2} concentrations in carbonate minerals and rocks. J. Sediment. Petrol. **39**, 486–508.

KINSMAN, D. J. J. (1970): Trace cations in aragonite (abst.). Ann. Meeting Geol. Soc. Am., p. 596–597.

KINSMAN, D. J. J., HOLLAND, H. D. (1969): The co-precipitation of cations with $CaCO_3$. IV. The co-precipitation of Sr^{+2} with aragonite between 16° and 96 °C. Geochim. Cosmochim. Acta **33**, 1–17.

KINSMAN, D.J.J., PARK, R.K. (1969): Studies in recent sedimentology and early diagenesis, Trucial Coast, Arabian Gulf. 2nd Regional Tech. Symposium Soc. Petroleum Eng. of AIME Saudi Arabia Section, Dhahran, 10 pp.

KITANO, Y. (1962): The behavior of various inorganic ions in the separation of calcium carbonate from a bicarbonate solution. Bull. Chem. Soc. Japan **35**, 1973–1980.

KITANO, Y., HOOD, D.W. (1962): Calcium carbonate crystal forms formed from sea water by inorganic processes. J. Ocean Soc. Japan **18**, 141–145.

KITANO, Y., KANAMORI, N. (1966): Synthesis of magnesian calcite at low temperatures and pressures. Geochem. J. **1**, 1–10.

KITANO, Y., KANAMORI, N., TOKUYAMA, A. (1969): Effects of organic matter on solubilities and crystal form of carbonates (abst.). Am. Zoologist **9**, 681–688.

KITANO, Y., PARK, K., HOOD, D.W. (1962): Pure aragonite synthesis. J. Geophys. Res. **67**, 4873–4874.

KLÄHN, H. (1932): Der quantitative Verlauf der Aufarbeitung von Sanden, Geröllen und Schalen in wässerigen Medium. Neues Jahrb. Mineral. Geol. Palaeontol. **67**, 313–412.

KLEMENT, M.C. (895): Sur l'origine de la dolomie dans les formations sédimentaires. Mem. Soc. Belge Geol. **9**, 3–23.

KLENOVA, M.V., ZENKEVITCH, N.L. (1962): Geological studies in the western part of the North Atlantic. Trudy morskogo gidrofizicheskogo instituta (Moskva) **25**, 142–186.

KLUG, H.P., ALEXANDER, L.E. (1954): Diffraction procedures for polycrystalline and amorphous materials, 716 pp. New York: John Wiley & Sons, Inc.

KOBAYASHI, I. (1969): Internal microstructure of the shell of bivalve mollusks. Am. Zoologist **9**, 663–672.

KOCZY, F.F. (1956): The specific alkalinity. Deep-Sea Res. **3**, 279–288.

KOHLMEYER, J. (1969): The role of marine fungi in the penetration of calcareous substrates. Am. Zoologist **9**, 741–746.

KOLDEWIJN, B.W. (1958): Sediments of the Paria-Trinidad shelf. Rept. Orinoco Shelf Exped. **3**, 109 pp.

KORNICKER, L.S. (1962): Evolutionary trends among mollusc fecal pellets. J. Paleontol. **36**, 829–834.

KORNICKER, L.S., BOYD, D.W. (1962): Shallow-water geology and environments of Alacran Reef complex, Campeche Bank, Mexico. Am. Assoc. Petrol. Geologists Bull. **46**, 640–673.

KORNICKER, L.S., BRYANT, W.R. (1969): Sedimentation on continental shelf of Guatemala and Honduras. In: A.R. MCBIRNEY (ed.), Tectonic relations of northern central America and the western Caribbean—the Bonancca expedition. Am. Assoc. Petrol. Geologists Mem. **11**, 244–257.

KORNICKER, L.S., PURDY, E.A. (1957): A Bahamian faecal-pellet sediment. J. Sediment. Petrol. **27**, 126–128.

KOSTER, J.T. (1939): Notes on Javanese calcicole Cyanophyceae. Blumea **3**, 243–247.

KRAUS, R.W., GALLOWAY, R.A. (1960): The role of algae in the formation of beachrock in certain islands of the Caribbean. Caribb. Beach Studies Tech. Rept., No. 11 (E), Coastal Studies Inst., Louisiana State Univ., 49 pp.

KRAUSKOPF, K.B. (1967): Introduction to geochemistry, 721 pp. New York: McGraw-Hill Book Co.

KRINSLEY, D. (1959): Manganese in modern and fossil gastropod shells. Nature **183**, 770–771.

KRINSLEY, D. (1960a): Trace elements in the tests of planktonic Foraminifera. Micropaleontology **6**, 297–300.

KRINSLEY, D. (1960b): Magnesium, strontium, and aragonite in the shells of certain littoral gastropods. J. Paleontol. **34**, 744–755.

KRINSLEY, D., BIERI, R. (1959): Changes in the chemical composition of pteropod shells after deposition on the sea floor. J. Paleontol. **33**, 682–684.

KRUMBEIN, W.C., PETTIJOHN, F.J. (1938): Manual of sedimentary petrography, 549 pp. New York: Appleton-Century-Crofts, Inc.

KUENEN, P.H. (1933): Geology of coral reefs. Snellius Exped. **5**, Part 2, 126 pp.

KUENEN, P.H. (1950): Marine geology, 568 pp. New York: John Wiley & Sons, Inc.

LABOREL, J., PÉRÈS, J.M., PICARD, J., VACELET, J. (1961): Étude directe des fonds des parages de Marseille de 30 à 300 m avec la Soucoupe plongeante Cousteau. Bull. Inst. Oceanog. Monaco, No. 1206, 16 pp.

LADD, H.S. (1961): Reef building. Science **134**, 703–715.

LADD, H.S., HEDGPETH, J.W., POST, R. (1957): Environments and facies of existing bays on the central Texas coast. In H.S. LADD (ed.), Treatise on Paleoecology, Geol. Soc. Amer. Mem. **67**(b), 599–640.

LADD, H.S., SCHLANGER, S.O. (1960): Drilling operations on Eniwetok Atoll. U. S. Geol. Surv. Profess. Paper **260-Y**, 863–903.

LADD, H.S., TRACEY, J.I., JR., GROSS, M.G. (1967): Drilling on Midway Atoll, Hawaii. Science **156**, 1088–1094.

LAGAAIJ, R., GAUTIER, Y.V. (1965): Bryozoan assemblages from marine sediments of the Rhone delta, France. Micropaleontology **11**, 39–58.

LAHOUD, J.A., MILLER, D.S., FRIEDMAN, G.M. (1966): Relationship between depositional environment and uranium concentrations of molluscan shells. J. Sediment. Petrol. **36**, 541–547.

LALOU, C. (1957a): Étude expérimentale de la production de carbonates par les bactéries des vases de la baie de Villefranche-sur-mer. Ann. Inst. Oceanog. Paris **33**, 201–266.

LALOU, C. (1957b): Studies of bacterial precipitation of carbonates on sea water. J. Sediment. Petrol. **27**, 190–195.

LAND, L.S. (1966): Diagenesis of metastable skeletal carbonates. Unpubl. Ph. D. Thesis, Lehigh Univ., 141 pp.

LAND, L.S. (1970): Carbonate mud: production by epibiont growth on *Thalassia testudinum*. J. Sediment. Petrol. **40**, 1361–1363.

LAND, L.S. (1972): Comparative geochemistry of Recent marine dolomites and Pleistocene (meteoric) dolomites. Unpubl. ms., 2 p.

LAND, L.S., GOREAU, T.F. (1970): Submarine lithification of Jamaican reefs. J. Sediment. Petrol. **40**, 457–462.

LANE, D.W. (1962): Improved acetate peel technique. J. Sediment. Petrol. **32**, 870.

LANGHUS, B.G., MEDIOLI, G., WATKINS, C. (1970): Foraminiferal tests as sedimentary particles (abs.). Ann. Meeting Geol. Soc. Am., p. 601.

LEES, A., BUTLER, A.T. (1971): Temperate water shallow marine carbonate sediments and their ancient equivalents (abst.). 8th Intern. Sediment. Congr., Heidelberg, p. 58.

LEES, A., BUTLER, A.T., SCOTT, J. (1969): Marine carbonate sedimentation processes, Connemara, Ireland. Reading Univ. Geol. Rept., No. 2, 64 pp.

LEIGHTON, M.W., PENDEXTER, C. (1962): Carbonate rock types. In: W.E. HAM (ed.), Classification of carbonate rocks. Am. Assoc. Petrol. Geologists Mem. **1**, p. 33–61.

LEMOINE, P. (1911): Structure anatomique des Melobesiees. Ann. Inst. Oceanog., Paris **2**, 213 pp.

LEMOINE, P. (1940): Les algues calcaires de la zone neritique. Contrib. a l'etudes de la repartition actuelle et passee des organismes dans la zone neritique, Sec. Biogeogr. Mem., No. 7, p. 75–138.

LERMAN, A. (1965): Strontium and magnesium in water and in *Crassostrea* calcite. Science **150**, 745–751.

LEWIN, J.C. (1962): Calcification. In: R.A. LEWIN (ed.), Physiology and biochemistry of algae, p. 457–465. New York: Academic Press.

LEWIS, J.B. (1960): The fauna of rocky shores of Barbados, West Indies. Can. J. Zool. **38**, 391–435.

LEWIS, J.B., AXELSEN, F., GOODBODY, I., PAGE, C., CHISLETT, G. (1969): Comparative growth rates of some reef corals in the Caribbean. McGill Univ. Marine Sciences Manuscript, Report 10, 26 pp.

LEWIS, M.S. (1968): The morphology of the fringing reefs along the east coast of Mahé, Seychelles. J. Geol. **76**, 140–153.

LI, Y.H. (1967): The degree of saturation of $CaCO_3$ in the oceans. Unpubl. Ph. D. Dissertation Columbia Univ., 176 pp.

LI, Y.H., TAKAHASHI, T., BROECKER, W.S. (1969): Degree of saturation of $CaCO_3$ in the oceans. J. Geophys. Res. **74**, 5507–5525.

LIEBERMANN, O. (1967): Synthesis of dolomite. Nature **213**, 241–245.

LIKINS, R.C., BERGY, E.G., POSNER, A.S. (1963): Comparative fixation of calcium and strontium by snail shell. Ann. N. Y. Acad. Sci. **109**, 269–277.

LINCK, G. (1903): Die Bildung der Oolithe und Rogensteine. Neues Jahrb. Mineral. Geol. Palaeontol. Abt. A **16**, 495–513.

LIPMAN, C.B. (1924): A critical and experimental study of Drew's bacterial hypothesis on $CaCO_3$ precipitation in the sea. Carnegie Inst. Wash. Publ. 340, p. 181–191.

LIPPMANN, F. (1960): Versuche zur Aufklärung der Bildungsbedingungen von Calcit und Aragonit. Fortschr. Mineral. **38**, 156.

LIPPMANN, F. (1968): Syntheses of $BaMg(CO_3)_2$ (Norsethite) at 20 °C and the formation of dolomite in sediments. In: G. MULLER and G.M. FIREDMAN (eds.), Recent developments in carbonate sedimentology in central Europe, p. 33–37. Berlin-Heidelberg-New York: Springer.

LIPPS, J. H., RIBBE, P. H. (1967): Electron-probe microanalysis of planktonic Foraminifera. J. Paleontol. **41**, 492–496.

LISITZIN, A. P. (1960): Bottom sediments of the Eastern Antarctic and the South Indian Ocean. Deep-Sea Res. **7**, 89–99.

LISITZIN, A. P. (1970): Sedimentation and geochemical considerations. In: Scientific exploration of the South Pacific. National Acad. Sci. Standard Book No. 309-1755-6, p. 89–132.

LISITZIN, A. P. (1971): Distribution of carbonate microfossils in suspension and in bottom sediments. In: B. M. FUNNELL and W. R. RIEDEL (eds.), The micropaleontology of oceans, p. 197–218. Cambridge Univ. Press.

LISITZIN, A. P., PETELIN, V. P. (1970): Distribution of $CaCO_3$ in the Pacific bottom sediments (in Russian). In: The Pacific Ocean. Sedimentation in the Pacific Ocean, vol. 2, p. 26–68. Moscow: Publ. House "Nauka".

LISTER, J. J. (1900): *Astrosclera willeyana*, the type of a new family of sponges. Zool. Results ... [by ARTHUR WILLEY] **4**, 459–482.

LIVINGSTON, H. D., THOMPSON, G. (1971): Trace element concentrations in some modern corals. Limnol. Oceanog. **16**, 786–796.

LIVINGSTONE, D. A. (1963): Chemical composition of rivers and lakes. U.S. Geol. Surv. Profess. Papers **440-G**, 61 pp.

LLOYD, R. M. (1964): Variation in the oxygen and carbon isotope ratios of Florida Bay mollusks and their environmental significance. J. Geol. **72**, 84–113.

LLOYD, R. M. (1966): Oxygen isotope enrichment of sea water by evaporation. Geochim. Cosmochim. Acta **30**, 801–814.

LLOYD, R. M. (1971): Some observations on recent sediment alteration ("micritization") and the possible role of algae in submarine lithification. In: O. P. BRICKER (ed.), Carbonate cements, The Johns Hopkins Univ. Studies in Geology, No. 19, p. 72–79.

LOEBLICH, A. R., JR., TAPPAN, H. (1964): Protista. Sarcodina, chiefly "Thecamobians" and Foraminiferida. In: R.C. MOORE (ed.), Treatise on invertebrate paleontology. Geol. Soc. Am. Part C, vol. 1, 510 pp.

LOEBLICH, A. R., JR., TAPPAN, H. (1966): Annotated index and bibliography of the calcareous nannoplankton. Phycologia **5**, 81–216.

LOGAN, B. W. (1961): Cryptozoon and associated stromatolites from the recent, Shark Bay, Western Australia. J. Geol. **69**, 517–533.

LOGAN, B. W. (1969): Coral reefs and banks, Yucatan shelf, Mexico. In: B. W. LOGAN et al. Carbonate sediments and reefs, Yucatan Shelf, Mexico. Am. Assoc. Petrol. Geologists Mem. **11**, 129–198.

LOGAN, B. W., CEBLUSKI, D. E. (1970): Sedimentary environments of Shark Bay, Western Australia. In: Carbonate sedimentation and environments, Shark Bay, Western Australia. Am. Assoc. Petrol. Geologists Mem. **13**, 1–37.

LOGAN, B. W., HARDING, J. L., AHR, W. M., WILLIAMS, J. D., SNEAD, R. G. (1969): Late Quaternary sediments of Yucatan Shelf, Mexico. In: B. W. LOGAN et al. Carbonate sediments and reefs, Yucatan Shelf, Mexico. Am. Assoc. Petrol. Geologists Mem. **11**, 1–128.

LOGAN, B. W., REZAK, R., GINSBURG, R. N. (1964): Classification and environmental significance of algal stromatolites. J. Geol. **72**, 68–83.

LOREAU, J. P. (1970): Ultrastructure de la phase carbonaté des oolithes marines actuelles. Compt. Rend. Ser. D., **271**, 816–819.

LOVERING, T. S. (1969): The origin of hydrothermal and low temperature dolomite. Econ. Geol. **64**, 743–754.

LOWENSTAM, H. A. (1954a): Factors affecting the aragonite: calcite ratios in carbonate-secreting marine organisms. J. Geol. **62**, 284–321.

LOWENSTAM, H. A. (1954b): Environmental relations of modification compositions of certain carbonate secreting marine invertebrates. Proc. Natl. Acad. Sci. **40**, 39–48.

LOWENSTAM, H. A. (1955): Aragonite needles secreted by algae and some sedimentary implications. J. Sediment. Petrol. **25**, 270–272.

LOWENSTAM, H. A. (1961): Mineralogy, O^{18}/O^{16} ratios and strontium and magnesium contents of Recent and fossil brachiopods and their bearing on the history of the oceans. J. Geol. **69**, 241–260.

LOWENSTAM, H. A. (1962a): Goethite in radular teeth of recent marine gastropods. Science **137**, 279–280.

LOWENSTAM, H. A. (1962b): Magnetite in denticle capping in recent chitons. Bull. Geol. Soc. Am. **73**, 435–438.

LOWENSTAM, H. A. (1963): Biological problems relating to the composition and diagenesis of sediments. In: T. W. DONNELLY (ed.), The earth sciences, p. 137–195. Chicago: The Univ. Chicago Press.

LOWENSTAM, H. A. (1964a): Sr/Ca ratio of skeletal aragonites. In: Isotopic and cosmic chemistry, p. 114–132. Amsterdam: North Holland Publ. Co.

LOWENSTAM, H. A. (1964b): Coexisting calcites and aragonites from skeletal carbonates of marine organisms and their strontium and magnesium contents. In: Recent researches in the fields of hydrosphere, atmosphere and nuclear geochemistry, p. 373–404. Tokyo: Maruzen Co., Ltd.

LOWENSTAM, H. A., EPSTEIN, S. (1957): On the origin of sedimentary aragonite needles of the Great Bahama Bank. J. Geol. **65**, 364–375.

LOWENSTAM, H. A., MCCONNELL, D. (1968): Biologic precipitation of fluorite. Science **162**, 1496–1498.

LUCAS, G. (1948): La sédimentation calcaire. Action du carbonate de sodium sur l'eau de mer. Compt. Rend. **226**, 937–939.

LUCAS, G. (1955): Oolithes marines actuelles et calcaires oolithiques recents sur la rivage Africain de la Méditerranée orientale (Egypt et sud Tunisien). Bull. sta. oceanogr. Salammbo, No. 52, p. 19–38.

LUCIA, F. J. (1968): Recent sediments and diagenesis of South Bonaire, Netherlands Antilles. J. Sediment. Petrol. **38**, 845–858.

LUDBROOK, N. H. (1960): Scaphopoda. In: J. B. KNIGHT et al., (eds.) Mollusca. 1. Treatise on invertebrate paleontology, Geol. Soc. Amer. Part I, p. 37–40.

LUDWICK, J. C., WALTON, W. R. (1957): Shelf-edge calcareous prominences in northeastern Gulf of Mexico. Am. Assoc. Petrol. Geologists Bull. **41**, 2054–2101.

LYMAN, J. (1956): Buffer mechanism of sea water. Ph. D. Thesis, Univ. Calif., Los Angeles.

LYNTS, G. W. (1966): Relationship of sediment-size distribution to ecologic factors in Buttonwood Sound, Florida Bay. J. Sediment. Petrol. **36**, 66–74.

MABESOONE, J. M. (1964): Origin and age of the sandstone reefs of Pernambuco (northeastern Brazil). J. Sediment. Petrol. **34**, 715–726.

MABESOONE, J. M. (1971): Recent marine limestones from the shelf of tropical Brazil. Geol. Mijnbouw **50**, 451–460.

MABESOONE, J. M., TINOCO, I. M. (1965/66): Shelf off Alagoas and Sergipe (northeastern Brazil) 2. Geology. Trab. Inst. Oceanog. Univ. Fed. Pe., Recife **7/8**, 151–186.

MACCLINTOCK, C. (1967): Shell structure of patelloid and bellerophontoid gastropods (Mollusca). Yale Univ., Peabody Mus. Nat. Hist. Bull. **22**, 1–140.

MACCLINTOCK, C., PANNELLA, G. (1969): Time calcification in the bivalve mollusk *Mercenaria mercenaria* (Linneaus) during the 24-hour period (abst.). Ann. Meeting Program Geol. Soc. Am., p. 140.

MACDONALD, G. F. (1956): Experimental determination of calcite-aragonite equilibrium relations at elevated temperatures and pressures. Am. Mineralogist **41**, 744–756.

MACFAYDEN, W. A. (1930): The undercutting of coral reef limestone on the coasts of some islands in the Red Sea. Geogr. J. **75**, 27–34.

MACINTYRE, I. G. (1967a): Recent sediments off the west coast of Barbados, W. I., Unpubl. Ph. D. Thesis, McGill Univ., 169 pp.

MACINTYRE, I. G. (1967b): Submerged coral reefs, west coast of Barbados, West Indies. Can. J. Earth Sci. **4**, 461–474.

MACINTYRE, I. G. (1972): Submerged reefs of eastern Caribbean. Am. Assoc. Petrol. Geologists, Bull. **56**, 720–738.

MACINTYRE, I. G., MILLIMAN, J. D. (1970): Physiographic features on the outer shelf and upper slope, Atlantic continental margin southeastern United States. Bull. Geol. Soc. Am. **81**, 2577–2598.

MACINTYRE, I. G., MILLIMAN, J. D. (1971): Limestones from the outer shelf and upper slope, continental margin, southeastern U. S. In: O. P. BRICKER (ed.), Carbonate cements. The Johns Hopkins Univ. Studies in Geology, No. 19, p. 103–110.

MACINTYRE, I. G., MOUNTJOY, E. W., D'ANGLEJAN, B. F. (1968): An occurrence of submarine cementation of carbonate sediments off the west coast of Barbados, West Indies. J. Sediment. Petrol. **38**, 660–664.

MACINTYRE, I. G., PILKEY, O. H. (1969): Tropical reef corals: tolerance to low temperatures on the North Carolina continental shelf. Science **166**, 374–375.

MACINTYRE, W. G. (1965): The temperature variation of the solubility product of $CaCO_3$ in sea water. Fish. Res. Board, Canada, Ms. Rept. No. 200, 153 pp.

MACKENZIE, F. T., GARRELS, R. M. (1966a): Chemical mass balance between rivers and oceans. Am. J. Sci. **264**, 507–525.

MacKenzie, F.T., Garrels, R.M. (1966 b): Silica-bicarbonate balance in the ocean and early diagenesis. J. Sediment. Petrol. **36**, 1075–1084.

MacKenzie, F.T., Kulm, L.D., Cooley, R.L., Barnhart, J.T. (1965): *Homotrema rubrum* (Lamarck), a sediment transport indicator. J. Sediment. Petrol. **35**, 265–272.

MacNeil, F.S. (1954): The shape of atolls: an inheritance from subaerial erosion forms. Am. J. Sci. **252**, 402–427.

Maiklem, W.R. (1967): Black and brown speckled foraminiferal sand from the southern part of the Great Barrier Reef. J. Sediment. Petrol. **37**, 1023–1030.

Majewske, O.P. (1969): Recognition of invertebrate fossil fragments in rocks and thin sections, 101 pp. Leiden: E.J. Brill.

Malone, P.G., Dodd, J.R. (1967): Temperature and salinity effects on calcification rate of *Mytilus edulis* and its paleoecological implications. Limnol. Oceanog. **12**, 432–436.

Manning, R.B., Kumpf, H.E. (1959): Preliminary investigation of the fecal pellets of certain invertebrates of the South Florida area. Bull. Marine Sci. Gulf and Caribbean **9**, 291–309.

Manton, S.M., Stephenson, T.A. (1935): Ecological surveys of coral reefs. Sci. Rept. Great Barrier Reef Exped. **3**, 273–312.

Marszalek, D.S., Wright, R.C., Hay, W.W. (1969): Foraminiferal test as an environmental buffer (abst.). Am. Assoc. Petrol. Geologists Bull. **53**, 730.

Margolis, S., Rex, R.W. (1971): Endolithic algae and micrite envelope formation in Bahamian oolites as revealed by scanning electron microscopy. Bull. Geol. Soc. Am. **82**, 843–852.

Marlowe, J.I. (1971 a): High-magnesian calcite cement in calcarinite from Aves Swell, Caribbean Sea. In: O.P. Bricker (ed.), Carbonate cements, The Johns Hopkins Univ. Studies in Geology, No. 19, p. 111–115.

Marlowe, J.I. (1971 b): Dolomite, phosphorite, and carbonate diagenesis on a Caribbean seamount. J. Sediment. Petrol. **41**, 809–827.

Marsden, J.R. (1962): A coral-eating polychaete. Nature **193**, 598.

Marsh, J.A., Jr. (1970): Primary productivity of reef-building calcareous red algae. Ecology **51**, 255–263.

Martin, E.L., Ginsburg, R.N. (1966): Radiocarbon ages of oolitic sands on Great Bahama Bank. Proc. 6th Intern. Conf. Radiocarbon and Tritium Dating. U. S. Atomic Energy Comm. Rept. Conf. 650652, p. 705–719.

Mason, B. (1962): Principles of geochemistry, 310 pp. New York: John Wiley & Sons.

Matheja, J., Degens, E.T. (1968): Molekulare Entwicklung mineralisationsfähiger organischer Matrizen. Neues Jahrb. Geol. Paleontol. Mh., No. 4, p. 215–229.

Matthews, A, A. L. (1930): Origin and growth of Great Salt Lake ooliths. J. Geol. **38**, 633–642.

Matthews, R.K. (1963): Continuous seismic profiles of a shelf-edge bathymetric prominence in northern Gulf of Mexico. Gulf Coast Assoc. Geol. Socs. Trans. **13**, 49–58.

Matthews, R,K. (1966): Genesis of Recent lime mud in southern British Honduras. J. Sediment. Petrol. **36**, 428–454.

Mauchline, J., Templeton, W.L. (1966): Strontium, calcium and barium in marine organisms from the Irish Sea. J. Conseil Perm. Intern. Exploration Mer. **30**, 161–170.

Mawson, D. (1929): Some South Australian algal limestone in process of formation Quart. J. Geol. Soc. London **85**, 613–623.

Maxwell, W.G.H. (1968 a): Atlas of the Great Barrier Reef, 258 pp. Amsterdam: Elsevier Publ. Co.

Maxwell, W.G.H. (1968 b): Relict sediments, Queensland continental shelf. Australian J. Sci. **31**, 85–86.

Mayer, A.G. (1915): The lower temperatures at which reef-corals lose their ability to capture food. Carnegie Inst. Wash. Yearbook **14**, 212.

Mayer, A.G. (1916): Submarine solution of limestone in relation to the Murray-Agassiz theory of atolls. Proc. Natl. Acad. Sci. **2**, 28–30.

Mayor, A.G. (1924): Causes which produce stable conditions in the depth of the floors of Pacific fringing reef-flats. Carnegie Inst. Washing. Publ. **340**, 27–36.

McCammon, H.M., Auld, J.A., Watson, J.A. (1969): Adsorption of iron to the shells of brachiopods. Bull. Geol. Soc. Am. **80**, 527–530.

McConnell, D. (1963): Inorganic constituents in the shell of the living brachiopod *Lingula*. Bull. Geol. Soc. Am. **74**, 363–364.

McCrea, J. M. (1950): On the isotopic chemistry of carbonates and a paleotemperature scale. J. Phys. Chem. **18**, 849.

McCrone, A. W. (1963): Quick preparation of peel-prints for sedimentary petrography. J. Sediment. Petrol. **33**, 228–230.

McFarlin, P. F. (1967): Aragonite vein fillings in marine manganese nodules. J. Sediment. Petrol. **37**, 68–72.

McIntosh, W. C. (1902): On the boring of *Polydora* in Australian oysters. Ann. Mag. Nat. Hist. Ser. F, p. 299–308.

McIntyre, A., Be, A. W. H. (1967): Modern coccolithophoridae of the Atlantic Ocean. I. Placoliths and cyrtholiths. Deep-Sea Res. **14**, 561–597.

McIntyre, A., McIntyre, R. (1971): Coccolith concentrations and differential solution in oceanic sediments. In: B. M. Funnell and W. R. Riedel (eds.), The paleontology of oceans. The micropaleontology of oceans, p. 253–261. Cambridge: Cambridge Univ. Press.

McIntyre, A., Ruddiman, W. F., Jantzen, R. (1972): Southward penetrations of the North Atlantic polar front: faunal and floral evidence of large-scale surface water mass movements over the last 225000 years. Deep-Sea Res. **19**, 61–77.

McKee, E. D. (1959): Storm sediments on a Pacific atoll. J. Sediment. Petrol. **29**, 354–364.

McKee, E. D., Chronic, J., Leopold, E. B. (1959): Sedimentary belts in lagoon of Kapingamarangi Atoll. Am. Assoc. Petrol. Geologists Bull. **43**, 501–562.

McLean, P. F. (1967a): Origin and development of ridge-furrow systems in beachrock in Barbados, West Indies. Marine Geol. **5**, 181–193.

McLean, P. F. (1967b): Measurements of beachrock erosion by some tropical marine gastropods. Bull. Marine Sci. **17**, 551–561.

McMaster, R. L., Conover, J. T. (1967): Recent algal stromatolites from the Canary Islands. J. Geol. **74**, 647–652.

McMaster, R. L., Lachance, T. P. (1969): Northwestern African continental shelf sediments. Marine Geol. **7**, 57–67.

McMaster, R. L., LaChance, T. P., Ashraf, A. (1970): Continental geomorphic features off Portuguese Guinea, Guinea, and Sierra Leone, West Africa. Marine Geol. **9**, 203–213.

McMaster, R. L., Milliman, J. D., Ashraf, A. (1971): Continental shelf and upper slope surface sediments off Portuguese Guinea, Guinea, and Sierra Leone, West Africa. J. Sediment. Petrol. **41**, 150–158.

Mecarini, G., Shimaoka, G., Krause, D. C. (1965): Submarine lithification of *Globigerina* ooze (abst.). N. E. Regional Meeting Geol. Soc. Am. p. 33.

Menard, H. W. (1964): Marine geology of the Pacific, 271 pp. New York: McGraw-Hill Book Co.

Menard, H. W., Smith, S. M. (1966): Hypsometry of ocean basin provinces. J. Geophys. Res. **71**, 4305–4325.

Menzies, R. J., Pilkey, O. H., Blackwelder, B. W., Dexter, D., Huling, P., McCloskey, L. (1966): A submerged reef off North Carolina. Intern. Rev. ges. Hydrobiol. **51**, 393–431.

Merrill, A. S., Emery, K. O., Rubin, M. (1965): Ancient oyster shells on the Atlantic continental shelf. Science **147**, 398–400.

Mesolella, K. J. (1967): Zonation of uplifted Pleistocene coral reefs on Barbados, West Indies. Science **156**, 633–640.

Miller, D. N., Jr. (1961): Early diagenetic dolomite associated with salt extraction process, Inagua, Bahamas. J. Sediment. Petrol. **31**, 473–476.

Milliman, J. D. (1965): An annotated bibliography of recent papers on corals and coral reefs. Atoll Res. Bull., No. 111, 58 pp.

Milliman, J. D. (1966): Submarine lithification of carbonate sediments. Science **153**, 994–997.

Milliman, J. D. (1967): Carbonate sedimentation on Hogsty Reef, a Bahamian atoll. J. Sediment. Petrol. **37**, 658–676.

Milliman, J. D. (1969a): Four southwestern Caribbean atolls: Courtown Cays, Albuquerque Cays, Roncador Bank and Serrana Bank. Atoll Res. Bull., No. 129, 41 pp.

Milliman, J. D. (1969b): Carbonate sedimentation on four southwestern Caribbean atolls and its relation to the "oolite problem". Trans. Gulf Coast Assoc. Geol. Soc. **19**, 195–206.

Milliman, J. D. (1971a): Examples of submarine lithification. In: O. P. Bricker (ed.), Carbonate cements. The Johns Hopkins Univ. Studies in Geology, No. 19, p. 95–102.

Milliman, J. D. (1971b): The role of calcium carbonate in continental shelf sedimentation. In:

D. J. STANLEY (ed.), The new concepts of continental margin sedimentation supplement. Amer. Geol. Inst., 20 pp.

MILLIMAN, J. D. (1972): Atlantic continental shelf and slope of the United States. Petrology of the sand fraction—northern New Jersey to southern Florida. U.S. Geol. Surv. Profess. Paper **529-J**, 40 pp.

MILLIMAN, J. D. (1973): Caribbean coral reefs. In: R. ENDEAN and O. A. JONES (eds.), Biology and geology of coral reefs. New York: Academic Press, in press.

MILLIMAN, J. D., EMERY, K. O. (1968): Sea levels during the past 35,000 years. Science **162**, 1121–1123.

MILLIMAN, J. D., GASTNER, M., MÜLLER, J. (1971): Utilization of magnesium in coralline algae. Bull. Geol. Soc. Am. **82**, 573–580.

MILLIMAN, J. D., MANHEIM, F. T., PRATT, R. M., ZARUDZKI, E. F. K. (1967): ALVIN dives on the continental margin off the southeastern United States. Woods Hole Oceanographic Institution Tech. Rept. Reference No. 67–80.

MILLIMAN, J. D., MÜLLER, J. (1973): Precipitation and lithification of magnesian calcite in the deep-sea sediments of the eastern Mediterranean Sea. Sedimentology, **20**, 29–46.

MILLIMAN, J. D., PILKEY, O. H., BLACKWELDER, B. W. (1968): Carbonate sedimentation on the continental shelf, Cape Hatteras to Cape Romain. Southeastern Geol. **9**, 245–267.

MILLIMAN, J. D., PILKEY, O. H., ROSS, D. A. (1972): Sediments of the continental margin off the eastern United States, Maine to Florida. Bull. Geol. Soc. Am. **82**, 1315–1334.

MILLIMAN, J. D., ROSS, D. A., KU, T. H. (1969): Precipitation and lithification of deep-sea carbonates in the Red Sea. J. Sediment. Petrol. **39**, 724–736.

MILLIMAN, J. D., WEILER, Y., STANLEY, D. J. (1973): Morphology and carbonate sedimentation on shallow banks in the Alboran Sea. In: D. J. STANLEY (ed.), The Mediterranean sea: a natural sedimentation laboratory. p. 241–259. Stroudsburg, Pa: Dowden, Hutchinson and Ross.

MINCHIN, E. A. (1908): Materials for a monograph of the Ascons. II. The formation of spicules in genus *Leucosolenia* with some notes on the histology of sponges. Quart. J. Microscop. Sci. **52**, 301–355.

MITTERER, R. M. (1968): Amino acid composition of organic matrix in calcareous oolites. Science **162**, 1498–1499.

MITTERER, R. M. (1971): Comparative amino acid composition of calcified and non-calcified polychaete worm tubes. Comp. Biochem. Physiol. **38 B**, 405–409.

MOBERLY, R., JR. (1968): Composition of magnesian calcites of algae and pelcypods by electron microprobe analysis. Sedimentology **11**, 61–82.

MOLINIER, R. (1955): Les plate-formes et corniches recifales de Vermets (*Vermetus cristatus* Biondi) en Mediterranee occidentale. Compt. Rend. **240**, 361–363.

MOLINIER, R., PICARD, J. (1952): Recherches sur les herbiers de phanerogames marines du littoral Mediterranean Francais. Ann. Inst. Oceanog. **27**, 157–234.

MOLNIA, B. F., PILKEY, O. H. (1972): Origin and distribution of calcareous fines on the Carolina continental shelf. Sedimentology. **18**, 293–310.

MONAGHAN, P. H., LYTLE, M. A. (1956): The origin of calcareous ooliths. J. Sediment. Petrol. **26**, 111–118.

MONTY, C. (1965): Recent algal stromatolites in the windward lagoon, Andros Island, Bahamas. Ann. Soc. Geol. Belg., Bull. **88**, 269–276.

MONTY, C. (1967): Distribution and structure of Recent stromatolitic algal mats, eastern Andros Island, Bahamas. Ann. Soc. Geol. Belg., Bull. 3, **90**, 55–100.

MOORE, C. H., JR. (1971): Beach rock cements, Grand Cayman, B. W. I. In: O. P. BRICKER (ed.), Carbonate cements. The Johns Hopkins Univ. Studies in Geology, No. 19, p. 9–12.

MOORE, C. H., JR., BILLINGS, G. K. (1971): Preliminary model of beach rock formation, Grand Cayman Island, B. W. I. In: O. P. BRICKER (ed.), Carbonate cements, The Johns Hopkins Univ. Studies in Geology, No. 19, p. 40–45.

MOORE, C. H., JR., BOLENEUS, D., MULTER, H. G. (1971): Recent carbonate cementation, Cotton Garden Bay, St. Croix, Virgin Islands (abst.). 1971 Ann. Meetings Geol. Soc. Am. p. 650.

MOORE, H. B. (1939): Faecal pellets in relation to marine deposits. In: P. D. TRASK (ed.), Recent marine sediments. Am. Assoc. Petrol. Geologists Special Publ. No. 4, p. 516–524.

MOORE, H. B., KRUSE, P. (1956): A review of present knowledge of faecal pellets. Marine Lab. Rept. Inst. Marine Sci., Univ. Miami, No. 13805, 25 pp.

MORELOCK, J., KOENIG, K. J. (1967): Terrigenous sedimentation in a shallow water coral reef environment. J. Sediment. Petrol. **37**, 1001–1005.

MORESBY, R. (1835): Extracts from Commander Moresby's report on the northern atolls of the Maldives. J. Roy. Geog. Soc. **5**, 398–404.

MORGAN, J. P., TREADWELL, R. C. (1954): Cemented sandstone slabs of the Chandeleur Islands, Louisiana. J. Sediment. Petrol. **24**, 71–75.

MORRIS, R. C., DICKEY, P. A. (1956): Modern evaporite deposition in Peru. Am. Assoc. Petrol. Geologists Bull. **41**, 2467–2474.

MORRIS, R. W., KITTLEMAN, L. R. (1967): Piezoelectric property of otoliths. Science **158**, 368–370.

MÜLLER, G. (1966): Grain size, carbonate content, and carbonate mineralogy of recent sediments of the Indian Ocean off the eastern coast of Somalia. Naturwissenschaften **21**, 547–550.

MÜLLER, G. (1967): Methods in sedimentary petrology, 283 pp. Stuttgart: Schweizerbart Verlag.

MÜLLER, G. (1970): Lakustreische und marine Karbonate. Geol. Rundschau, in press.

MÜLLER, G., BLASCHKE, R. (1969): Zur Entstehung des Tiefseekalkschlammes im Schwarzen Meer. Naturwissenschaften **56**, 561–562.

MÜLLER, G., IRION, G. (1969): Subaerial cementation and subsequent dolomitization of lacustrine carbonate muds and sands from Paleotuz Golu ("Salt Lake") Turkey. Sedimentology **12**, 193–204.

MÜLLER, G., MÜLLER, J. (1967): Mineralogisch-sedimentpetrographische und chemische Untersuchungen an einem Bank-Sediment (Cross Bank) der Florida Bay, USA. Neues Jahrb. Mineral. Abhandl. **106**, 257–286.

MÜLLER, G., TIETZ, G. (1966): Recent dolomitization of Quaternary biocalcarenites from Fuerteventura (Canary Islands). Contr. Mineral. and Petrol. **13**, 89–96.

MÜLLER, J. (1969): Mineralogisch-sedimentpetrographische Untersuchungen an Karbonatsedimenten aus dem Schelfbereich um Fuerteventura und Lanzarote (Kanarische Inseln). Unpubl. Ph. D. Thesis, Heidelberg Univ. (W. Germany), 99 pp.

MÜLLER, J., FABRICIUS, F. (1970): Carbonate mineralogy of deep sea sediments from the Ionian Sea. 22nd Congr. C. I. E. S. M. Rome.

MÜLLER, J., MILLIMAN, J. D. (1972): Relict carbonate-rich sediments on Grand Banks. Can. J. Earth Sci. in press.

MULLER, J., RUDOWSKI, S. (1967): Cementation of recent beach sediments of the Southern Baltic (in Polish, with English summary). Prace Museum Ziemi nr. **11**, Warszawa, p. 243–253.

MULTER, H. G. (1969): Field Guide to some carbonate rock environments: Florida Keys and Western Bahamas. Miami Geological Society, Miami.

MULTER, H. G. (1971): Holocene cementation of skeletal grains into beach rock, Dry Tortugas, Florida. In: O. P. BRICKER (ed.), Carbonate cements. The Johns Hopkins Univ. Studies in Geology, No. 19, p. 25–26.

MULTER, H. G., MILLIMAN, J. D. (1967): Geologic aspects of Sabellarian reefs, southeastern Florida. Bull. Marine Sci. **17**, 257–267.

MUNK, W. H., SARGENT, M. C. (1954): Adjustment of Bikini Atoll to ocean waves, Bikini and nearby atolls, Marshall Islands. U. S. Geol. Surv. Profess. Paper **260-C**, 275–280.

MURATA, K. J., ERD, R. C. (1964): Composition of sediments from the experimental Mohole project (Guadalupe Site). J. Sediment. Petrol. **34**, 633–655.

MURRAY, J., HJORT, J. (1912): The depths of the ocean, 821 pp. London: MacMillan & Co., Ltd.

MURRAY, J., IRVINE, R. (1891): On coral reefs and other carbonate of lime formations in modern seas. Roy. Soc. Edinburgh Proc. **17**, 79–109.

MURRAY, J., PHILIPPI, E. (1908): Die Grundproben der „Deutschen Tiefsee-Expedition“. Wissenschaftl. Ergebn. der Deutschen Tiefsee-Exped. "Valdivia", 1898–1899, vol. 10, p. 80–206.

MURRAY, J. W. (1966): The Foraminiferida of the Persian Gulf. 4. Khor al Bazam. Palaeogeog., Palaeoclimatol., Palaoecol. **2**, 153–169.

MURRAY, J. W. (1967): Production in benthic foraminiferids. J. Natl. Hist. **1**, 61–68.

MURRAY, R. C., PRAY, L. C. (1965): Dolomitization and limestone diagenesis: An introduction. In: L. C. PRAY and R. C. MURRAY (eds.), Dolomitization and limestone diagenesis. Soc. Econ. Paleontologist Mineralogists Special Publ. **13**, 1–2.

MUSCATINE, L. (1967): Glycerol excretion by symbiotic algae from corals and *Tridacna* and its control by the host. Science **156**, 516–519.

NACHTRIEB, N. H. (1950): Principles and practice spectrochemical analysis, 324 pp. New York: McGraw-Hill Book Co., Inc.

NADSON, G. (1927): Les algues perforantes, leur distribution et leur rôle dans la nature. Compt. Rend. **184**, 195–1017.

NAGLE, J. S. (1968): Distribution of epibiota of macroepibenthic plants. Contrib. Marine Sci., Univ. Texas Marine Sci. Inst. **13**, 103–144.

NAIDU, A. S. (1969): Some aspects of texture, mineralogy and geochemistry of modern deltaic sediments of the Godavari River, India. Unpubl. Ph. D. Thesis, Andhra Univ. Waltair, India.

NAIR, R. R. (1969): Phosphatized oolites on the western continental shelf of India. Proc. Natl. Inst. Sci. India **35**, A., 858–863.

NAIR, R. R., PYLEE, A. (1968): Size distribution and carbonate content of the sediments of the western shelf of India. Proc. Symposium on "Indian Ocean". Bull. Natl. Inst. Sci. India, No. 38, p. 411–420.

NAME, W. G. VAN (1930): The ascidians of Porto Rico and the Virgin Islands. Scientific Survey of Porto Rico and Virgin Islands. Ann. N. Y. Acad. Sci **10**, No. 4, 403–535.

NAME, W. G. VAN (1945): The North and South American ascidians. Bull. Am. Mus. Nat. Hist. **84**, 476 pp.

NATTERER, K. (1894): Chemische Untersuchungen im östlichen Mittelmeer. In: Reise S. M. Schiff „Pola" in Jahre 1890. Denkschr. Akad. Wiss. Wien, Math-naturw. **65**, 445–572.

NATTERER, K. (1898): Expedition S. M. Schiff „Pola" in das Rote Meer, Nördliche Hälfte (Oktober 1895—Mai 1896 Denkschr. der Kaiserl. Akad. Wiss. **65**, 445–572.

NEFF, J. M. (1967): Calcium carbonate tube formation by serpulid polychaete worms: physiology and ultrastructure. Unpubl. Ph. D. Dissertation, Duke Univ.

NEFF, J. M. (1969): Mineral regeneration by serpulid polychaete worms. Biol. Bull. **136**, 76–90.

NEHER, J., ROHRER, E. (1958): Dolomitbildung unter Mitwirkung von Bakterien. Ecologae Geol. Helv. **51**, 213–215.

NELSON, D. J. (1965): Strontium in calcite: new analyses. Publ. Inst. Marine Sci., Univ. Texas **10**, 76–79.

NELSON, H. F. (1971): Cementation in a Holocene chenier sand. In: O. P. BRICKER (ed.), Carbonate cements. The Johns Hopkins Univ. Studies in Geology, No. 19, p. 141–142.

NESTEROFF, W. D. (1954): Sur la formation des grès de plage ou "beach-rock" en Mer Rouge. Compt. Rend. **238**, 2547–2548.

NESTEROFF, W. D. (1956): De l'origine des oolithes. Compt. Rend. **242**, 1047–1049.

NEUMANN, A. C. (1965): Processes of recent carbonate sedimentation in Harrington Sound, Bermuda. Bull. Marine Sci. **15**, 987–1035.

NEUMANN, A. C. (1966): Observations on coastal erosion in Bermuda and measurements of the boring rate of the sponge, *Cliona lampa*. Limnol. Oceanog. **11**, 92–108.

NEUMANN, A. C., GEBELEIN, C. D., SCOFFIN, T. P. (1970): Composition, structure and erodability of subtidal mats, Abaco, Bahamas. J. Sediment. Petrol. **40**, 274–297.

NEUMANN, A. C., KELLER, G. H., KOFOED, J. W. (1972): Lithoherms. in the Straits of Florida (abst.). Program, Geol. Soc. Am. Annual Meeting **4**, 611.

NEUMANN, A. C., LAND, L. S. (1969): Algal production and lime mud deposition in the Bight of Abaco: A budget (abst.). Geol. Soc. Am. Special Paper **121**, 219.

NEWELL, N. D. (1956): Geological reconnaissance of the Raroia (Kon Tiki) Atoll, Tuamotu Archipelago. Bull. Am. Mus. Natl. Hist. **109**, 315–372.

NEWELL, N. D. (1959): The biology of coral reefs. Nat. Hist. **68**, 226–235.

NEWELL, N. D. (1971): An outline history of tropical organic reefs. Am. Mus. Novitates No. 2465, 37 pp.

NEWELL, N. D., IMBRIE, J., PURDY, E. G., THURBER, D. L. (1959): Organism communities and bottom facies, Great Bahama Bank. Bull. Amer. Mus. Nat. Hist. **117**, 177–228.

NEWELL, N. D., PURDY, E. G., IMBRIE, J. (1960): Bahamian oolitic sand. J. Geol. **68**, 481–497.

NEWELL, N. D., RIGBY, J. K. (1957): Geological studies on the Great Bahama Bank. In: Regional aspects of carbonate deposition. Soc. Econ. Paleontologists Mineralogists Spec. Publ. No. 5, p. 13–72.

NEWELL, N. D., RIGBY, J. K., WHITEMAN, A. J., BRADLEY, J. S. (1951): Shoal-water geology and environments, eastern Andros Island, Bahamas. Bull. Am. Mus. Nat. Hist. **97**, 1–29.

NEWHOUSE, J. (1954): Floristics and plant ecology of Raroia Atoll, Tuamotu. Part 2. Ecological and floristic notes on the Myxophyta of Raroia. Atoll Res. Bull., No. 33.

NEWMAN, W. A. (1960): On the paucity of intertidal barnacles in the tropical western Pacific. Veliger **2**, 89–94.

NEWMAN, W. A., ZULLO, V. A., WAINWRIGHT, S. A. (1967): A critique on recent concepts of growth in Balanomorpha (Cirripedia, Thoracia). Crustaceana **12**, 167–178.

NEWMAN, W. A., ZULLO, V. A., WITHERS, T. H. (1969): Cirripedia. In: R. C. MOORE (ed.), Treatise on invertebrate paleontology. Geol. Soc. Am. part R, (1), 206–295.

NICHOLLS, C. D., CURL, H., JR., BOWEN, V. T. (1959): Spectrographic analyses of marine plankton. Limnol. Oceanog. **4**, 472–478.

NICHOLS, D. (1962): Echinoderms. Hutchinson Univ. Lab. London, 200 pp.

NICOLS, D. (1967): Some characteristics of cold-water marine pelecypods. J. Paleontol. **41**, 1330–1340.

NIINO, H. (1931): Concretions and pseudoconcretions dredged from the sea bottom around Japan. J. Imp. Fish. Inst., Tokyo, **30**, No. 2.

NIINO, H., EMERY, K. O. (1961): Sediments of shallow portion of the East China Sea and South China Sea. Bull. Geol. Soc. Am. **72**, 731–762.

NIINO, H., EMERY, K. O. (1966): Continental shelf sediments off northeastern Asia. J. Sediment. Petrol. **36**, 152–161.

NOBREGA, C. P., ONOFREDE, M. J. (1971): Distribution de los sedimentos en la platforma continental norte-nordeste de Brasil. Symposium on Investigations and Resources of the Caribbean Sea and Adjacent Regions, UNESCO, p. 273–284.

NOLL, W. (1934): Geochemie strontiums. Mit Bemerkungen zur Geochemie des Bariums. Chem. Erde **8**, 507–534.

NORIN, E. (1956): The sediments of the central Tyrrhenian Sea. Rept. Swed. Deep-Sea Exped. **8**, 287–334.

NORTH, W. J. (1954): Size distribution, erosive activities and gross metabolic efficiency of the marine intertidal snails *Littorina planaxis* and *L. scutulata*. Biol. Bull. **106**, 185–197.

NORTH, W. J. (1968): Kelp habitat improvement project, annual report. 1 July, 1967—30 June 1968. Calif. Inst. Technology, 123 pp.

NOTA, D. J. G. (1958): Sediments of the western Guiana shelf. Mededel. Landbouwhogeschool Wagingen, Netherlands **58**, 98 pp.

OBA, T. (1969): Biostratigraphy and isotopic paleotemperature of some deep-sea cores from the Indian Ocean. Sci. Rept., Tohoku Univ., 2nd Ser., Geol., **41**, 2.

ODUM, H. T. (1957): Biochemical deposition of strontium. Inst. Marine Sci. **4**, 38–114.

ODUM, H. T., ODUM, E. T. (1955): Trophic structure and productivity of a windward coral reef community on Eniwetok Atoll. Ecol. Monographs **25**, 291–320.

OELTZSCHNER, H., SIGL, W. (1970): Sedimentologische Untersuchungen im Golf von Manfredonia (Sudadria). Geol. Rundschau **60**, 131–144.

OKADA, H., HONJO, S. (1973): The distribution of oceanic coccolithophorids in the Pacific. Deep-Sea Res., **20**, 355–374.

OKAZAKI, K. (1960): Skeletal formation of sea urchin larvae. II. Organic matrix of the spicule. Embryologia **5**, 283–320.

OLAUSSON, E. (1960): Description of sediment cores from the Indian Ocean. Rept. Swed. Deep-Sea Exped. **9**, 53–88.

OLAUSSON, E. (1965): Evidence of climatic changes in North Atlantic deep-sea cores, with remarks on isotopic paleotemperature analysis. In: M. SEARS (ed.), Progress in oceanography, vol. 3, p. 221–252. Oxford: Pergamon Press.

O'NEIL, J. R., EPSTEIN, S. (1966): Oxygen isotope fractionation in the system dolomite-calcite-carbon dioxide. Science **152**, 198–201.

OPPENHEIMER, C. H. (1960): Bacterial activity in sediments of shallow marine bays. Geochim. Cosmochim. Acta **19**, 244–260.

OPPENHEIMER, C. H., MASTER, I. M. (1964): Transition of silicate and carbonate crystal structure by photosynthesis and metabolism (abst.). Geol. Soc. Am. Special Paper **76**, 125.

ORR, W. N. (1967): Secondary calcification in the foraminiferal genus *Globorotalia*. Science **157**, 1554–1555.

ORTON, J. H. (1928): On rhythmic periods in shell growth in *Ostrea edulis*, with a note on fattening. J. Marine Biol. Assoc. **15**, 365–427.

OTTER, G. W. (1932): Rock-burrowing echinoids. Biol. Rev. **7**, 89–107.

OTTER, G. W. (1937): Rock-destroying organisms in relation to coral reefs. Rept. Great Barrier Reef Exped. **1**, 323–352.

OWEN, B. B., BRINKLEY, S. R., JR. (1941): Calculation of the effect of pressure upon ion equilibria in pure water and in salt conditions. Chem. Rev. **29**, 461–474.

PAASCHE, E. (1962): Coccolith formation. Nature **193**, 1094–1095.

PAASCHE, E. (1966): Adjustment to light and dark rates of coccolith formation. Physiol. Plantarum **19**, 271–278.

PAASCHE, E. (1968): Biology and physiology of coccolithophorids. Ann. Rev. Microbiol. **22**, 71–86.

PALACHE, C., BERMAN, H., FRENDEL, C. (1963): Dana's system of mineralogy (7th ed.), vol. II, 124 pp. New York: John Wiley & Sons, Inc.

PANNELLA, G. (1971): Fish otoliths: Daily growth layers and periodical patterns. Science **173**, 1124–1127.

PANTIN, H. M. (1958): Rate of formation of a diagenetic calcareous concretion. J. Sediment. Petrol. **28**, 366–371.

PARK, K. (1968): Seawater hydrogen-ion concentration: vertical distribution. Science **162**, 357–358.

PARK, K., HOOD, D. W., ODUM, H. T. (1958): Diurnal pH variation in Texas bays and its application to primary productivity estimation. Publ. Inst. Mar. Sci. Univ. Texas **5**, 47–64.

PARK, K., EPSTEIN, E. (1960): Carbon isotope fractionation during photosynthesis. Geochim. Cosmochim. Acta **21**, 110–126.

PARKER, F. L. (1971): Distribution of planktonic foraminifera in recent deep-sea sediments. In: B. M. FUNNEL and W. R. RIEDEL (eds.), The micropaleontology of oceans, p. 289–307. Cambridge: Cambridge Univ. Press.

PARKER, R. H. (1960): Ecology and distribution patterns of marine macro-invertebrates, northern Gulf of Mexico in Recent sediments, northwest Gulf of Mexico, Amer. Assoc. Petroleum Geol., p. 302–337.

PARKER, R. H., CURRAY, J. R. (1956): Fauna and bathymetry of banks on continental shelf, northwest Gulf of Mexico. Am. Assoc. Petrol. Geologists Bull. **40**, 2428–2439.

PATRIQUIN, D. G. (1972): Carbonate mud production by epibionts on *Thallasia*: an estimate based on leaf growth rate data. J. Sediment. Petrol. **42**, 687–689.

PEARSE, A. S., WILLIAMS, L. G. (1951): Biota of the reefs off the Carolinas. J. Elisha Mitchell Sci. Soc. **73**, 11–68.

PEARSE, V. B. (1970): Incorporation of metabolic CO_2 into coral skeleton. Nature **228**, 383.

PEARSE, V. B. (1972): Radioisotopic study of calcification in the articulated coralline alga *Bossiella orbigniana*. J. Phycol. **8**, 88–97.

PEARSE, V. B., MUSCATINE, L. (1971): Role of symbiotic algae (Zooxanthellae) in coral calcification. Biol. Bull. **141**, 350–363.

PEQUEGNAT, W. E., FREDERICKS, A. D. (1967): Organic production of epifaunal organisms. Texas A & M Univ. Dept. Oceanog., Ref. 67-15t, 45 pp.

PÉRÈS, J. M. (1967a): Les biocoenoses benthiques dans le système phytal. Rec. Trav. Sta. Marine End. Bull. **42**, 3–113.

PÉRÈS, J. M. (1967b): The Mediterranean benthos. In: H. BARNES (ed.), Oceanog. Marine Biol. Ann. Rev. **5**, 449–533.

PÉRÈS, J. M., PICARD, J. (1958): Manuel de bionomie benthique de la Mer Méditerranée. Rec. Trav. Sta. Marine End. Bull. **14**, 5–122.

PÉRÈS, J. M., PICARD, J. (1964): Nouveau manuel de bionomie benthique de la Mer Méditerranée. Rec. Trav. Sta. Marine End. Bull. **31**, 5–137.

PERKIN-ELMER (1966): Analytical methods for atomic absorption spectrophotometer. Norwalk, Conn.

PERKINS, R. D., ENOS, P. (1968): Hurricane Betsy in the Florida-Bahama area: Geologic effects and comparison with Hurricane Donna. J. Geol. **76**, 710–717.

PERKINS, R. D., HALSEY, S. D. (1971): Geologic significance of microboring fungi and algae in Carolina shelf sediments. J. Sediment. Petrol. **41**, 843–853.

PERKINS, R. D., MCKENZIE, M. D., BLACKWELDER, P. L. (1972): Aragonite crystals within Codiacean algae: distinctive morphology and sedimentary implications. Science **175**, 624–626.

PESSAGNO, E. A., MIYANO, K. (1968): Notes on the wall structure of the Globigerinacea. Micropaleontol. **14**, 38–50.

PETERSON, M. N. A. (1966): Calcite: rates of dissolution in a vertical profile in the central Pacific. Science **154**, 1542–1544.

PETERSON, M. N. A., BIEN, G. S., BERNER, R. A. (1963): Radiocarbon studies of Recent dolomite from Deep Spring Lake, California. J. Geophys. Res. **68**, 6493–6505.

PETERSON, M. N. A., BORCH, C. C. VON DER, BIEN, G. S. (1966): Growth of dolomite crystals. Am. J. Sci. **264**, 257–272.

PHILLIPS, A. H. (1922): Analytical search for metals in Tortugas marine organisms. Carnegie Inst. Wash. Publ. **312**, 95–99.

PHLEGER, F. B. (1960): Ecology and distribution of recent Foraminifera. 297 p. Baltimore: Johns Hopkins Press.

PHLEGER, F.B., EWING, G.C. (1962): Sedimentology and oceanography of coastal lagoons in Baja California. Bull. Geol. Soc. Am. **73**, 145–182.

PIA, J. (1926): Pflanzen als Gesteinsbildner, 355 pp. Berlin.

PIA, J. (1927): Thallophyta. In: M. HIRMER (ed.), Handbuch der Paläobotanik, vol. 1, p. 1–36.

PIA, J. (1933): Die rezenten Kalksteine, 420 pp. Leipzig: Academische Verlagsgesellschaft M. B. H.

PIERCE, J.W., MELSON, W.G. (1967): Dolomite from the continental slope off California. J. Sediment. Petrol. **37**, 963–966.

PILKEY, O.H. (1964): Mineralogy of the fine fraction of certain carbonate cores. Bull. Marine Sci. Gulf and Caribbean **14**, 126–139.

PILKEY, O.H., BLACKWELDER, B.W. (1968): Mineralogy of the sand size carbonate fraction of some recent marine terrigenous and carbonate sediments. J. Sediment. Petrol. **38**, 799–810.

PILKEY, O.H., GOODELL, H.G. (1963): Trace elements in recent mollusk shells. Limnol. Oceanog. **8**, 137–148.

PILKEY, O.H., GOODELL, H.G. (1964): Comparison of the composition of fossil and recent mollusk shells. Bull. Geol. Soc. Am. **75**, 217–228.

PILKEY, O.H., HARRISS, R.C. (1966): The effect of intertidal environment on the composition of calcareous skeletal material. Limnol. Oceanog. **11**, 381–385.

PILKEY, O.H., HOWER, J. (1960): The effect of environment on the concentration of skeletal magnesium and strontium in *Dendraster*. J. Geol. **68**, 203–216.

PILKEY, O.H., RUCKER, J.B. (1966): Mineralogy of Tongue of the Ocean sediments. J. Marine Res. **24**, 276–285.

PILKEY, O.H., SCHNITKER, D., PEVEAR, D.R. (1966): Oolites on the Georgia continental shelf edge. J. Sediment. Petrol. **36**, 462–467.

PIMM, A.C., GARRISON, R.E., BOYCE, R.E. (1971): Sedimentology synthesis: lithology, chemistry and physical properties of sediments in the northwestern Pacific Ocean. Initial Repts. Deep-Sea Drilling Project **6**, 1131–1252.

POBEGUIN, T. (1954): Contribution a l'étude des carbonates du calcium, précipitation du calcaire par les végétaux, comparison avec le monde animal. Ann. Sci. Nat., Botanique **15**, 29–109.

PORTER, C.L., ZEBROWSKI, G. (1937): Lime-loving molds from Australian sands. Mycologia **29**, 252–257.

POLUZZI, A., SARTORI, R. (1973): Carbonate mineralogy of some Bryozoa from Talbot Shoal. Ann. Museo Geol. Bolognia, Ser. 2a, **39**, 11–15.

PRATJE, O. (1924): Korallenbänke in tiefem und kühlem Wasser. Centralbl. Mineral. Geol. 410–415.

PRAY, L.C. (1966): Hurricane Betsy (1965) and nearshore carbonate sediments of the Florida Keys (abst.). Program Ann. Meeting Geol. Soc. Am., p. 168–169.

PRAY, L.C., MURRAY, R.C. (eds.) (1965): Dolomitization and limestone diagenesis: A Symposium. Soc. Econ. Paleontologists Mineralogists. Spec. Publ. No. 13, 180 p.

PRENANT, M. (1925): Contributions à l'étude cytologique du calcaire. II. Sur les conditions de formation des spicules chez les Didemnides. Bull. Biol. Franc. et Belg. **59**, 403–425.

PRESLEY, B.J., KAPLAN, I.R. (1968): Changes in dissolved sulfate, calcium and carbonate from interstitial water of nearshore sediments. Geochim. Cosmochim. Acta **32**, 1037–1048.

PRICE, N.B., HALLAM, A. (1967): Variation of strontium content within shells of recent *Nautilus* and *Sepia*. Nature **215**, 1272–1274.

PURDY, E.G. (1963): Recent calcium carbonate facies of the Great Bahama Bank. 1. Petrography and reaction groups. 2. Sedimentary facies. J. Geol. **71**, 334–355; 472–497.

PURDY, E.G. (1968): Carbonate diagenesis: An environmental survey. Geol. Romana **7**, 183–228.

PURDY, E.G., IMBRIE, J. (1964): Carbonate Sediments, Great Bahama Bank. Ann. Meeting Geol. Soc. Am. Guidebook Field Trip No. 2, 66 pp.

PURDY, E.G., KORNICKER, L.S. (1958): Algal disintegration of Bahamian limestone coasts. J. Geol. **66**, No. 1, 96–99.

PURI, H.S., COLLIER, A. (1967): Role of micro-organisms in formation of limestones. Trans. Gulf. Coast Assoc. Geol. Soc. **17**, 355–367.

PUSEY, W.C. (1964): Recent calcium carbonate sedimentation in northern British Honduras. Unpubl. Ph. D. Dissertation Rice Univ., 247 pp.

PYLE, T.E., TIEH, T.T. (1970): Strontium, vanadium and zinc in the shells of pteropods. Limnol. Oceanog. **15**, 153–154.

PYTKOWICZ, R. M. (1965): Calcium carbonate saturation in the ocean. Limnol. Oceanog. **10**, 220–225.

PYTKOWICZ, R. M. (1970): On the carbonate compensation depth in the Pacific Ocean. Geochim. Cosmochim. Acta **34**, 836–839.

PYTKOWICZ, R. M., CONNERS, D. N. (1964): High pressure solubility of calcium carbonate in seawater. Science **144**, 840–841.

QUIEVREUX, C. (1963): Secretion tupibare des larves de Spi ror binae (Annélides, polychètes). Cahier Biol., Mar. **4**, 399–406.

RAKESTRAW, N. W. (1949): The conception of alkalinity or excess base of sea water. J. Marine Res. **8**, 14–20.

RANDALL, J. E. (1967): Food habits of reef fishes of the West Indies. In: Proc. Internat. Conf. Trop. Oceanog. Univ. Miami Press, p. 665–847.

RANSON, G. (1955): Observations sur des facteurs biologiques de la dissolution du calcaire d'origine récifale dans les Tuamotu. Proc. 8th Pacific Sci. Congr. 3 A, p. 979–988.

RANSON, G. (1958): Coraux et recifs coralliens (Bibliographie). Bull. Inst. Oceanog. Monaco No. 1121, 80 pp.

RAUP, D. M. (1958): The relation between water temperature and morphology in *Dendraster*. J. Geol. **66**, 668–677.

RAUP, D. M. (1965): Crystal orientation in the echinoid apical system. J. Paleontol. **39**, 934–951.

RAUP, D. M. (1966): The endoskeleton. In: R. A. BOOLOOTIAN (ed.), The physiology of Echinodermata, p. 379–395. New York: John Wiley & Sons, Inc.

RAYMOND, P. E., HUTCHINS, F. (1932): A calcareous beach at John O'Groats, Scotland. J. Sediment. Petrol. **2**, 63–67.

RAYMOND, P. E., STETSON, H. C. (1932): A calcareous beach on the coast of Maine. J. Sediment. Petrol. **2**, 51–62.

REISS, Z. (1958): Classification of lamellar foraminifera. Micropaleontology **4**, 51–70.

REVELLE, R. (1934): Physico-chemical factors affecting the solubility of calcium carbonate in sea water. J. Sediment. Petrol. **4**, 103–110.

REVELLE, R., EMERY, K. O. (1957): Chemical erosion of beach rock and exposed reef rock. U. S. Geol. Surv. Profess. Paper **260-T**, 699–709.

REVELLE, R., FAIRBRIDGE, R. W. (1957): Carbonates and carbon dioxide. In: J. W. HEDGPETH (ed.), Treatise on marine ecology. Geol. Soc. Am. Mem. **67** (1), 239–296.

RHOADS, D. C. (1967): Biogenic reworking of intertidal and subtidal sediments in Barnstable Harbor and Buzzards Bay, Massachusetts. J. Geol. **75**, 461–476.

RHOADS, D. C., YOUNG, D. K. (1970): The influence of deposit-feeding organisms on sediment stability and community trophic structure. J. Marine Res. **28**, 150–176.

RICE, M. K. (1969): Possible boring structures of sipunculids. Am. Zoologist **9**, 803–812.

RICHARDS, F. A. (1965): Anoxic basins and fjords. In: J. P. RILEY and G. SKIRROW (eds.), Chemical oceanography, vol. 1, p. 611–645. London: Academic Press.

RIEDEL, W. R. (1963): The preserved record: paleontology of pelagic sediments. In: M. N. HILL (ed.), The sea, vol. 3, p. 866–887. New York: Interscience Publ.

RIEDEL, W. R., LADD, H. S., TRACEY, J. I., JR, BRAMLETTE, M. A. (1961): Preliminary drilling phase of Mohole Project. II. Summary of coring operations. Am. Assoc. Petrol. Geologists Bull. **45**, 1793–1798.

RILEY, J. P., SEGAR, D. A. (1970): The distribution of the major and some minor elements in marine animals. 1. Echinoderms and coelenterates. J. Marine Biol. Assoc. U. K. **50**, 721–730.

RITTENBERG, S. C., EMERY, K. O., HULSEMANN, J., DEGENS, E. T., FAY, R. C., REUTER, J. H., GRADY, J. R., RICHARDSON, S. H., BRAY, E. E. (1963): Biogeochemistry of sediments in experimental Mohole. J. Sediment. Petrol. **33**, 140–172.

ROBERTS, W. P., PIERCE, J. W. (1967): Outcrop of the Yorktown Formation (upper Miocene) in Onslow Bay, North Carolina. Southeastern Geol. **8**, 131–138.

ROBERTSON, J. D., PANTIN, C. F. E. (1938): Tube formation in *Pomatoceros triqueter*. Nature **141**, 648–649.

ROBERTSON, R. (1961): The feeding of *Strombus* and related herbivorous marine gastropods: with a review and field observations. Notulae Naturae No. 343, 9 pp.

ROBERTSON, R. (1970): Review of the predators and parasites of stony corals, with special reference to symbiotic prosobranch gastropods. Pacific. Sci. **24**, 43–54.

ROBINSON, M., CLAYTON, R. N. (1969): Carbon-14 fractionation between aragonite and calcite. Geochim. Cosmochim. Acta **33**, 997–1002.

ROONEY, W. S., PERKINS, R. D. (1972): Distribution and geologic significance of microboring organisms within sediments of the Arlington Reet complex, Australia. Bull. Geol. Soc. Am. **83**, 1139–1150.

ROSS, A. M., CERAME-VIVAS, M. J., MCCLOSKEY, L. (1964): New barnacle records for the coast of North Carolina. Crustaceana **7**, 312–313.

ROSS, D. A., DEGENS, E. T., MACILVAINE, J. (1970): Black Sea: Recent sedimentary history. Science **170**, 163–165.

ROTHPLETZ, A. (1892): Über die Oolithe. (English translation by CROGIN, F. W., 1892, Am. Geol. **10**, 279–282.)

ROY, K. J., SMITH, S. V. (1971): Sedimentation and coral reef development in turbid water: Fanning Lagoon. Pacific Sci. **25**, 234–248.

ROYSE, C. F., JR., WADELL, J. S., PETERSEN, L. E. (1971): X-ray determination of calcite-dolomite: an evaluation. J. Sediment. Petrol. **41**, 483–488.

RUCKER, J. B. (1967): Paleoecological analysis of cheilostome Bryozoa from Venezuela-British Guiana shelf sediments. Bull. Marine Sci. **17**, 787–839.

RUCKER, J. B. (1968): Carbonate mineralogy of sediments of Exuma Sound, Bahamas. J. Sediment. Petrol. **38**, 68–72.

RUCKER, J. B., CARVER, R. E. (1969): A survey of the carbonate mineralogy of cheilostome Bryozoa. J. Paleontol. **43**, 791–799.

RUCKER, J. B., VALENTINE, J. W. (1961): Salinity response of trace element concentration in *Crassostrea virginica*. Nature **190**, 1099–1100.

RUDDIMAN, W. F., HEEZEN, B. C. (1967): Differential solution of planktonic foraminifera. Deep-Sea Res. **14**, 801–808.

RUDWICK, M. J. S. (1965): Ecology and paleoecology—Brachiopoda. In: R. C. MOORE (ed.), Treatise on invertebrate paleontology. Geol. Soc. Amer., part H, p. 199–214.

RUNHAM, N. W. (1961): The histochemistry of the radula of *Patella vulgata*. Quart. J. Microscop. Sci. **102**, 371–380.

RUNNELLS, D. D. (1970): Errors in x-ray analysis of carbonates due to solid-solution variation in composition of component minerals. J. Sediment. Petrol. **40**, 1158–1166.

RUSNAK, G. A. (1960): Some observations of recent oolites. J. Sediment. Petrol. **30**, 471–480.

RUSNAK, G. A., NESTEROFF, W. D. (1964): Modern turbidites: terrigenous abyssal plain versus bioclastic basin. In: R. L. MILLER (ed.), Papers in marine geology (Shepard Commemorative Volume), p. 488–507. New York: MacMillan.

RUSSELL, K. L., DEFFEYES, K. S., FOWLER, G. A., LLOYD, R. M. (1967): Marine dolomite of unusual isotopic composition. Science **155**, 189–191.

RUSSELL, R. J. (1962a): Beachrock—geological observations. Guidebook Field Trip to Peninsula of Yucatan New Orleans Geol. Soc., p. 64–72.

RUSSELL, R. J. (1962b): Origin of beach rock. Z. Geomorphol. **6**, 1–16.

RUSSELL, R. J. (1968): Algal flats of Port Hedland, Western Australia. Coastal Studies Bull. Louisiana State Univ., No. 2, p. 45–55.

RUSSELL, R. J., MACINTYRE, W. G. (1965): Southern hemisphere beach rock. Geog. Rev. **55**, 17–45.

RYLAND, J. S. (1967): Polyzoa. In: H. BARNES (ed.), Oceanog. Marine Biol. Ann. Rev. **5**, 343–359.

SAITO, T. (1971): Distribution of carbonates in deep-sea sediments (abst.). Am. Assoc. Petrol. Geologists Bull. **55**, 363.

SAITO, T., BE, A. W. H. (1968): Paleontology of deep-sea deposits. International dictionary of geophysics. Oxford: Pergamon Press.

SAITO, T., EWING, M., BURCKLE, L. H. (1966): Tertiary sediment from the Mid-Atlantic Ridge. Science **151**, 1075–1079.

SANDBERG, P. A. (1970): Skeletal ultrastructure and development in cheilostome Bryozoa (abst.). Ann. Meeting Geol. Soc. Am., p. 672.

SANDERS, H. L., HESSLER, R. R. (1969): Ecology of the deep-sea benthos. Science **163**, 1419–1424.

SANZ-ECHEVERRIA, J. (1949): Identificacion de los peces de la familia Centrolophidae de Espana por medio de los otolitos. Real Soc. Espanola de Historia Nat., Madrid Tomo Extraort., p. 151–156.

SARNTHEIM, M. (1971): Oberflächensedimente im Persischen Golf und Golf von Oman. II. Quantitative Komponentenanalyse der Grobfraktion. „Meteor"-Forsch.-Ergeb. V. C. (5), 113 p.

SASS, E., WEILER, Y., KATZ, A. (1971): Recent lagoonal oolites in the Gulf of Suez (abst.). 8th Intern. Sediment. Congr., Heidelberg, p. 87.
SAUDRAY, Y., BOUFFANDEAU, M. (1958): Sur la composition chimique du systeme tegumentaire du quelques Bryozoaires. Bull. Inst. Oceanog., Monaco, No. 1119, p. 1–13.
SCHÄFER, W. (1962): Aktuo.-Paläontologie nach Studien in der Nordsee. 666 p. Frankfurt/Main: W. Kramer.
SCHÄFER, W. (1967): Biofazies-Bereiche im subfossilen Korallenriff Sarso (Rotes Meer). Seneken-bergiana Lethaea **48**, 107–133.
SCHLANGER, S.O. (1957): Dolomite growth in coralline algae. J. Sediment. Petrol. **27**, 181–187.
SCHLANGER, S.O. (1963): Subsurface geology of Eniwetok Atoll. U. S. Geol. Surv. Profess. Paper **260-BB**, 991–1066.
SCHLANGER, S.O. (1964): Petrology of the limestones of Guam. U. S. Geol. Surv. Profess. Paper **403-D**, 52 pp.
SCHLANGER, S.O., BROOKHART, J.W. (1955): Geology and water resources of Falalop Island, Ulithi Atoll, western Caroline Islands. Am. J. Sci. **253**, 553–573.
SCHLANGER, S.O., TRACEY, J.I., JR. (1970): Dolomitization related to recent emergence of Jarvis Island, Southern Line Islands (Pacific Ocean) (abst.). Ann. Meeting Geol. Soc. Am., p. 676.
SCHMALZ, R.F. (1965): Brucite in carbonate secreted by the red alga *Goniolithon* sp. Science **149**, 993–996.
SCHMALZ, R.F. (1970): Surface energy measurements for calcite, and textural equilibria of carbonate cements (abst.). N. E. Section Ann. Meeting Geol. Soc. Am., p. 34–35.
SCHMALZ, R.F. (1971): Beachrock formation on Eniwetok Atoll. In: O.P. BRICKER (ed.), Carbonate cements. The Johns Hopkins Univ. Studies in Geology, No. 19, p. 17–24.
SCHMALZ, R.F., CHAVE, K.E. (1963): Calcium carbonate factors affecting saturation in ocean waters off Bermuda. Science **139**, 1206–1207.
SCHMALZ, R.F., SWANSON, F.S. (1969): Diurnal variations in the carbonate saturation of seawater. J. Sediment. Petrol. **39**, 255–267.
SCHMIDT, W.J. (1924): Die Bausteine des Tierkörpers in polarisiertem Licht, 528 pp. Bonn: Friedrich Cohen.
SCHNEIDERMANN, N. (1971): Selective dissolution of recent coccoliths in the Atlantic Ocean (abst.). 1971 Ann. Meetings Geol. Soc. Am., p. 695.
SCHOLL, D.W. (1966): Florida Bay: a modern site of limestone formation. In: R.W. FAIRBRIDGE (ed.). The encyclopedia of oceanography, p. 282–288. New York: Reinhold Publ. Co.
SCHOLLE, P.A., KLING, S.A. (1971): Southern British Honduras: lagoonal coccolith ooze. J. Sediment. Petrol. **42**, 195–204.
SCHOPF, T.J.M., ALLAN, J.R. (1970): Phylum Ectoprocta, Order Cheilostomata: Microprobe analysis of calcium, magnesium, strontium, and phosphorus in skeletons. Science **169**, 280–282.
SCHOPF, T.J.M., MANHEIM, F.T. (1967): Chemical composition of Ectoprocta (Bryozoa). J. Paleontol. **41**, 1197–1225.
SCHROEDER, J.H., DWORNIK, E.J., PAPIKE, J.J. (1969): Primary protodolomite in echinoid skeletons. Bull. Geol. Soc. Am. **80**, 1613–1616.
SCHROEDER, J.H., GINSBURG, R.N. (1971): Calcified algae filaments in reefs: criterion of early dia-genesis (abst.). Am. Assoc. Petrol. Geologists Bull. **55**, 364.
SCHROEDER, J.H., MILLER, D.S., FRIEDMAN, G.M. (1970): Uranium distributions in recent skeletal carbonates. J. Sediment. Petrol. **40**, 672–681.
SCOFFIN, T.P. (1970): The trapping and binding of subtidal carbonate sediments by marine vegetation in Bimini lagoon, Bahamas. J. Sediment. Petrol. **40**, 249–273.
SCOTT, H.W. (1961): Shell morphology of Ostracoda. In: R.C. MOORE (ed.), Treatise on invertebrate paleontology, Ged. Soc. Amer. Part Q(3), p. 21–37.
SEGAR, D.A., COLLINS, J.D., RILEY, J.P. (1971): The distribution of the major and some minor elements in marine animals. Part II. Molluscs. J. Marine Biol. Assoc., U. K. **51**, 131–136.
SEIBOLD, E. (1962): Untersuchungen zur Kalkfällung und Kalklösung am Westrand der Great Bahama Bank. Sedimentology **1**, 50–74.
SHACKLETON, N. (1967): Oxygen isotope analyses and Pleistocene temperatures reassessed. Nature **215**, 15–17.
SHARMA, T., CLAYTON, R.N. (1965): Measurement of O^{18}/O^{16} ratios of total oxygen of carbonates. Geochim. Cosmochim. Acta **29**, 1347–1353.

SHARP, J.H. (1969): Blue-green algae and carbonates—*Schizothrix calcicola* and algal stromatolites from Bermuda. Limnol. Oceanog. **14**, 568–578.

SHEARMAN, D.J., SHIPWITH, P.A. D'E. (1965): Organic matter in recent and ancient limestones and its rôle in their diagenesis. Nature **208**, 1310–1311.

SHEARMAN, D.J., TYMAN, J., ZAND KARIMI, M. (1970): The genesis and diagenesis of oolites. Proc. Geol. Assoc. **81**, 561–575.

SHEPARD, F.P. (1970): Lagoonal topography of Caroline and Marshall Islands. Bull. Geol. Soc. Am. **81**, 1905–1914.

SHIER, D.E. (1965): Vermetid reefs and coastal development in southwest Florida. Unpubl. Ph. D. Thesis, Florida State Univ., Tallahassee, Fla., 138 pp.

SHIER, D.E. (1969): Vermetid reefs and coastal development in the Ten Thousand Islands, southwest Florida. Bull. Geol. Soc. Am. **80**, 485–508.

SHINN, E.A. (1963): Spur and groove formation on the Florida reef tract. J. Sediment. Petrol. **33**, 291–303.

SHINN, E.A. (1966): Coral growth-rate, an environmental indicator. J. Paleontol. **40**, 233–240.

SHINN, E.A. (1968a): Practical significance of birdseye structures in carbonate rocks. J. Sediment. Petrol. **38**, 215–223.

SHINN, E.A. (1968b): Selective dolomitization of recent sedimentary structures. J. Sediment. Petrol. **38**, 611–616.

SHINN, E.A. (1968c): Burrowing in recent lime sediments of Florida and the Bahamas. J. Paleontol. **42**, 879–894.

SHINN, E.A. (1969): Submarine lithification of Holocene carbonate sediments in the Persian Gulf. Sedimentology **12**, 109–144.

SHINN, E.A., GINSBURG, R.N., LLOYD, R.M. (1965): Recent supratidal dolomite from Andros Island, Bahamas. In: L.C. PRAY and R.C. MURRAY (eds.), Dolomitization and limestone diagenesis. Soc. Econ. Paleontologists Mineralogists Special Publ. No. 13, p. 89–111.

SHINN, E.A., LLOYD, R.M., GINSBURG, R.N. (1969): Anatomy of a modern carbonate tidal-flat, Andros Island, Bahamas. J. Sediment. Petrol. **39**, 1202–1228.

SIBLEY, D.F., MURRAY, R.C. (1972): Marine diagenesis of carbonate sediment, Bonaire, Netherlands antilles (abst.). Am. Assoc. Petrol. Geol. Bull. **56**, 653.

SIDDIQUIE, H.N. (1967): Recent sediments of the Bay of Bengal. Marine Geol. **5**, 259–291.

SIEGEL, F.R. (1960): The effect of strontium on the aragonite-calcite ratios of Pleistocene corals. J. Sediment. Petrol. **30**, 297–304.

SIESSER, W.G., ROGERS, J. (1971): An investigation of the suitability of four methods used in routine carbonate analysis of marine sediments. Deep-Sea Res. **18**, 135–139.

SIEVER, R. (1968): Sedimentological consequences of a steady-state ocean atmosphere. Sedimentology **11**, 5–29.

SIEVER, R., BECK, K.C., BERNER, R.A. (1965): Composition of interstitial waters of modern sediments. J. Geol. **73**, 39–73.

SILEN, L. (1946): On two new groups of Bryozoa living in shells of molluscs. Archiv. Zool. **40B**, No. 1, 1–7.

SILLÉN, L.G. (1961): The physical chemistry of sea water. In: M. SEARS (ed.), Oceanography, p. 549–581. Washington, D. C.: Am. Assoc. Adv. Sci.

SILLÉN, L.G. (1967): The ocean as a chemical system. Science **156**, 1189–1197.

SIMKISS, K. (1964): Variations in the crystalline form of calcium carbonate precipitated from sea water. Nature **201**, 492–493.

SKINNER, H.C.W. (1963): Precipitation of calcian dolomites and magnesian calcites in the southeast of South Australia. Am. J. Sci. **261**, 449–472.

SKINNER, H.C.W., SKINNER, B.J., RUBIN, M. (1963): Age and accumulation rate of dolomite-bearing carbonate sediments in South Australia. Science **139**, 335–336.

SKIRROW, G. (1965): The dissolved gases-carbon dioxide. In: J.P. RILEY and G. SKIRROW (eds.), Chemical oceanography, p. 227–322. London: Academic Press.

SLACK-SMITH, R.J. (1959): An investigation of coral deaths at Peel Island, Moreton Bay, in early 1956. Papers Dept. Zool. Univ. Queensland **1**, 211–222.

SLAVIN, W. (1968): Atomic absorption spectroscopy, 307 pp. New York: Interscience Publ.

SMITH, A.G. (1960): Amphineura. In: J.B. KNIGHT et al. (eds.), Mollusca 1, Treatise on invertebrate paleontology, Geol. Soc. Am. Part I. p. 41–76.

SMITH, C.L. (1940): The Great Bahama Bank. 1. General hydrographical and chemical features. 2. Calcium carbonate precipitation. J. Marine Res. **3**, 147–189.

SMITH, C.L. (1941): The solubility of calcium carbonate in tropical sea water. J. Marine Biol. Assoc. **25**, 235–242.

SMITH, N.R. (1926): Report on a bacteriological examination of "Chalky Mud" and sea-water from the Bahama Banks. Carnegie Inst. Wash. Publ. **344**, 69–72.

SMITH, R.A., WRIGHT, E.R. (1962): Elemental composition of oyster shell. Texas J. Sci. **14**, 222–224.

SMITH, S.V. (1970): Calcium carbonate budget of the southern California continental borderland. Unpubl. Ph. D. Thesis Univ. Hawaii, 174 pp.

SMITH, S.V., DYGAS, J.A., CHAVE, K.E. (1969): Distribution of calcium carbonate in pelagic sediments. Marine Geol. **6**, 391–400.

SMITH, S.V., HADERLIE, E.C. (1969): Growth and longevity of some calcareous fouling organisms, Monterey Bay, California. Pacific Sci. **23**, 447–451.

SOHN, I.G. (1958): Chemical constituents of ostracodes: some applications to paleontology and paleoecology. J. Paleontol. **32**, 730–736.

SOHN, I.G., KORNICKER, L.S. (1969): Significance of calcareous nodules in myodocopid ostracod carapaces. In: J.W. NEALE (ed.), The taxonomy, morphology and ecology of recent ostracoca, p. 99–108. Edinburgh: Oliver & Boyd.

SOLIMAN, G.N. (1969): Ecological aspects of some coral-boring gastropods and bivalves of the northwestern Red Sea. Am. Zoologist **9**, 887–894.

SORBY, H.C. (1879): Anniversary Address of the President. Proc. Quart. J. Geol. Soc. London **35**, 56–95.

SOULE, J.D., SOULE, D.F. (1969): Systematics and biogeography of burrowing bryozoans. Am. Zoologist **9**, 791–802.

SOUTHWARD, A.J. (1964): Limpet grazing and the control of vegetation on rocky shorelines. In: D.J. CRISP (ed.), Grazing in terrestrial and marine environments, p. 265–273. Oxford: Blackwell Scientific Publ.

SPENCER, C.P. (1965): The carbon dioxide system in sea water: a critical appraisal. Oceanography and Marine Biology Rev. **3**, 31–57.

SPENCER, M. (1967): Bahamas deep test. Am. Assoc. Petrol. Geologists Bull. **51**, 263–268.

SPENDER, M.A. (1930): Island-reefs of the Queensland coast. Geograph. J. **76**, 194–214.

SPIRO, B.F. (1971): Ultrastructure and chemistry of the skeleton of *Tubipora musica* Linne. Med. Fra Dansk Geol. Forening, **20**, 279–284.

SPOTTS, J.H., SILVERMAN, S.R. (1966): Organic dolomite from Point Fermin, California. Am. Mineralogist **51**, 1144–1155.

SQUIRES, D.F. (1959): Deep sea corals collected by the Lamont Geological Observatory. 1. Atlantic Corals. Am. Mus. Novitates, No. 1965, 42 pp.

SQUIRES, D.F. (1962): Corals at the mouth of the Rewa River, Viti Levu, Fiji. Nature **195**, 361–362.

SQUIRES, D.F. (1963): Calcareous shelf-edge prominences off the Orinoco River of South America (abst.). Geol. Soc. Am. Special Paper **76**, 155.

STACKELBERG, V.U. VON (1970): Faziesverteilung in Sedimenten des indischpakistanischen Kontinentalrandes (Arabisches Meer). Geol. Rundschau **60**, 268–270.

STANLEY, D.J., SWIFT, D.J.P. (1967): Bermuda's southern aeolianite reef tract. Science **157**, 677–681.

STANLEY, D.J., SWIFT, D.J.P., RICHARDS, H.G. (1967): Fossiliferous concretions on Georges Bank. J. Sediment. Petrol. **37**, 1070–1083.

STARK, J.T., DAPPLES, E.C. (1941): Near shore and coral lagoon sediments from Raiatea, Society Islands. J. Sediment. Petrol. **11**, 21–27.

STARK, L.M., ALMODOVAR, L., KRAUSS, R.W. (1969): Factors affecting the rate of calcification in *Halimeda opuntia* (L.) Lamouroux and *Halimeda discoidea* Decaisne. J. Phycol. **5**, 305–312.

STEERS, J.A., CHAPMAN, V.J., COLMAN, J., LOFTHOUSE, J.A. (1940): Sand cays and mangroves in Jamaica. Geograph. J. **96**, 305–328.

STEHLI, F.G., HOWER, J. (1961): Mineralogy and early diagenesis of carbonate sediments. J. Sediment. Petrol. **31**, 358–371.

STENZEL, H.B. (1963): Aragonite and calcite as constituents of adult oyster shells. Science **142**, 232–233.

STEPHENSON, T.A., STEPHENSON, A. (1950): Life between tide marks in North America. I. The Florida Keys. J. Ecol. **38**, 354–402.

STEPHENSON, T.A., STEPHENSON, A. (1954): The Bermuda Islands. Endeavour **13**, 72–80.

STEPHENSON, W. (1961): Experimental studies on the ecology of intertidal environments at Heron Island. II. The effect of substratum. Australian J. Marine Freshwater Res. **12**, 164–176.
STEPHENSON, W., ENDEAN, R., BENNETT, I. (1958): An ecological survey of the marine fauna of Low Isles, Queensland. Australian J. Marine Freshwater Res. **9**, 262–318.
STEPHENSON, W., SEARLES, R.B. (1960): Experimental studies on the ecology of intertidal environments at Heron Island. I. Exclusion of fish from beachrock. Australian J. Marine Freshwater Res. **11**, 241–267.
STETSON, H.C. (1938): The sediments of the continental shelf off the eastern coast of the United States. Mass. Inst. Tech. and Woods Hole Oceanogr. Inst. Papers Phys. Oceanog. Meteor. **5**, 5–48.
STETSON, H.C. (1953): The sediments of the western Gulf of Mexico. Papers Phys. Oceanog. Meteor. **12**, 1–45.
STETSON, T.R., SQUIRES, D.F., PRATT, R.M. (1962): Coral banks occurring in deep water on the Blake Plateau. Am. Mus. Novitates, No. 2114, 39 pp.
STETSON, T.R., UCHUPI, E., MILLIMAN, J.D. (1969): Surface and subsurface morphology of two small areas of the Blake Plateau. Trans. Gulf Coast Assoc. Geol. Soc. **19**, 131–142.
STIEGLITZ, R.D. (1972): Scanning electron microscopy of the fine fraction of recent carbonate sediments from Bimini, Bahamas. J. Sediment. Petrol. **42**, 211–226.
STOCKMAN, K.W., GINSBURG, R.N., SHINN, E.A. (1967): The production of lime mud by algae in South Florida. J. Sediment. Petrol. **37**, 633–648.
STODDART, D.R. (1962): Three Caribbean atolls: Turneffe Islands, Lighthouse Reef and Glovers Reef, British Honduras. Atoll Res. Bull., No. 87, 147 pp.
STODDART, D.R. (1963): Effects of Hurricane Hattie on the British Honduras reefs and cays, October 30–31, 1961. Atoll Res. Bull., No. 95, 142 pp.
STODDART, D.R. (1964): Carbonate sediments of Half Moon Cay, British Honduras. Atoll Res. Bull., No. 104, 16 pp.
STODDART, D.R. (1969): Ecology and morphology of recent coral reefs. Biol. Rev. **44**, 433–498.
STODDART, D.R., CANN, J.R. (1965): Nature and origin of beach rock. J. Sediment. Petrol. **35**, 243–247.
STODDART, D.R., YONGE, M. (eds.) (1971): Regional variation in Indian Ocean coral reefs. Symp. Zool. Soc. London, No. 28, 584 p.
STOLKOWSKI, J. (1951): Essai sur le déterminisme des formes mineralogiques du calcaire chez les êtres vivants (calcaires coquilliers). Ann. Inst. Oceanog. **26**, 1–115.
STONE, S.W. (1956): Some ecologic data relating to pelagic foraminifera. Micropaleontology **2**, No. 4, 361–370.
STORR, J.F. (1964): Ecology and oceanography of the coral-reef tract, Abaco Island, Bahamas. Geol. Soc. Am. Special Paper **79**, 98 pp.
STRAATEN, L.M.J.U. VAN (1967): Solution of aragonite in a core from the southeastern Adriatic Sea. Marine Geol. **5**, 241–248.
STUBBINGS, H.G. (1937): Pteropoda. John Murray Exped. Rept., **5**, No. 2, 15–33.
STUBBINGS, H.G. (1939): The marine deposits of the Arabian Sea. John Murray Exped. Rept. **3**, No. 2, 32–158.
SUBBA RAO, M. (1958): Distribution of calcium carbonate in the shelf sediments off east coast of India. J. Sediment. Petrol. **28**, 274–285.
SUBBA RAO, M. (1964): Some aspects of continental shelf sediments off the east coast of India. Marine Geol. **1**, 59–87.
SUESS, E. (1970): Interaction of organic compounds with calcium carbonate—I. Association phenomena and geochemical implications. Geochim. Cosmochim. Acta **34**, 157–168.
SUMMERHAYES, C.P. (1970): Phosphate deposits on the northwest African continental shelf and slope. Unpubl. Ph. D. Thesis, Univ. London, 282 pp.
SVERDRUP, H.U., JOHNSON, M.W., FLEMING, R.H. (1942): The oceans, their physics, chemistry and general biology, 1087 pp. Englewood Cliffs, N. J.: Prentice-Hall, Inc.
SWAN, E.F. (1950): The calcareous tube secreting glands of the serpulid polychaetes. J. Morphol. **86**, 285–314.
SWAN, E.F. (1952): The growth of the clam *Mya arenaria* as affected by the substratum. Ecology **33**, 530–534.
SWAN, E.F. (1956): The meaning of strontium-calcium ratios. Deep-Sea Res. **4**, 71.
SWAN, E.F. (1966): Growth, autonomy and regeneration. In: R.A. BOOLOOTIAN (ed.), The physiology of echinodermata, p. 397–434. New York: John Wiley & Sons, Inc.

SWINCHATT, J.P. (1965): Significance of constituent composition, texture, and skeletal breakdown in some recent carbonate sediments. J. Sediment. Petrol. **35**, 71–90.

TAFT, W.H. (1961): Authigenic dolomite in modern carbonate sediments along the southern coast of Florida. Science **134**, 561–562.

TAFT, W.H., ARRINGTON, F., HAIMOVITZ, A., MACDONALD, C., WOOLHEATER, C. (1968): Lithification of modern marine carbonate sediments at Yellow Bank, Bahamas. Bull. Marine Sci. **18**, 762–828.

TAFT, W.H., HARBAUGH, J.W. (1964): Modern carbonate sediments of southern Florida, Bahamas, and Espiritu Santo Island, Baja California: A comparison of their mineralogy and chemistry. Stanford Univ. Publ. Geol. Sci. **8**, No. 2, 133 pp.

TALBOT, F.H. (1965): A description of the coral structure of Tutia Reef (Tanganyika Territory, East Africa), and its fish fauna. Proc. Zool. Soc. London **145**, 431–470.

TAPPAN, H., LOEBLICH, A.R., JR. (1968): Lorica composition of modern and fossil Tintinnida (ciliate Protozoa), systematics, geologic distribution and some new Tertiary taxa. J. Paleontol. **42**, 1378–1394.

TARUTANI, T., CLAYTON, R.N., MAYEDA, T.K. (1969): The effect of polymorphism and magnesium substitution on oxygen isotope formation between calcium carbonate and water. Geochim. Cosmochim. Acta **33**, 987–996.

TATSUMOTO, M., GOLDBERG, E.D. (1959): Some aspects of the marine geochemistry of uranium. Geochim. Cosmochim. Acta **17**, 201–208.

TAYLOR, J.C.M., ILLING, L.V. (1969): Holocene intertidal calcium carbonate cementation, Qatar, Persian Gulf. Sedimentology **12**, 69–107.

TAYLOR, J.C.M., ILLING, L.V. (1971): Development of Recent cemented layers within intertidal sand-flats, Qatar, Persian Gulf. In: O.P. BRICKER (ed.), Carbonate cementation. The Johns Hopkins Univ. Studies in Geology, No. 19, p. 27–31.

TAYLOR, J.D. (1968): Coral reef and associated invertebrate communities (mainly molluscan) around Mahe, Seychelles. Phil. Trans. Roy. Soc., Ser. B **254**, 129–206.

TAYLOR, J.D., KENNEDY, W.J. (1969): The influence of periostracum on bivalve shell structure. Calc. Tiss. Res. **3**, 274–283.

TAYLOR, J.D., LEWIS, M.S. (1970): The flora, fauna and sediments of the marine grass beds of Mahe, Seychelles. J. Nat. Hist. **4**, 199–220.

TEICHERT, C. (1958): Cold- and deep-water coral banks. Am. Assoc. Petrol. Geologists Bull. **42**, 1064–1082.

TEICHERT, C. (1970): Oolite, oolith, ooid: Discussion. Am. Assoc. Petrol. Geologists Bull. **54**, 1748–1749.

TENNANT, C.B., BERGER, R.W. (1957): X-ray determination of dolomite-calcite ratio of a carbonate rock. Am. Mineralogist **42**, 23–29.

TEODOROVICH, G.I. (1946): On the genesis of dolomite in sedimentary deposits (in Russian). Dokl. Acad. Nauka S.S.S.R. **53**, 817–820.

TERLECKY, P.M. (1967): The nature and distribution of oolites on the Atlantic continental shelf of the southeastern United States. Unpubl. M. S. Thesis, Duke Univ., 40 pp.

THODE, H.G., WANLESS, R.K., WALLOUCH, R. (1954): The origin of native sulfur deposits from isotope fractionation studies. Geochim. Cosmochim. Acta **5**, 286–298.

THOMAS, L.P., MOORE, D.R., WORK, R. (1961): Effects of Hurricane Donna on the turtle grass beds of Biscayne Bay, Florida. Bull. Marine Sci. **11**, 191–197.

THOMPSON, G. (1972): A Geochemical study of some lithified carbonate sediments from the deep sea. Geochim. Cosmochim. Acta. **36**, 1237–1253.

THOMPSON, G., BOWEN, V.T., MELSON, W.G., CIFELLI, R. (1968): Lithified carbonates from the deep-sea of the equatorial Atlantic. J. Sediment. Petrol. **38**, 1305–1312.

THOMPSON, G., BOWEN, V.T. (1969): Analyses of coccolith ooze from the deep tropical Atlantic. J. Marine Res. **27**, 32–38.

THOMPSON, G., LIVINGSTON, H.D. (1970): Strontium and uranium concentrations in aragonite precipitated by some modern corals. Earth and Planet. Sci. Letters **8**, 439–442.

THOMPSON, T.G., CHOW, T.J. (1955): The strontium-calcium atom ratio in carbonate secreting marine organisms. Papers in Marine Biol. and Oceanogr., Deep-Sea Res. Suppl. to **3**, 20–39.

THORP, E.M. (1936a): Calcareous shallow-water marine deposits of Florida and the Bahamas. Carnegie Inst. Wash. Publ. 452, Papers Tortugas Lab., **29**, 37–120.

THORP, E.M. (1936b): The sediments of the Pearl and Hermes Reef. J. Sediment. Petrol. **6**, 109–118.

TILL, R. (1970): The relationship between environment and sediment composition (geochemistry and petrology) in the Bimini Lagoon, Bahamas. J. Sediment. Petrol. **40**, 367–385.

TODD, R., LOW, D. (1971): Foraminifera from the Bahama Bank west of Andros Island. U. S. Geol. Surv. Profess. Paper **683-C**, 22 pp.

TOMLINSON, J.T. (1969): Shell-burrowing barnacles. Am. Zoologist **9**, 837–840.

TOURTELOT, H.A., RYE, R.O. (1969): Distribution of oxygen and carbon isotopes in fossils of late Cretaceous age, western interior region of North America. Bull. Geol Soc. Am. **80**, 1904–1922.

TOWE, K.M. (1967): Echinoderm calcite: Single crystal or polycrystalline aggregates. Science **157**, 1048–1050.

TOWE, K.M., CIFELLI, R. (1967): Wall ultrastructure in calcareous foraminifera: crystallographic aspects and a model for calcification. J. Paleontol. **41**, 742–762.

TOWE, K.M., HAMILTON, G.H. (1968): Ultrastructure and inferred calcification of the mature and developing nacre in bivalve mollusks. Calc. Tiss. Res. **1**, 306–318.

TOWE, K.M., MALONE, P.G. (1970): Precipitation of metastable carbonate phases from seawater. Nature **226**, 348–349.

TRACEY, J.I., JR., ABOTT, D.P., ARNOW, T. (1961): Natural history of Ifaluk Atoll: physical environment. Bull. B. P. Bishop Mus. No. 222, 75 pp.

TRACEY, J.I., JR., SCHLANGER, S.O., STARK, J.T., DOAN, D.B., MAY, H.G. (1964): General geology of Guam. U. S. Geol. Surv. Profess. Paper **403-A**, 104 pp.

TRAGANZA, E.D. (1967): Dynamics of the carbon dioxide system on the Great Bahama Bank. Bull. Marine Sci. **17**, 348–366.

TRAVIS, D.F., FRANCOIS, C.J., BONAR, L.C., GLIMCHER, M.J. (1967): Comparative studies of the organic matrices of invertebrate mineralized tissues. J. Ultrastruct. Res. **18**, 519–550.

TRAVIS, D.F., GONSALVES, M. (1969): Comparative ultrastructure and organization of the prismatic region of two bivalves, and its possible relation to the chemical mechanism of boring. Am. Zoologist **9**, 635–661.

TREFZ, S.M. (1958): The physiology of digestion of *Holothuria atra* Jager with special reference to its ecology on coral reefs. Ph. D. thesis, Univ. Hawaii, 149 pp.

TUREKIAN, K.K., ARMSTRONG, R.L. (1960): Magnesium, strontium and barium concentrations and calcite/aragonite ratios of some recent molluscan shells. J. Marine Res. **18**, 133–151.

TUYL, F.M. VAN (1916): The origin of dolomite. Iowa Geol. Surv. Ann. Rept. **25**, 251–422.

TWENHOFEL, W.H. (1942): The rate of deposition of sediments: a major factor connected with alteration of sediments after deposition. J. Sediment. Petrol. **12**, 99–110.

TWENHOFEL, W.H., TYLER, S.A. (1941): Methods of study of sediments, 183pp. New York: McGraw-Hill.

UCHUPI, E., MILLIMAN, J.D., LUYENDYK, B.P., BOWIN, C.O., EMERY, K.O. (1971): Structure and origin of Southeastern Bahamas. Am. Assoc. Petrol.Geologists Bull. **55**, 687–704.

UMBGROVE, J.H.F. (1947): Coral reefs of the East Indies. Bull. Geol. Soc. Am. **58**, 729–778.

UPSHAW, C.F., CREATH, W.B., BROOKS, F.L. (1966): Sediments and microfauna off the coasts of Mississippi and adjacent states. Miss. Geol. Econ. Topog. Surv. Bull., No. 106, 127 pp.

UREY, H.C. (1947): The thermodynamic properties of isotopic substances. J. Chem. Soc. 562–581.

UREY, H.C., LOWENSTAM, H.A., EPSTEIN, S., MCKINNEY, C.R. (1951): Measurements of paleotemperatures and temperatures of the upper Cretaceous of England, Denmark, and the southeastern United States. Bull. Geol. Soc. Am. **62**, 399–416.

USDOWSKI, H.E. (1963): Der Rogenstein des norddeutschen unteren Buntsandsteins, ein Kalkoölith des marinen Faxiesbereichs. Fortschr. Geol. Rheinland Westfalen. **10**, 337–342.

UTINOMI, H. (1953): Coral-dwelling organisms as destructive agents of corals. Proc. 7th Pacific Sci. Congr. **4**, 533–536.

VANNEY, J.R. (1965): Etude sedimentologique du Mor Bras, Bretagne. Marine Geol. **3**, 195–222.

VAUGHAN, T.W. (1914): Preliminary remarks on the geology of the Bahamas: origin of oolite. Carnegie Inst. Wash. Publ. **182**, 47–54.

VAUGHAN, T.W. (1919a): Corals and the formation of coral reefs. Ann. Rept. Smithsonian Inst. **17**, 189–238.

VAUGHAN, T.W. (1919b): Fossil corals from Central America, Cuba, and Puerto Rico, with an account of the American Tertiary, Pleistocene, and Recent coral reefs. Bull. U.S. Nat. Mus. **103**, 189–524.

VAUGHAN, T.W. (1933): The oceanographic point of view. Contributions to Marine Biology, p. 40–56.

VAUGHAN, T.W. (1940): Ecology of modern marine organisms with reference to paleogeography. Bull. Geol. Soc. Am. **51**, 433–468.

VAUGHAN, T.W., WELLS, J.W. (1943): Revision of the suborders, families, and genera of the Scleractinia. Geol. Soc. Am. Special Paper **44**, 363 pp.

VEEH, H.H., VEEVERS, J.H. (1970): Sea level at −175 m off the Great Barrier Reef 13,600 to 17,000 years ago. Nature **226**, 536–537.

VERGNAUD-GRAZZINI, C., HERMAN-ROSENBERG, Y. (1969): Étude paleoclimatique d'une carotte de Méditerranée Orientale. Rev. Geogr. Phys. Geol. Dynam. **9**, 279–292.

VERMEER, D.E. (1963): Effects of Hurricane Hattie, 1961, on the cays of British Honduras. Z. Geomorphol. **7**, 332–354.

VERRILL, A.E. (1900): Notes on the geology of the Bermudas. Am. J. Sci. **9**, 313–340.

VINOGRADOV, A.P. (1953): The elementary chemical composition of marine organisms. Sears Found. Marine Res., Mem. II, 647 pp.

VIRLET-D'AOUST (1857): Sur les oeufs d'insectes donnant lieu à la formation d'oolithes dans des calcaires lacustres au Mexique. Compt. Rend. **45**, 865.

VITA-FINZI, C., CORNELIUS, P.F.S. (1973): Cliff sapping by molluscs in Oman. J. Sediment. Petrol. **43**, 31–32.

VOGEL, J.C. (1959): Über den Isotopengehalt des Kohlenstoffs in Süßwasser-Kalkablagerungen. Geochim. Cosmochim. Acta **16**, 236–242

VOGEL, J.C. (1970): Groningen radiocarbon dates IX. Radiocarbon, **12**, 444–471.

WAHLSTROM, E. (1955): Petrographic mineralogy, 408 pp. New York: John Wiley & Sons, Inc.

WAINWRIGHT, S.A. (1963): Skeletal organization in the coral, *Pocillopora damicornis*. Quart. J. Microscop. Sci. **104**, 169–184.

WAINWRIGHT, S.A. (1967): Diurnal activity of hermatypic gorgonians. Nature **216**, 1041.

WAINWRIGHT, S.A. (1969): Stress and design in bivalved mollusc shell. Nature **224**, 777–779.

WAINWRIGHT, S.A., DILLON, J.R. (1969): On the orientation of sea fans (genus *Gorgonia*). Biol. Bull. **136**, 130–139.

WALL, D., DALE, B. (1968): Quaternary calcareous dinoflagellates (Calciodinellideae) and their natural affinities. J. Paleontol. **42**, 1395–1408.

WALL, D., GUILLARD, R.R.L., DALE, B., SWIFT, E., WATABE, N. (1970): Calcitic resting cysts in *Peridinium trochoideum* (Stein) Lemmermann, an autotrophic marine dinoflagellate. Phycologia **9**, 151–156.

WALTHER, J. (1888): Die Korallenriffe der Sinaihalbinsel. Bandes Abhand. math.-phys. Classe Königl. Sächsischen Gesell. Wissensch., Leipzig No. 10, 439–505.

WALTON, W.R. (1955): Ecology of living benthonic foraminifera, Todes Santos, Baja California. J. Paleontol. **29**, 952–990.

WALTON, W.R. (1964): Recent foraminiferal ecology and paleoecology. In: J. IMBRIE and N.D. NEWELL (eds.), Approaches to paleoecology, p. 151–237. New York: John Wiley & Sons.

WANGERSKY, P.J. (1969): Distribution of suspended carbonate with depth in the ocean. Limnol. Oceanog. **14**, 929–933.

WANGERSKY, P.J., JOENSUU, O. (1964): Strontium, magnesium and manganese in foraminiferal carbonates. J. Geol. **72**, 477–483.

WANLESS, H.R. (1969): Sediments of Biscayne Bay—Distribution and depositional history. Unpubl. M.S. Thesis, Inst. Marine Sciences, Univ. Miami (Fla.), 260 pp.

WANTLAND, K.F. (1967): Recent benthonic foraminifera in the sediments of the British Honduras shelf. Ph.D. Thesis, Rice Univ.

WARBURTON, F.E. (1958): The manner in which the sponge *Cliona* bores in calcareous objects. Can. J. Zool. **36**, 555–562.

WARME, J.E., MARSHALL, N.F. (1969): Marine borers in calcareous terrigenous rocks of Pacific coast. Am. Zoologist **9**, 765–774.

WARNE, S.ST.J. (1962): A quick laboratory staining scheme for the differentiation of the major carbonate minerals. J. Sediment. Petrol. **32**, 29–38.

WASKOWIAK, R. (1962): Geochimische Untersuchungen an rezenten Molluskenschalen mariner Herkunft. Freiberger Forschungsh. **136**, 1–155.

WASS, R.E., CONOLLY, J.R., MACINTYRE, R.J. (1970): Bryozoan carbonate sand continuous along southern Australia. Marine Geol. **9**, 63–73.

WATABE, N. (1967): Crystallographic analysis of the coccolith of *Coccolithus huxleyi*. Calc. Tiss. Res. **1**, 114–121.

WATABE, N., WILBUR, K.M. (1960): Influence of the organic matrix on crystal types in molluscs. Nature **188**, 334.

WATTENBERG, H. (1936): Kohlensäure und Kalziumkarbonat im Meer. Fortschr. Mineral. **20**, 168–195.

WATTENBERG, H., TIMMERMANN, E. (1936): Über die Sättigung des Seewassers an $CaCO_3$ und die anorganogene Bildung von Kalksedimenten. Ann. Hydrog. Marit. Meteorol. **64**, 23–31.

WEBER, J. N. (1964): Oxygen isotope fractionation between coexisting calcite and dolomite. Science **145**, 1303–1305.

WEBER, J. N. (1968): Fractionation of the stable isotopes of carbon and oxygen in calcareous invertebrates—the Asteroida, Ophiuroidea and Crinoidea. Geochim. Cosmochim. Acta **32**, 33–70.

WEBER, J. N. (1973): Deep-sea ahermatypic scleractinian corals: isotopic composition of the skeleton. Deep-sea Res. (in press).

WEBER, J. N., KAUFMAN, J. W. (1965): Brucite in the calcareous alga *Goniolithon*. Science **149**, 996–997.

WEBER, J. N., RAUP, D. M. (1966): Fractionation of the stable isotopes of carbon and oxygen in marine calcareous organisms—the Echinoidea. Part I. Variation of C^{13} and O^{18} content within individuals. Geochim. Cosmochim. Acta **30**, 681–703.

WEBER, J. N., RAUP, D. M. (1968): Comparison of C^{13}/C^{12} and O^{18}/O^{16} in the skeletal calcite of recent and fossil echinoids. J. Paleontol. **42**, 37–50.

WEBER, J. N., WOODHEAD, P. M. J. (1970): Carbon and oxygen isotope fractionation in the skeletal carbonate of reef-building corals. Chem. Geol. **6**, 93–117.

WEBER, J. N., WOODHEAD, P. M. J. (1972a): Temperature dependence of oxygen-18 concentrations in reef coral carbonates. J. Geophys. Res. **77**, 463–473.

WEBER, J. N., WOODHEAD, P. M. J. (1972b): Stable isotope variations in nonscleractinian coelenterate carbonates as a function of temperature. Marine Biol. **15**, 293–297.

WELLMAN, H. W., WILSON, A. T. (1965): Salt weathering, a neglected geological erosive agent in coastal and arid environments. Nature **205**, 1097–1098.

WELLS, A. J. (1962): Recent dolomite in the Persian Gulf. Nature **194**, 274–275.

WELLS, A. J., ILLING, L. V. (1963): Present-day precipitation of calcium carbonate in the Persian Gulf. In: L. M. J. U. VAN STRAATEN (ed.), Deltaic and shallow marine deposits. Developments in sedimentology, **1**, 429–435. Amsterdam: Elsevier Publ. Co.

WELLS, J. W. (1951): The coral reefs of Arno Atoll, Marshall Islands. Atoll Res. Bull. No. 9, 14 pp.

WELLS, J. W. (1954): Recent corals of the Marshall Islands, Bikini and nearby atolls. U.S. Geol. Surv. Profess. Paper **260-I**, 385–486.

WELLS, J. W. (1956): Scleractinia. In: R. C. MOORE (ed.), Treatise on invertebrate paleontology, Geol. Soc. Am. part F, Coelenterata, p. 328–444.

WELLS, J. W. (1957a): Coral Reefs. In: J. HEDGPETH (ed.), Treatise on marine ecology, Geol. Soc. Am. Mem. **67**, (1), 609–631.

WELLS, J. W. (1957b): Corals. In: J. HEDGPETH (ed.), Treatise on marine ecology. Geol. Soc. Am. Mem. **67**, (1), 1087–1104.

WELLS, J. W. (1963): Coral growth and geochronometry. Nature **197**, 948–950.

WENTWORTH, C. K. (1938): Marine bench-forming processes—Part I. Water level weathering. J. Geomorphol. **1**, 5–32.

WENTWORTH, C. K. (1944): Potholes, pits and pans: subaerial and marine. J. Geol. **52**, 117–130.

WERNER, W. E., JR. (1967): The distribution and ecology of the barnacle *Balanus trigonus*. Bull. Marine Sci. **17**, 64–84.

WETHERED, E. B. (1895): The formation of oolite. Quart. J. Geol. Soc. London **51**, 196–206.

WEYL, P. K. (1958): The solution kinetics of calcite. J. Geol. **66**, 163–176.

WEYL, P. K. (1967): The solution behavior of carbonate materials in sea water. Studies in Tropical Oceanogr. Univ. Miami, **5**, 178–228.

WIEDEMANN, H. U. (1972): Shell deposits and shell preservation in Quaternary and Tertiary estuarine sediments in Georgia, U.S.A. Sedimentary Geol. **7**, 103–115.

WIENS, H. J. (1962): Atoll environment and ecology, 532 pp. New Haven: Yale Univ. Press.

WILBUR, K. M. (1964): Shell formation and regeneration. In: K. M. WILBUR and C. M. YONGE (eds.), Physiology of mollusca, p. 243–282. New York: Academic Press.

WILBUR, K. M., COLINVAUX, L. H., WATABE, N. (1969): Electron microscope study of calcification in the alga *Halimeda* (order Siphonales). Physiologia **8**, 27–35.

WILBUR, K. M., WATABE, N. (1963): Experimental studies on calcification in molluscs and the alga *Coccolithus huxleyi*. Ann. N.Y. Acad. Sci. **109**, 82–112.

WILBUR, K. M., WATABE, N. (1967): Mechanisms of calcium carbonate deposition in coccolithophorids and molluscs. Studies in Trop. Oceanog., Univ. Miami **5**, 133–154.

WILLIAMS, A., ROWELL, A. J. (1965): Morphology (of Brachiopoda). In: R. C. MOORE (ed.), Treatise on invertebrate paleontology, Geol. Soc. Am., Part H, p. 57–155.
WINLAND, H. D. (1968): The role of high Mg calcite in the preservation of micrite envelopes and textural features of aragonite sediments. J. Sediment. Petrol. **38**, 1320–1325.
WINLAND, H. D. (1969): Stability of calcium carbonate polymorphs in warm, shallow seawater. J. Sediment. Petrol. **39**, 1579–1587.
WINLAND, H. D. (1971): Non-skeletal deposition of high Mg calcite in the marine environment and its role in the retention of textures. In: O. P. BRICKER (ed.), Carbonate cements. The Johns Hopkins Univ. Studies in Geology, No. 19, p. 278–284.
WINLAND, H. D., MATTHEWS, R. K. (1969): Origin of recent grapestone grains, Bahama Islands (abst.). Ann. Meeting Program Geol. Soc. Am., p. 239.
WISE, S. W., JR. (1969): Organization and formation of fasciculi in the skeletons of scleractinian corals (abst.). Ann. Meeting Program Geol. Soc. Am., p. 241.
WISE, S. W., JR., KELTS, K. R. (1971): Submarine lithification of middle Tertiary chalks in the South Pacific Ocean basin (abst.). Program, 8th Intern. Sedimentological Congr., Heidelberg, p. 110.
WISEMAN, J. D. H. (1956): The rates of accumulation of nitrogen and calcium carbonate on the equatorial Atlantic floor. Advan. Sci. **12**, 579–582.
WISEMAN, J. D. H. (1965): Calcium and magnesium carbonates in some Indian Ocean sediments. In: M. SEARS (ed.), Progr. in Ocean **3**, 373–383.
WODINSKY, J. (1969): Penetration of the shell and feeding on gastropods by *Octopus*. Am. Zoologist **9**, 997–1010.
WOLF, K. H. (1962): The importance of calcareous algae in limestone genesis and sedimentation. Neues Jahrb. Geol. Paleontol. Monatsh., 245–261.
WOLF, K. H. (1965): "Grain-diminution" of algal colonies to micrite. J. Sediment. Petrol. **35**, 420–427.
WOMERSLEY, H. B., BAILEY, A. (1969): The marine algae of the Solomon Islands and their place in biotic reefs. Phil. Tians. Roy. Soc. Ser. B **255**, 433–442.
WOOD, A. (1949): The structure of the wall of the test in the foraminifera, its value in classification. Quart. J. Geol. Soc. London **104**, 229–252.
WRAY, J. L., DANIELS, F. (1957): Precipitation of calcite and aragonite. J. Am. Chem. Soc. **79**, 2031–2034.
YAMANOUTI, T. (1939): Ecological and physiological studies on the holothurians in coral reefs of Palao Islands. Palao Trop. Biol. Sta. Studies **1**, No. 4, 603–636.
YONGE, C. M. (1940): The biology of reef-building corals. Sci. Rept. Great Barrier Reef Exped. **1**, 353–391.
YONGE, C. M. (1958): Ecology and physiology of reef-building corals. In: A. A. BUZZATI-TRAVERSO (ed.), Perspectives in marine biology, p. 117–135. Berkely: Univ. Calif. Press.
YONGE, C. M. (1963a): The biology of coral reefs. Advan. Marine Biol. **1**, 209–260.
YONGE, C. M. (1963b): Rock-boring organisms. In R. F. SOGNNAES (ed.). Mechanisms of hard tissue distruction, Publ. 75. Am. Assoc. Advan. Sci., p. 1–24.
YONGE, C. M. (1971): THOMAS F. GOREAN: A tribute in D. R. STODDART and M. YONGE (eds.). Regional variations in Indian Ocean Reefs. Symp. Zool. Soc. London No. 28, p. XXI–XXXV.
ZALMANSON, E. S. (1951): Sediment formation in Lake Balkhash. Bull. Moskov. Obshchestva Ispytatel. Prirody, Otdel. Geol. **26**, 41–59.
ZANS, V. A. (1958): The Pedro Cays and Pedro Bank. Report on the survey of the cays, 1955–57. Kingston Geol. Surv. Dept. Bull. **3**, 1–47.
ZELLER, E. J., SAUNDERS, D. F., SIEGEL, F. R. (1959): Laboratory precipitation of dolomitic carbonate (abst.). Bull. Geol. Soc. Am. **70**, 1704.
ZELLER, E. J., WRAY, J. (1956): Factors influencing precipitation of calcium carbonate. Am. Assoc. Petrol. Geologists Bull. **40**, 140–152.
ZEMBRUSCKI, S. G. (1968): Geologia e magnetometria submarine na platforma continental de Alagoas, Sergipe, Bahia e Espirito Santo. Marinha Brasil, Hydrog. e. Naveg. Publ. DG 26-11, 46 pp.
ZEN, E-AN (1959): Mineralogy and petrography of marine bottom sediment samples off the coast of Peru and Chile. J. Sediment. Petrol. **29**, 513–539.
ZENKOVITCH, V. P. (1967): Processes of coastal development. 738 p., New York, Intersci. Publ.
ZHIRMUNSKIY, A. V., ZADOROZHNYY, I. K., NAYDIN, P. D., SAKS, V. N., TEYS, R. V. (1967): Determination of temperatures of growth of modern and fossil mollusks by O^{18}/O^{16} ratio of their shells. Geochem. Internat. **4**, 459–468.
ZULLO, V. A. (1966): Thoracid cirripedia from the continental shelf off South Carolina, U.S.A. Crustaceana **11**, 229–244.

Subject Index